PRINCIPLES OF
STRUCTURAL DESIGN

PRINCIPLES OF STRUCTURAL DESIGN

Edited by
Wai-Fah Chen
Eric M. Lui

CRC Press
Taylor & Francis Group
Boca Raton London New York

CRC Press is an imprint of the
Taylor & Francis Group, an **informa** business
A TAYLOR & FRANCIS BOOK

This material was previously published in the *Handbook of Structural Engineering, Second Edition*. © CRC Press LLC, 2005.

CRC Press
Taylor & Francis Group
6000 Broken Sound Parkway NW, Suite 300
Boca Raton, FL 33487-2742

First issued in paperback 2019

© 2006 by Taylor & Francis Group, LLC
CRC Press is an imprint of Taylor & Francis Group, an Informa business

No claim to original U.S. Government works

ISBN-13: 978-0-8493-7235-3 (hbk)
ISBN-13: 978-0-367-39194-2 (pbk)
Library of Congress Card Number 2005051478

Library of Congress Cataloging-in-Publication Data

Principles of structural design / edited by Wai-Fah Chen, Eric M. Lui.
    p. cm.
  Includes bibliographical references and index.
  ISBN 0-8493-7235-6 (alk. paper)
  1. Structural design. I. Chen, Wai-Fah, 1936- II. Lui, E. M.

TA658.P745 2006
624.1'771--dc22
                                           2005051478

Visit the Taylor & Francis Web site at
http://www.taylorandfrancis.com

and the CRC Press Web site at
http://www.crcpress.com

Publisher's Preface

In scientific publishing, two types of books provide essential cornerstones to a field of study: the textbook and the handbook. CRC Press is best known for its handbooks, a tradition dating back to 1913 with publication of the first edition of the *Handbook of Chemistry and Physics*.

In recent years, we have had an increasing number of requests for reprintings of portions of our handbooks to fit a narrower scope of interest than the handbook.

Because each chapter is written by an expert, these derivative works fill a niche between the general textbook and comprehensive handbook, and are suitable as supplemental reading for upper-level university courses or, in some cases, even as primary textbooks. We believe that researchers and professional engineers will also find this smaller and more affordable format useful when their requirements do not merit purchase of the entire handbook.

This book is comprised of ten chapters reprinted from the *Handbook of Structural Engineering, Second Edition*, edited by Wai-Fah Chen and Eric M. Lui.

The Editors

Wai-Fah Chen is presently dean of the College of Engineering at University of Hawaii at Manoa. He was a George E. Goodwin Distinguished Professor of Civil Engineering and head of the Department of Structural Engineering at Purdue University from 1976 to 1999.

He received his B.S. in civil engineering from the National Cheng-Kung University, Taiwan, in 1959, M.S. in structural engineering from Lehigh University, Pennsylvania, in 1963, and Ph.D. in solid mechanics from Brown University, Rhode Island, in 1966.

Dr. Chen received the Distinguished Alumnus Award from National Cheng-Kung University in 1988 and the Distinguished Engineering Alumnus Medal from Brown University in 1999.

Dr. Chen is the recipient of numerous national engineering awards. Most notably, he was elected to the U.S. National Academy of Engineering in 1995, was awarded the Honorary Membership in the American Society of Civil Engineers in 1997, and was elected to the Academia Sinica (National Academy of Science) in Taiwan in 1998.

A widely respected author, Dr. Chen has authored and coauthored more than 20 engineering books and 500 technical papers. He currently serves on the editorial boards of more than 10 technical journals. He has been listed in more than 30 *Who's Who* publications.

Dr. Chen is the editor-in-chief for the popular 1995 *Civil Engineering Handbook*, the 1997 *Structural Engineering Handbook*, the 1999 *Bridge Engineering Handbook*, and the 2002 *Earthquake Engineering Handbook*. He currently serves as the consulting editor for the McGraw-Hill's *Encyclopedia of Science and Technology*.

He has worked as a consultant for Exxon Production Research on offshore structures, for Skidmore, Owings and Merrill in Chicago on tall steel buildings, for the World Bank on the Chinese University Development Projects, and for many other groups.

Eric M. Lui is currently chair of the Department of Civil and Environmental Engineering at Syracuse University. He received his B.S. in civil and environmental engineering with high honors from the University of Wisconsin at Madison in 1980 and his M.S. and Ph.D. in civil engineering (majoring in structural engineering) from Purdue University, Indiana, in 1982 and 1985, respectively.

Dr. Lui's research interests are in the areas of structural stability, structural dynamics, structural materials, numerical modeling, engineering computations, and computer-aided analysis and design of building and bridge structures. He has authored and coauthored numerous journal papers, conference proceedings, special publications, and research reports in these areas. He is also a contributing author to a number of engineering monographs and handbooks, and is the coauthor of two books on the subject of structural stability. In addition to conducting research, Dr. Lui teaches a variety of undergraduate and graduate courses at Syracuse University. He was a recipient of the College of Engineering and Computer Science Crouse Hinds Award for Excellence in Teaching in 1997. Furthermore, he has served as the faculty advisor of Syracuse University's chapter of the American Society of Civil Engineers (ASCE) for more than a decade and was recipient of the ASCE Faculty Advisor Reward Program from 2001 to 2003.

Dr. Lui has been a longtime member of the ASCE and has served on a number of ASCE publication, technical, and educational committees. He was the associate editor (from 1994 to 1997) and later the book editor (from 1997 to 2000) for the ASCE *Journal of Structural Engineering*. He is also a member of many other professional organizations such as the American Institute of Steel Construction, American Concrete Institute, American Society of Engineering Education, American Academy of Mechanics, and Sigma Xi.

He has been listed in more than 10 *Who's Who* publications and has served as a consultant for a number of state and local engineering firms.

Contributors

Wai-Fah Chen
College of Engineering
University of Hawaii at Manoa
Honolulu, Hawaii

J. Daniel Dolan
Department of Civil and Environmental
    Engineering
Washington State University
Pullman, Washington

Achintya Haldar
Department of Civil Engineering and
    Engineering Mechanics
The University of Arizona
Tucson, Arizona

S. E. Kim
Department of Civil Engineering
Sejong University
Seoul, South Korea

Richard E. Klingner
Department of Civil Engineering
University of Texas
Austin, Texas

Yoshinobu Kubo
Department of Civil Engineering
Kyushu Institute of Technology
Tobata, Kitakyushu, Japan

Eric M. Lui
Department of Civil and
    Environmental Engineering
Syracuse University
Syracuse, New York

Edward G. Nawy
Department of Civil and
    Environmental Engineering
Rutgers University — The State University
    of New Jersey
Piscataway, New Jersey

Austin Pan
T.Y. Lin International
San Francisco, California

Maurice L. Sharp
Consultant — Aluminum
Structures
Avonmore, Pennsylvania

Wei-Wen Yu
Department of Civil Engineering
University of Missouri
Rolla, Missouri

Contents

1

Steel Structures

Eric M. Lui
*Department of Civil and
Environmental Engineering,
Syracuse University,
Syracuse, NY*

1-1

1.1 Materials

1.1.1 Stress–Strain Behavior of Structural Steel

Structural steel is a construction material that possesses attributes such as *strength, stiffness, toughness,* and *ductility* that are desirable in modern constructions. Strength is the ability of a material to resist stress. It is measured in terms of the material's yield strength F_y and ultimate or tensile strength F_u. Steel used in ordinary constructions normally have values of F_y and F_u that range from 36 to 50 ksi (248 to 345 MPa) and from 58 to 70 ksi (400 to 483 MPa), respectively, although higher-strength steels are becoming more common. Stiffness is the ability of a material to resist deformation. It is measured in terms of the modulus of elasticity E and modulus of rigidity G. With reference to Figure 1.1, in which several uniaxial engineering stress–strain curves obtained from coupon tests for various grades of steels are shown, it is seen that the modulus of elasticity E does not vary appreciably for the different steel grades. Therefore, a value of 29,000 ksi (200 GPa) is often used for design. Toughness is the ability of a material to absorb energy before failure. It is measured as the area under the material's stress–strain curve. As shown in Figure 1.1, most (especially the lower grade) steels possess high toughness that made them suitable for both static and seismic applications. Ductility is the ability of a material to undergo large inelastic (or plastic) deformation before failure. It is measured in terms of percent elongation or percent reduction in area of the specimen tested in uniaxial tension. For steel, percent elongation ranges from around 10 to 40 for a 2-in. (5-cm) gage length specimen. Ductility generally decreases with increasing steel strength. Ductility is a very important attribute of steel. The ability of structural steel to deform considerably before failure by fracture allows an indeterminate structure to undergo stress redistribution. Ductility also enhances the energy absorption characteristic of the structure, which is extremely important in seismic design.

1.1.2 Types of Steel

Structural steels used for construction are designated by the American Society of Testing and Materials (ASTM) as follows:

ASTM designation*	Steel type
A36/A36M	Carbon structural steel
A131/A131M	Structural steel for ships
A242/A242M	High-strength low-alloy structural steel
A283/A283M	Low and intermediate tensile strength carbon steel plates
A328/A328M	Steel sheet piling
A514/A514M	High-yield strength, quenched and tempered alloy steel plate suitable for welding
A529/A529M	High-strength carbon–manganese steel of structural quality
A572/A572M	High-strength low-alloy columbium–vanadium steel
A573/A573M	Structural carbon steel plates of improved toughness
A588/A588M	High-strength low-alloy structural steel with 50 ksi (345 MPa) minimum yield point to 4 in. [100 mm] thick
A633/A633M	Normalized high-strength low-alloy structural steel plates
A656/A656M	Hot-rolled structural steel, high-strength low-alloy plate with improved formability
A678/A678M	Quenched and tempered carbon and high-strength low-alloy structural steel plates
A690/A690M	High-strength low-alloy steel H-Piles and sheet piling for use in marine environments
A709/A709M	Carbon and high-strength low-alloy structural steel shapes, plates, and bars and quenched and tempered alloy structural steel plates for bridges

ASTM designation*	Steel type
A710/A710M	Age-hardening low-carbon nickel–copper–chromium–molybdenum–columbium alloy structural steel plates
A769/A769M	Carbon and high-strength electric resistance welded steel structural shapes
A786/A786M	Rolled steel floor plates
A808/A808M	High-strength low-alloy carbon, manganese, columbium, vanadium steel of structural quality with improved notch toughness
A827/A827M	Plates, carbon steel, for forging and similar applications
A829/A829M	Plates, alloy steel, structural quality
A830/A830M	Plates, carbon steel, structural quality, furnished to chemical composition requirements
A852/A852M	Quenched and tempered low-alloy structural steel plate with 70 ksi [485 MPa] minimum yield strength to 4 in. [100 mm] thick
A857/A857M	Steel sheet piling, cold formed, light gage
A871/A871M	High-strength low-alloy structural steel plate with atmospheric corrosion resistance
A913/A913M	High-strength low-alloy steel shapes of structural quality, produced by quenching and self-tempering process (QST)
A945/A945M	High-strength low-alloy structural steel plate with low carbon and restricted sulfur for improved weldability, formability, and toughness
A992/A992M	Steel for structural shapes (W-sections) for use in building framing

*The letter M in the designation stands for Metric.

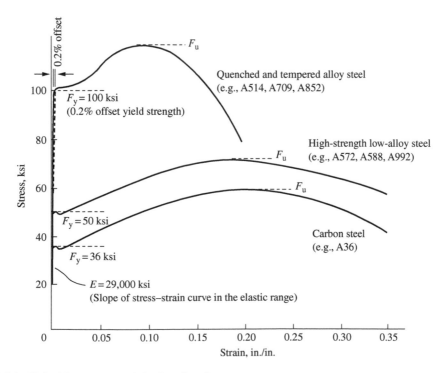

FIGURE 1.1 Uniaxial stress–strain behavior of steel.

TABLE 1.1 Steel Types and General Usages

ASTM designation	F_y (ksi)[a]	F_u (ksi)[a]	Plate thickness (in.)[b]	General usages
A36/A36M	36	58–80	To 8	Riveted, bolted, and welded buildings and bridges
A529/A529M	50	65–100	To 2.5	Similar to A36. The higher yield stress for
	55	70–100	To 1.5	A529 steel allows for savings in weight. A529 supersedes A441
A572/A572M				Grades 60 and 65 not suitable for welded
Grade 42	42	60	To 6	bridges
Grade 50	50	65	To 4	
Grade 55	55	70	To 2	
Grade 60	60	75	To 1.25	
Grade 65	65	80	To 1.25	
A242/A242M	42	63	1.5–5	Riveted, bolted, and welded buildings and
	46	67	0.75–1.5	bridges. Used when weight savings
	50	70	0.5–0.75	and enhanced atmospheric corrosion resistance are desired. Specific instructions must be provided for welding
A588/A588M	42	63	5–8	Similar to A242. Atmospheric corrosion
	46	67	4–5	resistance is about four times that of
	50	70	To 4	A36 steel
A709/A709M				Primarily for use in bridges
Grade 36	36	58–80	To 4	
Grade 50	50	65	To 4	
Grade 50W	50	70	To 4	
Grade 70W	70	90–110	To 4	
Grade 100 and 100W	90	100–130	2.5–4	
Grade 100 and 100W	100	110–130	To 2.5	
A852/A852M	70	90–110	To 4	Plates for welded and bolted construction where atmospheric corrosion resistance is desired
A514/A514M	90–100	100–130	2.5–6	Primarily for welded bridges. Avoid usage if
		110–130		ductility is important
A913/A913M	50–65	65 (Max. $F_y/F_u = 0.85$)	To 4	Used for seismic applications
A992/A992M	50–65	65 (Max. $F_y/F_u = 0.85$)	To 4	Hot-rolled wide flange shapes for use in building frames

[a] 1 ksi = 6.895 MPa.
[b] 1 in. = 25.4 mm.

A summary of the specified minimum yield stresses F_y, the specified minimum tensile strengths F_u, and general usages for some commonly used steels are given in Table 1.1.

1.1.3 High-Performance Steel

High-performance steel (HPS) is a name given to a group of high-strength low-alloy (HSLA) steels that exhibit high strength, higher yield to tensile strength ratio, enhanced toughness, and improved weldability. Although research is still underway to develop and quantify the properties of a number of HPS, one HPS that is currently in use especially for bridge construction is HPS70W. HPS70W is a derivative of ASTM A709 Grade 70W steel (see Table 1.1). Compared to ASTM A709 Grade 70W, HPS70W has improved mechanical properties and is more resistant to postweld cracking even without preheating before welding.

1.1.4 Fireproofing of Steel

Although steel is an incombustible material, its strength (F_y, F_u) and stiffness (E) reduce quite noticeably at temperatures normally reached in fires when other materials in a building burn. Exposed steel members that may be subjected to high temperature in a fire should be fireproofed to conform to the fire ratings set forth in city codes. Fire ratings are expressed in units of time (usually hours) beyond which the structural members under a standard ASTM Specification (E119) fire test will fail under a specific set of criteria. Various approaches are available for fireproofing steel members. Steel members can be fireproofed by encasement in concrete if a minimum cover of 2 in. (5.1 mm) of concrete is provided. If the use of concrete is undesirable (because it adds weight to the structure), a lath and plaster (gypsum) ceiling placed underneath the structural members supporting the floor deck of an upper story can be used. In lieu of such a ceiling, spray-on materials, such as mineral fibers, perlite, vermiculite, gypsum, etc., can also be used for fireproofing. Other means of fireproofing include placing steel members away from the source of heat, circulating liquid coolant inside box or tubular members, and the use of insulative paints. These special paints foam and expand when heated, thus forming a shield for the members (Rains 1976). For a more detailed discussion of structural steel design for fire protection, refer to the latest edition of AISI publication No. FS3, *Fire-Safe Structural Steel — A Design Guide*. Additional information on fire-resistant standards and fire protection can be found in the AISI booklets on *Fire Resistant Steel Frame Construction, Designing Fire Protection for Steel Columns*, and *Designing Fire Protection for Steel Trusses* as well as in the *Uniform Building Code*.

1.1.5 Corrosion Protection of Steel

Atmospheric corrosion occurs when steel is exposed to a continuous supply of water and oxygen. The rate of corrosion can be reduced if a barrier is used to keep water and oxygen from contact with the surface of bare steel. Painting is a practical and cost-effective way to protect steel from corrosion. The Steel Structures Painting Council issues specifications for the surface preparation and the painting of steel structures for corrosion protection of steel. In lieu of painting, the use of other coating materials such as epoxies or other mineral and polymeric compounds can be considered. The use of corrosion resistance steels such as ASTM A242, A588 steel, or galvanized or stainless steel is another alternative. Corrosion resistant steels such as A588 retard corrosion by the formation of a layer of deep reddish-brown to black patina (an oxidized metallic film) on the steel surface after a few wetting–drying cycles, which usually take place within 1 to 3 years. Galvanized steel has a zinc coating. In addition to acting as a protective cover, zinc is anodic to steel. The steel, being cathodic, is therefore protected from corrosion. Stainless steel is more resistant to rusting and staining than ordinary steel primarily because of the presence of chromium as an alloying element.

1.1.6 Structural Steel Shapes

Steel sections used for construction are available in a variety of shapes and sizes. In general, there are three procedures by which steel shapes can be formed: hot rolled, cold formed, and welded. All steel shapes must be manufactured to meet ASTM standards. Commonly used steel shapes include the wide flange (W) sections, the American Standard beam (S) sections, bearing pile (HP) sections, American Standard channel (C) sections, angle (L) sections, tee (WT) sections, as well as bars, plates, pipes, and hollow structural sections (HSS). Sections that, by dimensions, cannot be classified as W or S shapes are designated as miscellaneous (M) sections and C sections that, by dimensions, cannot be classified as American Standard channels are designated as miscellaneous channel (MC) sections.

Hot-rolled shapes are classified in accordance with their tensile property into five size groups by the American Society of Steel Construction (AISC). The groupings are given in the AISC Manuals (1989,

2001). Groups 4 and 5 shapes and group 3 shapes with flange thickness exceeding $1\frac{1}{2}$ in. are generally used for application as *compression members*. When weldings are used, care must be exercised to minimize the possibility of cracking in regions at the vicinity of the welds by carefully reviewing the material specification and fabrication procedures of the pieces to be joined.

1.1.7 Structural Fasteners

Steel sections can be fastened together by rivets, bolts, and welds. While rivets were used quite extensively in the past, their use in modern steel construction has become almost obsolete. Bolts have essentially replaced rivets as the primary means to connect nonwelded structural components.

1.1.7.1 Bolts

Four basic types of bolts are commonly in use. They are designated by ASTM as A307, A325, A490, and A449 (ASTM 2001a–d). A307 bolts are called common, unfinished, machine, or rough. They are made from low-carbon steel. Two grades (A and B) are available. They are available in diameters from $\frac{1}{4}$ to 4 in. (6.4 to 102 mm) in $\frac{1}{8}$ in. (3.2 mm) increments. They are used primarily for low-stress connections and for secondary members. A325 and A490 bolts are called high-strength bolts. A325 bolts are made from a heat-treated medium-carbon steels. They are available in two types: Type 1 — bolts made of medium-carbon steel. Type 3 — bolts having atmospheric corrosion resistance and weathering characteristics comparable to A242 and A588 steels. A490 bolts are made from quenched and tempered alloy steel and thus have higher strength than A325 bolts. Like A325 bolts, two types (Types 1 and 3) are available. Both A325 and A490 bolts are available in diameters from $\frac{1}{2}$ to $1\frac{1}{2}$ in. (13 to 38 mm) in $\frac{1}{8}$ in. (3.2 mm) increments. They are used for general construction purposes. A449 bolts are made from quenched and tempered steels. They are available in diameters from $\frac{1}{4}$ to 3 in. (6.4 to 76 mm). Because A449 bolts are not produced to the same quality requirements nor have the same heavy-hex head and nut dimensions as A325 or A490 bolts, they are not to be used for slip critical connections. A449 bolts are used primarily when diameters over $1\frac{1}{2}$ in. (38 mm) are needed. They are also used for anchor bolts and threaded rod.

   High-strength bolts can be tightened to two conditions of tightness: snug tight and fully tight. The snug-tight condition can be attained by a few impacts of an impact wrench or the full effort of a worker using an ordinary spud wrench. The snug-tight condition must be clearly identified in the design drawing and is permitted in bearing-type connections where slip is permitted, or in tension or combined shear and tension applications where loosening or fatigue due to vibration or load fluctuations are not design considerations. Bolts used in slip-critical conditions (i.e., conditions for which the integrity of the connected parts is dependent on the frictional force developed between the interfaces of the joint) and in conditions where the bolts are subjected to direct tension are required to be tightened to develop a pretension force equal to about 70% of the minimum tensile stress F_u of the material from which the bolts are made. This can be accomplished by using the turn-of-the-nut method, the calibrated wrench method, or by the use of alternate design fasteners or direct tension indicator (RCSC 2000).

1.1.7.2 Welds

Welding is a very effective means to connect two or more pieces of materials together. The four most commonly used welding processes are shielded metal arc welding (SMAW), submerged arc welding (SAW), gas metal arc welding (GMAW), and flux core arc welding (FCAW) (AWS 2000). Welding can be done with or without filler materials although most weldings used for construction utilize filler materials. The filler materials used in modern-day welding processes are electrodes. Table 1.2 summarizes the electrode designations used for the aforementioned four most commonly used welding processes. In general, the strength of the electrode used should equal or exceed the strength of the steel being welded (AWS 2000).

TABLE 1.2 Electrode Designations

Welding processes	Electrode designations	Remarks
Shielded metal arc welding (SMAW)	E60XX E70XX E80XX E100XX E110XX	The "E" denotes electrode. The first two digits indicate tensile strength in ksi.[a] The two "X"s represent numbers indicating the electrode usage
Submerged arc welding (SAW)	F6X-EXXX F7X-EXXX F8X-EXXX F10X-EXXX F11X-EXXX	The "F" designates a granular flux material. The digit(s) following the "F" indicate the tensile strength in ksi (6 means 60 ksi, 10 means 100 ksi, etc.). The digit before the hyphen gives the Charpy V-notched impact strength. The "E" and the "X"s that follow represent numbers relating to the electrode usage
Gas metal arc welding (GMAW)	ER70S-X ER80S ER100S ER110S	The digits following the letters "ER" represent the tensile strength of the electrode in ksi
Flux cored arc welding (FCAW)	E6XT-X E7XT-X E8XT E10XT E11XT	The digit(s) following the letter "E" represent the tensile strength of the electrode in ksi (6 means 60 ksi, 10 means 100 ksi, etc.)

[a] 1 ksi = 6.895 MPa.

Finished welds should be inspected to ensure their quality. Inspection should be performed by qualified welding inspectors. A number of inspection methods are available for weld inspections, including visual inspection, the use of liquid penetrants, magnetic particles, ultrasonic equipment, and radiographic methods. Discussion of these and other welding inspection techniques can be found in the *Welding Handbook* (AWS 1987).

1.1.8 Weldability of Steel

Weldability is the capacity of a material to be welded under a specific set of fabrication and design conditions and to perform as expected during its service life. Generally, weldability is considered very good for low-carbon steel (carbon level < 0.15% by weight), good for mild steel (carbon levels 0.15 to 0.30%), fair for medium-carbon steel (carbon levels 0.30 to 0.50%), and questionable for high-carbon steel (carbon levels 0.50 to 1.00%). Because weldability normally decreases with increasing carbon content, special precautions such as preheating, controlling heat input, and post-weld heat treating are normally required for steel with carbon content reaching 0.30%. In addition to carbon content, the presence of other alloying elements will have an effect on weldability. Instead of more accurate data, the table below can be used as a guide to determine the weldability of steel (Blodgett, undated).

Element	Range for satisfactory weldability	Level requiring special care (%)
Carbon	0.06–0.25%	0.35
Manganese	0.35–0.80%	1.40
Silicon	0.10% max.	0.30
Sulfur	0.035% max.	0.050
Phosphorus	0.030% max.	0.040

A quantitative approach for determining weldability of steel is to calculate its *carbon equivalent value.* One definition of the carbon equivalent value C_{eq} is

$$C_{eq} = \text{Carbon} + \frac{(\text{manganese} + \text{silicon})}{6} + \frac{(\text{copper} + \text{nickel})}{15}$$
$$+ \frac{(\text{chromium} + \text{molybdenum} + \text{vanadium} + \text{columbium})}{5} \tag{1.1}$$

A steel is considered weldable if $C_{eq} \leq 0.50\%$ for steel in which the carbon content does not exceed 0.12% and if $C_{eq} \leq 0.45\%$ for steel in which the carbon content exceeds 0.12%.

Equation 1.1 indicates that the presence of alloying elements decreases the weldability of steel. An example of high-alloy steels is stainless steel. There are three types of stainless steel: austenitic, martensitic, or ferritic. Austenitic stainless steel is the most weldable, but care must be exercised to prevent thermal distortion because heat dissipation is only about one third as fast as in plain carbon steel. Martensitic steel is also weldable but prone to cracking because of its high hardenability. Preheating and maintaining interpass temperature are often needed, especially when the carbon content is above 0.10%. Ferritic steel is weldable but decreased ductility and toughness in the weld area can present a problem. Preheating and postweld annealing may be required to minimize these undesirable effects.

1.2 Design Philosophy and Design Formats

1.2.1 Design Philosophy

Structural design should be performed to satisfy the criteria for strength, serviceability, and economy. *Strength* pertains to the general integrity and safety of the structure under extreme load conditions. The structure is expected to withstand occasional overloads without severe distress and damage during its lifetime. *Serviceability* refers to the proper functioning of the structure as related to its appearance, maintainability, and durability under normal, or service load, conditions. Deflection, vibration, permanent deformation, cracking, and corrosion are some design considerations associated with serviceability. *Economy* concerns with the overall material, construction, and labor costs required for the design, fabrication, erection, and maintenance processes of the structure.

1.2.2 Design Formats

At present, steel design in the United States is being performed in accordance with one of the following three formats.

1.2.2.1 Allowable Stress Design (ASD)

ASD has been in use for decades for steel design of buildings and bridges. It continues to enjoy popularity among structural engineers engaged in steel building design. In allowable stress (or working stress) design, member stresses computed under service (or working) loads are compared to some predesignated stresses called allowable stresses. The allowable stresses are often expressed as a function of the yield stress (F_y) or tensile stress (F_u) of the material divided by a factor of safety. The factor of safety is introduced to account for the effects of overload, understrength, and approximations used in structural analysis. The general format for an allowable stress design has the form

$$\frac{R_n}{FS} \geq \sum_{i=1}^{m} Q_{ni} \tag{1.2}$$

where R_n is the nominal resistance of the structural component expressed in unit of stress (i.e., the allowable stress), Q_{ni} is the service or working stresses computed from the applied working load of type i, FS is the factor of safety; i is the load type (dead, live, wind, etc.), and m is the number of load types considered in the design.

1.2.2.2 Plastic Design (PD)

PD makes use of the fact that steel sections have reserved strength beyond the first yield condition. When a section is under flexure, yielding of the cross-section occurs in a progressive manner, commencing with the fibers farthest away from the neutral axis and ending with the fibers nearest the neutral axis. This phenomenon of progressive yielding, referred to as *plastification*, means that the cross-section does not fail at first yield. The additional moment that a cross-section can carry in excess of the moment that corresponds to first yield varies depending on the shape of the cross-section. To quantify such reserved capacity, a quantity called *shape factor*, defined as the ratio of the *plastic moment* (moment that causes the entire cross-section to yield, resulting in the formation of a *plastic hinge*) to the *yield moment* (moment that causes yielding of the extreme fibers only) is used. The shape factor for hot-rolled I-shaped sections bent about the strong axes has a value of about 1.15. The value is about 1.50 when these sections are bent about their weak axes.

For an indeterminate structure, failure of the structure will not occur after the formation of a plastic hinge. After complete yielding of a cross-section, force (or, more precisely, moment) redistribution will occur in which the unyielded portion of the structure continues to carry some additional loadings. Failure will occur only when enough cross-sections have yielded rendering the structure unstable, resulting in the formation of a *plastic collapse mechanism.*

In PD, the factor of safety is applied to the applied loads to obtain *factored loads.* A design is said to have satisfied the strength criterion if the load effects (i.e., forces, shears, and moments) computed using these factored loads do not exceed the nominal plastic strength of the structural component. PD has the form

$$R_n \geq \gamma \sum_{i=1}^{m} Q_{ni} \tag{1.3}$$

where R_n is the nominal plastic strength of the member, Q_{ni} is the nominal load effect from loads of type i, γ is the load factor, i is the load type, and m is the number of load types.

In steel building design, the load factor is given by the AISC Specification as 1.7 if Q_n consists of dead and live gravity loads only, and as 1.3 if Q_n consists of dead and live gravity loads acting in conjunction with wind or earthquake loads.

1.2.2.3 Load and Resistance Factor Design (LRFD)

LRFD is a probability-based limit state design procedure. A *limit state* is defined as a condition in which a structure or structural component becomes unsafe (i.e., a violation of the strength limit state) or unsuitable for its intended function (i.e., a violation of the serviceability limit state). In a limit state design, the structure or structural component is designed in accordance to its limits of usefulness, which may be strength related or serviceability related. In developing the LRFD method, both load effects and resistance are treated as random variables. Their variabilities and uncertainties are represented by frequency distribution curves. A design is considered satisfactory according to the strength criterion if the resistance exceeds the load effects by a comfortable margin. The concept of safety is represented schematically in Figure 1.2. Theoretically, the structure will not fail unless the load effect Q exceeds the resistance R as shown by the shaded portion in the figure. The smaller this shaded area, the less likely that the structure will fail. In actual design, a resistance factor ϕ is applied to the nominal resistance of the structural component to account for any uncertainties associated with the determination of its strength and a load factor γ is applied to each load type to account for the uncertainties and difficulties associated with determining its actual load magnitude. Different load factors are used for different load types to reflect the varying degree of uncertainties associated with the determination of load magnitudes. In general, a lower load factor is used for a load that is more predicable and a higher load factor is used for a load that is less predicable. Mathematically, the LRFD format takes the form

$$\phi R_n \geq \sum_{i=1}^{m} \gamma_i Q_{ni} \tag{1.4}$$

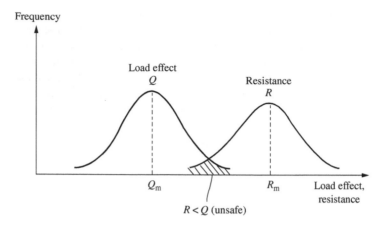

FIGURE 1.2   Frequency distribution of load effect and resistance.

TABLE 1.3   Load Factors and Load Combinations

$1.4(D+F)$
$1.2(D+F+T)+1.6(L+H)+0.5(L_r \text{ or } S \text{ or } R)$
$1.2D+1.6(L_r \text{ or } S \text{ or } R)+(L \text{ or } 0.8W)$
$1.2D+1.6W+L+0.5(L_r \text{ or } S \text{ or } R)$
$1.2D+1.0E+L+0.2S$
$0.9D+1.6W+1.6H$
$0.9D+1.0E+1.6H$

Notes: D is the dead load, *E* is the earthquake load, *F* is the load due to fluids with well-defined pressures and maximum heights, *H* is the load due to the weight and lateral pressure of soil and water in soil, *L* is the live load, L_r is the roof live load, *R* is the rain load, *S* is the snow load, *T* is the self-straining force, and *W* is the wind load.

The load factor on *L* in the third, fourth, and fifth load combinations shown above can be set to 0.5 for all occupancies (except for garages or areas occupied as places of public assembly) in which the design live load per square foot of area is less than or equal to 100 psf (4.79 kN/m²). The load factor on *H* in the sixth and seventh load combinations shall be set to zero if the structural action due to *H* counteracts that due to *W* or *E*.

where ϕR_n represents the design (or usable) strength and $\sum \gamma_i Q_{ni}$ represents the required strength or load effect for a given load combination. Table 1.3 shows examples of load combinations (ASCE 2002) to be used on the right-hand side of Equation 1.4. For a safe design, all load combinations should be investigated and the design is based on the worst-case scenario.

1.3   Tension Members

Tension members are designed to resist tensile forces. Examples of tension members are hangers, truss members, and bracing members that are in tension. Cross-sections that are used most often for tension members are solid and hollow circular rods, bundled bars and cables, rectangular plates, single and double angles, channels, WT- and W-sections, and a variety of built-up shapes.

1.3.1   Tension Member Design

Tension members are to be designed to preclude the following possible failure modes under normal load conditions: yielding in gross section, fracture in effective net section, block shear, shear rupture along plane through the fasteners, bearing on fastener holes, prying (for lap- or hanger-type joints). In addition, the fasteners' strength must be adequate to prevent failure in the fasteners. Also, except for rods in tension, the slenderness of the tension member obtained by dividing the length of the member by its least radius of gyration should preferably not exceed 300.

1.3.1.1 Allowable Stress Design

The computed tensile stress f_t in a tension member shall not exceed the allowable stress for tension, F_t, given by $0.60F_y$ for yielding on the gross area and by $0.50F_u$ for fracture on the effective net area. While the gross area is just the nominal cross-sectional area of the member, the *effective net area* is the smallest cross-sectional area accounting for the presence of fastener holes and the effect of *shear lag*. It is calculated using the equation

$$A_e = UA_n = U\left[A_g - \sum_{i=1}^{m} d_{ni}t_i + \sum_{j=1}^{k}\left(\frac{s^2}{4g}\right)_j t_j \right] \qquad (1.5)$$

where U is a reduction coefficient given by (Munse and Chesson 1963)

$$U = 1 - \frac{\bar{x}}{l} \le 0.90 \qquad (1.6)$$

in which l is the length of the connection and $\bar{x}$ is the larger of the distance measured from the centroid of the cross-section to the contact plane of the connected pieces or to the fastener lines. In the event that the cross-section has two symmetrically located planes of connection, $\bar{x}$ is measured from the centroid of the nearest one-half the area (Figure 1.3). This reduction coefficient is introduced to account for the shear lag effect that arises when some component elements of the cross-section in a joint are not connected, rendering the connection less effective in transmitting the applied load. The terms in brackets in Equation 1.5 constitute the so-called net section A_n. The various terms are defined as follows: A_g is the gross cross-sectional area, d_n is the nominal diameter of the hole (bolt cutout) taken as the nominal bolt diameter plus $\frac{1}{8}$ in. (3.2 mm), t is the thickness of the component element, s is the longitudinal center-to-center spacing (pitch) of any two consecutive fasteners in a chain of staggered holes, and g is the transverse center-to-center spacing (gage) between two adjacent fasteners gage lines in a chain of staggered holes.

The second term inside the brackets of Equation 1.5 accounts for loss of material due to bolt cutouts; the summation is carried for all bolt cutouts lying on the failure line. The last term inside the brackets of Equation 1.5 indirectly accounts for the effect of the existence of a combined stress state (tensile and shear) along an inclined failure path associated with staggered holes; the summation is carried for all staggered paths along the failure line. This term vanishes if the holes are not staggered. Normally, it is necessary to investigate different failure paths that may occur in a connection; the critical failure path is the one giving the smallest value for A_e.

To prevent block shear failure and shear rupture, the allowable strengths for block shear and shear rupture are specified as follows:

Block shear:

$$R_{BS} = 0.30A_v F_u + 0.50A_t F_u \qquad (1.7)$$

Shear rupture:

$$F_v = 0.30F_u \qquad (1.8)$$

where A_v is the net area in shear, A_t is the net area in tension, and F_u is the specified minimum tensile strength.

The tension member should also be designed to possess adequate thickness and the fasteners should be placed within a specific range of spacings and edge distances to prevent failure due to bearing and failure by prying action (see Section 1.11).

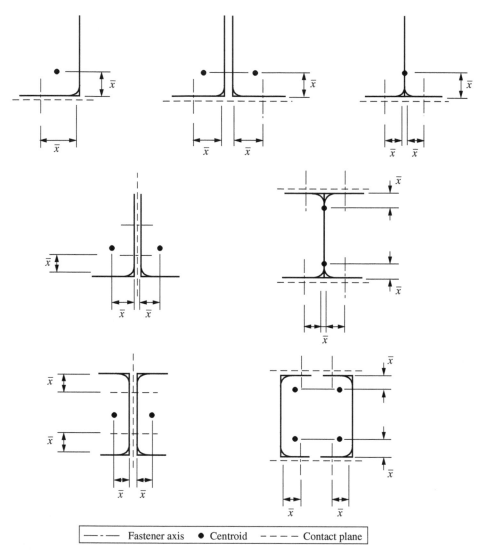

FIGURE 1.3   Definition of $\bar{x}$ for selected cross-sections.

1.3.1.2 Load and Resistance Factor Design

According to the LRFD Specification (AISC 1999), tension members designed to resist a factored axial force of P_u calculated using the load combinations shown in Table 1.3 must satisfy the condition of

$$\phi_t P_n \geq P_u \tag{1.9}$$

The design strength $\phi_t P_n$ is evaluated as follows:

Yielding in gross section:

$$\phi_t P_n = 0.90[F_y A_g] \tag{1.10}$$

where 0.90 is the resistance factor for tension, F_y is the specified minimum yield stress of the material, and A_g is the gross cross-sectional area of the member.

Fracture in effective net section:

$$\phi_t P_n = 0.75[F_u A_e] \tag{1.11}$$

where 0.75 is the resistance factor for fracture in tension, F_u is the specified minimum tensile strength, and A_e is the effective net area given in Equation 1.5.

Block shear: If $F_u A_{nt} \geq 0.6 F_u A_{nv}$ (i.e., shear yield–tension fracture)

$$\phi_t P_n = 0.75[0.60 F_y A_{gv} + F_u A_{nt}] \leq 0.75[0.6 F_u A_{nv} + F_u A_{nt}] \tag{1.12a}$$

and if $F_u A_{nt} < 0.6 F_u A_{nv}$ (i.e., shear fracture–tension yield)

$$\phi_t P_n = 0.75[0.60 F_u A_{nv} + F_y A_{gt}] \leq 0.75[0.60 F_u A_{nv} + F_u A_{nt}] \tag{1.12b}$$

where 0.75 is the resistance factor for block shear, F_y, F_u are the specified minimum yield stress and tensile strength, respectively, A_{gv} is the gross shear area, A_{nt} is the net tension area, A_{nv} is the net shear area, and A_{gt} is the gross tension area.

EXAMPLE 1.1

Using LRFD, select a double-channel tension member shown in Figure 1.4a to carry a dead load D of 40 kip and a live load L of 100 kip. The member is 15 ft long. Six 1-in. diameter A325 bolts in standard size holes are used to connect the member to a $\frac{3}{8}$-in. gusset plate. Use A36 steel ($F_y = 36$ ksi, $F_u = 58$ ksi) for all the connected parts.

Load combinations: From Table 1.3, the applicable load combinations are

$$1.4D = 1.4(40) = 56 \text{ kip}$$
$$1.2D + 1.6L = 1.2(40) + 1.6(100) = 208 \text{ kip}$$

The design of the tension member is to be based on the larger of the two, that is, 208 kip and so *each* channel is expected to carry 104 kip.

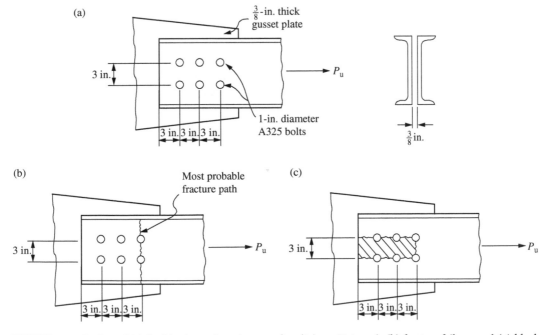

FIGURE 1.4 Design of (a) double-channel tension member (1 in. = 25.4 mm); (b) fracture failure; and (c) block shear failure.

Yielding in gross section: Using Equations 1.9 and 1.10, the gross area required to prevent cross-section yielding is

$$0.90[F_y A_g] \geq P_u$$
$$0.90[(36)(A_g)] \geq 104 \text{ kip}$$
$$(A_g)_{\text{req'd}} \geq 3.21 \text{ in.}^2$$

From the section properties table contained in the AISC-LRFD Manual, one can select the following trial sections: C8 × 11.5 ($A_g = 3.38$ in.²), C9 × 13.4 ($A_g = 3.94$ in.²), and C8 × 13.75 ($A_g = 4.04$ in.²).

Check for the limit state of fracture on effective net area: The above sections are checked for the limiting state of fracture in the following table:

Section	A_g (in.²)	t_w (in.)	$\bar{x}$ (in.)	U[a]	A_e[b] (in.²)	$\phi_t P_n$ (kip)
C8 × 11.5	3.38	0.220	0.571	0.90	2.6	113.1
C9 × 13.4	3.94	0.233	0.601	0.90	3.07	133.5
C8 × 13.75	4.04	0.303	0.553	0.90	3.02	131.4

[a] Equation 1.6.
[b] Equation 1.5, Figure 1.4b.

From the last column of the above table, it can be seen that fracture is not a problem for any of the trial sections.

Check for the limit state of block shear: Figure 1.4c shows a possible block shear failure mode. To avoid block shear failure the required strength of $P_u = 104$ kip should not exceed the design strength, $\phi_t P_n$, calculated using Equations 1.12a or 1.12b, whichever is applicable.

For the C8 × 11.5 section:

$$A_{gv} = 2(9)(0.220) = 3.96 \text{ in.}^2$$
$$A_{nv} = A_{gv} - 5\left(1 + \frac{1}{8}\right)(0.220) = 2.72 \text{ in.}^2$$
$$A_{gt} = (3)(0.220) = 0.66 \text{ in.}^2$$
$$A_{nt} = A_{gt} - 1\left(1 + \frac{1}{8}\right)(0.220) = 0.41 \text{ in.}^2$$

Substituting the above into Equation 1.12b, since ($F_u A_{nt} = 23.8$ kip) is smaller than ($0.6 F_u A_{nv} = 94.7$ kip), we obtain $\phi_t P_n = 88.8$ kip, which is less than $P_u = 104$ kip. The C8 × 11.5 section is therefore not adequate. A significant increase in block shear strength is not expected from the C9 × 13.4 section because its web thickness t_w is just slightly over that of the C8 × 11.5 section. As a result, we shall check the adequacy of the C8 × 13.75 section instead.

For the C8 × 13.75 section:

$$A_{gv} = 2(9)(0.303) = 5.45 \text{ in.}^2$$
$$A_{nv} = A_{gv} - 5\left(1 + \frac{1}{8}\right)(0.303) = 3.75 \text{ in.}^2$$
$$A_{gt} = (3)(0.303) = 0.91 \text{ in.}^2$$
$$A_{nt} = A_{gt} - 1\left(1 + \frac{1}{8}\right)(0.303) = 0.57 \text{ in.}^2$$

Substituting the above into Equation 1.12b, since ($F_u A_{nt} = 33.1$ kip) is smaller than ($0.6 F_u A_{nv} = 130.5$ kip), we obtain $\phi_t P_n = 122$ kip, which exceeds the required strength P_u of 104 kip. Therefore, block shear will not be a problem for the C8 × 13.75 section.

Check for the limiting slenderness ratio: Using parallel axis theorem, the least radius of gyration of the double-channel cross-section is calculated to be 0.96 in. Therefore, $L/r = (15 \text{ ft})(12 \text{ in./ft})/0.96 \text{ in.} = 187.5$, which is less than the recommended maximum value of 300.

Check for the adequacy of the connection: The calculations are shown in an example in Section 1.11.

Longitudinal spacing of connectors: According to Section J3.5 of the LRFD Specification, the maximum spacing of connectors in built-up tension members shall not exceed:

- Twenty-four times the thickness of the thinner plate or 12 in. (305 mm) for painted members or unpainted members not subject to corrosion.
- Fourteen times the thickness of the thinner plate or 7 in. (180 mm) for unpainted members of weathering steel subject to atmospheric corrosion.

Assuming the first condition applies, a spacing of 6 in. is to be used. Use 2C8 × 13.75 connected intermittently at 6-in. interval.

1.3.2 Pin-Connected Members

Pin-connected members shall be designed to preclude the following failure modes:

- Tension yielding in the gross section
- Tension fracture on the effective net area
- Longitudinal shear on the effective area
- Bearing on the projected pin area (Figure 1.5)

1.3.2.1 Allowable Stress Design

The allowable stresses for tension yield, tension fracture, and shear rupture are $0.60F_y$, $0.45F_y$, and $0.30F_u$, respectively. The allowable stresses for bearing are given in Section 1.11.

1.3.2.2 Load and Resistance Factor Design

The design tensile strength $\phi_t P_n$ for a pin-connected member are given as follows:

Tension on gross area: see Equation 1.10.
Tension on effective net area:

$$\phi_t P_n = 0.75[2tb_{eff}F_u] \tag{1.13}$$

Shear on effective area:

$$\phi_{sf} P_n = 0.75[0.6A_{sf}F_u] \tag{1.14}$$

Bearing on projected pin area: see Section 1.11.

The terms in Figure 1.5 and the above equations are defined as follows: a is the shortest distance from the edge of the pin hole to the edge of the member measured in the direction of the force, A_{pb} is the projected bearing area $= dt$, $A_{sf} = 2t(a + d/2)$, $b_{eff} = 2t + 0.63$, in. (or, $2t + 16$, mm) but not more than the actual distance from the edge of the hole to the edge of the part measured in the direction normal to the applied force, d is the pin diameter, and t is the plate thickness.

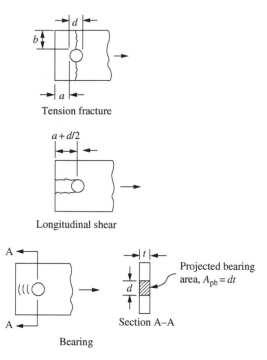

Tension fracture

Longitudinal shear

Bearing

FIGURE 1.5   Failure modes of pin-connected members.

1.3.3   Threaded Rods

1.3.3.1   Allowable Stress Design

Threaded rods under tension are treated as bolts subject to tension in allowable stress design. These allowable stresses are given in Section 1.11.

1.3.3.2   Load and Resistance Factor Design

Threaded rods designed as tension members shall have an gross area A_b given by

$$A_b \geq \frac{P_u}{\phi\, 0.75 F_u} \tag{1.15}$$

where A_b is the gross area of the rod computed using a diameter measured to the outer extremity of the thread, P_u is the factored tensile load, ϕ is the resistance factor given as 0.75, and F_u is the specified minimum tensile strength.

1.4   Compression Members

Members under compression can fail by yielding, inelastic buckling, or elastic buckling depending on the slenderness ratio of the members. Members with low slenderness ratios tend to fail by yielding while members with high slenderness ratio tend to fail by elastic buckling. Most compression members used in construction have intermediate slenderness ratios and so the predominant mode of failure is inelastic buckling. Overall member buckling can occur in one of three different modes: flexural, torsional, and flexural–torsional. Flexural buckling occurs in members with doubly symmetric or doubly antisymmetric cross-sections (e.g., I or Z sections) and in members with singly symmetric sections (e.g., channel, tee, equal-legged angle, double-angle sections) when such sections are buckled about an axis that is *perpendicular* to the axis of symmetry. Torsional buckling occurs in members with doubly symmetric sections such as cruciform or built-up shapes with very thin walls. Flexural–torsional buckling occurs in members with singly symmetric cross-sections (e.g., channel, tee, equal-legged angle, double-angle sections) when such sections are buckled about the axis of symmetry and in members with unsymmetric cross-sections (e.g., unequal-legged L). Normally, torsional buckling of symmetric shapes is not particularly important in the design of hot-rolled compression members. It either does not govern or its buckling strength does not differ significantly from the corresponding weak axis flexural buckling strengths. However, torsional buckling may become important for open sections with relatively thin component plates. It should be noted that for a given cross-sectional area, a closed section is much stiffer torsionally than an open section. Therefore, if torsional deformation is of concern, a closed section should be used. Regardless of the mode of buckling, the governing effective slenderness ratio (Kl/r) of the compression member preferably should not exceed 200.

In addition to the slenderness ratio and cross-sectional shape, the behavior of compression members is affected by the relative thickness of the component elements that constitute the cross-section. The relative thickness of a component element is quantified by the width–thickness ratio (b/t) of the element. The width–thickness ratios of some selected steel shapes are shown in Figure 1.6. If the width–thickness ratio falls within a limiting value [denoted by the LRFD specification (AISC 1999) as λ_r] as shown in Table 1.4, the section will not experience local buckling prior to overall buckling of the member. However, if the width–thickness ratio exceeds this limiting width–thickness value, consideration of local buckling in the design of the compression member is required.

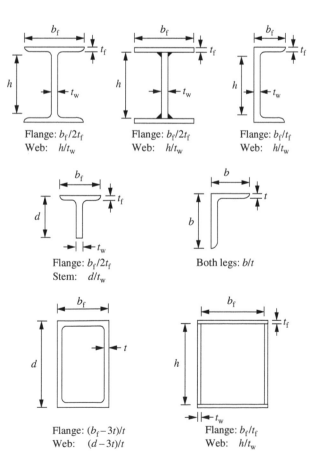

FIGURE 1.6 Definition of width–thickness ratio of selected cross-sections.

TABLE 1.4 Limiting Width–Thickness Ratios for Compression Elements Under Pure Compression

Component element	Width–thickness ratio	Limiting value, λ_r
Flanges of I-shaped sections; plates projecting from compression elements; outstanding legs of pairs of angles in continuous contact; flanges of channels	b/t	$0.56\sqrt{E/F_y}$
Flanges of square and rectangular box and HSS of uniform thickness; flange cover plates and diaphragm plates between lines of fasteners or welds	b/t	$1.40\sqrt{E/F_y}$
Unsupported width of cover plates perforated with a succession of access holes	b/t	$1.86\sqrt{E/F_y}$
Legs of single-angle struts; legs of double-angle struts with separators; unstiffened elements (i.e., elements supported along one edge)	b/t	$0.45\sqrt{E/F_y}$
Flanges projecting from built-up members	b/t	$0.64\sqrt{E/(F_y/k_c)}$
Stems of tees	d/t	$0.75\sqrt{E/F_y}$
All other uniformly compressed stiffened elements (i.e., elements supported along two edges)	b/t h/t_w	$1.49\sqrt{E/F_y}$
Circular hollow sections	D/t D is the outside diameter and t is the wall thickness	$0.11E/F_y$

Note: E is the modulus of elasticity, F_y is the specified minimum yield stress, $k_c = 4/\sqrt{(h/t_w)}$, and $0.35 \leq k_c \leq 0.763$ for I-shaped sections, k_c is equal to 0.763 for other sections, where h is the web depth and t_w is the web thickness.

To facilitate the design of compression members, column tables for W, tee, double angle, square/rectangular tubular, and circular pipe sections are available in the AISC Manuals for both allowable stress design (AISC 1989) and load and resistance factor design (AISC 2001).

1.4.1 Compression Member Design

1.4.1.1 Allowable Stress Design

The computed compressive stress f_a in a compression member shall not exceed its allowable value given by

$$F_a = \begin{cases} \dfrac{\left[1 - \left((Kl/r)^2/2C_c^2\right)\right]F_y}{(5/3) + (3(Kl/r)/8C_c) - ((Kl/r)^3/8C_c^3)}, & \text{if } Kl/r \leq C_c \\[2ex] \dfrac{12\pi^2 E}{23(Kl/r)^2}, & \text{if } Kl/r > C_c \end{cases} \tag{1.16}$$

where Kl/r is the slenderness ratio, K is the effective length factor of the compression member in the plane of buckling, l is the unbraced member length in the plane of buckling, r is the radius of gyration of the cross-section about the axis of buckling, E is the modulus of elasticity, and $G_c = \sqrt{(2\pi^2 E/F_y)}$ is the slenderness ratio that demarcates between inelastic member buckling from elastic member buckling. Kl/r should be evaluated for both buckling axes, and the larger value used in Equation 1.16 to compute F_a.

The first part of Equation 1.16 is the allowable stress for inelastic buckling and the second part is the allowable stress for elastic buckling. In ASD, no distinction is made between flexural, torsional, and flexural–torsional buckling.

1.4.1.2 Load and Resistance Design

Compression members are to be designed so that the design compressive strength $\phi_c P_n$ will exceed the required compressive strength P_u. $\phi_c P_n$ is to be calculated as follows for the different types of overall buckling modes.

Flexural buckling (with width–thickness ratio $\leq \lambda_r$):

$$\phi_c P_n = \begin{cases} 0.85[A_g(0.658^{\lambda_c^2})F_y], & \text{if } \lambda_c \leq 1.5 \\[2mm] 0.85\left[A_g\left(\dfrac{0.877}{\lambda_c^2}\right)F_y\right], & \text{if } \lambda_c > 1.5 \end{cases} \tag{1.17}$$

where $\lambda_c = (KL/r\pi)\sqrt{(F_y/E)}$ is the slenderness parameter, A_g is the gross cross-sectional area, F_y is the specified minimum yield stress, E is the modulus of elasticity, K is the effective length factor, l is the unbraced member length in the plane of buckling, and r is the radius of gyration of the cross-section about the axis of buckling.

The first part of Equation 1.17 is the design strength for inelastic buckling and the second part is the design strength for elastic buckling. The slenderness parameter $\lambda_c = 1.5$ is the slenderness parameter that demarcates between inelastic behavior from elastic behavior.

Torsional buckling (with width–thickness ratio $\leq \lambda_r$): $\phi_c P_n$ is to be calculated from Equation 1.17, but with λ_c replaced by λ_e and given by

$$\lambda_e = \sqrt{\frac{F_y}{F_e}} \tag{1.18}$$

where

$$F_e = \left[\frac{\pi^2 E C_w}{(K_z L)^2} + GJ\right]\frac{1}{I_x + I_y} \tag{1.19}$$

in which C_w is the warping constant, G is the shear modulus $= 11{,}200$ ksi (77,200 MPa), I_x, I_y are the moments of inertia about the major and minor principal axes, respectively, J is the torsional constant, and K_z is the effective length factor for torsional buckling.

The warping constant C_w and the torsional constant J are tabulated for various steel shapes in the AISC-LRFD Manual (AISC 2001). Equations for calculating approximate values for these constants for some commonly used steel shapes are shown in Table 1.5.

Flexural–torsional buckling (with width–thickness ratio $\leq \lambda_r$): Same as for torsional buckling except F_e is now given by

For singly symmetric sections:

$$F_e = \frac{F_{es} + F_{ez}}{2H}\left[1 - \sqrt{1 - \frac{4F_{es}F_{ez}H}{(F_{es} + F_{ez})^2}}\right] \tag{1.20}$$

where

$F_{es} = F_{ex}$ if the x-axis is the axis of symmetry of the cross-section, or
$F_{es} = F_{ey}$ if the y-axis is the axis of symmetry of the cross-section
$F_{ex} = \pi^2 E/(Kl/r)_x^2$
$F_{ey} = \pi^2 E/(Kl/r)_x^2$
$H = 1 - (x_o^2 + y_o^2)/r_o^2$

in which K_x, K_y are the effective length factors for buckling about the x and y axes, respectively, l is the unbraced member length in the plane of buckling, r_x, r_y are the radii of gyration about the x and y axes, respectively, x_o, y_o are shear center coordinates with respect to the centroid (Figure 1.7), and $r_o^2 = x_o^2 + y_o^2 + r_x^2 + r_y^2$.

TABLE 1.5 Approximate Equations for C_w and J

Structural shape	Warping constant, C_w	Torsional constant, J	
I	$h'^2 I_c I_t/(I_c + I_t)$	$\sum C_i (b_i t_i^3/3)$	
C	$(b' - 3E_0)h'^2 b'^2 t_f/6 + E_0^2 I_x$	where	
	where	b_i = width of component element i	
	$E_0 = b'^2 t_f/(2b' t_f + h' t_w/3)$	t_i = thickness of component element i	
		C_i = correction factor for component element i	
		(see values below)	
T	$(b_f^3 t_f^3/4 + h''^3 t_w^3)/36$ (≈ 0 for small t)	b_i/t_i	C_i
L	$(l_1^3 t_1^3 + l_2^3 t_2^3)/36$ (≈ 0 for small t)	1.00	0.423
		1.20	0.500
		1.50	0.588
		1.75	0.642
		2.00	0.687
		2.50	0.747
		3.00	0.789
		4.00	0.843
		5.00	0.873
		6.00	0.894
		8.00	0.921
		10.00	0.936
		∞	1.000

Note: b' is the distance measured from toe of flange to centerline of web, h' is the distance between centerline lines of flanges, h'' is the distance from centerline of flange to tip of stem, l_1, l_2 are the length of the legs of the angle, t_1, t_2 are the thickness of the legs of the angle, b_f is the flange width, t_f is the average thickness of flange, t_w is the thickness of web, I_c is the moment of inertia of compression flange taken about the axis of the web, I_t is the moment of inertia of tension flange taken about the axis of the web, and I_x is the moment of inertia of the cross-section taken about the major principal axis.

Numerical values for r_o and H are given for hot-rolled W, channel, tee, single-angle, and double-angle sections in the AISC-LRFD Manual (AISC 2001).

For unsymmetric sections: F_e is to be solved from the cubic equation

$$(F_e - F_{ex})(F_e - F_{ey})(F_e - F_{ez}) - F_e^2(F_e - F_{ey})\left(\frac{x_o}{r_o}\right)^2 - F_e^2(F_e - F_{ex})\left(\frac{y_o}{r_o}\right)^2 = 0 \qquad (1.21)$$

The definitions of the terms in the above equation are as in Equation 1.20.

Local Buckling (with width–thickness ratio $\geq \lambda_r$): local buckling in the component element of the cross-section is accounted for in design by introducing a reduction factor Q in Equation 1.17 as follows:

$$\phi_c P_n = \begin{cases} 0.85[A_g Q(0.658^{Q\lambda^2})F_y], & \text{if } \lambda\sqrt{Q} \leq 1.5 \\ 0.85\left[A_g\left(\dfrac{0.877}{\lambda^2}\right)F_y\right], & \text{if } \lambda\sqrt{Q} > 1.5 \end{cases} \qquad (1.22)$$

where $\lambda = \lambda_c$ for flexural buckling and $\lambda = \lambda_e$ for flexural–torsional buckling.

The Q factor is given by

$$Q = Q_s Q_a \qquad (1.23)$$

where Q_s is the reduction factor for unstiffened compression elements of the cross-section (see Table 1.6) and Q_a is the reduction factor for stiffened compression elements of the cross-section (see Table 1.7).

1.4.2 Built-Up Compression Members

Built-up members are members made by bolting and/or welding together two or more standard structural shapes. For a built-up member to be fully effective (i.e., if all component structural

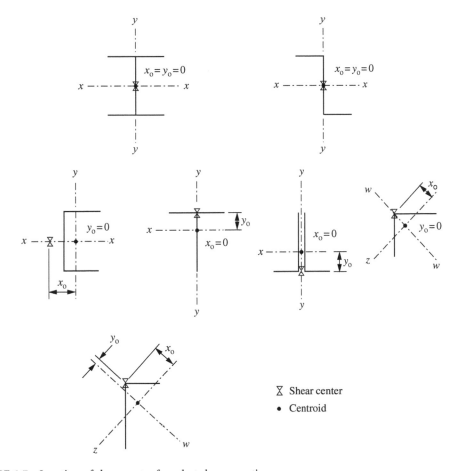

FIGURE 1.7 Location of shear center for selected cross-sections.

TABLE 1.6 Formulas for Q_s

Structural element	Range of b/t or d/t	Q_s
Single-angles	$0.45\sqrt{(E/F_y)} < b/t < 0.91\sqrt{(E/F_y)}$ $b/t \geq 0.91\sqrt{(E/F_y)}$	$1.340 - 0.76(b/t)\sqrt{(F_y/E)}$ $0.53E/[F_y(b/t)^2]$
Flanges, angles, and plates projecting from columns or other compression members	$0.56\sqrt{(E/F_y)} < b/t < 1.03\sqrt{(E/F_y)}$ $b/t \geq 1.03\sqrt{(E/F_y)}$	$1.415 - 0.74(b/t)\sqrt{(F_y/E)}$ $0.69E/[F_y(b/t)^2]$
Flanges, angles, and plates projecting from built-up columns or other compression members	$0.64\sqrt{[E/(F_y/k_c)]} < b/t < 1.17\sqrt{[E/(F_y/k_c)]}$ $b/t \geq 1.17\sqrt{[E/(F_y/k_c)]}$	$1.415 - 0.65(b/t)\sqrt{(F_y/k_cE)}$ $0.90Ek_c/[F_y(b/t)^2]$
Stems of tees	$0.75\sqrt{(E/F_y)} < d/t < 1.03\sqrt{(E/F_y)}$ $d/t \geq 1.03\sqrt{(E/F_y)}$	$1.908 - 1.22(b/t)\sqrt{(F_y/E)}$ $0.69E/[F_y(b/t)^2]$

Notes: k_c is defined in the footnote of Table 1.4, E is the modulus of elasticity, F_y is the specified minimum yield stress, b is the width of the component element, and t is the thickness of the component element.

TABLE 1.7   Formula for Q_a

$$Q_s = \frac{\text{effective area}}{\text{actual area}}$$

The effective area is equal to the summation of the effective areas of the stiffened elements of the cross-section. The effective area of a stiffened element is equal to the product of its thickness t and its effective width b_e given by

For flanges of square and rectangular sections of uniform thickness, when $b/t \geq 1.40\sqrt{(E/f)}$[a]

$$b_e = 1.91t\sqrt{\frac{E}{f}}\left[1 - \frac{0.38}{(b/t)}\sqrt{\frac{E}{f}}\right] \leq b$$

For other noncircular uniformly compressed elements, when $b/t \geq 1.49\sqrt{(E/f)}$[a]

$$b_e = 1.91t\sqrt{\frac{E}{f}}\left[1 - \frac{0.34}{(b/t)}\sqrt{\frac{E}{f}}\right] \leq b$$

For axially loaded circular sections with $0.11E/F_y < D/t < 0.45E/F_y$

$$Q_a = \frac{0.038E}{F_y(D/t)} + \frac{2}{3}$$

where b is the actual width of the stiffened element, t is the wall thickness, E is the modulus of elasticity, f is the computed elastic compressive stress in the stiffened elements, and D is the outside diameter of circular sections.

[a] $b_e = b$ otherwise.

shapes are to act as one unit rather than as individual units), the following conditions must be satisfied:

1. Slippage of component elements near the ends of the built-up member must be prevented.
2. Adequate fasteners must be provided along the length of the member.
3. The fasteners must be able to provide sufficient gripping force on all component elements.

Condition 1 is satisfied if all component elements in contact near the ends of the built-up member are connected by a weld having a length not less than the maximum width of the member or by bolts spaced longitudinally not more than four diameters apart for a distance equal to one and a half times the maximum width of the member. Condition 2 is satisfied if continuous welds are used throughout the length of the built-up compression member. Condition 3 is satisfied if either welds or fully tightened bolts are used as the fasteners. While condition 1 is mandatory, conditions 2 and 3 can be violated in design. If condition 2 or 3 is violated, the built-up member is not fully effective and slight slippage among component elements may occur. To account for the decrease in capacity due to slippage, a modified slenderness ratio is used to compute the design compressive strength when buckling of the built-up member is about an axis *coinciding* or *parallel* to at least one plane of contact for the component shapes. The modified slenderness ratio $(KL/r)_m$ is given as follows:

If condition 2 is violated

$$\left(\frac{KL}{r}\right)_m = \sqrt{\left(\frac{KL}{r}\right)_o^2 + \frac{0.82\alpha^2}{(1+\alpha^2)}\left(\frac{a}{r_{ib}}\right)^2} \tag{1.24}$$

If condition 3 is violated

$$\left(\frac{KL}{r}\right)_m = \sqrt{\left(\frac{KL}{r}\right)_o^2 + \left(\frac{a}{r_i}\right)^2} \tag{1.25}$$

In the above equations, $(KL/r)_o = (KL/r)_x$ if the buckling axis is the x-axis and at least one plane of contact between component elements is parallel to that axis; $(KL/r)_o = (KL/r)_y$ if the buckling axis is the

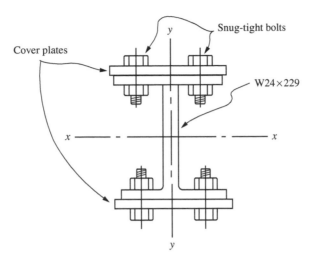

FIGURE 1.8 Design of cover plates for a compression member.

y-axis and at least one plane of contact is parallel to that axis. a is the longitudinal spacing of the fasteners, r_i is the minimum radius of gyration of any component element of the built-up cross-section, r_{ib} is the radius of gyration of individual component relative to its centroidal axis parallel to the axis of buckling of the member, and h is the distance between centroids of components elements measured perpendicularly to the buckling axis of the built-up member.

No modification to (KL/r) is necessary if the buckling axis is perpendicular to the planes of contact of the component shapes. Modifications to both $(KL/r)_x$ and $(KL/r)_y$ are required if the built-up member is so constructed that planes of contact exist in both the x and y directions of the cross-section.

Once the modified slenderness ratio is computed, it is to be used in the appropriate equation to calculate F_a in allowable stress design, or $\phi_c P_n$ in load and resistance factor design.

An additional requirement for the design of built-up members is that the effective slenderness ratio, Ka/r_{ib} of each component element, where K is the effective length factor of the component element between adjacent fasteners, does not exceed $\frac{3}{4}$ of the governing slenderness ratio of the built-up member. This provision is provided to prevent component element buckling between adjacent fasteners from occurring prior to overall buckling of the built-up member.

EXAMPLE 1.2

Using LRFD, determine the size of a pair of cover plates to be bolted, using fully tightened bolts, to the flanges of a W24 × 229 section as shown in Figure 1.8 so that its design strength, $\phi_c P_n$, will be increased by 20%. Also determine the spacing of the bolts along the longitudinal axis of the built-up column. The effective lengths of the section about the major $(KL)_x$ and minor $(KL)_y$ axes are both equal to 20 ft. A992 steel is to be used.

Determine the design strength for the W24 × 229 section: Since $(KL)_x = (KL)_y$ and $r_x > r_y$, $(KL/r)_y$ will exceed $(KL/r)_x$ and the design strength will be controlled by flexural buckling about the minor axis. Using section properties, $r_y = 3.11$ in. and $A = 67.2$ in.^{2}, obtained from the AISC-LRFD Manual (AISC 2001), the slenderness parameter λ_c about the minor axis can be calculated as follows:

$$(\lambda_c)_y = \frac{1}{\pi}\left(\frac{KL}{r}\right)_y \sqrt{\frac{F_y}{E}} = \frac{1}{3.142}\left(\frac{20 \times 12}{3.11}\right)\sqrt{\frac{50}{29,000}} = 1.02$$

Substituting $\lambda_c = 1.02$ into Equation 1.17, the design strength of the section is

$$\phi_c P_n = 0.85\lfloor 67.2(0.658^{1.02^2})50\rfloor = 1848 \text{ kip}$$

Determine the design strength for the built-up section: The built-up section is expected to possess a design strength that is 20% in excess of the design strength of the W24 × 229 section, so

$$(\phi_c P_n)_{\text{req'd}} = (1.20)(1848) = 2218 \text{ kip}$$

Determine the size of the cover plates: After the cover plates are added, the resulting section is still doubly symmetric. Therefore, the overall failure mode is still flexural buckling. For flexural buckling about the minor axis (y–y), no modification to (KL/r) is required since the buckling axis is perpendicular to the plane of contact of the component shapes and so no relative movement between the adjoining parts is expected. However, for flexural buckling about the major (x–x) axis, modification to (KL/r) is required since the buckling axis is parallel to the plane of contact of the adjoining structural shapes and slippage between the component pieces will occur. We shall design the cover plates assuming flexural buckling about the minor axis will control and check for flexural buckling about the major axis later.

A W24 × 229 section has a flange width of 13.11 in.; so, as a trial, use cover plates with widths of 14 in. as shown in Figure 1.8. Denoting t as the thickness of the plates, we have

$$(r_y)_{\text{built-up}} = \sqrt{\frac{(I_y)_{\text{W-shape}} + (I_y)_{\text{plates}}}{A_{\text{W-shape}} + A_{\text{plates}}}} = \sqrt{\frac{651 + 457.3t}{67.2 + 28t}}$$

and

$$(\lambda_c)_{y,\text{built-up}} = \frac{1}{\pi}\left(\frac{KL}{r}\right)_{y,\text{built-up}} \sqrt{\frac{F_y}{E}} = 3.17\sqrt{\frac{67.2 + 28t}{651 + 457.3t}}$$

Assuming $(\lambda)_{y,\text{built-up}}$ is less than 1.5, one can substitute the above expression for λ_c in Equation 1.17. When $\phi_c P_n$ equals 2218, we can solve for t. The result is $t \approx \frac{3}{8}$ in. Back-substituting $t = \frac{3}{8}$ into the above expression, we obtain $(\lambda)_{c,\text{built-up}} = 0.975$, which is indeed < 1.5. So, try 14 in. × $\frac{3}{8}$ in. cover plates.

Check for local buckling: For the I-section:

$$\text{Flange:} \quad \left[\frac{b_f}{2t_f} = 3.8\right] < \left[0.56\sqrt{\frac{E}{F_y}} = 0.56\sqrt{\frac{29{,}000}{50}} = 13.5\right]$$

$$\text{Web:} \quad \left[\frac{h_c}{t_w} = 22.5\right] < \left[1.49\sqrt{\frac{E}{F_y}} = 1.49\sqrt{\frac{29{,}000}{50}} = 35.9\right]$$

For the cover plates, if $\frac{3}{4}$ in. diameter bolts are used and assuming an edge distance of 2 in., the width of the plate between fasteners will be $13.11 - 4 = 9.11$ in. Therefore, we have

$$\left[\frac{b}{t} = \frac{9.11}{3/8} = 24.3\right] < \left[1.40\sqrt{\frac{E}{F_y}} = 1.40\sqrt{\frac{29{,}000}{50}} = 33.7\right]$$

Since the width–thickness ratios of all component shapes do not exceed the limiting width–thickness ratio for local buckling, local buckling is not a concern.

Check for flexural buckling about the major (x–x) axis: Since the built-up section is doubly symmetric, the governing buckling mode will be flexural buckling regardless of the axes. Flexural buckling will occur about the major axis if the modified slenderness ratio $(KL/r)_m$ about the major axis exceeds $(KL/r)_y$. Therefore, as long as $(KL/r)_m$ is less than $(KL/r)_y$, buckling will occur about the minor axis and flexural buckling about the major axis will not be controlled. In order to arrive at an optimal design, we shall determine the longitudinal fastener spacing, a, such that the modified slenderness ratio $(KL/r)_m$ about the major axis will be equal to $(KL/r)_y$. That is, we shall solve for a from the equation

$$\left[\left(\frac{KL}{r}\right)_m = \sqrt{\left(\frac{KL}{r}\right)_x^2 + \left(\frac{a}{r_i}\right)^2}\right] = \left[\left(\frac{KL}{r}\right)_y = 73.8\right]$$

In the above equation, $(KL/r)_x$ is the slenderness ratio about the major axis of the built-up section, r_i is the least radius of gyration of the component shapes, which in this case is the cover plate.

Substituting $(KL/r)_x = 21.7$, $r_i = r_{\text{cover plate}} = \sqrt{(I/A)}_{\text{cover plate}} = \sqrt{[(\frac{3}{8})^2/12]} = 0.108$ into the above equation, we obtain $a = 7.62$ in. Since $(KL) = 20$ ft, we shall use $a = 6$ in. for the longitudinal spacing of the fasteners.

Check for component element buckling between adjacent fasteners:

$$\left[\frac{Ka}{r_i} = \frac{1 \times 6}{0.108} = 55.6\right] \approx \left[\frac{3}{4}\left(\frac{KL}{r}\right)_y = \frac{3}{4}(73.8) = 55.4\right]$$

so, the component element buckling criterion is not a concern.

Use 14 in. $\times \frac{3}{8}$ in. cover plates bolted to the flanges of the W24 $\times$ 229 section by $\frac{3}{4}$ in. diameter fully tightened bolts spaced 6 in. longitudinally.

1.4.3 Column Bracing

The design strength of a column can be increased if lateral braces are provided at intermediate points along its length in the buckled direction of the column. The AISC-LRFD Specification (1999) identifies two types of bracing systems for columns. A relative bracing system is one in which the movement of a braced point with respect to other adjacent braced points is controlled, for example, the diagonal braces used in buildings. A nodal (or discrete) brace system is one in which the movement of a braced point with respect to some fixed points is controlled, for example, the guy wires of guyed towers. A bracing system is effective only if the braces are designed to satisfy both stiffness and strength requirements. The following equations give the required stiffness and strength for the two bracing systems.

Required braced stiffness:

$$\beta_{cr} = \begin{cases} \dfrac{2.67 P_u}{L_{br}} & \text{for relative bracing} \\ \dfrac{10.7 P_u}{L_{br}} & \text{for nodal bracing} \end{cases} \qquad (1.26)$$

where P_u is the required compression strength of the column and L_{br} is the distance between braces (for ASD, replace P_u in Equation 1.26 by $1.5P_a$, where P_a is the required compressive strength based on ASD load combinations).

If L_{br} is less than L_q (the maximum unbraced length for P_u), L_{br} can be replaced by L_q in the above equations.

Required braced strength:

$$P_{br} = \begin{cases} 0.004 P_u & \text{for relative bracing} \\ 0.01 P_u & \text{for nodal bracing} \end{cases} \qquad (1.27)$$

where P_u is defined as in Equation 1.26.

1.5 Flexural Members

Depending on the width–thickness ratios of the component elements, steel sections used as *flexural members* are classified as compact, noncompact, and slender element sections. Compact sections are sections that can develop the cross-section plastic moment (M_p) under flexure and sustain that moment through a large hinge rotation without fracture. Noncompact sections are sections that either cannot develop the cross-section full plastic strength or cannot sustain a large hinge rotation at M_p, probably due to local buckling of the flanges or web. Slender elements are sections that fail by local buckling of component elements long before M_p is reached. A section is considered compact if all its component elements have width–thickness ratios less than a limiting value (denoted as λ_p in LRFD). A section is considered noncompact if one or more of its component elements have width–thickness ratios that fall in between λ_p and λ_r. A section is considered a slender element if one or more of its component elements have width–thickness ratios that exceed λ_r. Expressions for λ_p and λ_r are given in the Table 1.8.

In addition to the compactness of the steel section, another important consideration for beam design is the lateral unsupported (unbraced) length of the member. For beams bent about their strong axes, the failure modes, or limit states, vary depending on the number and spacing of lateral supports provided to brace the compression flange of the beam. The compression flange of a beam behaves somewhat like a compression member. It buckles if adequate lateral supports are not provided in a phenomenon called *lateral torsional buckling*. Lateral torsional buckling may or may not be accompanied by yielding, depending on the lateral unsupported length of the beam. Thus, lateral torsional buckling can be inelastic or elastic. If the lateral unsupported length is large, the limit state is elastic lateral torsional buckling. If the lateral unsupported length is smaller, the limit state is inelastic lateral torsional buckling. For compact section beams with adequate lateral supports, the limit state is full yielding of the cross-section (i.e., plastic hinge formation). For noncompact section beams with adequate lateral supports, the limit state is flange or web local buckling. For beams bent about their weak axes, lateral torsional buckling will not occur and so the lateral unsupported length has no bearing on the design. The limit states for such beams will be the formation of the plastic hinge if the section is compact and the limit state will be a flange or web local buckling if the section is noncompact.

Beams subjected to high shear must be checked for possible web shear failure. Depending on the width–thickness ratio of the web, failure by shear yielding or web shear buckling may occur. Short, deep beams with thin webs are particularly susceptible to web shear failure. If web shear is of concern, the use of thicker webs or web reinforcements such as stiffeners is required.

Beams subjected to concentrated loads applied in the plane of the web must be checked for a variety of possible flange and web failures. Failure modes associated with concentrated loads include local flange bending (for tensile concentrated load), local web yielding (for compressive concentrated load), web crippling (for compressive load), side-sway web buckling (for compressive load), and compression buckling of the web (for a compressive load pair). If one or more of these conditions is critical, transverse stiffeners extending at least one-half the beam depth (use full depth for compressive buckling of the web) must be provided adjacent to the concentrated loads.

Long beams can have deflections that may be too excessive, leading to problems in serviceability. If deflection is excessive, the use of intermediate supports or beams with higher flexural rigidity is required.

The design of flexural members should satisfy, at the minimum, the following criteria:

- Flexural strength criterion
- Shear strength criterion
- Criteria for concentrated loads
- Deflection criterion

To facilitate beam design, a number of beam tables and charts are given in the AISC Manuals (1989, 2001) for both allowable stress and load and resistance factor design.

TABLE 1.8 λ_p and λ_r for Members Under Flexural Compression

Component element	Width–thickness ratio[a]	λ_p	λ_r
Flanges of I-shaped rolled beams and channels	b/t	$0.38\sqrt{(E/F_y)}$	$0.83\sqrt{(E/F_L)}$[b] F_L = smaller of $(F_{yf} - F_r)$ or F_{yw} F_{yf} = flange yield strength F_{yw} = web yield strength F_r = flange compressive residual stress (10 ksi for rolled shapes, 16.5 ksi for welded shapes)
Flanges of I-shaped hybrid or welded beams	b/t	$0.38\sqrt{(E/F_{yf})}$ for nonseismic application $0.31\sqrt{(E/F_{yf})}$ for seismic application F_{yf} = flange yield strength	$0.95\sqrt{[E/(F_L/k_c)]}$[c] F_L = as defined above k_c = as defined in the footnote of Table 1.4
Flanges of square and rectangular box and HSS of uniform thickness; flange cover plates and diaphragm plates between lines of fasteners or welds	b/t	$0.939\sqrt{(E/F_y)}$ for plastic analysis	$1.40/\sqrt{(E/F_y)}$
Unsupported width of cover plates perforated with a succession of access holes	b/t	NA	$1.86\sqrt{(E/F_y)}$
Legs of single-angle struts; legs of double-angle struts with separators; unstiffened elements	b/t	NA	$0.45/\sqrt{(E/F_y)}$
Stems of tees	d/t	NA	$0.75\sqrt{(E/F_y)}$
Webs in flexural compression	h_c/t_w	$3.76\sqrt{(E/F_y)}$ for nonseismic application $3.05\sqrt{(E/F_y)}$ for seismic application	$5.70\sqrt{(E/F_y)}$[d]
Webs in combined flexural and axial compression	h_c/t_w	For $P_u/\phi_b P_y \leq 0.125$: $3.76(1 - 2.75 P_u/\phi_b P_y)\sqrt{(E/F_y)}$ for nonseismic application $3.05(1 - 1.54 P_u/\phi_b P_y)\sqrt{(E/F_y)}$ for seismic application For $P_u/\phi_b P_y > 0.125$: $1.12(2.33 - P_u/\phi_b P_y)\sqrt{(E/F_y)} \geq$ $1.49\sqrt{(E/F_y)}$ $\phi_b = 0.90$ P_u = factored axial force $P_y = A_g F_y$ $0.07E/F_y$	$5.70(1 - 0.74 P_u/\phi_b P_y)\sqrt{(E/F_y)}$
Circular hollow sections	D/t D = outside diameter t = wall thickness		$0.31E/F_y$

[a] See Figure 1.5 for definition of b, h_c, and t.
[b] For ASD, this limit is $0.56\sqrt{(E/F_y)}$.
[c] For ASD, this limit is $0.56\sqrt{[E/(F_{yf}/k_c)]}$, where $k_c = 4.05/(h/t)^{0.46}$ if $h/t > 70$, otherwise $k_c = 1.0$.
[d] For ASD, this limit is $4.46\sqrt{(E/F_b)}$, where F_b = allowable bending stress.
Note: In all the equations, E is the modulus of elasticity and is F_y is the minimum specified yield strength.

1.5.1 Flexural Member Design

1.5.1.1 Allowable Stress Design

1.5.1.1.1 Flexural Strength Criterion

The computed flexural stress, f_b, shall not exceed the allowable flexural stress, F_b, as given below (in all equations, the minimum specified yield stress, F_y, can not exceed 65 ksi).

1.5.1.1.1.1 Compact Section Members Bent about Their Major Axes — For $L_b \leq L_c$

$$F_b = 0.66F_y \tag{1.28}$$

where

L_c = smaller of $\{76b_f/\sqrt{F_y}, 20,000/(d/A_f)F_y\}$, for I and Channel shapes

$= [1,950 + 1,200(M_1/M_2)](b/F_y) \geq 1,200(b/F_y)$, for box sections, rectangular and circular tubes

in which b_f is the flange width (in.), d is the overall depth of the section (ksi), A_f is the area of the compression flange (in.^{2}), b is the width of the cross-section (in.), and M_1/M_2 is the ratio of the smaller to larger moment at the ends of the unbraced length of the beam; M_1/M_2 is positive for reverse curvature bending and negative for single curvature bending.

For the above sections to be considered as compact, in addition to having the width–thickness ratios of their component elements falling within the limiting value of λ_p shown in Table 1.8, the flanges of the sections must be continuously connected to the webs. For box-shaped sections, the following requirements must also be satisfied: the depth-to-width ratio should not exceed six, and the flange-to-web thickness ratio should not exceed two.

For $L_b > L_c$, the allowable flexural stress in tension is given by

$$F_b = 0.60F_y \tag{1.29}$$

and the allowable flexural stress in compression is given by the larger value calculated from Equations 1.30 and 1.31. While Equation 1.30 normally controls for deep, thin-flanged sections where warping restraint torsional resistance dominates, Equation 1.31 normally controls for shallow, thick-flanged sections where St Venant torsional resistance dominates:

$$F_b = \begin{cases} \left[\dfrac{2}{3} - \dfrac{F_y(l/r_T)^2}{1,530 \times 10^3 C_b}\right]F_y \leq 0.60F_y, & \text{if } \sqrt{\dfrac{102,000C_b}{F_y}} \leq \dfrac{l}{r_T} < \sqrt{\dfrac{510,000C_b}{F_y}} \\ \dfrac{170,000C_b}{(l/r_T)^2} \leq 0.60F_y, & \text{if } \dfrac{l}{r_T} \geq \sqrt{\dfrac{510,000C_b}{F_y}} \end{cases} \tag{1.30}$$

$$F_b = \frac{12,000C_b}{ld/A_f} \leq 0.60F_y \tag{1.31}$$

where l is the distance between cross-sections braced against twist or lateral displacement of the compression flange (in.), r_T is the radius of gyration of a section comprising the compression flange plus $\frac{1}{3}$ of the compression web area taken about an axis in the plane of the web (in.), A_f is the compression flange area (in.^{2}), d is the depth of cross-section (in.), $C_b = 12.5M_{max}/(2.5M_{max} + 3M_A + 4M_B + 3M_C)$, where M_{max}, M_A, M_B, M_C are absolute values of the maximum moment, quarter-point moment, midpoint moment, and three-quarter point moment along the unbraced length of the member, respectively. (For simplicity in design, C_b can conservatively be taken as unity.)

It should be cautioned that Equations 1.30 and 1.31 are applicable only to I and Channel shapes with an axis of symmetry in, and loaded in the plane of the web. In addition, Equation 1.31 is applicable only if the compression flange is solid and approximately rectangular in shape and its area is not less than the tension flange.

1.5.1.1.1.2 Compact Section Members Bent about Their Minor Axes — Since lateral torsional buckling will not occur for bending about the minor axes, regardless of the value of L_b, the allowable flexural stress is

$$F_b = 0.75F_y \qquad (1.32)$$

1.5.1.1.1.3 Noncompact Section Members Bent about Their Major Axes — For $L_b \leq L_c$

$$F_b = 0.60F_y \qquad (1.33)$$

where L_c is defined as in Equation 1.28.

For $L_b > L_c$, F_b is given in Equations 1.29 to 1.31.

1.5.1.1.1.4 Noncompact Section Members Bent about Their Minor Axes — Regardless of the value of L_b,

$$F_b = 0.60F_y \qquad (1.34)$$

1.5.1.1.1.5 Slender Element Sections — Refer to Section 1.10.

1.5.1.1.2 Shear Strength Criterion

For practically all structural shapes commonly used in constructions, the shear resistance from the flanges is small compared to the webs. As a result, the shear resistance for flexural members is normally determined on the basis of the webs only. The amount of web shear resistance is dependent on the width–thickness ratio h/t_w of the webs. If h/t_w is small, the failure mode is web yielding. If h/t_w is large, the failure mode is web buckling. To avoid web shear failure, the computed shear stress, f_v, shall not exceed the allowable shear stress, F_v, given by

$$F_v = \begin{cases} 0.40F_y, & \text{if } \dfrac{h}{t_w} \leq \dfrac{380}{\sqrt{F_y}} \\ \dfrac{C_v}{2.89}F_y \leq 0.40F_y, & \text{if } \dfrac{h}{t_w} > \dfrac{380}{\sqrt{F_y}} \end{cases} \qquad (1.35)$$

where

$C_v = 45{,}000k_v/[F_y(h/t_w)^2]$, if $C_v \leq 0.8$

$\quad = [190/(h/t_w)]\sqrt{(k_v/F_y)}$, if $C_v > 0.8$

$k_v = 4.00 + 5.34/(a/h)^2$, if $a/h \leq 1.0$

$\quad = 5.34 + 4.00/(a/h)^2$, if $a/h > 1.0$

t_w = web thickness (in.)

a = clear distance between transverse stiffeners (in.)

h = clear distance between flanges at section under investigation (in.)

1.5.1.1.3 Criteria for Concentrated Loads

1.5.1.1.3.1 Local Flange Bending — If the concentrated force that acts on the beam flange is tensile, the beam flange may experience excessively bending, leading to failure by fracture. To preclude this type of failure, transverse stiffeners are to be provided opposite the tension flange unless the length of the load when measured across the beam flange is less than 0.15 times the flange width, or if the flange thickness, t_f, exceeds

$$0.4\sqrt{\dfrac{P_{bf}}{F_y}} \qquad (1.36)$$

where P_{bf} is the computed tensile force multiplied by $\frac{5}{3}$ if the force is due to live and dead loads only or by $\frac{4}{3}$ if the force is due to live and dead loads in conjunction with wind or earthquake loads (kip) and F_y is the specified minimum yield stress (ksi).

1.5.1.1.3.2 Local Web Yielding — To prevent local web yielding, the concentrated compressive force, R, should not exceed $0.66R_n$, where R_n is the web yielding resistance given in Equation 1.54 or 1.55, whichever applies.

1.5.1.1.3.3 Web Crippling — To prevent web crippling, the concentrated compressive force, R, should not exceed $0.50R_n$, where R_n is the web crippling resistance given in Equations 1.56, 1.57, or 1.58, whichever applies.

1.5.1.1.3.4 Sideways Web Buckling — To prevent sideways web buckling, the concentrated compressive force, R, should not exceed R_n, where R_n is the sideways web buckling resistance given in Equation 1.59 or 1.60, whichever applies, except the term $C_r t_w^3 t_f/h^2$ is replaced by $6800 t_w^3/h$.

1.5.1.1.3.5 Compression Buckling of the Web — When the web is subjected to a pair of concentrated force acting on both flanges, buckling of the web may occur if the web depth clear of fillet, d_c, is greater than

$$\frac{4100 t_w^3 \sqrt{F_y}}{P_{bf}} \tag{1.37}$$

where t_w is the web thickness, F_y is the minimum specified yield stress, and P_{bf} is as defined in Equation 1.36.

1.5.1.1.4 Deflection Criterion

Deflection is a serviceability consideration. Since most beams are fabricated with a camber that somewhat offsets the dead load deflection, consideration is often given to deflection due to live load only. For beams supporting plastered ceilings, the service live load deflection preferably should not exceed $L/360$ where L is the beam span. A larger deflection limit can be used if due considerations are given to ensure the proper functioning of the structure.

EXAMPLE 1.3

Using ASD, determine the amount of increase in flexural capacity of a W24 × 55 section bent about its major axis if two 7 in. × $\frac{1}{2}$ in. (178 mm × 13 mm) cover plates are bolted to its flanges as shown in Figure 1.9. The beam is laterally supported at every 5-ft (1.52-m) interval. Use A36 steel. Specify the type, diameter, and longitudinal spacing of the bolts used if the maximum shear to be resisted by the cross-section is 100 kip (445 kN).

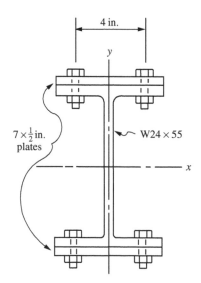

FIGURE 1.9 Beam section with cover plates.

Section properties: A W24 × 55 section has the following section properties:

$$b_f = 7.005 \text{ in.}, \quad t_f = 0.505 \text{ in.}, \quad d = 23.57 \text{ in.}, \quad t_w = 0.395 \text{ in.}, \quad I_x = 1350 \text{ in.}^4, \quad S_x = 114 \text{ in.}^3$$

Check compactness: Refer to Table 1.8, and assuming that the transverse distance between the two bolt lines is 4 in., we have

$$\text{Beam flanges} \quad \left[\frac{b_f}{2t_f} = 6.94\right] < \left[0.38\sqrt{\frac{E}{F_y}} = 10.8\right]$$

$$\text{Beam web} \quad \left[\frac{d}{t_w} = 59.7\right] < \left[3.76\sqrt{\frac{E}{F_y}} = 107\right]$$

$$\text{Cover plates} \quad \left[\frac{4}{1/2} = 8\right] < \left[0.939\sqrt{\frac{E}{F_y}} = 26.7\right]$$

Therefore, the section is compact.

Determine the allowable flexural stress, F_b: Since the section is compact and the lateral unbraced length, $L_b = 60$ in., is less than $L_c = 83.4$ in., the allowable bending stress from Equation 1.28 is $0.66F_y = 24$ ksi.

Determine the section modulus of the beam with cover plates:

$$S_{x,\text{combination section}} = \frac{I_{x,\text{combination section}}}{c}$$

$$= \frac{1350 + 2\left[\left(\frac{1}{12}\right)(7)\left(\frac{1}{2}\right)^3 + (7)\left(\frac{1}{2}\right)(12.035)^2\right]}{\left[\left(\frac{23.57}{2}\right) + \left(\frac{1}{2}\right)\right]}$$

$$= 192 \text{ in.}^3$$

Determine flexural capacity of the beam with cover plates:

$$M_{x,\text{combination section}} = S_{x,\text{combination section}} F_b = (192)(24) = 4608 \text{ k in.}$$

Since the flexural capacity of the beam without cover plates is

$$M_x = S_x F_b = (114)(24) = 2736 \text{ k in.}$$

the increase in flexural capacity is 68.4%.

Determine diameter and longitudinal spacing of bolts: From Mechanics of Materials, the relationship between the shear flow, q, the number of bolts per shear plane, n, the allowable bolt shear stress, F_v, the cross-sectional bolt area, A_b, and the longitudinal bolt spacing, s, at the interface of two component elements of a combination section is given by

$$\frac{nF_v A_b}{s} = q$$

Substituting $n = 2$, $q = VQ/I = (100)[(7)(\frac{1}{2})(12.035)]/2364 = 1.78$ k/in. into the above equation, we have

$$\frac{F_v A_b}{s} = 0.9 \text{ k/in.}$$

If $\frac{1}{2}$ in. diameter A325-N bolts are used, we have $A_b = \pi(\frac{1}{2})^2/4 = 0.196$ in.^{2} and $F_v = 21$ ksi (from Table 1.12), from which s can be solved from the above equation to be 4.57 in. However, for ease of installation, use $s = 4.5$ in.

In calculating the section properties of the combination section, no deduction is made for the bolt holes in the beam flanges nor the cover plates, which is allowed provided that the following condition is satisfied:

$$0.5F_u A_{fn} \geq 0.6F_y A_{fg}$$

where F_y and F_u are the minimum specified yield strength and tensile strength, respectively. A_{fn} is the net flange area and A_{fg} is the gross flange area. For this problem

Beam flanges $[0.5F_u A_{fn} = 0.5(58)(7.005 - 2 \times 1/2)(0.505) = 87.9\,\text{kip}]$
$> [0.6F_y A_{fg} = 0.6(36)(7.005)(0.505) = 76.4\,\text{kip}]$

Cover plates $[0.5F_u A_{fn} = 0.5(58)(7 - 2 - 1/2)(1/2) = 87\,\text{kip}]$
$> [0.6F_y A_{fg} = 0.6(36)(7)(1/2) = 75.6\,\text{kip}]$

so the use of gross cross-sectional area to compute section properties is justified. In the event that the condition is violated, cross-sectional properties should be evaluated using an effective tension flange area A_{fe} given by

$$A_{fe} = \frac{5}{6}\frac{F_u}{F_y} A_{fn}$$

So, use $\frac{1}{2}$-in. diameter A325-N bolts spaced 4.5 in. apart longitudinally in two lines 4 in. apart to connect the cover plates to the beam flanges.

1.5.1.2 Load and Resistance Factor Design

1.5.1.2.1 *Flexural Strength Criterion*
Flexural members must be designed to satisfy the flexural strength criterion of

$$\phi_b M_n \geq M_u \tag{1.38}$$

where $\phi_b M_n$ is the design flexural strength and M_u is the required strength. The design flexural strength is determined as given below.

1.5.1.2.1.1 Compact Section Members Bent about Their Major Axes — For $L_b \leq L_p$ (plastic hinge formation)

$$\phi_b M_n = 0.90 M_p \tag{1.39}$$

For $L_p < L_b \leq L_r$ (inelastic lateral torsional buckling)

$$\phi_b M_n = 0.90 C_b \left[M_p - (M_p - M_r)\left(\frac{L_b - L_p}{L_r - L_p}\right)\right] \leq 0.90 M_p \tag{1.40}$$

For $L_b > L_r$ (elastic lateral torsional buckling)

For I-shaped members and channels:

$$\phi_b M_n = 0.90 C_b \left[\frac{\pi}{L_b}\sqrt{EI_y GJ + \left(\frac{\pi E}{L_b}\right)^2 I_y C_w}\right] \leq 0.90 M_p \tag{1.41}$$

For solid rectangular bars and symmetric box sections:

$$\phi_b M_n = 0.90 C_b \frac{57,000\sqrt{JA}}{L_b/r_y} \leq 0.90 M_p \tag{1.42}$$

The variables used in the above equations are defined in the following:

L_b = lateral unsupported length of the member
L_p, L_r = limiting lateral unsupported lengths given in the following table:

Structural shape	L_p	L_r
I-shaped sections, channels	$1.76 r_y / \sqrt{(E/F_{yf})}$ where r_y = radius of gyration about minor axis E = modulus of elasticity F_{yf} = flange yield strength	$[r_y X_1 / F_L]\{\sqrt{[1 + \sqrt{(1 + X_2 F_L^2)}]}\}$ where r_y = radius of gyration about minor axis, in. $X_1 = (\pi/S_x)\sqrt{(EGJA/2)}$ $X_2 = (4C_w/I_y)(S_x/GJ)^2$ F_L = smaller of $(F_{yf} - F_r)$ or F_{yw} F_{yf} = flange yield stress, ksi F_{yw} = web yield stress, ksi F_r = 10 ksi for rolled shapes, 16.5 ksi for welded shapes S_x = elastic section modulus about the major axis, in.^{3} (use S_{xc}, the elastic section modulus about the major axis with respect to the compression flange if the compression flange is larger than the tension flange) I_y = moment of inertia about the minor axis, in.^{4} J = torsional constant, in.^{4} C_w = warping constant, in.^{6} E = modulus of elasticity, ksi G = shear modulus, ksi
Solid rectangular bars, symmetric box sections	$[0.13 r_y E\sqrt{(JA)}]/M_p$ where r_y = radius of gyration about minor axis E = modulus of elasticity J = torsional constant A = cross-sectional area M_p = plastic moment capacity = $F_y Z_x$ F_y = yield stress Z_x = plastic section modulus about the major axis	$[2 r_y E\sqrt{(JA)}]/M_r$ where r_y = radius of gyration about minor axis J = torsional constant A = cross-sectional area $M_r = F_{yf} S_x$ F_y = yield stress F_{yf} = flange yield strength S_x = elastic section modulus about the major axis

Note: L_p given in this table are valid only if the bending coefficient C_b is equal to unity. If $C_b > 1$, the value of L_p can be increased. However, using the L_p expressions given above for $C_b > 1$ will give conservative value for the flexural design strength.

and

$M_p = F_y Z_x$
$M_r = F_L S_x$ for I-shaped sections and channels, $F_{yf} S_x$ for solid rectangular bars and box sections
F_L = smaller of $(F_{yf} - F_r)$ or F_{yw}
F_{yf} = flange yield stress, ksi
F_{yw} = web yield stress, ksi
F_r = 10 ksi for rolled sections, 16.5 ksi for welded sections
F_y = specified minimum yield stress

S_x = elastic section modulus about the major axis
Z_x = plastic section modulus about the major axis
I_y = moment of inertia about the minor axis
J = torsional constant
C_w = warping constant
E = modulus of elasticity
G = shear modulus
C_b = $12.5M_{max}/(2.5M_{max} + 3M_A + 4M_B + 3M_C)$
M_{max}, M_A, M_B, M_C = absolute value of maximum moment, quarter-point moment, midpoint moment, and three-quarter point moment along the unbraced length of the member, respectively.

C_b is a factor that accounts for the effect of moment gradient on the lateral torsional buckling strength of the beam. Lateral torsional buckling strength increases for a steep moment gradient. The worst loading case as far as lateral torsional buckling is concerned is when the beam is subjected to a uniform moment resulting in single curvature bending. For this case $C_b = 1$. Therefore, the use of $C_b = 1$ is conservative for the design of beams.

1.5.1.2.1.2 Compact Section Members Bent about Their Minor Axes — Regardless of L_b, the limit state will be plastic hinge formation

$$\phi_b M_n = 0.90M_{py} = 0.90F_y Z_y \tag{1.43}$$

1.5.1.2.1.3 Noncompact Section Members Bent about Their Major Axes — For $L_b \leq L_p'$ (flange or web local buckling)

$$\phi_b M_n = \phi_b M_n' = 0.90\left[M_p - (M_p - M_r)\left(\frac{\lambda - \lambda_p}{\lambda_r - \lambda_p}\right)\right] \tag{1.44}$$

where

$$L_p' = L_p + (L_r - L_p)\left(\frac{M_p - M_n'}{M_p - M_r}\right) \tag{1.45}$$

L_p, L_r, M_p, M_r are defined as before for compact section members, and

For flange local buckling:
   $\lambda = b_f/2t_f$ for I-shaped members, b_f/t_f for channels
   λ_p, λ_r are defined in Table 1.8
For web local buckling:
   $\lambda = h_c/t_w$
   λ_p, λ_r are defined in Table 1.8

in which b_f is the flange width, t_f is the flange thickness, h_c is twice the distance from the neutral axis to the inside face of the compression flange less the fillet or corner radius, and t_w is the web thickness.

For $L_p' < L_b \leq L_r$ (inelastic lateral torsional buckling), $\phi_b M_n$ is given by Equation 1.40 except that the limit $0.90M_p$ is to be replaced by the limit $0.90M_n'$.

For $L_b > L_r$ (elastic lateral torsional buckling), $\phi_b M_n$ is the same as for compact section members as given in Equation 1.41 or 1.42.

1.5.1.2.1.4 Noncompact Section Members Bent about Their Minor Axes — Regardless of the value of L_b, the limit state will be either flange or web local buckling, and $\phi_b M_n$ is given by Equation 1.42.

1.5.1.2.1.5 Slender Element Sections — Refer to Section 1.10.

1.5.1.2.1.6 Tees and Double Angle Bent about Their Major Axes — The design flexural strength for tees and double-angle beams with flange and web slenderness ratios less than the corresponding limiting slenderness ratios λ_r shown in Table 1.8 is given by

$$\phi_b M_n = 0.90 \left[\frac{\pi \sqrt{EI_y GJ}}{L_b} \left(B + \sqrt{1 + B^2} \right) \right] \leq 0.90(\beta M_y) \tag{1.46}$$

where

$$B = \pm 2.3 \left(\frac{d}{L_b} \right) \sqrt{\frac{I_y}{J}} \tag{1.47}$$

Use the plus sign for B if the *entire* length of the stem along the unbraced length of the member is in tension. Otherwise, use the minus sign. β equals 1.5 for stems in tension and equals 1.0 for stems in compression. The other variables in Equation 1.46 are defined as before in Equation 1.41.

1.5.1.2.2 Shear Strength Criterion

For a satisfactory design, the design shear strength of the webs must exceed the factored shear acting on the cross-section, that is

$$\phi_v V_n \geq V_u \tag{1.48}$$

Depending on the slenderness ratios of the webs, three limit states can be identified: shear yielding, inelastic shear buckling, and elastic shear buckling. The design shear strength that corresponds to each of these limit states are given as follows:

For $h/t_w \leq 2.45\sqrt{(E/F_{yw})}$ (shear yielding of web)

$$\phi_v V_n = 0.90[0.60 F_{yw} A_w] \tag{1.49}$$

For $2.45\sqrt{(E/F_{yw})} < h/t_w \leq 3.07\sqrt{(E/F_{yw})}$ (inelastic shear buckling of web)

$$\phi_v V_n = 0.90 \left[0.60 F_{yw} A_w \frac{2.45\sqrt{(E/F_{yw})}}{h/t_w} \right] \tag{1.50}$$

For $3.07\sqrt{(E/F_{yw})} < h/t_w \leq 260$ (elastic shear buckling of web)

$$\phi_v V_n = 0.90 A_w \left[\frac{4.52 E}{(h/t_w)^2} \right] \tag{1.51}$$

The variables used in the above equations are defined in the following, where h is the clear distance between flanges less the fillet or corner radius, t_w is the web thickness, F_{yw} is the yield stress of web, $A_w = d t_w$, and d is the overall depth of the section.

1.5.1.2.3 Criteria for Concentrated Loads

When concentrated loads are applied normal to the flanges in planes parallel to the webs of flexural members, the flanges and webs must be checked to ensure that they have sufficient strengths ϕR_n to withstand the concentrated forces R_u, that is

$$\phi R_n \geq R_u \tag{1.52}$$

The design strengths for a variety of limit states are given below.

1.5.1.2.3.1 Local Flange Bending — The design strength for local flange bending is given by

$$\phi R_n \geq 0.90[6.25 t_f^2 F_{yf}] \tag{1.53}$$

where t_f is the flange thickness of the loaded flange and F_{yf} is the flange yield stress.

The design strength in Equation 1.53 is applicable only if the length of load across the member flange exceeds $0.15b$, where b is the member flange width. If the length of load is less than $0.15b$, the limit state of local flange bending need not be checked. Also, Equation 1.53 shall be reduced by a factor of half if the concentrated force is applied less than $10 t_f$ from the beam end.

1.5.1.2.3.2 Local Web Yielding — The design strength for yielding of the beam web at the toe of the fillet under tensile or compressive loads acting on one or both flanges are

If the load acts at a distance from the beam end which exceeds the depth of the member

$$\phi R_n = 1.00[(5k + N)F_{yw} t_w] \tag{1.54}$$

If the load acts at a distance from the beam end which does not exceed the depth of the member

$$\phi R_n = 1.00[(2.5k + N)F_{yw} t_w] \tag{1.55}$$

where k is the distance from the outer face of the flange to the web toe of the fillet, N is the length of bearing on the beam flange, F_{yw} is the web yield stress, and t_w is the web thickness.

1.5.1.2.3.3 Web Crippling — The design strength for crippling of beam web under compressive loads acting on one or both flanges are

If the load acts at a distance from the beam end which exceeds half the depth of the beam

$$\phi R_n = 0.75 \left\{ 0.80 t_w^2 \left[1 + 3 \left(\frac{N}{d} \right) \left(\frac{t_w}{t_f} \right)^{1.5} \right] \sqrt{\frac{E F_{yw} t_f}{t_w}} \right\} \tag{1.56}$$

If the load acts at a distance from the beam end which does not exceed half the depth of the beam and if $N/d \leq 0.2$

$$\phi R_n = 0.75 \left\{ 0.40 t_w^2 \left[1 + 3 \left(\frac{N}{d} \right) \left(\frac{t_w}{t_f} \right)^{1.5} \right] \sqrt{\frac{E F_{yw} t_f}{t_w}} \right\} \tag{1.57}$$

If the load acts at a distance from the beam end which does not exceed half the depth of the beam and if $N/d > 0.2$

$$\phi R_n = 0.75 \left\{ 0.40 t_w^2 \left[1 + \left(\frac{4N}{d} - 0.2 \right) \left(\frac{t_w}{t_f} \right)^{1.5} \right] \sqrt{\frac{E F_{yw} t_f}{t_w}} \right\} \tag{1.58}$$

where d is the overall depth of the section, t_f is the flange thickness, and the other variables are the same as those defined in Equations 1.54 and 1.55.

1.5.1.2.3.4 Sideways Web Buckling — Sideways web buckling may occur in the web of a member if a compressive concentrated load is applied to a flange not restrained against relative movement by stiffeners or lateral bracings. The sideways web buckling design strength for the member is

If the loaded flange is restrained against rotation about the longitudinal member axis and $(h/t_w)(l/b_f)$ is less than 2.3

$$\phi R_n = 0.85 \left\{ \frac{C_r t_w^2 t_f}{h^2} \left[1 + 0.4 \left(\frac{h/t_w}{l/b_f} \right)^3 \right] \right\} \tag{1.59}$$

If the loaded flange is not restrained against rotation about the longitudinal member axis and $(d_c/t_w)(l/b_f)$ is less than 1.7

$$\phi R_n = 0.85 \left\{ \frac{C_r t_w^2 t_f}{h^2} \left[0.4 \left(\frac{h/t_w}{l/b_f} \right)^3 \right] \right\} \tag{1.60}$$

where t_f is the flange thickness (in.), t_w is the web thickness (in.), h is the clear distance between flanges less the fillet or corner radius for rolled shapes; distance between adjacent lines of fasteners or clear distance between flanges when welds are used for built-up shapes (in.), b_f is the flange width (in.), l is the largest laterally unbraced length along either flange at the point of load (in.), $C_r = 960,000$ ksi if $M_u/M_y < 1$ at the point of load and $C_r = 480,000$ ksi if $M_u/M_y \geq 1$ at the point of load, and M_y is the yield moment.

1.5.1.2.3.5 Compression Buckling of the Web — This limit state may occur in members with unstiffened webs when both flanges are subjected to compressive forces. The design strength for this limit state is

$$\phi R_n = 0.90 \left[\frac{24 t_w^3 \sqrt{EF_{yw}}}{h} \right] \tag{1.61}$$

This design strength shall be reduced by a factor of half if the concentrated forces are acting at a distance less than half the beam depth from the beam end. The variables in Equation 1.61 are the same as those defined in Equations 1.58 to 1.60.

Stiffeners shall be provided in pairs if any one of the above strength criteria is violated. If the local flange bending or the local web yielding criterion is violated, the stiffener pair to be provided to carry the excess R_u need not extend more than one-half the web depth. The stiffeners shall be welded to the loaded flange if the applied force is tensile. They shall either bear on or be welded to the loaded flange if the applied force is compressive. If the web crippling or the compression web buckling criterion is violated, the stiffener pair to be provided shall extend the full height of the web. They shall be designed as axially loaded compression members (see Section 1.4) with an effective length factor $K = 0.75$, a cross-section A_g composed of the cross-sectional areas of the stiffeners plus $25 t_w^2$ for interior stiffeners and $12 t_w^2$ for stiffeners at member ends.

1.5.1.2.4 Deflection Criterion
The deflection criterion is the same as that for ASD. Since deflection is a serviceability limit state, service (rather than factored) loads are used in deflection computations.

1.5.2 Continuous Beams

Continuous beams shall be designed in accordance with the criteria for flexural members given in the preceding section. However, a 10% reduction in negative moments due to gravity loads is permitted at the supports provided that

- The maximum positive moment between supports is increased by $\frac{1}{10}$ the average of the negative moments at the supports.
- The section is compact.
- The lateral unbraced length does not exceed L_c (for ASD) or L_{pd} (for LRFD) where L_c is as defined in Equation 1.26 and L_{pd} is given by

$$L_{pd} = \begin{cases} \left[0.12 + 0.076 \left(\frac{M_1}{M_2} \right) \right] \left(\frac{E}{F_y} \right) r_y, & \text{for I-shaped members} \\ \left[0.17 + 0.10 \left(\frac{M_1}{M_2} \right) \right] \left(\frac{E}{F_y} \right) r_y \geq 0.10 \left(\frac{E}{F_y} \right) r_y, & \text{for solid rectangular and box sections} \end{cases} \tag{1.62}$$

in which, F_y is the specified minimum yield stress of the compression flange, M_1/M_2 is the ratio of smaller to larger moment within the unbraced length, taken as positive if the moments cause reverse

curvature and negative if the moments cause single curvature, and r_y is the radius of gyration about the minor axis.
- The beam is not a hybrid member.
- The beam is not made of high-strength steel.
- The beam is continuous over the supports (i.e., not cantilevered).

EXAMPLE 1.4

Using LRFD, select the lightest W-section for the three-span continuous beam shown in Figure 1.10a to support a uniformly distributed dead load of 1.5 k/ft (22 kN/m) and a uniformly distributed live load of 3 k/ft (44 kN/m). The beam is laterally braced at the supports A, B, C, and D. Use A36 steel.

Load combinations: The beam is to be designed based on the worst load combination of Table 1.3. By inspection, the load combination $1.2D + 1.6L$ will control the design. Thus, the beam will be designed to support a factored uniformly distributed dead load of $1.2 \times 1.5 = 1.8$ k/ft and a factored uniformly distributed live load of $1.6 \times 3 = 4.8$ k/ft.

Placement of loads: The uniform dead load is to be applied over the entire length of the beam as shown in Figure 1.10b. The uniform live load is to be applied to spans AB and CD as shown in Figure 1.10c to obtain the maximum positive moment and it is to be applied to spans AB and BC as shown in Figure 1.10d to obtain the maximum negative moment.

Reduction of negative moment at supports: Assuming the beam is compact and $L_b < L_{pd}$ (we shall check these assumptions later), a 10% reduction in support moment due to gravity load is allowed provided that the maximum moment is increased by $\frac{1}{10}$ the average of the negative support moments. This reduction is shown in the moment diagrams as solid lines in Figure 1.10b and d. (The dotted lines in these figures represent the unadjusted moment diagrams.) This provision for support moment reduction takes into consideration the beneficial effect of moment redistribution in continuous beams, and it allows for the selection of a lighter section if the design is governed by negative moments. Note that no reduction in negative moments is made to the case when only spans AB and CD are loaded. This is because for this load case, the negative support moments are less than the positive in-span moments.

Determination of the required flexural strength, M_u: Combining load case 1 and load case 2, the maximum positive moment is found to be 256 kip ft. Combining load case 1 and load case 3, the maximum negative moment is found to be 266 kip ft. Thus, the design will be controlled by the negative moment and so $M_u = 266$ kip ft.

Beam selection: A beam section is to be selected based on Equation 1.38. The critical segment of the beam is span AB. For this span, the lateral unsupported length, $L_b = 20$ ft. For simplicity, the bending coefficient, C_b, is conservatively taken as 1. The selection of a beam section is facilitated by the use of a series of beam charts contained in the AISC-LRFD Manual (AISC 2001). Beam charts are plots of flexural design strength $\phi_b M_n$ of beams as a function of the lateral unsupported length L_b based on Equations 1.39 to 1.41. A beam is considered satisfactory for the limit state of flexure if the beam strength curve envelopes the required flexural strength for a given L_b.

For the present example, $L_b = 20$ ft and $M_u = 266$ kip ft, the lightest section (the first solid curve that envelopes $M_u = 266$ kip ft for $L_b = 20$ ft) obtained from the chart is a W16 × 67 section. Upon adding the factored dead weight of this W16 × 67 section to the specified loads, the required flexural strength increases from 266 to 269 kip ft. Nevertheless, the beam strength curve still envelopes this required strength for $L_b = 20$ ft; therefore, the section is adequate.

Check for compactness: For the W16 × 67 section

$$\text{Flange:} \quad \left[\frac{b_f}{2t_f} = 7.7\right] < \left[0.38\sqrt{\frac{E}{F_y}} = 10.8\right]$$

$$\text{Web:} \quad \left[\frac{h_c}{t_w} = 35.9\right] < \left[3.76\sqrt{\frac{E}{F_y}} = 106.7\right]$$

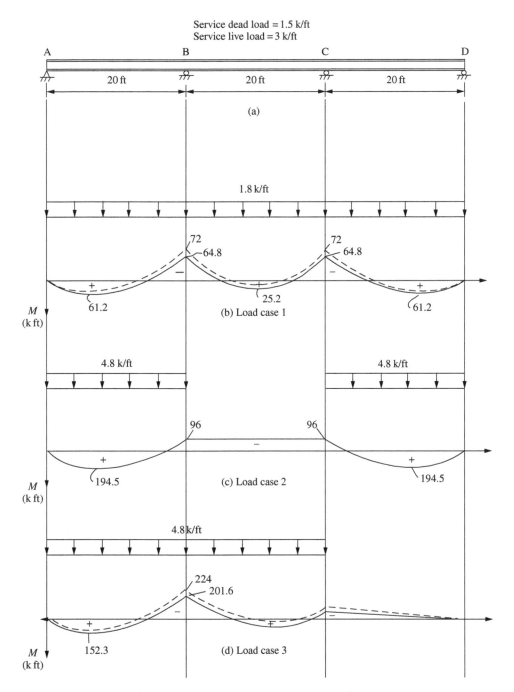

FIGURE 1.10 Design of a three-span continuous beam (1 k = 4.45 kN, 1 ft = 0.305 m).

Therefore, the section is compact.

Check whether $L_b < L_{pd}$: Using Equation 1.62, with $M_1/M_2 = 0$, $r_y = 2.46$ in., and $F_y = 36$ ksi, we have $L_{pd} = 246$ in. (or 20.5 ft). Since $L_b = 20$ ft is less than $L_{pd} = 20.5$ ft, the assumption made earlier is validated.

Check for the limit state of shear: The selected section must satisfy the shear strength criterion of Equation 1.48. From structural analysis, it can be shown that maximum shear occurs just to the left of support B under load case 1 (for dead load) and load case 3 (for live load). It has a magnitude of 81.8 kip. For the W16 × 67 section, $h/t_w = 35.9$, which is less than $2.45\sqrt{(E/F_{yw})} = 69.5$, so the design shear strength is given by Equation 1.47. We have, for $F_{yw} = 36$ ksi and $A_w = dt_w = (16.33)(0.395)$

$$[\phi_v V_n = 0.90(0.60 F_{yw} A_w) = 125 \text{ kip}] > [V_u = 81.8 \text{ kip}]$$

Therefore, shear is not a concern. Normally, the limit state of shear will not control unless for short beams subjected to heavy loads.

Check for limit state of deflection: Deflection is a serviceability limit state. As a result, a designer should use service (not factored) load for deflection calculations. In addition, most beams are cambered to offset deflection caused by dead loads, so only live loads are considered in deflection calculations. From structural analysis, it can be shown that maximum deflection occurs in spans AB and CD when (service) live loads are placed on those two spans. The magnitude of the deflection is 0.297 in. Assuming the maximum allowable deflection is $L/360$ where L is the span length between supports, we have an allowable deflection of $20 \times 12/360 = 0.667$ in. Since the calculated deflection is less than the allowable deflection, deflection is not a problem.

Check for the limit state of web yielding and web crippling at points of concentrated loads: From structural analysis, it can be shown that maximum support reaction occurs at support B when the beam is subjected to loads shown as load case 1 (for dead load) and load case 3 (for live load). The magnitude of the reaction R_u is 157 kip. Assuming point bearing, that is, $N=0$, we have for $d=16.33$ in., $k=1.375$ in., $t_f = 0.665$ in., and $t_w = 0.395$ in.

Web yielding: $[\phi R_n = \text{Equation } 4.54 = 97.8 \text{ kip}] < [R_u = 157 \text{ kip}]$
Web crippling: $[\phi R_n = \text{Equation } 4.56 = 123 \text{ kip}] < [R_u = 157 \text{ kip}]$

Thus, both the web yielding and web crippling criteria are violated. As a result, we need to provide web stiffeners or bearing plate at support B. Suppose we choose the latter, the size of the bearing plate can be determined by solving Equations 1.54 and 1.56 for N using $R_u = 157$ kip. The result is $N = 4.2$ in. and 3.3 in., respectively. So, use $N = 4.25$ in. The width of the plate, B, should conform with the flange width, b_f, of the W-section. The W16 × 67 section has a flange width of 10.235 in., so use $B = 10.5$ in. The thickness of the bearing plate is to be calculated from the following equation (AISC 2001):

$$t = \sqrt{\frac{2.22 R_u [(B - 2k)/2]^2}{A F_y}}$$

where R_u is the factored concentrated load at the support (kip), B is the width of the bearing plate (in.), k is the distance from the web toe of the fillet to the outer surface of the flange (in.), A is the area of the bearing plate (in.²), and F_y is the yield strength of the bearing plate.

Substituting $R_u = 157$ kip, $B = 10.5$ in., $k = 1.375$ in., $A = 42$ in.², and $F_y = 36$ ksi into the above equation, we obtain $t = 1.86$ in. Therefore, use a $1\frac{7}{8}$ in. plate.

For uniformity, use the same size plate at all the supports. The bearing plates are to be welded to the supporting flange of the W-section.

Use a W16 × 67 section. Provide bearing plates of size $1\frac{7}{8}$ in. × 4 in. × $10\frac{1}{2}$ in. at the supports.

1.5.3 Beam Bracing

The design strength of beams that are bent about their major axes depends on their lateral unsupported length L_b. The manner a beam is braced against out-of-plane deformation affects its design. Bracing can be provided by various means such as cross-frames, cross-beams or diaphragms, or encasement of the beam flange in the floor slab (Yura 2001). Two types of bracing systems are identified in the AISC-LRFD Specification — relative and nodal. A relative brace controls the movement of a braced point with respect to adjacent braced points along the span of the beam. A nodal (or discrete) brace controls the movement of a braced point without regard to the movement of adjacent braced points. Regardless of the type of bracing system used, braces must be designed with sufficient strength and stiffness to prevent out-of-plane movement of the beam at the braced points. Out-of-plane movement consists of lateral deformation of the beam and twisting of cross-sections. Lateral stability of beams can be achieved by lateral bracing, torsional bracing, or a combination of the two. For lateral bracing, bracing shall be attached near the compression flange for members bend in single curvature (except cantilevers). For cantilevers, bracing shall be attached to the tension flange at the free end. For members bend in double curvature, bracing shall be attached to both flanges near the inflection point. For torsional bracing, bracing can be attached at any cross-sectional location.

1.5.3.1 Stiffness Requirement for Lateral Bracing

The required brace stiffness of the bracing assembly in a direction perpendicular to the longitudinal axis of the braced member, in the plane of buckling, is given by

$$\beta_{br} = \begin{cases} \dfrac{5.33 M_u C_d}{L_{br} h_o} & \text{for relative bracing} \\[3mm] \dfrac{13.3 M_u C_d}{L_{br} h_o} & \text{for nodal bracing} \end{cases} \tag{1.63}$$

where M_u is the required flexural strength, $C_d = 1.0$ for single curvature bending; 2.0 for double curvature bending near the inflection point, L_{br} is the distance between braces, and h_o is the distance between flange centroids.

L_{br} can be replaced by L_q (the maximum unbraced length for M_u) if $L_{br} < L_q$. (For ASD, replace M_u in Equation 1.63 by $1.5 M_a$, where M_a is the required flexural strength based on ASD load combinations.)

1.5.3.2 Strength Requirement for Lateral Bracing

In addition to the stiffness requirement as stipulated above, braces must be designed for a required brace strength given by

$$P_{br} = \begin{cases} 0.008 \dfrac{M_u C_d}{h_o} & \text{for relative bracing} \\[3mm] 0.02 \dfrac{M_u C_d}{h_o} & \text{for nodal bracing} \end{cases} \tag{1.64}$$

The terms in Equation 1.64 are defined as in Equation 1.63.

1.5.3.3 Stiffness Requirement for Torsional Bracing

The required bracing stiffness is

$$\beta_{Tbr} = \frac{\beta_T}{\left(1 - \dfrac{\beta_T}{\beta_{sec}}\right)} \geq 0 \tag{1.65}$$

where

$$\beta_T = \begin{cases} \dfrac{3.2 L M_u^2}{n E I_y C_b^2} & \text{for nodal bracing} \\[3mm] \dfrac{3.2 M_u^2}{E I_y C_b^2} & \text{for continuous bracing} \end{cases} \tag{1.66}$$

and

$$\beta_{sec} = \begin{cases} \dfrac{3.3 E}{h_o} \left(\dfrac{1.5 h_o t_w^3}{12} + \dfrac{t_s b_s^3}{12} \right) & \text{for nodal bracing} \\[3mm] \dfrac{3.3 E t_w^3}{12 h_o} & \text{for continuous bracing} \end{cases} \tag{1.67}$$

in which L is the span length, M_u is the required moment, n is the number of brace points within the span, E is the modulus of elasticity, I_y is the moment of inertia of the minor axis, C_b is the bending coefficient as defined earlier, h_o is the distance between the flange centroids, t_w is the thickness of the beam web, t_s is the thickness of the web stiffener, and b_s is the width of the stiffener (or, for pairs of stiffeners, b_s = total width of stiffeners). (For ASD, replace M_u in Equation 1.66 by $1.5M_a$, where M_a is the required flexural strength based on ASD load combinations.)

1.5.3.4 Strength Requirement for Torsional Bracing

The connection between a torsional brace and the beam being braced must be able to withstand a moment given by

$$M_{Tbr} = \frac{0.024 M_u L}{n C_b L_{br}} \qquad (1.68)$$

where L_{br} is the distance between braces (if $L_{br} < L_q$, where L_q is the maximum unbraced length for M_u, then use L_q). The other terms in Equation 1.68 are defined in Equation 1.66.

EXAMPLE 1.5

Design an I-shaped cross-beam 12 ft (3.7 m) in length to be used as lateral braces to brace a 30-ft (9.1 m) long simply supported W30 × 90 girder at every third point. The girder was designed to carry a moment of 8000 kip in. (904 kN m). A992 steel is used.

Because a brace is provided at every third point, $L_{br} = 10$ ft $= 120$ in., $M_u = 8000$ kip in. as stated. $C_d = 1$ for single curvature bending. $h_o = d - t_f = 29.53 - 0.610 = 28.92$ in. for the W30 × 90 section. Substituting these values into Equations 1.63 and 1.64 for nodal bracing, we obtain $\beta_{br} = 30.7$ kip/in., and $P_{br} = 5.53$ kip.

As the cross-beam will be subject to compression, its slenderness ratio, l/r, should not exceed 200. Let us try the smallest size W-section, a W4 × 13 section, with $A = 3.83$ in.^{2}, $r_y = 1.00$ in., $\phi_c P_n = 25$ kip.

$$\text{Stiffness:} \quad \frac{EA}{l} = \frac{(29,000)(3.83)}{12 \times 12} = 771 \text{ kip/in.} > 30.7 \text{ kip/in.}$$

$$\text{Strength:} \quad \phi_c P_n = 25 \text{ kip} > 5.53 \text{ kip}$$

$$\text{Slenderness:} \quad \frac{l}{r_y} = \frac{12 \times 12}{1.00} = 144 < 200$$

Since all criteria are satisfied, the W4 × 13 section is adequate. Use W4 × 13 as cross-beams to brace the girder.

1.6 Combined Flexure and Axial Force

When a member is subject to the combined action of bending and axial force, it must be designed to resist stresses and forces arising from both bending and axial actions. While a tensile axial force may induce a stiffening effect on the member, a compressive axial force tends to destabilize the member, and the instability effects due to member instability (P–δ effect) and frame instability (P–Δ effect) must be properly accounted for. The P–δ effect arises when the axial force acts through the lateral deflection of the member relative to its chord. The P–Δ effect arises when the axial force acts through the relative displacements of the two ends of the member. Both effects tend to increase member deflection and moment, and so they must be considered in the design. A number of approaches are available in the literature to handle these so-called P–Δ effects (see, e.g., Chen and Lui 1991; Galambos 1998). The design of members subject to combined bending and axial force is facilitated by the use interaction equations. In these equations, the effects of bending and axial actions are combined in a certain manner to reflect the capacity demand on the member.

1.6.1 Design for Combined Flexure and Axial Force

1.6.1.1 Allowable Stress Design

The interaction equations are

If the axial force is tensile:

$$\frac{f_a}{F_t} + \frac{f_{bx}}{F_{bx}} + \frac{f_{by}}{F_{by}} \leq 1.0 \tag{1.69}$$

where f_a is the computed axial tensile stress, f_{bx}, f_{by} are computed bending tensile stresses about the major and minor axes, respectively, F_t is the allowable tensile stress (see Section 1.3), and F_{bx}, F_{by} are allowable bending stresses about the major and minor axes, respectively (see Section 1.5).

If the axial force is compressive:
Stability requirement

$$\frac{f_a}{F_a} + \left[\frac{C_{mx}}{1 - (f_a/F'_{ex})}\right]\frac{f_{bx}}{F_{bx}} + \left[\frac{C_{my}}{1 - (f_a/F'_{ey})}\right]\frac{f_{by}}{F_{by}} \leq 1.0 \tag{1.70}$$

Yield requirement

$$\frac{f_a}{0.66F_y} + \frac{f_{bx}}{F_{bx}} + \frac{f_{by}}{F_{by}} \leq 1.0 \tag{1.71}$$

However, if the axial force is small (when $f_a/F_a \leq 0.15$), the following interaction equation can be used in lieu of the above equations.

$$\frac{f_a}{F_a} + \frac{f_{bx}}{F_{bx}} + \frac{f_{by}}{F_{by}} \leq 1.0 \tag{1.72}$$

The terms in Equations 1.70 to 1.72 are defined as follows:

f_a, f_{bx}, f_{by} = Computed axial compressive stress, computed bending stresses about the major and minor axes, respectively. These stresses are to be computed based on a *first-order analysis*.

F_y = Minimum specified yield stress.

F'_{ex}, F'_{ey} = Euler stresses about the major and minor axes $(\pi^2 E/(Kl/r)_x, \pi^2 E/(Kl/r)_y)$ divided by a factor of safety of $\frac{23}{12}$.

C_m = a coefficient to account for the effect of moment gradient on member and frame instabilities (C_m is defined in Section 1.6.1.2).

The other terms are defined as in (Equation 1.69).

The terms in brackets in (Equation 1.70) are moment magnification factors. The computed bending stresses f_{bx}, f_{by} are magnified by these magnification factors to account for the P–δ effects in the member.

1.6.1.2 Load and Resistance Factor Design

Doubly or singly symmetric members subject to combined flexure and axial forces shall be designed in accordance with the following interaction equations:

For $P_u/\phi P_n \geq 0.2$

$$\frac{P_u}{\phi P_n} + \frac{8}{9}\left(\frac{M_{ux}}{\phi_b M_{nx}} + \frac{M_{uy}}{\phi_b M_{ny}}\right) \leq 1.0 \tag{1.73}$$

1-44

For $P_u/\phi P_n < 0.2$

$$\frac{P_u}{2\phi P_n} + \left(\frac{M_{ux}}{\phi_b M_{nx}} + \frac{M_{uy}}{\phi_b M_{ny}}\right) \le 1.0 \tag{1.74}$$

where, if P is tensile,

P_u = factored tensile axial force
P_n = design tensile strength (see Section 1.3)
M_u = factored moment (preferably obtained from a second-order analysis)
M_n = design flexural strength (see Section 1.5)
$\phi = \phi_t$ = resistance factor for tension = 0.90
ϕ_b = resistance factor for flexure = 0.90

and, if P is compressive,

P_u = factored compressive axial force
P_n = design compressive strength (see Section 1.4)
M_u = required flexural strength (see discussion below)
M_n = design flexural strength (see Section 1.5)
$\phi = \phi_c$ = resistance factor for compression = 0.85
ϕ_b = resistance factor for flexure = 0.90

The required flexural strength M_u shall be determined from a second-order elastic analysis. In lieu of such an analysis, the following equation may be used:

$$M_u = B_1 M_{bt} + B_2 M_{lt} \tag{1.75}$$

where

M_{nt} = factored moment in member assuming the frame does not undergo lateral translation (see Figure 1.11)
M_{lt} = factored moment in member as a result of lateral translation (see Figure 1.11)
$B_1 = C_m/(1 - P_u/P_e) \ge 1.0$ is the $P-\delta$ moment magnification factor
$P_e = \pi^2 EI/(KL)^2$, with $K \le 1.0$ in the plane of bending
C_m = a coefficient to account for moment gradient (see discussion below)
$B_2 = 1/[1 - (\sum P_u \Delta_{oh}/\sum HL)]$ or $B_2 = 1/[1 - (\sum P_u/\sum P_e)]$
$\sum P_u$ = sum of all factored loads acting on and above the story under consideration
Δ_{oh} = first-order interstory translation
$\sum H$ = sum of all lateral loads acting on and above the story under consideration
L = story height
$P_e = \pi^2 EI/(KL)^2$

For end-restrained members that do not undergo relative joint translation and not subject to transverse loading between their supports in the plane of bending, C_m is given by

$$C_m = 0.6 - 0.4\left(\frac{M_1}{M_2}\right)$$

where M_1/M_2 is the ratio of the smaller to larger member end moments. The ratio is positive if the member bends in reverse curvature and negative if the member bends in single curvature.

For members that do not undergo relative joint translation and subject to transverse loading between their supports in the plane of bending, C_m can conservatively be taken as 1.

For purpose of design, Equations 1.73 and 1.74 can be rewritten in the form (Aminmansour 2000):
For $P_u/\phi_c P_n > 0.2$

$$bP_u + mM_{ux} + nM_{uy} \le 1.0 \tag{1.76}$$

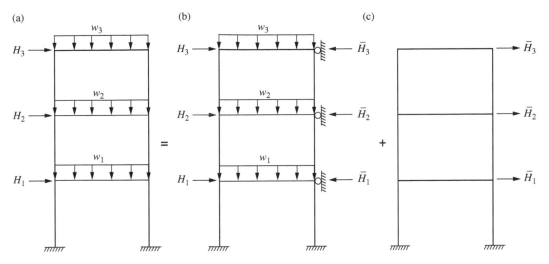

FIGURE 1.11 Calculation of M_{nt} and M_{lt}: (a) original frame; (b) nonsway frame analysis for M_{nt}; and (c) sway frame analysis M_{lt}.

For $P_u/\phi_c P_n \leq 0.2$

$$\frac{b}{2}P_u + \frac{9}{8}mM_{ux} + \frac{9}{8}nM_{uy} \leq 1.0 \qquad (1.77)$$

where

$b = 1/(\phi_c P_n)$
$m = 8/(9\phi_b M_{nx})$
$n = 8/(9\phi_b M_{ny})$

Numerical values for b, m, and n are provided in the AISC Manual (AISC 2001).

1.7 Biaxial Bending

Members subjected to bending about both principal axes (e.g., purlins on an inclined roof) should be designed for *biaxial bending*. Since both moment about the major axis M_{ux} and moment about the minor axis M_{uy} create flexural stresses over the cross-section of the member, the design must take into consideration these stress combinations.

1.7.1 Design for Biaxial Bending

1.7.1.1 Allowable Stress Design

The following interaction equation is often used for the design of beams subject to biaxial bending

$$f_{bx} + f_{by} \leq 0.60F_y \quad \text{or} \quad \frac{M_x}{S_x} + \frac{M_y}{S_y} \leq 0.60F_y \qquad (1.78)$$

where M_x, M_y are service load moments about the major and minor beam axes, respectively, S_x, S_y are elastic section moduli about the major and minor axes, respectively, and F_y is the specified minimum yield stress.

EXAMPLE 1.6

Using ASD, select a W-section to carry dead load moments $M_x = 20$ k ft (27 kN m) and $M_y = 5$ k ft (6.8 kN m) and live load moments $M_x = 50$ k ft (68 kN m) and $M_y = 15$ k ft (20 kN m). Use A992 steel.

Calculate service load moments:

$$M_x = M_{x,\text{dead}} + M_{x,\text{live}} = 20 + 50 = 70 \text{ k ft}$$
$$M_y = M_{y,\text{dead}} + M_{y,\text{live}} = 5 + 15 = 20 \text{ k ft}$$

Select section: Substituting the above service load moments into Equation 1.78, we have

$$\frac{70 \times 12}{S_x} + \frac{20 \times 12}{S_y} \le 0.60(50) \quad \text{or} \quad 840 + 240\frac{S_x}{S_y} \le 30 S_x$$

For W-sections with depth below 14 in. the value of S_x/S_y normally falls in the range 3 to 8, and for W-sections with depth above 14 in. the value of S_x/S_y normally falls in the range 5 to 12. Assuming $S_x/S_y = 10$, we have from the above equation, $S_x \ge 108$ in.^{3} Using the ASD Selection Table in the AISC-ASD Manual, let us try a W24 × 55 section ($S_x = 114$ in.^{3}, $S_y = 8.30$ in.^{3}). For the W24 × 55 section

$$\left[840 + 240\frac{114}{8.30} = 4136\right] > [30 S_x = 30(114) = 3420] \text{ (Therefore, NG)}$$

The next lightest section is W21 × 62 ($S_x = 127$ in.^{3}, $S_y = 13.9$ in.^{3}). For this section

$$\left[840 + 240\frac{127}{13.9} = 3033\right] < [30 S_x = 30(127) = 3810] \text{ (Therefore, Okay)}$$

Therefore, use a W21 × 62 section.

1.7.1.2 Load and Resistance Factor Design

To avoid distress at the most severely stressed point, the following equation for the limit state of yielding must be satisfied:

$$f_{\text{un}} \le \phi_b F_y \tag{1.79}$$

where $f_{\text{un}} = (M_{ux}/S_x + M_{uy}/S_y)$ is the flexural stress under factored loads, S_x, S_y are elastic section moduli about the major and minor axes, respectively, $\phi_b = 0.90$, and F_y is the specified minimum yield stress.

In addition, the limit state for lateral torsional buckling about the major axis should also be checked, that is,

$$\phi_b M_{nx} \ge M_{ux} \tag{1.80}$$

$\phi_b M_{nx}$ is the design flexural strength about the major axis (see Section 1.5).

To facilitate design for biaxial bending, Equation 1.79 can be rearranged to give

$$S_x \ge \frac{M_{ux}}{\phi_b F_y} + \frac{M_{uy}}{\phi_b F_y}\left(\frac{S_x}{S_y}\right) \approx \frac{M_{ux}}{\phi_b F_y} + \frac{M_{uy}}{\phi_b F_y}\left(3.5\frac{d}{b_f}\right) \tag{1.81}$$

In the above equation, d is the overall depth and b_f the flange width of the section. The approximation $(S_x/S_y) \approx (3.5d/b_f)$ was suggested by Gaylord et al. (1992) for doubly symmetric I-shaped sections.

1.8 Combined Bending, Torsion, and Axial Force

Members subjected to the combined effect of bending, torsion, and axial force should be designed to satisfy the following limit states:

Yielding under normal stress:

$$\phi F_y \ge f_{\text{un}} \tag{1.82}$$

where $\phi = 0.90$, F_y is the specified minimum yield stress, and f_{un} is the maximum normal stress determined from an elastic analysis under factored loads.

Yielding under shear stress:

$$\phi(0.6F_{\mathrm{y}}) \geq f_{\mathrm{uv}} \qquad (1.83)$$

where $\phi = 0.90$, F_{y} is the specified minimum yield stress, and f_{uv} is the maximum shear stress determined from an elastic analysis under factored loads.

Buckling:

$$\phi_{\mathrm{c}} F_{\mathrm{cr}} \geq f_{\mathrm{un}} \quad \text{or} \quad \phi_{\mathrm{c}} F_{\mathrm{cr}} \geq f_{\mathrm{uv}}, \qquad \text{whichever is applicable} \qquad (1.84)$$

where $\phi_{\mathrm{c}} F_{\mathrm{cr}} = \phi_{\mathrm{c}} P_{\mathrm{n}}/A_{\mathrm{g}}$, in which $\phi_{\mathrm{c}} P_{\mathrm{n}}$ is the design compressive strength of the member (see Section 1.4), A_{g} is the gross cross-section area, and f_{un}, f_{uv} are normal and shear stresses as defined in Equations 1.82 and 1.83.

1.9 Frames

Frames are designed as a collection of structural components such as beams, beam–columns (columns), and connections. According to the restraint characteristics of the connections used in the construction, frames can be designed as Type I (rigid framing), Type II (simple framing), Type III (semirigid framing) in ASD, or fully restrained (rigid) and partially restrained (semirigid) in LRFD.

- The design of rigid frames necessitates the use of connections capable of transmitting the full or a significant portion of the moment developed between the connecting members. The rigidity of the connections must be such that the angles between intersecting members should remain virtually unchanged under factored loads.
- The design of simple frames is based on the assumption that the connections provide no moment restraint to the beam insofar as gravity loads are concerned, but these connections should have adequate capacity to resist wind moments.
- The design of semirigid frames is permitted upon evidence of the connections to deliver a predicable amount of moment restraint. Over the past two decades, a large body of work has been published in the literature on semirigid connection and frame behavior (see, e.g., Chen 1987, 2000; CTBUH 1993; Chen et al. 1996; Faella et al. 2000). However, because of the vast number of semirigid connections that can exhibit an appreciable difference in joint behavior, no particular approach has been recommended by any building specifications as of this writing. The design of semirigid or partially restrained frames is dependent on the sound engineering judgment by the designer.

Semirigid and simple framings often incur inelastic deformation in the connections. The connections used in these constructions must be proportioned to possess sufficient ductility to avoid overstress of the fasteners or welds.

Regardless of the types of constructions used, due consideration must be given to account for member and frame instability (P–δ and P–Δ) effects either by the use of a second-order analysis or by other means such as moment magnification factors or notional loads (ASCE 1997). The end-restrained effect on the member should also be accounted for by the use of the effective length factor K.

1.9.1 Frame Design

Frames can be designed as *side-sway inhibited* (braced) or *side-sway uninhibited* (unbraced). In side-sway inhibited frames, frame *drift* is controlled by the presence of a bracing system (e.g., shear walls, diagonal, cross, K-braces, etc.). In side-sway uninhibited frames, frame drift is limited by the flexural rigidity of the connected members and the diaphragm action of the floors. Most side-sway uninhibited frames are designed as Type I or Type FR frames using moment connections. Under normal circumstances, the amount of interstory drift under service loads should not exceed $h/500$ to $h/300$ where h is the story height. A higher value of interstory drift is allowed only if it does not create serviceability concerns.

Beams in side-sway inhibited frames are often subject to high axial forces. As a result, they should be designed as beam–columns using beam–column interaction equations. Furthermore, vertical bracing systems should be provided for braced multistory frames to prevent vertical buckling of the frames under gravity loads.

When designing members of a frame, a designer should consider a variety of loading combinations and load patterns, and the members are designed for the most severe load cases. Preliminary sizing of members can be achieved by the use of simple behavioral models such as the simple beam model, cantilever column model, and portal and cantilever method of frame analysis (see, e.g., Rossow 1996).

1.9.2 Frame Bracing

The subject of frame bracing is discussed in a number of references; see, for example, SSRC (1993) and Galambos (1998). According to the LRFD Specification (AISC 1999), the required story or panel bracing shear stiffness in side-sway inhibited frames is

$$\beta_{cr} = \frac{2.67 \sum P_u}{L} \qquad (1.85)$$

where $\sum P_u$ is the sum of all factored gravity loads acting on and above the story or panel supported by the bracing and L is the story height or panel spacing (for ASD, replace P_u in Equation 1.85 by $1.5P_a$, where P_a is the required compressive strength based on ASD load combinations).

The required story or panel bracing force is

$$P_{br} = 0.004 \sum P_u \qquad (1.86)$$

1.10   Plate Girders

Plate girders are built-up beams. They are used as flexural members to carry extremely large lateral loads. A flexural member is considered a plate girder if the width–thickness ratio of the web, h_c/t_w, exceeds $760/\sqrt{F_b}$ (F_b is the allowable flexural stress) according to ASD, or λ_r (see Table 1.8) according to LRFD. Because of the large web slenderness, plate girders are often designed with transverse stiffeners to reinforce the web and to allow for postbuckling (shear) strength (i.e., *tension field action*) to develop. Table 1.9 summarizes the requirements of transverse stiffeners for plate girders based on the web slenderness ratio h/t_w. Two types of transverse stiffeners are used for plate girders: bearing stiffeners and intermediate stiffeners. Bearing stiffeners are used at unframed girder ends, and at concentrated load points where the web yielding or web crippling criterion is violated. Bearing stiffeners extend the full depth of the web from the bottom of the top flange to the top of the bottom flange. Intermediate stiffeners are used when the width–thickness ratio of the web, h/t_w, exceeds 260, when the shear criterion is violated, or when tension field action is considered in the design. Intermediate stiffeners need not extend to the full depth of the web but must be in contact with the compression flange of the girder.

Normally, the depths of plate girder sections are so large that simple beam theory, which postulates that plane sections before bending remain in plane after bending, does not apply. As a result, a different set of design formulae are required for plate girders.

TABLE 1.9   Web Stiffeners Requirements

Range of web slenderness	Stiffeners requirements
$\dfrac{h}{t_w} \leq 260$	Plate girder can be designed without web stiffeners
$260 < \dfrac{h}{t_w} \leq \dfrac{0.48E}{\sqrt{F_{yf}(F_{yf} + 16.5)}}$	Plate girder must be designed with web stiffeners. The spacing of stiffeners, a, can exceed $1.5h$. The actual spacing is determined by the shear criterion
$\dfrac{0.48E}{\sqrt{F_{yf}(F_{yf} + 16.5)}} < \dfrac{h}{t_w} \leq 11.7\sqrt{\dfrac{E}{F_{yf}}}$	Plate girder must be designed with web stiffeners. The spacing of stiffeners, a, cannot exceed $1.5h$

Note: a is the clear distance between stiffeners, h is the clear distance between flanges when welds are used or the distance between adjacent lines of fasteners when bolts are used, t_w is the web thickness, and F_{yf} is the compression flange yield stress (ksi).

1.10.1 Plate Girder Design

1.10.1.1 Allowable Stress Design

1.10.1.1.1 Allowable Bending Stress
The maximum bending stress in the compression flange of the girder computed using the flexure formula shall not exceed the allowable value, F_b', given by

$$F_b' = F_b R_{PG} R_e \tag{1.87}$$

where

F_b = applicable allowable bending stress as discussed in Section 1.5 (ksi)
R_{PG} = plate girder stress reduction factor = $1 - 0.0005(A_w/A_f)(h/t_w - 760/\sqrt{F_b}) \leq 1.0$
R_e = hybrid girder factor = $[12 + (A_w/A_f)(3\alpha - \alpha^3)]/[12 + 2(A_w/A_f)] \leq 1.0$, $R_e = 1$ for a nonhybrid girder
A_w = area of the web
A_f = area of the compression flange
α = $0.60 F_{yw}/F_b \leq 1.0$
F_{yw} = yield stress of the web

1.10.1.1.2 Allowable Shear Stress
Without tension field action: The allowable shear stress is the same as that for beams given in Equation 1.35.

With tension field action: The allowable shear stress is given by

$$F_v = \frac{F_y}{2.89}\left[C_v + \frac{1 - C_v}{1.15\sqrt{1 + (a/h)^2}}\right] \leq 0.40 F_y \tag{1.88}$$

Note that the tension field action can be considered in the design only for nonhybrid girders. If the tension field action is considered, transverse stiffeners must be provided and spaced at a distance so that the computed average web shear stress, f_v, obtained by dividing the total shear by the web area does not exceed the allowable shear stress, F_v, given by Equation 1.88. In addition, the computed bending tensile stress in the panel where tension field action is considered cannot exceed $0.60 F_y$, nor $(0.825 - 0.375 f_v/F_v) F_y$ where f_v is the computed average web shear stress and F_v is the allowable web shear stress given in Equation 1.88. The shear transfer criterion given by Equation 1.91 must also be satisfied.

1.10.1.1.3 Transverse Stiffeners
Transverse stiffeners must be designed to satisfy the following criteria.

Moment of inertia criterion: With reference to an axis in the plane of the web, the moment of inertia of the stiffeners, in in.[4], shall satisfy the condition

$$I_{st} \geq \left(\frac{h}{50}\right)^4 \tag{1.89}$$

where h is the clear distance between flanges, in inches.

Area criterion: The total area of the stiffeners, in in.[2], shall satisfy the condition

$$A_{st} \geq \frac{1 - C_v}{2}\left[\frac{a}{h} - \frac{(a/h)^2}{\sqrt{1 + (a/h)^2}}\right] YDht_w \tag{1.90}$$

where C_v is the shear buckling coefficient as defined in Equation 1.35, a is the stiffeners' spacing, h is the clear distance between flanges, t_w is the web thickness, Y is the ratio of web yield stress to stiffener yield stress, D is equal to 1.0 for stiffeners furnished in pairs, 1.8 for single angle stiffeners, and 2.4 for single plate stiffeners.

Shear transfer criterion: If tension field action is considered, the total shear transfer, in kip/in., of the stiffeners shall not be less than

$$f_{vs} = h\sqrt{\left(\frac{F_{yw}}{340}\right)^3} \tag{1.91}$$

where F_{yw} is the web yield stress (ksi) and h is the clear distance between flanges (in.).

The value of f_{vs} can be reduced proportionally if the computed average web shear stress, f_v, is less than F_v given in Equation 1.88.

1.10.1.2 Load and Resistance Factor Design

1.10.1.2.1 Flexural Strength Criterion

Doubly or singly symmetric, single web plate girders loaded in the plane of the web should satisfy the flexural strength criterion of Equation 1.38. The plate girder design flexural strength is given by

For the limit state of tension flange yielding:

$$\phi_b M_n = 0.90[S_{xt} R_e F_{yt}] \tag{1.92}$$

For the limit state of compression flange buckling:

$$\phi_b M_n = 0.90[S_{xc} R_{PG} R_e F_{cr}] \tag{1.93}$$

where

S_{xt} = section modulus referred to the tension flange = I_x/c_t
S_{xc} = section modulus referred to the compression flange = I_x/c_c
I_x = moment of inertia about the major axis
c_t = distance from neutral axis to extreme fiber of the tension flange
c_c = distance from neutral axis to extreme fiber of the compression flange
R_{PG} = plate girder bending strength reduction factor
    = $1 - a_r[h_c/t_w - 5.70\sqrt{E/F_{cr}})]/[1200 + 300a_r] \le 1.0$
R_e = hybrid girder factor
    = $[12 + a_r(3m - m^3)]/[12 + 2a_r] \le 1.0$ ($R_e = 1$ for nonhybrid girder)
a_r = ratio of web area to compression flange area
m = ratio of web yield stress to flange yield stress or ratio of web yield stress to F_{cr}
F_{yt} = tension flange yield stress
F_{cr} = critical compression flange stress calculated as follows:

Limit state	Range of slenderness	F_{cr} (ksi)
Flange local buckling	$\dfrac{b_f}{2t_f} \le 0.38\sqrt{\dfrac{E}{F_{yf}}}$	F_{yf}
	$0.38\sqrt{\dfrac{E}{F_{yf}}} < \dfrac{b_f}{2t_f} \le 1.35\sqrt{\dfrac{E}{F_{yf}/k_c}}$	$F_{yf}\left[1 - \dfrac{1}{2}\left(\dfrac{(b_f/2t_f) - 0.38\sqrt{E/F_{yf}}}{1.35\sqrt{E/(F_{yf}/k_c)} - 0.38\sqrt{E/F_{yf}}}\right)\right] \le F_{yf}$
	$\dfrac{b_f}{2t_f} > 1.35\sqrt{\dfrac{E}{F_{yf}/k_c}}$	$\dfrac{26{,}200k_c}{(b_f/2t_f)^2}$
Lateral torsional buckling	$\dfrac{L_b}{r_T} \le 1.76\sqrt{\dfrac{E}{F_{yf}}}$	F_{yf}
	$1.76\sqrt{\dfrac{E}{F_{yf}}} < \dfrac{L_b}{r_T} \le 4.44\sqrt{\dfrac{E}{F_{yf}}}$	$C_b F_{yf}\left[1 - \dfrac{1}{2}\left(\dfrac{(L_b/r_T) - 1.76\sqrt{E/F_{yf}}}{4.44\sqrt{E/F_{yf}} - 1.76\sqrt{E/F_{yf}}}\right)\right] \le F_{yf}$
	$\dfrac{L_b}{r_T} > 4.44\sqrt{\dfrac{E}{F_{yf}}}$	$\dfrac{286{,}000C_b}{(L_b/r_T)^2}$

$k_c = 4/\sqrt{(h/t_w)}, 0.35 \leq k_c \leq 0.763$
b_f = compression flange width
t_f = compression flange thickness
L_b = lateral unbraced length of the girder
$r_T = \sqrt{[(t_f b_f^3/12 + h_c t_w^3/72)/(b_f t_f + h_c t_w/6)]}$
h_c = twice the distance from the neutral axis to the inside face of the compression flange less the fillet
t_w = web thickness
F_{yf} = yield stress of compression flange (ksi)
C_b = Bending coefficient (see Section 1.5)

F_{cr} must be calculated for both flange local buckling and lateral torsional buckling. The smaller value of F_{cr} is used in Equation 1.93.

The plate girder bending strength reduction factor R_{PG} is a factor to account for the nonlinear flexural stress distribution along the depth of the girder. The hybrid girder factor is a reduction factor to account for the lower yield strength of the web when the nominal moment capacity is computed assuming a homogeneous section made entirely of the higher yield stress of the flange.

1.10.1.2.2 Shear Strength Criterion
Plate girders can be designed with or without the consideration of tension field action. If tension field action is considered, intermediate web stiffeners must be provided and spaced at a distance, a, such that a/h is smaller than 3 or $[260/(h/t_w)]^2$, whichever is smaller. Also, one must check the flexure–shear interaction of Equation 1.96, if appropriate. Consideration of tension field action is not allowed if

- The panel is an end panel
- The plate girder is a hybrid girder
- The plate girder is a web tapered girder
- a/h exceeds 3 or $[260/(h/t_w)]^2$, whichever is smaller

The design shear strength, $\phi_v V_n$, of a plate girder is determined as follows:

If tension field action is not considered: $\phi_v V_n$ are the same as those for beams as given in Equations 1.49 to 1.51.

If tension field action is considered and $h/t_w \leq 1.10\sqrt{(k_v E/F_{yw})}$:

$$\phi_v V_n = 0.90[0.60 A_w F_{yw}] \tag{1.94}$$

and if $h/t_w > 1.10\sqrt{(k_v E/F_{yw})}$:

$$\phi_v V_n = 0.90 \left[0.60 A_w F_{yw} \left(C_v + \frac{1 - C_v}{1.15\sqrt{1 + (a/h)^2}} \right) \right] \tag{1.95}$$

where

$k_v = 5 + 5/(a/h)^2$ (k_v shall be taken as 5.0 if a/h exceeds 3.0 or $[260/(h/t_w)]^2$, whichever is smaller)
$A_w = d t_w$ (where d is the section depth and t_w is the web thickness)
F_{yw} = web yield stress
C_v = shear coefficient, calculated as follows:

Range of h/t_w	C_v
$1.10\sqrt{\dfrac{k_v E}{F_{yw}}} \leq \dfrac{h}{t_w} \leq 1.37\sqrt{\dfrac{k_v E}{F_{yw}}}$	$\dfrac{1.10\sqrt{k_v E/F_{yw}}}{h/t_w}$
$\dfrac{h}{t_w} > 1.37\sqrt{\dfrac{k_v E}{F_{yw}}}$	$\dfrac{1.51 k_v E}{(h/t_w)^2 F_{yw}}$

1.10.1.2.3 Flexure–Shear Interaction

Plate girders designed for tension field action must satisfy the flexure–shear interaction criterion in regions where $0.60\phi V_n \leq V_u \leq \phi V_n$ and $0.75\phi M_n \leq M_u \leq \phi M_n$

$$\frac{M_u}{\phi M_n} + 0.625\,\frac{V_u}{\phi V_n} \leq 1.375 \tag{1.96}$$

where $\phi = 0.90$.

1.10.1.2.4 Bearing Stiffeners

Bearing stiffeners must be provided for a plate girder at unframed girder ends and at points of concentrated loads where the web yielding or the web crippling criterion is violated (see Section 1.5.1.1.3). Bearing stiffeners shall be provided in pairs and extend from the upper flange to the lower flange of the girder. Denoting b_{st} as the width of one stiffener and t_{st} as its thickness, bearing stiffeners shall be portioned to satisfy the following limit states:

For the limit state of local buckling:

$$\frac{b_{st}}{t_{st}} \leq 0.56\sqrt{\frac{E}{F_y}} \tag{1.97}$$

For the limit state of compression: The design compressive strength, $\phi_c P_n$, must exceed the required compressive force acting on the stiffeners. $\phi_c P_n$ is to be determined based on an effective length factor K of 0.75 and an effective area, A_{eff}, equal to the area of the bearing stiffeners plus portion of the web. For end bearing, this effective area is equal to $2(b_{st}t_{st}) + 12t_w^2$; and for interior bearing, this effective area is equal to $2(b_{st}t_{st}) + 25t_w^2$, where t_w is the web thickness. The slenderness parameter, λ_c, is to be calculated using a radius of gyration, $r = \sqrt{(I_{st}/A_{eff})}$, where $I_{st} = t_{st}(2b_{st} + t_w)^3/12$.

For the limit state of bearing: The bearing strength, ϕR_n, must exceed the required compression force acting on the stiffeners. ϕR_n is given by

$$\phi R_n \geq 0.75[1.8F_y A_{pb}] \tag{1.98}$$

where F_y is the yield stress and A_{pb} is the bearing area.

1.10.1.2.5 Intermediate Stiffeners

Intermediate stiffeners shall be provided if

- The shear strength capacity is calculated based on tension field action.
- The shear criterion is violated (i.e., when the V_u exceeds $\phi_v V_n$).
- The web slenderness h/t_w exceeds $2.45(\sqrt{E/F_{yw}})$.

Intermediate stiffeners can be provided in pairs or on one side of the web only in the form of plates or angles. They should be welded to the compression flange and the web but they may be stopped short of the tension flange. The following requirements apply to the design of intermediate stiffeners:

Local buckling: The width–thickness ratio of the stiffener must be proportioned so that Equation 1.97 is satisfied to prevent failure by local buckling.

Stiffener area: The cross-section area of the stiffener must satisfy the following criterion:

$$A_{st} \geq \frac{F_{yw}}{F_y}\left[0.15\,Dht_w(1 - C_v)\frac{V_u}{\phi_v V_n} - 18t_w^2\right] \geq 0 \tag{1.99}$$

where F_y is the yield stress of stiffeners, $D = 1.0$ for stiffeners in pairs, $D = 1.8$ for single angle stiffeners, and $D = 2.4$ for single plate stiffeners.

The other terms in Equation 1.99 are defined as before in Equations 1.94 and 1.95.

Stiffener moment of inertia: The moment of inertia for stiffener pairs taken about an axis in the web center or for single stiffeners taken about the face of contact with the web plate must satisfy the following criterion:

$$I_{st} \geq at_w^3 \left[\frac{2.5}{(a/h)^2} - 2 \right] \geq 0.5at_w^3 \qquad (1.100)$$

Stiffener length: The length of the stiffeners l_{st}, should fall within the range

$$h - 6\,t_w < l_{st} < h - 6t_w \qquad (1.101)$$

where h is the clear distance between the flanges less the widths of the flange-to-web welds and t_w is the web thickness.

If intermittent welds are used to connect the stiffeners to the girder web, the clear distance between welds shall not exceed $16t_w$ by not more than 10 in. (25.4 cm). If bolts are used, their spacing shall not exceed 12 in. (30.5 cm).

Stiffener spacing: The spacing of the stiffeners, a, shall be determined from the shear criterion $\phi_v V_n \geq V_u$. This spacing shall not exceed the smaller of $3h$ and $[260/(h/t_w)]^2 h$.

EXAMPLE 1.7

Using LRFD, design the cross-section of an I-shaped plate girder shown in Figure 1.12a to support a factored moment M_u of 4600 kip ft (6240 kN m); dead weight of the girder is included. The girder is a 60-ft (18.3-m) long simply supported girder. It is laterally supported at every 20 ft (6.1 m) interval. Use A36 steel.

Proportion of the girder web: Ordinarily, the overall depth to span ratio d/L of a building girder is in the range $\frac{1}{12}$ to $\frac{1}{10}$. So, let us try $h = 70$ in.

Also, because h/t_w of a plate girder is normally in the range $5.70\sqrt{(E/F_{yf})}$ to $11.7\sqrt{(E/F_{yf})}$, using $E = 29,000$ ksi and $F_{yf} = 36$ ksi, let us try $t_w = \frac{5}{16}$ in.

Proportion of the girder flanges: For a preliminary design, the required area of the flange can be determined using the flange area method

$$A_f \approx \frac{M_u}{F_y h} = \frac{4600 \text{ kip ft} \times 12 \text{ in./ft}}{(36 \text{ ksi})(70 \text{ in.})} = 21.7 \text{ in.}^2$$

So, let $b_f = 20$ in. and $t_f = 1\frac{1}{8}$ in. giving $A_f = 22.5$ in.^{2}
Determine the design flexural strength $\phi_b M_n$ of the girder:
Calculate I_x:

$$\begin{aligned} I_x &= \sum [I_i + A_i\, y_i^2] \\ &= [8,932 + (21.88)(0)^2] + 2[2.37 + (22.5)(35.56)^2] \\ &= 65,840 \text{ in.}^4 \end{aligned}$$

Calculate S_{xt}, S_{xc}:

$$S_{xt} = S_{xc} = \frac{I_x}{c_t} = \frac{I_x}{c_c} = \frac{65,840}{35 + 1.125} = 1,823 \text{ in.}^3$$

Calculate r_T: Refer to Figure 1.12b,

$$r_T = \sqrt{\frac{I_T}{A_f + (1/6)A_w}} = \sqrt{\frac{(1.125)(20)^3/12 + (11.667)(5/16)^3/12}{22.5 + (1/6)(21.88)}} = 5.36 \text{ in.}$$

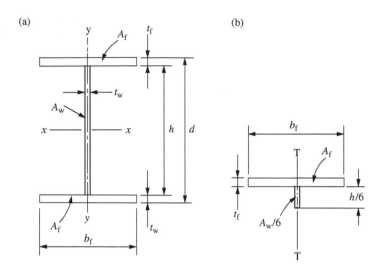

FIGURE 1.12   Design of a plate girder cross-section: (a) plate girder nomenclature and (b) calculation of r_T.

Calculate F_{cr}:
  For flange local buckling

$$\left[\frac{b_f}{2t_f} = \frac{20}{2(1.125)} = 8.89\right] < \left[0.38\sqrt{\frac{E}{F_{yf}}} = 10.8\right], \quad \text{so } F_{cr} = F_{yf} = 36 \text{ ksi}$$

  For lateral torsional buckling

$$\left[\frac{L_b}{r_T} = \frac{20 \times 12}{5.36} = 44.8\right] < \left[1.76\sqrt{\frac{E}{F_{yf}}} = 50\right], \quad \text{so } F_{cr} = F_{yf} = 36 \text{ ksi}$$

Calculate R_{PG}:

$$R_{PG} = 1 - \frac{a_r(h_c/t_w - 5.70\sqrt{(E/F_{cr})})}{(1{,}200 + 300a_r)} = 1 - \frac{0.972[70/(5/16) - 5.70\sqrt{(29{,}000/36)}]}{[1{,}200 + 300(0.972)]} = 0.96$$

Calculate $\phi_b M_n$:

$$\phi_b M_n = \text{smaller of } \begin{cases} 0.90 S_{xt} R_e F_{yt} & = (0.90)(1{,}823)(1)(36) = 59{,}065 \text{ kip in.} \\ 0.90 S_x R_{PG} R_e F_{cr} & = (0.90)(1{,}823)(0.96)(1)(36) = 56{,}700 \text{ kip in.} \end{cases}$$
$$= 56{,}700 \text{ kip in.}$$
$$= 4{,}725 \text{ kip ft.}$$

Since $[\phi_b M_n = 4{,}725 \text{ kip ft}] > [M_u = 4{,}600 \text{ kip ft}]$, the cross-section is acceptable. Use web plate $\frac{5}{16}$ in. × 70 in. and two flange plates $1\frac{1}{8}$ in. × 20 in. for the girder cross-section.

EXAMPLE 1.8

Design bearing stiffeners for the plate girder of the preceding example for a factored end reaction of 260 kip.
  Since the girder end is unframed, bearing stiffeners are required at the supports. The size of the stiffeners must be selected to ensure that the limit states of local buckling, compression, and bearing are not violated.

  Limit state of local buckling: Refer to Figure 1.13, try $b_{st} = 8$ in. To avoid problems with local buckling $b_{st}/2t_{st}$ must not exceed $0.56\sqrt{(E/F_y)} = 15.8$. Therefore, try $t_{st} = \frac{1}{2}$ in. So, $b_{st}/2t_{st} = 8$, which is less than 15.8.

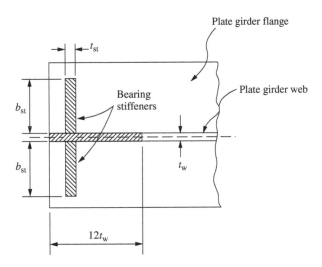

FIGURE 1.13 Design of bearing stiffeners.

Limit state of compression:

$$A_{eff} = 2(b_{st}t_{st}) + 12t_w^2 = 2(8)(0.5) + 12(5/16)^2 = 9.17 \text{ in.}^2$$
$$I_{st} = t_{st}(2b_{st} + t_w)^3/12 = 0.5[2(8) + 5/16]^3/12 = 181 \text{ in.}^4$$
$$r_{st} = \sqrt{(I_{st}/A_{eff})} = \sqrt{(181/9.17)} = 4.44 \text{ in.}$$
$$Kh/r_{st} = 0.75(70)/4.44 = 11.8$$
$$\lambda_c = (Kh/\pi r_{st})\sqrt{(F_y/E)} = (11.8/3.142)\sqrt{(36/29,000)} = 0.132$$

and from Equation 1.17

$$\phi_c P_n = 0.85(0.658^{\lambda^2})F_y A_{st} = 0.85(0.658)^{0.132^2}(36)(9.17) = 279 \text{ kip}$$

Since $\phi_c P_n > 260$ kip, the design is satisfactory for compression.

Limit state of bearing: Assuming there is a $\frac{1}{4}$-in. weld cutout at the corners of the bearing stiffeners at the junction of the stiffeners and the girder flanges, the bearing area for the stiffener pairs is: $A_{pb} = (8 - 0.25)(0.5)(2) = 7.75 \text{ in.}^2$. Substituting this into Equation 1.98, we have $\phi R_n = 0.75(1.8)(36)(7.75) = 377$ kip, which exceeds the factored reaction of 260 kip. So bearing is not a problem.

Use two $\frac{1}{2}$ in. × 8 in. plates for bearing stiffeners.

1.11 Connections

Connections are structural elements used for joining different members of a framework. Connections can be classified according to

- *The type of connecting medium used.* Bolted connections, welded connections, bolted–welded connections, riveted connections.
- *The type of internal forces the connections are expected to transmit.* Shear (semirigid, simple) connections, moment (rigid) connections.
- *The type of structural elements that made up the connections.* Single plate angle connections, double web angle connections, top and seated angle connections, seated beam connections, etc.
- *The type of members the connections are joining.* Beam-to-beam connections (beam splices), column-to-column connections (column splices), beam-to-column connections, hanger connections, etc.

To properly design a connection, a designer must have a thorough understanding of the behavior of the joint under loads. Different modes of failure can occur depending on the geometry of the connection and the relative strengths and stiffness of the various components of the connection. To ensure that the connection can carry the applied loads, a designer must check for all perceivable modes of failure pertinent to each component of the connection and the connection as a whole.

1.11.1 Bolted Connections

Bolted connections are connections whose components are fastened together primarily by bolts. The four basic types of bolts are discussed in Section 1.1.7. Depending on the direction and line of action of the loads relative to the orientation and location of the bolts, the bolts may be loaded in tension, shear, or a combination of tension and shear. For bolts subjected to shear forces, the design shear strength of the bolts also depends on whether or not the threads of the bolts are excluded from the shear planes. A letter X or N is placed at the end of the ASTM designation of the bolts to indicate whether the threads are excluded or not excluded from the shear planes, respectively. Thus, A325-X denotes A325 bolts whose threads are excluded from the shear planes and A490-N denotes A490 bolts whose threads are not excluded from the shear planes. Because of the reduced shear areas for bolts whose threads are not excluded from the shear planes, these bolts have lower design shear strengths than their counterparts whose threads are excluded from the shear planes.

Bolts can be used in both bearing-type connections and slip-critical connections. Bearing-type connections rely on bearing between the bolt shanks and the connecting parts to transmit forces. Some slippage between the connected parts is expected to occur for this type of connections. Slip-critical connections rely on the frictional force that develops between the connecting parts to transmit forces. No slippage between connecting elements is expected for this type of connection. Slip-critical connections are used for structures designed for vibratory or dynamic loads such as bridges, industrial buildings, and buildings in regions of high seismicity. Bolts used in slip-critical connections are denoted by the letter F after their ASTM designation, for example A325-F, A490-F.

Bolt holes. Holes made in the connected parts for bolts may be standard sized, oversized, short slotted, or long slotted. Table 1.10 gives the maximum hole dimension for ordinary construction usage.

Standard holes can be used for both bearing-type and slip-critical connections. Oversized holes shall be used only for slip-critical connections and hardened washers shall be installed over these holes in an outer ply. Short-slotted and long-slotted holes can be used for both bearing-type and slip-critical connections provided that when such holes are used for bearing, the direction of slot is transverse to the direction of loading. While oversized and short-slotted holes are allowed in any or all plies of the connection, long-slotted holes are allowed in only one of the connected parts. In addition, if long-slotted

TABLE 1.10   Nominal Hole Dimensions (in.)

Bolt diameter, d (in.)	Hole dimensions			
	Standard (dia.)	Oversize (dia.)	Short slot (width × length)	Long slot (width × length)
$\frac{1}{2}$	$\frac{9}{16}$	$\frac{5}{8}$	$\frac{9}{16} \times \frac{11}{16}$	$\frac{9}{16} \times 1\frac{1}{4}$
$\frac{5}{8}$	$\frac{11}{16}$	$\frac{13}{16}$	$\frac{11}{16} \times \frac{7}{8}$	$\frac{11}{16} \times 1\frac{9}{16}$
$\frac{3}{4}$	$\frac{13}{16}$	$\frac{15}{16}$	$\frac{13}{16} \times 1$	$\frac{13}{16} \times 1\frac{7}{8}$
$\frac{7}{8}$	$\frac{15}{16}$	$1\frac{1}{16}$	$\frac{15}{16} \times 1\frac{1}{8}$	$\frac{15}{16} \times 2\frac{3}{16}$
1	$1\frac{1}{16}$	$1\frac{1}{4}$	$1\frac{1}{16} \times 1\frac{5}{16}$	$1\frac{1}{16} \times 2\frac{1}{2}$
$\geq 1\frac{1}{8}$	$d + \frac{1}{16}$	$d + \frac{5}{16}$	$\left(d + \frac{1}{16}\right) \times \left(d + \frac{3}{8}\right)$	$\left(d + \frac{1}{16}\right) \times (2.5d)$

Note: 1 in. = 25.4 mm.

holes are used in an outer ply, plate washers, or a continuous bar with standard holes having a size sufficient to cover the slot shall be provided.

1.11.1.1 Bolts loaded in tension

If a tensile force is applied to the connection such that the direction of load is parallel to the longitudinal axes of the bolts, the bolts will be subjected to tension. The following conditions must be satisfied for bolts under tensile stresses.

Allowable stress design:

$$f_t \leq F_t \tag{1.102}$$

where f_t is the computed tensile stress in the bolt and F_t is the allowable tensile stress in bolt (see Table 1.11).

Load and resistance factor design:

$$\phi_t F_t \geq f_t \tag{1.103}$$

where $\phi_t = 0.75$, f_t is the tensile stress produced by factored loads (ksi), and F_t is the nominal tensile strength given in Table 1.11.

1.11.1.2 Bolts Loaded in Shear

When the direction of load is perpendicular to the longitudinal axes of the bolts, the bolts will be subjected to shear. The conditions that need to be satisfied for bolts under shear stresses are as follows:

Allowable stress design: For bearing-type and slip-critical connections, the condition is

$$f_v \leq F_v \tag{1.104}$$

where f_v is the computed shear stress in the bolt (ksi) and F_v is the allowable shear stress in bolt (see Table 1.12).

Load and resistance factor design: For bearing-type connections designed at factored loads, and for slip-critical connections designed at service loads, the condition is

$$\phi_v F_v \geq f_v \tag{1.105}$$

where $\phi_v = 0.75$ (for bearing-type connections design for factored loads), $\phi_v = 1.00$ (for slip-critical connections designed at service loads), f_v is the shear stress produced by factored loads (for bearing-type connections) or produced by service loads (for slip-critical connections), (ksi), and F_v is the

TABLE 1.11 F_t of Bolts (ksi)

	ASD		LRFD	
Bolt type	F_t, ksi (static loading)	F_t, ksi (fatigue loading)	F_t, ksi (static loading)	F_t, ksi (fatigue loading)
A307	20	Not allowed	45	Not allowed
A325	44	If $N \leq 20,000$: F_t = same as for	90	F_t is not defined, but the
A490	54	static loading	113	condition $F_t < F_{SR}$ needs
		If $20,000 < N \leq 500,000$:		to be satisfied
		$F_t = 40$ (A325) $= 49$ (A490)		where f_t = tensile stress caused
		If $N > 500,000$:		by service loads calculated
		$F_t = 31$ (A325) $= 38$ (A490)		using a net tensile area
		where N = number of stress		given by
		range fluctuations in design		$A_t = \dfrac{\pi}{4}\left(d_b - \dfrac{0.9743}{n}\right)^2$
		life F_u = minimum specified		F_{SR} is the design stress range
		tensile strength, ksi		given by
				$F_{SR} = \left(\dfrac{3.9 \times 10^8}{N}\right)^{1/3} \geq 7$ ksi
				in which d_b = nominal bolt diameter, n = threads per inch, N = number of stress range fluctuations in the design life

Note: 1 ksi = 6.895 MPa.

TABLE 1.12 F_v or F_n of Bolts (ksi)

Bolt type	F_v (for ASD)	F_v (for LRFD)
A307	10.0[a] (regardless of whether or not threads are excluded from shear planes)	24.0[a] (regardless of whether or not threads are excluded from shear planes)
A325-N	21.0[a]	48.0[a]
A325-X	30.0[a]	60.0[a]
A325-F[b]	17.0 (for standard size holes) 15.0 (for oversized and short-slotted holes) 12.0 (for long-slotted holes when direction of load is transverse to the slots) 10.0 (for long-slotted holes when direction of load is parallel to the slots)	17.0 (for standard size holes) 15.0 (for oversized and short-slotted holes) 12.0 (for long-slotted holes when direction of load is transverse to the slots) 10.0 (for long-slotted holes when direction of load is parallel to the slots)
A490-N	28.0[a]	60.0[a]
A490-X	40.0[a]	75.0[a]
A490-F[b]	21.0 (for standard size holes) 18.0 (for oversized and short-slotted holes) 15.0 (for long-slotted holes when direction of load is transverse to the slots) 13.0 (for long-slotted holes when direction of load is parallel to the slots)	21.0 (for standard size holes) 18.0 (for oversized and short-slotted holes) 15.0 (for long-slotted holes when direction of load is transverse to the slots) 13.0 (for long-slotted holes when direction of load is parallel to the slots)

[a] Tabulated values shall be reduced by 20% if the bolts are used to splice tension members having a fastener pattern whose length, measured parallel to the line of action of the force, exceeds 50 in.
[b] Tabulated values are applicable only to Class A surface, that is, unpainted clean mill surface and blast cleaned surface with Class A coatings (with slip coefficient = 0.33). For design strengths with other coatings, see "Load and Resistance Factor Design Specification to Structural Joints Using ASTM A325 or A490 Bolts" (RCSC 2000).
Note: 1 ksi = 6.895 MPa.

nominal shear strength given in Table 1.12.

For slip-critical connections designed at factored loads, the condition is

$$\phi r_{str} \geq r_u \tag{1.106}$$

where

$\phi = 1.0$ (for standard holes)
$\quad = 0.85$ (for oversized and short-slotted holes)
$\quad = 0.70$ (for long-slotted holes transverse to the direction of load)
$\quad = 0.60$ (for long-slotted holes parallel to the direction of load)
$r_{str} =$ design slip resistance per bolt $= 1.13\mu T_b N_s$
$\quad \mu = 0.33$ (for Class A surfaces, i.e., unpainted clean mill surfaces or blast-cleaned surfaces with Class A coatings)
$\quad = 0.50$ (for Class B surfaces, i.e., unpainted blast-cleaned surfaces or blast-cleaned surfaces with Class B coatings)
$\quad = 0.35$ (for Class C surfaces, i.e., hot-dip galvanized and roughened surfaces)
$\quad T_b =$ minimum fastener tension given in Table 1.13
$\quad N_s =$ number of slip planes
$r_u =$ required force per bolt due to factored loads

1.11.1.3 Bolts Loaded in Combined Tension and Shear

If a tensile force is applied to a connection such that its line of action is at an angle with the longitudinal axes of the bolts, the bolts will be subjected to combined tension and shear. The conditions that need to be satisfied are given below:

Allowable stress design: The conditions are

$$f_v \leq F_v \quad \text{and} \quad f_t \leq F_t \tag{1.107}$$

TABLE 1.13   Minimum Fastener Tension (kip)

Bolt diameter (in.)	A325 bolts	A490 bolts
$\frac{1}{2}$	012	015
$\frac{5}{8}$	019	024
$\frac{3}{4}$	028	035
$\frac{7}{8}$	039	049
1	051	064
$1\frac{1}{8}$	056	080
$1\frac{1}{4}$	071	102
$1\frac{3}{8}$	085	121
$1\frac{1}{2}$	103	148

Note: 1 kip $=4.45$ kN.

TABLE 1.14 F_t for Bolts under Combined Tension and Shear (ksi)

	Bearing-type connections			
	ASD		LRFD	
Bolt type	Threads not excluded from the shear plane	Threads excluded from the shear plane	Threads not excluded from the shear plane	Threads excluded from the shear plane
A307	$26 - 1.8f_v \leq 20$	$26 - 1.8f_v \leq 20$	$59 - 2.5f_v \leq 45$	$59 - 2.5f_v \leq 45$
A325	$\sqrt{(44^2 - 4.39f_v^2)}$	$\sqrt{(44^2 - 2.15f_v^2)}$	$117 - 2.5f_v \leq 90$	$117 - 2.0f_v \leq 90$
A490	$\sqrt{(54^2 - 3.75f_v^2)}$	$\sqrt{(54^2 - 1.82f_v^2)}$	$147 - 2.5f_v \leq 113$	$147 - 2.0f_v \leq 113$

Slip-critical connections

For ASD:
Only $f_v \leq F_v$ needs to be checked
where
f_v = computed shear stress in the bolt, ksi
$F_v = [1 - (f_t A_b / T_b)] \times$ (values of F_v given in Table 1.12)
f_t = computed tensile stress in the bolt, ksi
A_b = nominal cross-sectional area of bolt, in.^{2}
T_b = minimum pretension load given in Table 1.13.

For LRFD:
Only $\phi_v F_v \geq f_v$ needs to be checked
where
$\phi_v = 1.0$
f_v = shear stress produced by service load
$F_v = [1 - (T/0.8 T_b N_b)] \times$ (values of F_v given in Table 1.12)
T = service tensile force in the bolt, kip
T_b = minimum pretension load given in Table 1.13
N_b = number of bolts carrying the service-load tension T

Note: 1 ksi $=6.895$ MPa.

where f_v, F_v are defined in Equation 1.104, f_t is the computed tensile stress in the bolt (ksi), and F_t is the allowable tensile stress given in Table 1.14.

Load and resistance factor design: For bearing-type connections designed at factored loads and slip-critical connections designed at service loads, the conditions are

$$\phi_v F_v \geq f_v \quad \text{and} \quad \phi_t F_t \geq f_t \tag{1.108}$$

TABLE 1.15 Bearing Capacity

	ASD	LRFD
Conditions	Allowable bearing stress, F_p, ksi	Design bearing strength, ϕR_n, kip
1. For standard, oversized, or short-slotted holes loaded in any direction	$L_e F_u/2d \leq 1.2F_u$	$0.75[1.2L_c tF_u] \leq 0.75[2.4dtF_u]^a$
2. For long-slotted holes with direction of slot perpendicular to the direction of bearing	$L_e F_u/2d \leq 1.0F_u$	$0.75[1.0L_c tF_u] \leq 0.75[2.0dtF_u]$
3. If hole deformation at service load is not a design consideration	$L_e F_u/2d \leq 1.5F_u$	$0.75[1.5L_c tF_u] \leq 0.75[3.0dtF_u]$

[a] This equation is also applicable to long-slotted holes when the direction of slot is parallel to the direction of bearing force.
Note: L_e is the distance from free edge to center of the bolt; L_c is the clear distance, in the direction of force, between the edge of the hole and the edge of the adjacent hole or edge of the material; d is the nominal bolt diameter; t is the thickness of the connected part; and F_u is the specified minimum tensile strength of the connected part.

where ϕ_v, F_v, f_v are defined in Equation 1.105, ϕ_t is equal to 0.75, f_t is the tensile stress due to factored loads (for bearing-type connection) or due to service loads (for slip-critical connections) (ksi), and F_t is the nominal tension stress limit for combined tension and shear given in Table 1.14.

For slip-critical connections designed at factored loads, the condition is given in Equation 1.106, except that the design slip resistance per bolt ϕr_{str} shall be multiplied by a reduction factor given by $1 - (T_u/(1.13T_b N_b))$, where T_u is the factored tensile load on the connection, T_b is given in Table 1.13, and N_b is the number of bolts carrying the factored-load tension T_u.

1.11.1.4 Bearing Strength at Fastener Holes

Connections designed on the basis of bearing rely on the bearing force developed between the fasteners and the holes to transmit forces and moments. The limit state for bearing must therefore be checked to ensure that bearing failure will not occur. Bearing strength is independent of the type of fastener. This is because the bearing stress is more critical on the parts being connected than on the fastener itself. The AISC specification provisions for bearing strength are based on preventing excessive hole deformation. As a result, bearing capacity is expressed as a function of the type of holes (standard, oversized, slotted), bearing area (bolt diameter times the thickness of the connected parts), bolt spacing, edge distance (L_e), strength of the connected parts (F_u), and the number of fasteners in the direction of the bearing force. Table 1.15 summarizes the expressions and conditions used in ASD and LRFD for calculating the bearing strength of both bearing-type and slip-critical connections.

1.11.1.5 Minimum Fastener Spacing

To ensure safety, efficiency, and to maintain clearances between bolt nuts as well as to provide room for wrench sockets, the fastener spacing, s, should not be less than $3d$ where d is the nominal fastener diameter.

1.11.1.6 Minimum Edge Distance

To prevent excessive deformation and shear rupture at the edge of the connected part, a minimum edge distance L_e must be provided in accordance with the values given in Table 1.16 for standard holes. For oversized and slotted hole, the values shown must be incremented by C_2 given in Table 1.17.

1.11.1.7 Maximum Fastener Spacing

A limit is placed on the maximum value for the spacing between adjacent fasteners to prevent the possibility of gaps forming or buckling from occurring in between fasteners when the load to be transmitted by the connection is compressive. The maximum fastener spacing measured in the direction of the force is given as follows:

For painted members or unpainted members not subject to corrosion: smaller of $24t$ where t is the thickness of the thinner plate and 12 in. (305 mm).

TABLE 1.16 Minimum Edge Distance for Standard Holes (in.)

Nominal fastener diameter (in.)	At sheared edges	At rolled edges of plates, shapes, and bars or gas cut edges
$\frac{1}{2}$	$\frac{7}{8}$	$\frac{3}{4}$
$\frac{5}{8}$	$1\frac{1}{8}$	$\frac{7}{8}$
$\frac{3}{4}$	$1\frac{1}{4}$	1
$\frac{7}{8}$	$1\frac{1}{2}$	$1\frac{1}{8}$
1	$1\frac{3}{4}$	$1\frac{1}{4}$
$1\frac{1}{8}$	2	$1\frac{1}{2}$
$1\frac{1}{4}$	$2\frac{1}{4}$	$1\frac{5}{8}$
Over $1\frac{1}{4}$	$1\frac{3}{4}$ × Fastener diameter	$1\frac{1}{4}$ × Fastener diameter

Note: 1 in. = 25.4 mm.

TABLE 1.17 Values of Edge Distance Increment, C_2 (in.)

Nominal diameter of fastener (in.)	Oversized holes	Slotted holes		Slot parallel to edge
		Slot transverse to edge		
		Short slot	Long slot[a]	
$\leq\frac{7}{8}$	$\frac{1}{16}$	$\frac{1}{8}$	$3d/4$	0
1	$\frac{1}{8}$ 0	$\frac{1}{8}$		
$\geq 1\frac{1}{8}$	$\frac{1}{8}$ 0	$\frac{3}{16}$		

[a] If the length of the slot is less than the maximum shown in Table 1.10, the value shown may be reduced by one-half the difference between the maximum and the actual slot lengths.

Note: 1 in. = 25.4 mm.

For unpainted members of weathering steel subject to atmospheric corrosion: smaller of 14t where t is the thickness of the thinner plate and 7 in. (178 mm).

1.11.1.8 Maximum Edge Distance

A limit is placed on the maximum value for edge distance to prevent prying action from occurring. The maximum edge distance shall not exceed the smaller of 12t where t is the thickness of the connected part and 6 in. (15 cm).

EXAMPLE 1.9

Check the adequacy of the connection shown in Figure 1.4a. The bolts are 1-in. diameter A325-N bolts in standard holes. The connection is a bearing-type connection.

Check bolt capacity: All bolts are subjected to double shear. Therefore, the design shear strength of the bolts will be twice that shown in Table 1.12. Assuming each bolt carries an equal share of the factored applied load, we have from Equation 1.105

$$[\phi_v F_v = 0.75(2 \times 48) = 72 \text{ ksi}] > \left[f_v = \frac{208}{(6)(\pi d^2/4)} = 44.1 \text{ ksi} \right]$$

The shear capacity of the bolt is therefore adequate.

Check bearing capacity of the connected parts: With reference to Table 1.15, it can be seen that condition 1 applies for the present problem. Therefore, we have

$$[\phi R_n = 0.75(1.2 L_c t F_u) = 0.75(1.2)(3 - 1\tfrac{1}{8})(\tfrac{3}{8})(58) = 36.7 \text{ kip}$$
$$< 0.75(2.4 d t F_u) = 0.75(2.4)(1)(\tfrac{3}{8})(58) = 39.2 \text{ kip}$$
$$> [R_u = \frac{208}{6} = 34.7 \text{ kip}]$$

and so bearing is not a problem. Note that bearing on the gusset plate is more critical than bearing on the webs of the channels because the thickness of the gusset plate is less than the combined thickness of the double channels.

Check bolt spacing: The minimum bolt spacing is $3d = 3(1) = 3$ in. The maximum bolt spacing is the smaller of $14t = 14(0.303) = 4.24$ in. or 7 in. The actual spacing is 3 in., which falls within the range of 3 to 4.24 in., so bolt spacing is adequate.

Check edge distance: From Table 1.16, it can be determined that the minimum edge distance is 1.25 in. The maximum edge distance allowed is the smaller of $12t = 12(0.303) = 3.64$ in. or 6 in. The actual edge distance is 3 in., which falls within the range of 1.25 to 3.64 in., so edge distance is adequate.

The connection is therefore adequate.

1.11.1.9 Bolted Hanger-Type Connections

A typical hanger connection is shown in Figure 1.14. In the design of such connections, the designer must take into account the effect of *prying action*. Prying action results when flexural deformation occurs in the tee flange or angle leg of the connection (Figure 1.15). Prying action tends to increase the tensile force, called prying force, in the bolts. To minimize the effect of prying, the fasteners should be placed as close to the tee stem or outstanding angle leg as the wrench clearance will permit [see Tables on Entering and Tightening Clearances in Volume II — Connections of the AISC-LRFD Manual (AISC 2001)]. In addition, the flange and angle thickness should be proportioned so that the full tensile capacities of the bolts can be developed.

Two failure modes can be identified for hanger-type connections: formation of plastic hinges in the tee flange or angle leg at cross-sections 1 and 2, and tensile failure of the bolts when the tensile force including prying action B_c $(= T + Q)$ exceeds the tensile capacity of the bolt B. Since the determination of the actual prying force is rather complex, the design equation for the required thickness for the tee flange or angle leg is semiempirical in nature. It is given by

If ASD is used:

$$t_{\text{req'd}} = \sqrt{\frac{8 T b'}{p F_y (1 + \delta \alpha')}} \tag{1.109}$$

where T is the tensile force per bolt due to service load exclusive of initial tightening and prying force (kip). The other variables are as defined in Equation 1.110, except that B in the equation for α' is defined as the allowable tensile force per bolt. A design is considered satisfactory if the thickness of the tee flange or angle leg t_f exceeds $t_{\text{req'd}}$ and $B > T$.

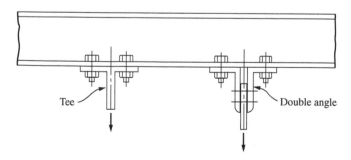

FIGURE 1.14   Hanger connections.

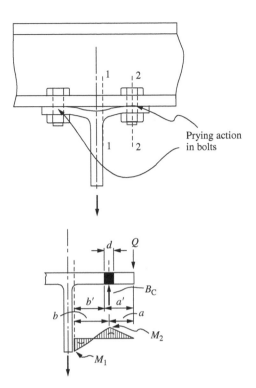

FIGURE 1.15   Prying action in hanger connections.

If LRFD is used:

$$t_{\text{req'd}} = \sqrt{\frac{4T_u b'}{\phi_b p F_y (1 + \delta \alpha')}} \tag{1.110}$$

where

$\phi_b = 0.90$

T_u = factored tensile force per bolt exclusive of initial tightening and prying force, kip

p = length of flange tributary to each bolt measured along the longitudinal axis of the tee or double-angle section, in.

δ = ratio of net area at bolt line to gross area at angle leg or stem face = $(p - d')/p$

   d' = diameter of bolt hole = bolt diameter + $\frac{1}{8}$ in.

$\alpha' = \dfrac{(B/T_u - 1)(a'/b')}{\delta[1 - (B/T_u - 1)(a'/b')]} \leq 1$ (if α' is less than zero, use $\alpha' = 1$)

   B = design tensile strength of one bolt = $\phi F_t A_b$ (kip) (ϕF_t is given in Table 1.11 and A_b is the nominal area of the bolt)

   $a' = a + d/2$

$b' = b - d/2$

   a = distance from bolt centerline to edge of tee flange or angle leg but not more than $1.25b$ in.

   b = distance from bolt centerline to face of tee stem or outstanding leg, in.

   d = nominal bolt diameter, in.

   A design is considered satisfactory if the thickness of the tee flange or angle leg t_f exceeds $t_{\text{req'd}}$ and $B > T_u$.
   Note that if t_f is much larger than $t_{\text{req'd}}$, the design will be too conservative. In this case α' should be recomputed using the equation

$$\alpha' = \frac{1}{\delta} \left[\frac{4T_u b'}{\phi_b p t_f^2 F_y} - 1 \right] \tag{1.111}$$

As before, the value of α' should be limited to the range $0 \leq \alpha' \leq 1$. This new value of α' is to be used in Equation 1.110 to recalculate $t_{req'd}$.

1.11.1.10 Bolted Bracket-Type Connections

Figure 1.16 shows three commonly used bracket-type connections. The bracing connection shown in Figure 1.16a should preferably be designed so that the line of action of the force will pass through the centroid of the bolt group. It is apparent that the bolts connecting the bracket to the column flange are subjected to combined tension and shear. As a result, the combined tensile-shear capacities of the bolts should be checked in accordance with Equation 1.107 in ASD or Equation 1.108 in LRFD. For simplicity, f_v and f_t are to be computed assuming that both the tensile and shear components of the force are distributed evenly to all bolts. In addition to checking for the bolt capacities, the bearing capacities of the column flange and the bracket should also be checked. If the axial component of the force is significant, the effect of prying should also be considered.

In the design of the eccentrically loaded connections shown in Figure 1.16b, it is assumed that the neutral axis of the connection lies at the center of gravity of the bolt group. As a result, the bolts above the neutral axis will be subjected to combined tension and shear and so Equation 1.107 or 1.108 needs to be checked. The bolts below the neutral axis are subjected to shear only and so Equation 1.104 or 1.105 applies. In calculating f_v, one can assume that all bolts in the bolt group carry an equal share of the shear force. In calculating f_t, one can assume that the tensile force varies linearly from a value of zero at the neutral axis to a maximum value at the bolt farthest away from the neutral axis. Using this assumption, f_t can be calculated from the equation Pey/I where y is the distance from the neutral axis to the location of the bolt above the neutral axis and $I = \sum A_b y^2$ is the moment of inertia of the bolt areas where A_b is the cross-sectional area of each bolt. The capacity of the connection is determined by the capacities of the bolts and the bearing capacity of the connected parts.

For the eccentrically loaded bracket connection shown in Figure 1.16c, the bolts are subjected to shear. The shear force in each bolt can be obtained by adding vectorally the shear caused by the applied

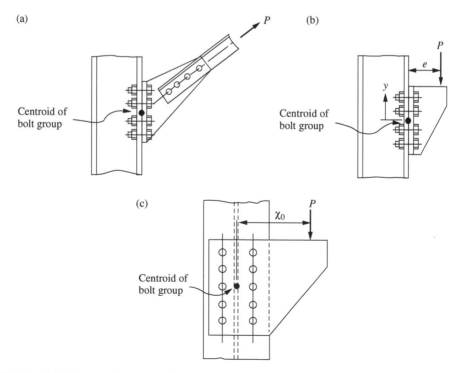

FIGURE 1.16 Bolted bracket-type connections.

load P and the moment $P\chi_0$. The design of this type of connections is facilitated by the use of tables contained in the AISC Manuals for Allowable Stress Design and Load and Resistance Factor Design (AISC 1986, 2001).

In addition to checking for bolt shear capacity, one needs to check the bearing and shear rupture capacities of the bracket plate to ensure that failure will not occur in the plate.

1.11.1.11 Bolted Shear Connections

Shear connections are connections designed to resist shear force only. They are used in Type 2 or Type 3 construction in ASD, and Type PR construction in LRFD. These connections are not expected to provide appreciable moment restraint to the connection members. Examples of these connections are shown in Figure 1.17. The framed beam connection shown in Figure 1.17a consists of two web angles that are often

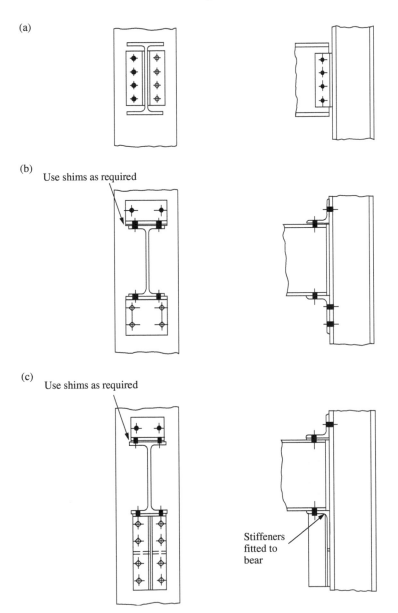

FIGURE 1.17 Bolted shear connections: (a) bolted frame beam connection; (b) bolted seated beam connection; and (c) bolted stiffened seated beam connection.

shop-bolted to the beam web and then field-bolted to the column flange. The seated beam connection shown in Figure 1.17b consists of two flange angles often shop-bolted to the beam flange and field-bolted to the column flange. To enhance the strength and stiffness of the seated beam connection, a stiffened seated beam connection shown in Figure 1.17c is sometimes used to resist large shear force. Shear connections must be designed to sustain appreciable deformation and yielding of the connections is expected. The need for ductility often limits the thickness of the angles that can be used. Most of these connections are designed with angle thickness not exceeding $\frac{5}{8}$ in. (16 mm).

The design of the connections shown in Figure 1.17 is facilitated by the use of design tables contained in the AISC-ASD and AISC-LRFD Manuals. These tables give design loads for the connections with specific dimensions based on the limit states of bolt shear, bearing strength of the connection, bolt bearing with different edge distances, and block shear (for coped beams).

1.11.1.12 Bolted Moment-Resisting Connections

Moment-resisting connections are connections designed to resist both moment and shear. They are used in Type 1 construction in ASD, and Type FR construction in LRFD. These connections are often referred to as rigid or fully restrained connections as they provide full continuity between the connected members and are designed to carry the full factored moments. Figure 1.18 shows some examples of moment-resisting connections. Additional examples can be found in the AISC-ASD and AISC-LRFD Manuals and Chapter 4 of the AISC Manual on Connections (AISC 1992).

1.11.1.13 Design of Moment-Resisting Connections

An assumption used quite often in the design of moment connections is that the moment is carried solely by the flanges of the beam. The moment is converted to a couple F_f given by $F_f = M/(d - t_f)$ acting on the beam flanges as shown in Figure 1.19.

The design of the connection for moment is considered satisfactory if the capacities of the bolts and connecting plates or structural elements are adequate to carry the flange force F_f. Depending on the geometry of the bolted connection, this may involve checking: (a) the shear and tensile capacities of the bolts; (b) the yield and fracture strength of the moment plate; (c) the bearing strength of the connected parts; and (d) bolt spacing and edge distance as discussed in the foregoing sections.

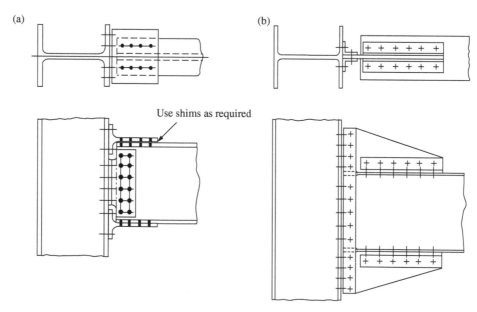

FIGURE 1.18   Bolted moment connections.

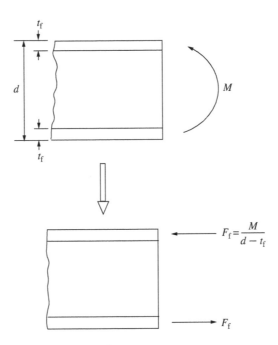

$$F_f = \frac{M}{d - t_f}$$

FIGURE 1.19 Flange forces in moment connections.

As for shear, it is common practice to assume that all the shear resistance is provided by the shear plates or angles. The design of the shear plates or angles is governed by the limit states of bolt shear, bearing of the connected parts, and shear rupture.

If the moment to be resisted is large, the flange force may cause bending of the column flange, or local yielding, crippling, or buckling of the column web. To prevent failure due to bending of the column flange or local yielding of the column web (for a tensile F_f) as well as local yielding, crippling, or buckling of the column web (for a compressive F_f), column stiffeners should be provided if any one of the conditions discussed in Section 1.5.1.1.3 is violated.

Following is a set of guidelines for the design of column web stiffeners (AISC 1989, 2001):

1. If local web yielding controls, the area of the stiffeners (provided in pairs) shall be determined based on any excess force beyond that which can be resisted by the web alone. The stiffeners need not extend more than one-half the depth of the column web if the concentrated beam flange force F_f is applied at only one column flange.
2. If web crippling or compression buckling of the web is controlled, the stiffeners shall be designed as axially loaded compression members (see Section 1.4). The stiffeners shall extend the entire depth of the column web.
3. The welds that connect the stiffeners to the column shall be designed to develop the full strength of the stiffeners.

   In addition, the following recommendations are given:

1. The width of the stiffener plus one-half of the column web thickness should not be less than one-half the width of the beam flange nor the moment connection plate that applies the force.
2. The stiffener thickness should not be less than one-half the thickness of the beam flange.
3. If only one flange of the column is connected by a moment connection, the length of the stiffener plate does not have to exceed one-half the column depth.
4. If both flanges of the column are connected by moment connections, the stiffener plate should extend through the depth of the column web and welds should be used to connect the stiffener

plate to the column web with sufficient strength to carry the unbalanced moment on opposite sides of the column.

5. If column stiffeners are required on both the tension and compression sides of the beam, the size of the stiffeners on the tension side of the beam should be equal to that on the compression size for ease of construction.

In lieu of stiffener plates, a stronger column section should be used to preclude failure in the column flange and web.

For a more thorough discussion of bolted connections, the readers are referred to the book by Kulak et al. (1987). Examples on the design of a variety of bolted connections can be found in the AISC-LRFD Manual (AISC 2001) and the AISC Manual on Connections (AISC 1992).

1.11.2 Welded Connections

Welded connections are connections whose components are joined together primarily by welds. The four most commonly used welding processes are discussed in Section 1.1.7. Welds can be classified according to

- *The types of welds:* groove welds, fillet welds, plug welds, and slot welds.
- *The positions of the welds:* horizontal welds, vertical welds, overhead welds, and flat welds.
- *The types of joints:* butt, lap, corner, edge, and tee.

Although fillet welds are generally weaker than groove welds, they are used more often because they allow for larger tolerances during erection than groove welds. Plug and slot welds are expensive to make and they do not provide much reliability in transmitting tensile forces perpendicular to the faying surfaces. Furthermore, quality control of such welds is difficult because inspection of the welds is rather arduous. As a result, plug and slot welds are normally used just for stitching different parts of the members together.

1.11.2.1 Welding Symbols

A shorthand notation giving important information on the location, size, length, etc. for the various types of welds was developed by the American Welding Society (AWS 1987) to facilitate the detailing of welds. This system of notation is reproduced in Figure 1.20.

1.11.2.2 Strength of Welds

In ASD, the strength of welds is expressed in terms of allowable stress. In LRFD, the design strength of welds is taken as the smaller of the design strength of the base material ϕF_{BM} (expressed as a function of the yield stress of the material) and the design strength of the weld electrode ϕF_W (expressed as a function of the strength of the electrode F_{EXX}). These allowable stresses and design strengths are summarized in Table 1.18 (ASD 1989; AISC 1999). During design using ASD, the computed stress in the weld shall not exceed its allowable value. During design using LRFD, the design strength of welds should exceed the required strength obtained by dividing the load to be transmitted by the effective area of the welds.

1.11.2.3 Effective Area of Welds

The effective area of groove welds is equal to the product of the width of the part joined and the effective throat thickness. The effective throat thickness of a full-penetration groove weld is taken as the thickness of the thinner part joined. The effective throat thickness of a partial-penetration groove weld is taken as the depth of the chamfer for J, U, bevel, or V (with bevel $\geq 60°$) joints and it is taken as the depth of the chamfer minus $\frac{1}{8}$ in. (3 mm) for bevel or V joints if the bevel is between 45° and 60°. For flare bevel groove welds the effective throat thickness is taken as $5R/16$ and for flare V-groove the effective throat thickness is taken as $R/2$ (or $3R/8$ for GMAW process when $R \geq 1$ in. or 25.4 mm). R is the radius of the bar or bend.

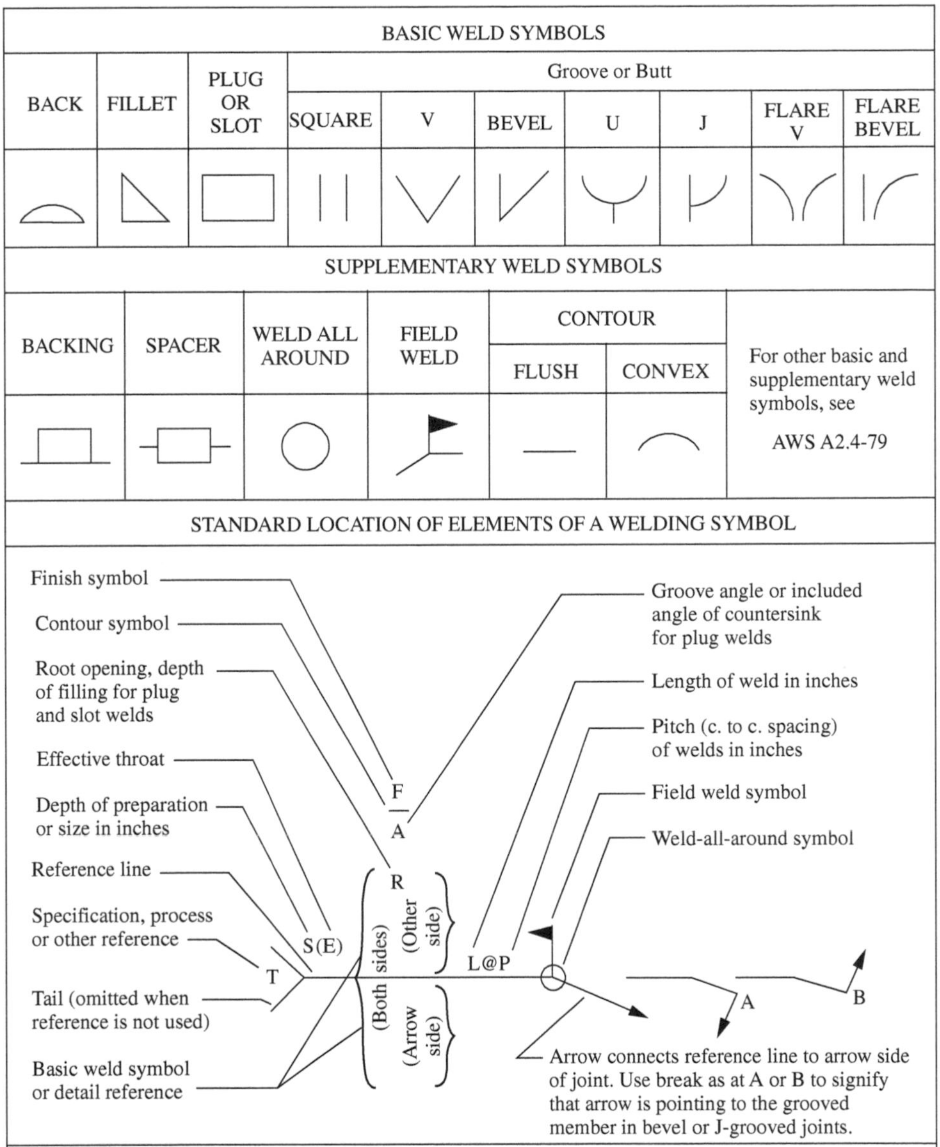
BASIC WELD SYMBOLS
BACK
FILLET
PLUG OR SLOT
Groove or Butt
SQUARE
V
BEVEL
U
J
FLARE V
FLARE BEVEL
SUPPLEMENTARY WELD SYMBOLS
BACKING
SPACER
WELD ALL AROUND
FIELD WELD
CONTOUR
FLUSH
CONVEX
For other basic and supplementary weld symbols, see
AWS A2.4-79
STANDARD LOCATION OF ELEMENTS OF A WELDING SYMBOL
Finish symbol
Contour symbol
Root opening, depth of filling for plug and slot welds
Effective throat
Depth of preparation or size in inches
Reference line
Specification, process or other reference
Tail (omitted when reference is not used)
Basic weld symbol or detail reference
Groove angle or included angle of countersink for plug welds
Length of weld in inches
Pitch (c. to c. spacing) of welds in inches
Field weld symbol
Weld-all-around symbol
F
A
R
S(E)
T
L@P
(Both sides)
(Other side)
(Arrow side)
A
B
Arrow connects reference line to arrow side of joint. Use break as at A or B to signify that arrow is pointing to the grooved member in bevel or J-grooved joints.

Note:

Size, weld symbol, length of weld and spacing must read in that order from left to right along the reference line. Neither orientation of reference line nor location of the arrow alters this rule.

The perpendicular leg of 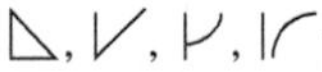 weld symbols must be at left.

Arrow and Other Side welds are of the same size unless otherwise shown. Dimensions of fillet welds must be shown on both the Arrow Side and the Other Side Symbol.

The point of the field weld symbol must point toward the tail.

Symbols apply between abrupt changes in direction of welding unless governed by the "all around" symbol or otherwise dimensioned.

These symbols do not explicitly provide for the case that frequently occurs in structural work, where duplicate material (such as stiffeners) occurs on the far side of a web or gusset plate. The fabricating industry has adopted this convention: that when the billing of the detail material discloses the existence of a member on the far side as well as on the near side, the welding shown for the near side shall be duplicated on the far side.

FIGURE 1.20 Basic weld symbols.

TABLE 1.18   Strength of Welds

Types of weld and stress[a]	Material	ASD allowable stress	LRFD ϕF_{BM} or ϕF_W	Required weld strength level[b,c]
		Full penetration groove weld		
Tension normal to effective area	Base	Same as base metal	$0.90F_y$	"Matching" weld must be used
Compression normal to effective area	Base	Same as base metal	$0.90F_y$	Weld metal with a strength level equal to or less than "matching" must be used
Tension or compression parallel to axis of weld	Base	Same as base metal	$0.90F_y$	
Shear on effective area	Base	$0.30 \times$ nominal tensile strength of weld metal	$0.90[0.60F_y]$	
	Weld electrode		$0.80[0.60F_{EXX}]$	
		Partial penetration groove welds		
Compression normal to effective area	Base	Same as base metal	$0.90F_y$	Weld metal with a strength level equal to or less than "matching" weld metal may be used
Tension or compression parallel to axis of weld[d]				
Shear parallel to axis of weld	Base	$0.30 \times$ nominal tensile strength of weld metal	$0.75[0.60F_{EXX}]$	
	Weld electrode			
Tension normal to effective area	Base	$0.30 \times$ nominal tensile strength of weld metal $\leq 0.60 \times$ yield stress of base metal	$0.90F_y$	
	Weld electrode		$0.80[0.60F_{EXX}]$	
		Fillet welds		
Stress on effective area	Base	$0.30 \times$ nominal tensile strength of weld metal	$0.75[0.60F_{EXX}]$	Weld metal with a strength level equal to or less than "matching" weld metal may be used
	Weld electrode			
Tension or compression parallel to axis of weld[d]	Base	Same as base metal	$0.90F_y$	
		Plug or slot welds		
Shear parallel to faying surfaces (on effective area)	Base	$0.30 \times$ nominal tensile strength of weld metal	$0.75[0.60F_{EXX}]$	Weld metal with a strength level equal to or less than "matching" weld metal may be used
	Weld electrode			

[a] See below for effective area.
[b] See AWS D1.1 for "matching" weld material.
[c] Weld metal one strength level stronger than "matching" weld metal will be permitted.
[d] Fillet welds and partial-penetration groove welds joining component elements of built-up members such as flange-to-web connections may be designed without regard to the tensile or compressive stress in these elements parallel to the axis of the welds.

The effective area of fillet welds is equal to the product of length of the fillets including returns and the effective throat thickness. The effective throat thickness of a fillet weld is the shortest distance from the root of the joint to the face of the diagrammatic weld as shown in Figure 1.21. Thus, for an equal leg fillet weld, the effective throat is given by 0.707 times the leg dimension. For fillet weld made by the SAW process, the effective throat thickness is taken as the leg size (for $\frac{3}{8}$ in. or 9.5 mm and smaller fillet welds) or as the theoretical throat plus 0.11 in. or 3 mm (for fillet weld over $\frac{3}{8}$ in. or 9.5 mm). A larger value for the effective throat thickness is permitted for welds made by the SAW process to account for the inherently superior quality of such welds.

The effective area of plug and slot welds is taken as the nominal cross-sectional area of the hole or slot in the plane of the faying surface.

1.11.2.4 Size and Length Limitations of Welds

To ensure effectiveness, certain size and length limitations are imposed for welds. For partial-penetration groove welds, minimum values for the effective throat thickness are given in the Table 1.19.

For plug welds, the hole diameter shall not be less than the thickness of the part that contains the weld plus $\frac{5}{16}$ in. (8 mm), rounded to the next larger odd $\frac{1}{16}$ in. (or even mm), nor greater than the minimum diameter plus $\frac{1}{8}$ in. (3 mm) or $2\frac{1}{4}$ times the thickness of the weld. The center-to-center spacing of plug welds shall not be less than four times the hole diameter. The thickness of a plug weld in material less than $\frac{5}{8}$ in. (16 mm) thick shall be equal to the thickness of the material. In material over $\frac{5}{8}$ in. (16 mm) thick, the thickness of the weld shall be at least one-half the thickness of the material but not less than $\frac{5}{8}$ in. (16 mm).

For slot welds, the slot length shall not exceed 10 times the thickness of the weld. The slot width shall not be less than the thickness of the part that contains the weld plus $\frac{5}{16}$ in. (8 mm) rounded to the nearest

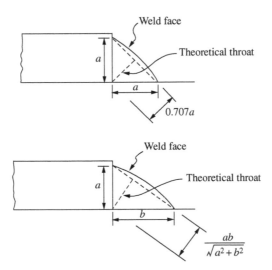

FIGURE 1.21   Effective throat of fillet welds.

TABLE 1.19   Minimum Effective Throat Thickness of Partial-Penetration Groove Welds

Thickness of the thicker part joined, t (in.)	Minimum effective throat thickness (in.)
$t \leq \frac{1}{4}$	$\frac{1}{8}$
$\frac{1}{4} < t \leq \frac{1}{2}$	$\frac{3}{16}$
$\frac{1}{2} < t \leq \frac{3}{4}$	$\frac{1}{4}$
$\frac{3}{4} < t \leq 1\frac{1}{2}$	$\frac{5}{16}$
$1\frac{1}{2} < t \leq 2\frac{1}{4}$	$\frac{3}{8}$
$2\frac{1}{4} < t \leq 6$	$\frac{1}{2}$
>6	$\frac{5}{8}$

Note: 1 in. $= 25.4$ mm.

larger odd $\frac{1}{16}$ in. (or even mm), nor larger than $2\frac{1}{4}$ times the thickness of the weld. The spacing of lines of slot welds in a direction transverse to their length shall not be less than four times the width of the slot. The center-to-center spacing of two slot welds on any line in the longitudinal direction shall not be less than two times the length of the slot. The thickness of a slot weld in material less than $\frac{5}{8}$ in. (16 mm) thick shall be equal to the thickness of the material. In material over $\frac{5}{8}$ in. (16 mm) thick, the thickness of the weld shall be at least one-half the thickness of the material but not less than $\frac{5}{8}$ in. (16 mm).

For fillet welds, the following size and length limitations apply:

- *Minimum leg size.* The minimum leg size is given in Table 1.20.
- *Maximum leg size.* Along the edge of a connected part less than $\frac{1}{4}$ in. (6 mm) thickness, the maximum leg size is equal to the thickness of the connected part. For thicker parts, the maximum leg size is t minus $\frac{1}{16}$ in. (2 mm), where t is the thickness of the part.
- *Length limitations.* The minimum effective length of a fillet weld is four times its nominal size. If a shorter length is used, the leg size of the weld shall be taken as $\frac{1}{4}$ in. (6 mm) its effective length for the purpose of stress computation. The effective length of end-loaded fillet welds with a length up to 100 times the leg dimension can be set equal to the actual length. If the length exceeds 100 times the weld size, the effective length shall be taken as the actual length multiplied by a reduction factor given by $[1.2 - 0.002(L/w)] \leq 1.0$, where L is the actual length of the end-loaded fillet weld, and w is the leg size. The length of longitudinal fillet welds used alone for flat bar tension members shall not be less than the perpendicular distance between the welds. The effective length of any segment of an intermittent fillet weld shall not be less than four times the weld size or $1\frac{1}{2}$ in. (38 mm).

TABLE 1.20 Minimum Leg Size of Fillet Welds

Thickness of thicker part joined, t (in.)	Minimum leg size (in.)
$t \leq \frac{1}{4}$	$\frac{1}{8}$
$\frac{1}{4} < t \leq \frac{1}{2}$	$\frac{3}{16}$
$\frac{1}{2} < t \leq \frac{3}{4}$	$\frac{1}{4}$
$t > \frac{3}{4}$	$\frac{5}{16}$

Note: 1 in. = 25.4 mm.

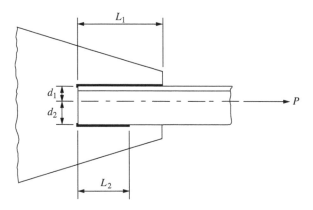

FIGURE 1.22 An eccentrically loaded welded tension connection.

1.11.2.5 Welded Connections for Tension Members

Figure 1.22 shows a tension angle member connected to a gusset plate by fillet welds. The applied tensile force P is assumed to act along the center of gravity of the angle. To avoid eccentricity, the lengths of the two fillet welds must be proportioned so that their resultant will act along the center of gravity of the angle. For example, if LRFD is used, the following equilibrium equations can be written:

Summing force along the axis of the angle:

$$(\phi F_{\mathrm{M}}) t_{\mathrm{eff}} L_1 + (\phi F_{\mathrm{M}}) t_{\mathrm{eff}} L_2 = P_{\mathrm{u}} \tag{1.112}$$

Summing moment about the center of gravity of the angle:

$$(\phi F_{\mathrm{M}}) t_{\mathrm{eff}} L_1 d_1 = (\phi F_{\mathrm{M}}) t_{\mathrm{eff}} L_2 d_2 \tag{1.113}$$

where P_{u} is the factored axial force, ϕF_{M} is the design strength of the welds as given in Table 1.18, t_{eff} is the effective throat thickness, L_1, L_2 are the lengths of the welds, and d_1, d_2 are the transverse distances from the center of gravity of the angle to the welds. The two equations can be used to solve for L_1 and L_2.

1.11.2.6 Welded Bracket-Type Connections

A typical welded bracket connection is shown in Figure 1.23. Because the load is eccentric with respect to the center of gravity of the weld group, the connection is subjected to both moment and shear. The welds must be designed to resist the combined effect of direct shear for the applied load and any additional shear from the induced moment. The design of a welded bracket connection is facilitated by the use of design tables in the AISC-ASD and AISC-LRFD Manuals. In both ASD and LRFD, the load capacity for the connection is given by

$$P = CC_1 Dl \tag{1.114}$$

where P is the allowable load (in ASD), or factored load, P_{u} (in LRFD), kip; l is the length of the vertical weld, in.; D is the number of sixteenths of an inch in fillet weld size; C are the coefficients tabulated in the AISC-ASD and AISC-LRFD Manuals. In the tables, values of C for a variety of weld geometries and dimensions are given; C_1 are the coefficients for the electrode used (see following table).

Electrode		E60	E70	E80	E90	E100	E110
ASD	F_{v} (ksi)	18	21	24	27	30	33
	C_1	0.857	1.0	1.14	1.29	1.43	1.57
LRFD	F_{EXX} (ksi)	60	70	80	90	100	110
	C_1	0.857	1.0	1.03	1.16	1.21	1.34

1.11.2.7 Welded Connections With Welds Subjected to Combined Shear and Flexure

Figure 1.24 shows a welded framed connection and a welded seated connection. The welds for these connections are subjected to combined shear and flexure. For purpose of design, it is common practice to assume that the shear force per unit length, R_{S}, acting on the welds is a constant and is given by

$$R_{\mathrm{S}} = \frac{P}{2l} \tag{1.115}$$

where P is the allowable load (in ASD), or factored load, P_{u} (in LRFD), and l is the length of the vertical weld.

In addition to shear, the welds are subjected to flexure as a result of load eccentricity. There is no general agreement on how the flexure stress should be distributed on the welds. One approach is to assume that the stress distribution is linear with half the weld subjected to tensile flexure stress and the other half is

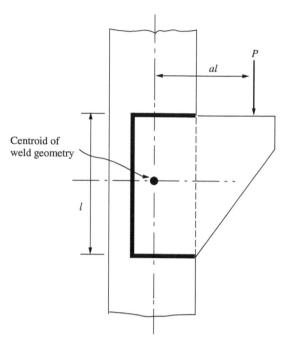

FIGURE 1.23 An eccentrically loaded welded bracket connection.

subjected to compressive flexure stress. Based on this stress distribution and ignoring the returns, the flexure tension force per unit length of weld, R_F, acting at the top of the weld can be written as

$$R_F = \frac{Mc}{I} = \frac{Pe(l/2)}{2l^3/12} = \frac{3Pe}{l^2} \tag{1.116}$$

where e is the load eccentricity.

The resultant force per unit length acting on the weld, R, is then

$$R = \sqrt{R_S^2 + R_F^2} \tag{1.117}$$

For a satisfactory design, the value R/t_{eff}, where t_{eff} is the effective throat thickness of the weld should not exceed the allowable values or design strengths given in Table 1.18.

1.11.2.8 Welded Shear Connections

Figure 1.25 shows three commonly used welded shear connections: a framed beam connection, a seated beam connection, and a stiffened seated beam connection. These connections can be designed by using the information presented in the earlier sections on welds subjected to eccentric shear and welds subjected to combined tension and flexure. For example, the welds that connect the angles to the beam web in the framed beam connection can be considered as eccentrically loaded welds and so Equation 1.114 can be used for their design. The welds that connect the angles to the column flange can be considered as welds subjected to combined tension and flexure and so Equation 1.117 can be used for their design. Like bolted shear connections, welded shear connections are expected to exhibit appreciable ductility and so the use of angles with thickness in excess of $\frac{5}{8}$ in. should be avoided. To prevent shear rupture failure, the shear rupture strength of the critically loaded connected parts should be checked.

To facilitate the design of these connections, the AISC-ASD and AISC-LRFD Manuals provide design tables by which the weld capacities and shear rupture strengths for different connection dimensions can be checked readily.

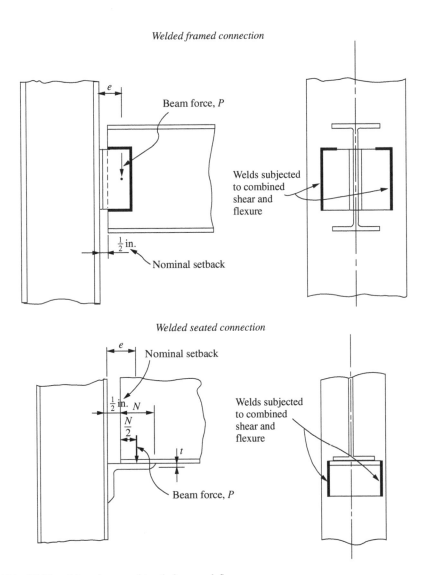

FIGURE 1.24   Welds subjected to combined shear and flexure.

1.11.2.9 Welded Moment-Resisting Connections

Welded moment-resisting connections (Figure 1.26), like bolted moment-resisting connections, must be designed to carry both moment and shear. To simplify the design procedure, it is customary to assume that the moment, to be represented by a couple F_f as shown in Figure 1.19, is to be carried by the beam flanges and that the shear is to be carried by the beam web. The connected parts (the moment plates, welds, etc.) are then designed to resist the forces F_f and shear. Depending on the geometry of the welded connection, this may include checking: (a) the yield and fracture strength of the moment plate; (b) the shear and tensile capacity of the welds; and (c) the shear rupture strength of the shear plate.

   If the column to which the connection is attached is weak, the designer should consider the use of column stiffeners to prevent failure of the column flange and web due to bending, yielding crippling, or buckling (see section on Design of moment-resisting connections).

   Examples on the design of a variety of welded shear and moment-resisting connections can be found in the AISC Manual on Connections (AISC 1992) and the AISC-LRFD Manual (AISC 2001).

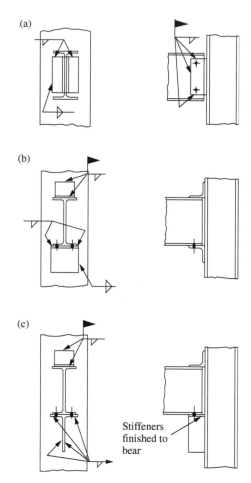

FIGURE 1.25 Welded shear connections: (a) framed beam connection; (b) seated beam connection; and (c) stiffened seated beam connection.

1.11.3 Shop Welded–Field Bolted Connections

A large percentage of connections used for construction are shop welded and field bolted types. These connections are usually more cost effective than fully welded connections and their strength and ductility characteristics often rival those of fully welded connections. Figure 1.27 shows some of these connections. The design of shop welded–field bolted connections is also covered in the AISC Manual on Connections and the AISC-LRFD Manual. In general, the following should be checked: (a) shear/tensile capacities of the bolts and/or welds; (b) bearing strength of the connected parts; (c) yield and/or fracture strength of the moment plate; and (d) shear rupture strength of the shear plate. Also, as for any other type of moment connections, column stiffeners shall be provided if any one of the following criteria — column flange bending, local web yielding, crippling and compression buckling of the column web — is violated.

1.11.4 Beam and Column Splices

Beam and column splices (Figure 1.28) are used to connect beam or column sections of different sizes. They are also used to connect beam or column of the same size if the design calls for extraordinary long span. Splices should be designed for both moment and shear unless it is the intention of the designer to utilize the splices as internal hinges. If splices are used for internal hinges, provisions must be made to ensure that the connections possess adequate ductility to allow for large hinge rotation.

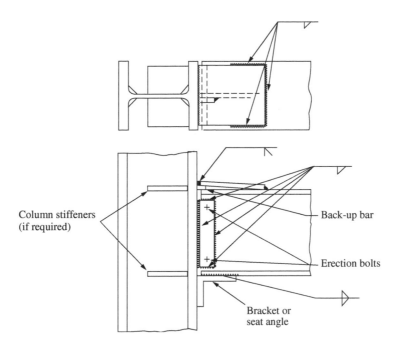

FIGURE 1.26 Welded moment connections.

Splice plates are designed according to their intended functions. Moment splices should be designed to resist the flange force $F_f = M/(d - t_f)$ (Figure 1.19) at the splice location. In particular, the following limit states need to be checked: yielding of gross area of the plate, fracture of net area of the plate (for bolted splices), bearing strengths of connected parts (for bolted splices), shear capacity of bolts (for bolted splices), and weld capacity (for welded splices). Shear splices should be designed to resist the shear forces acting at the locations of the splices. The limit states that are needed to be checked include: shear rupture of the splice plates, shear capacity of bolts under an eccentric load (for bolted splices), bearing capacity of the connected parts (for bolted splices), shear capacity of bolts (for bolted splices), weld capacity under an eccentric load (for welded splices). Design examples of beam and column splices can be found in the AISC Manual of Connections (AISC 1992) and the AISC-LRFD Manuals (AISC 2001).

1.12 Column Base Plates and Beam Bearing Plates (LRFD Approach)

1.12.1 Column Base Plates

Column base plates are steel plates placed at the bottom of columns whose function is to transmit column loads to the concrete pedestal. The design of column base plate involves two major steps:

- Determining the size $N \times B$ of the plate.
- Determining the thickness t_p of the plate.

Generally, the size of the plate is determined based on the limit state of bearing on concrete and the thickness of the plate is determined based on the limit state of plastic bending of critical sections in the plate. Depending on the types of forces (axial force, bending moment, shear force) the plate will be subjected to, the design procedures differ slightly. In all cases, a layer of grout should be placed between the base plate and its support for the purpose of leveling and anchor bolts should be provided to stabilize the column during erection or to prevent uplift for cases involving a large bending moment.

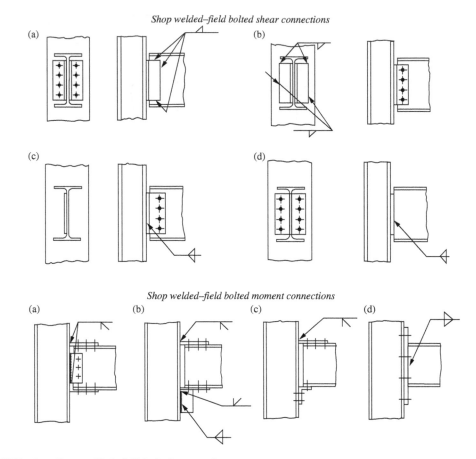

FIGURE 1.27   Shop welded–field bolted connections.

1.12.1.1 Axially Loaded Base Plates

Base plates supporting concentrically loaded columns in frames in which the column bases are assumed pinned are designed with the assumption that the column factored load P_u is distributed uniformly to the area of concrete under the base plate. The size of the base plate is determined from the limit state of bearing on concrete. The design bearing strength of concrete is given by the equation

$$\phi_c\, P_p = 0.60\left[0.85\, f_c'\, A_1 \sqrt{\frac{A_2}{A_1}}\,\right] \tag{1.118}$$

where f_c' is the compressive strength of concrete, A_1 is the area of the base plate, and A_2 is the area of the concrete pedestal that is geometrically similar to and concentric with the loaded area, $A_1 \le A_2 \le 4A_1$.

From Equation 1.118, it can be seen that the bearing capacity increases when the concrete area is greater than the plate area. This accounts for the beneficial effect of confinement. The upper limit of the bearing strength is obtained when $A_2 = 4A_1$. Presumably, the concrete area in excess of $4A_1$ is not effective in resisting the load transferred through the base plate.

Setting the column factored load, P_u, equal to the bearing capacity of the concrete pedestal, $\phi_c P_p$, and solving for A_1 from Equation 1.118, we have

$$A_1 = \frac{1}{A_2}\left[\frac{P_u}{0.6(0.85\, f_c')}\right]^2 \tag{1.119}$$

Beam splices

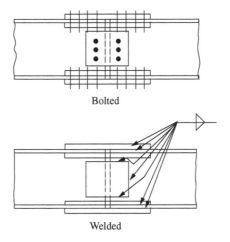

Bolted

Welded

Column splices

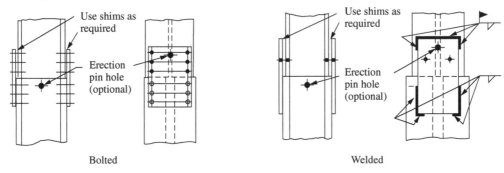

Bolted   Welded

FIGURE 1.28 Bolted and welded beam and column splices.

The length, N, and width, B, of the plate should be established so that $N \times B > A_1$. For an efficient design, the length can be determined from the equation

$$N \approx \sqrt{A_1} + 0.50(0.95d - 0.80b_f) \tag{1.120}$$

where $0.95d$ and $0.80b_f$ define the so-called effective load bearing area shown cross-hatched in Figure 1.29a. Once N is obtained, B can be solved from the equation

$$B = \frac{A_1}{N} \tag{1.121}$$

Both N and B should be rounded up to the nearest full inches.

The required plate thickness, $t_{\text{req'd}}$, is to be determined from the limit state of yield line formation along the most severely stressed sections. A yield line develops when the cross-section moment capacity is equal to its plastic moment capacity. Depending on the size of the column relative to the plate and the magnitude of the factored axial load, yield lines can form in various patterns on the plate. Figure 1.29 shows three models of plate failure in axially loaded plates. If the plate is large compared to the column, yield lines are assumed to form around the perimeter of the effective load bearing area (the cross-hatched area) as shown in Figure 1.29a. If the plate is small and the column factored load is light, yield lines are assumed to form around the inner perimeter of the I-shaped area as shown in Figure 1.29b. If the plate is small and the column factored load is heavy, yield lines are assumed to form around the inner edge of the

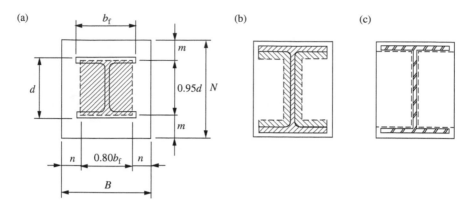

FIGURE 1.29 Failure models for centrally loaded column base plates: (a) plate with large *m, n*; (b) lightly loaded plate with small *m, n*; and (c) heavily loaded plate with small *m, n*.

column flanges and both sides of the column web as shown in Figure 1.29c. The following equation can be used to calculate the required plate thickness:

$$t_{\text{req'd}} = l\sqrt{\frac{2P_u}{0.90F_yBN}} \tag{1.122a}$$

where *l* is the larger of *m, n,* and $\lambda n'$ given by

$$m = \frac{(N - 0.95d)}{2} \tag{1.122b}$$

$$n = \frac{(B - 0.80b_f)}{2} \tag{1.122c}$$

$$n' = \sqrt{\frac{db_f}{4}} \tag{1.122d}$$

and

$$\lambda = \frac{2\sqrt{X}}{1 + \sqrt{1 - X}} \le 1 \tag{1.122e}$$

in which

$$X = \left(\frac{4db_f}{(d + b_f)^2}\right)\frac{P_u}{\phi_c P_p} \tag{1.122f}$$

1.12.1.2 Base Plates for Tubular and Pipe Columns

The design concept for base plates discussed above for I-shaped sections can be applied to the design of base plates for rectangular tubes and circular pipes. The critical section used to determine the plate thickness should be based on 0.95 times the outside column dimension for rectangular tubes and 0.80 times the outside dimension for circular pipes (Dewolf and Ricker 1990).

1.12.1.3 Base Plates with Moments

For columns in frames designed to carry moments at the base, base plates must be designed to support both axial forces and bending moments. If the moment is small compared to the axial force, the base plate can be designed without consideration of the tensile force that may develop in the anchor bolts. However, if the moment is large, this effect should be considered. To quantify the relative magnitude of this moment, an eccentricity $e = M_u/P_u$ is used. The general procedures for the design of base plates for different values of e will be given in the following (Dewolf and Ricker 1990).

1.12.1.3.1 Small Eccentricity, $e \leq N/6$

If e is small, the bearing stress is assumed to distribute linearly over the entire area of the base plate (Figure 1.30). The maximum bearing stress is given by

$$f_{max} = \frac{P_u}{BN} + \frac{M_u c}{I} \qquad (1.123)$$

where c is equal to $N/2$ and I is equal to $BN^3/12$.

The size of the plate is to be determined by a trial and error process. The size of the base plate should be such that the bearing stress calculated using Equation 1.123 does not exceed $\phi_c P_p/A_1$, that is

$$0.60 \left[0.85 f_c' \sqrt{\frac{A_2}{A_1}} \right] \leq 0.60[1.7 f_c'] \qquad (1.124)$$

The thickness of the plate is to be determined from

$$t_p = \sqrt{\frac{4 M_{plu}}{0.90 F_y}} \qquad (1.125)$$

where M_{plu} is the moment per unit width of critical section in the plate. M_{plu} is to be determined by assuming that the portion of the plate projecting beyond the critical section acts as an inverted cantilever loaded by the bearing pressure. The moment calculated at the critical section divided by the length of the critical section (i.e., B) gives M_{plu}.

1.12.1.3.2 Moderate Eccentricity, $N/6 < e \leq N/2$

For plates subjected to moderate moments, only portion of the plate will be subjected to bearing stress (Figure 1.31). Ignoring the tensile force in the anchor bolt in the region of the plate where no bearing occurs and denoting A as the length of the plate in bearing, the maximum bearing stress can be calculated from force equilibrium consideration as

$$f_{max} = \frac{2 P_u}{AB} \qquad (1.126)$$

where A is equal to $3(N/2 - e)$ is determined from moment equilibrium. The plate should be proportioned such that f_{max} does not exceed the value calculated using Equation 1.124. t_p is to be determined from Equation 1.125.

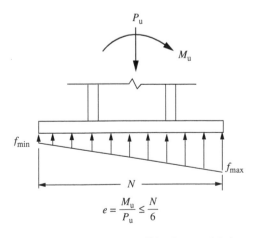

FIGURE 1.30 Eccentrically loaded column base plate (small load eccentricity).

1.12.1.3.3 Large Eccentricity, $e > N/2$

For plates subjected to large bending moments so that $e > N/2$, one needs to take into consideration the tensile force develops in the anchor bolts (Figure 1.32). Denoting T as the resultant force in the anchor bolts, A as the depth of the compressive stress block, N' as the distance from the line of action of the tensile force to the extreme compression edge of the plate, force equilibrium requires that

$$T + P_u = \frac{f_{max} A B}{2} \tag{1.127}$$

and moment equilibrium requires that

$$P_u \left(N' - \frac{N}{2} \right) + M = \frac{f_{max} A B}{2} \left(N' - \frac{A}{3} \right) \tag{1.128}$$

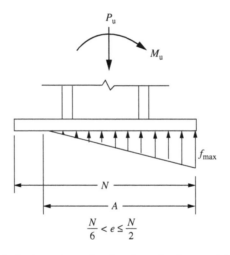

FIGURE 1.31 Eccentrically loaded column base plate (moderate load eccentricity).

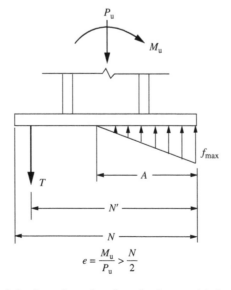

FIGURE 1.32 Eccentrically loaded column base plate (large load eccentricity).

The above equations can be used to solve for A and T. The size of the plate is to be determined using a trial and error process. The size should be chosen such that f_{max} does not exceed the value calculated using Equation 1.124, A should be smaller than N', and T should not exceed the tensile capacity of the bolts.

Once the size of the plate is determined, the plate thickness t_p is to be calculated using Equation 1.125. Note that there are two critical sections on the plate, one on the compression side of the plate and the other on the tension side of the plate. Two values of M_{plu} are to be calculated, and the larger value should be used to calculate t_p.

1.12.1.4 Base Plates with Shear

Under normal circumstances, the factored column base shear is adequately resisted by the frictional force developed between the plate and its support. Additional shear capacity is also provided by the anchor bolts. For cases in which exceptionally high shear force is expected such as in a bracing connection or in which uplift occurs which reduces the frictional resistance, the use of shear lugs may be necessary. Shear lugs can be designed based on the limit states of bearing on concrete and bending of the lugs. The size of the lug should be proportioned such that the bearing stress on concrete does not exceed $0.60(0.85f'_c)$. The thickness of the lug can be determined from Equation 1.125. M_{plu} is the moment per unit width at the critical section of the lug. The critical section is taken to be at the junction of the lug and the plate (Figure 1.33).

1.12.2 Anchor Bolts

Anchor bolts are provided to stabilize the column during erection and to prevent uplift for cases involving large moments. Anchor bolts can be cast-in-place bolts or drilled-in bolts. The latter are placed after the concrete is set and are not used often. Their design is governed by the manufacturer's specifications. Cast-in-place bolts are hooked bars, bolts, or threaded rods with nuts (Figure 1.34) placed before the concrete is set. Anchor rods and threaded rods shall conform to one of the following ASTM specifications: A36/A36M, A193/A193M, A354, A572/A572M, A588/A588M, or F1554.

Of the three types of cast-in-place anchors shown in the figure, the hooked bars are recommended for use only in axially loaded base plates. They are not normally relied upon to carry significant tensile force. Bolts and threaded rods with nuts can be used for both axially loaded base plates or base plates with moments. Threaded rods with nuts are used when the length and size required for the specific design exceed those of standard size bolts.

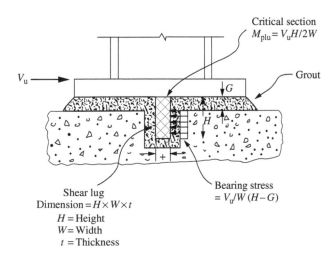

FIGURE 1.33 Column base plate subjected to shear.

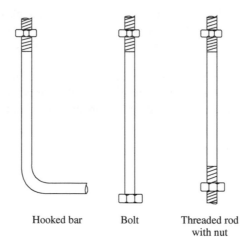

Hooked bar   Bolt   Threaded rod
                                   with nut

FIGURE 1.34 Base plate anchors.

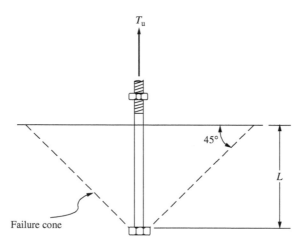

FIGURE 1.35 Cone pull-out failure.

Failure of bolts or threaded rods with nuts occurs when the tensile capacities of the bolts are reached. Failure is also considered to occur when a cone of concrete is pulled out from the pedestal. This cone pull-out type of failure is depicted schematically in Figure 1.35. The failure cone is assumed to radiate out from the bolt head or nut at an angle of 45° with tensile failure occurring along the surface of the cone at an average stress of $4\sqrt{f_c'}$ where f_c' is the compressive strength of concrete in psi. The load that will cause this cone pull-out failure is given by the product of this average stress and the projected area of the cone A_p (Marsh and Burdette 1985a,b). The design of anchor bolts is thus governed by the limit states of tensile fracture of the anchors and cone pull-out.

1.12.2.1 Limit State of Tensile Fracture

The area of the anchor should be such that

$$A_g \geq \frac{T_u}{\phi_t 0.75 F_u} \tag{1.129}$$

where A_g is the required gross area of the anchor, F_u is the minimum specified tensile strength, and $\phi_t = 0.75$ is the resistance factor for tensile fracture.

1.12.2.2 Limit State of Cone Pull-out

From Figure 1.35, it is clear that the size of the cone is a function of the length of the anchor. Provided that there are sufficient edge distance and spacing between adjacent anchors, the amount of tensile force required to cause cone pull-out failure increases with the embedded length of the anchor. This concept can be used to determine the required embedded length of the anchor. Assuming that the failure cone does not intersect with another failure cone or the edge of the pedestal, the required embedded length can be calculated from the equation

$$L \geq \sqrt{\frac{A_p}{\pi}} = \sqrt{\frac{(T_u/\phi_t 4\sqrt{f_c'})}{\pi}} \tag{1.130}$$

where A_p is the projected area of the failure cone, T_u is the required bolt force in pounds, f_c' is the compressive strength of concrete in psi, and ϕ_t is the resistance factor assumed to be equal to 0.75. If failure cones from adjacent anchors overlap one another or intersect with the pedestal edge, the projected area A_p must be adjusted according (see, e.g., Marsh and Burdette 1985a,b).

The length calculated using the above equation should not be less than the recommended values given by Shipp and Haninger (1983). These values are reproduced in the following table. Also shown in the table are the recommended minimum edge distances for the anchors.

Bolt type (material)	Minimum embedded length	Minimum edge distance
A307 (A36)	$12d$	$5d > 4$ in.
A325 (A449)	$17d$	$7d > 4$ in.

Note: d is the nominal diameter of the anchor.

1.12.3 Beam Bearing Plates

Beam bearing plates are provided between main girders and concrete pedestals to distribute the girder reactions to the concrete supports (Figure 1.36). Beam bearing plates may also be provided between cross-beams and girders if the cross-beams are designed to sit on the girders.

Beam bearing plates are designed based on the limit states of web yielding, web crippling, bearing on concrete, and plastic bending of the plate. The dimension of the plate along the beam axis, that is, N, is determined from the web yielding or web crippling criterion (see Section 1.5.1.1.3), whichever is more critical. The dimension B of the plate is determined from Equation 1.121 with A_1 calculated using Equation 1.119. P_u in Equation 1.119 is to be replaced by R_u, the factored reaction at the girder support.

Once the size $B \times N$ is determined, the plate thickness t_p can be calculated using the equation

$$t_p = \sqrt{\frac{2R_u n^2}{0.90 F_y BN}} \tag{1.131}$$

where R_u is the factored girder reaction, F_y is the yield stress of the plate, and $n = (B - 2k)/2$ in which k is the distance from the web toe of the fillet to the outer surface of the flange. The above equation was developed based on the assumption that the critical sections for plastic bending in the plate occur at a distance k from the centerline of the web.

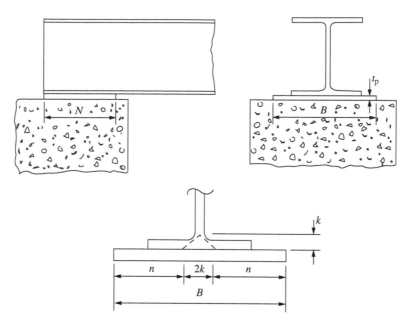

FIGURE 1.36   Beam bearing plate.

1.13   Composite Members (LRFD Approach)

Composite members are structural members made from two or more materials. The majority of composite sections used for building constructions are made from steel and concrete, although in recent years the use of fiber-reinforced polymer has been rising especially in the area of structural rehabilitation. Composite sections made from steel and concrete utilize the strength provided by steel and the rigidity provided by concrete. The combination of the two materials often results in efficient load-carrying members. Composite members may be concrete encased or concrete filled. For concrete encased members (Figure 1.37a), concrete is cast around steel shapes. In addition to enhancing strength and providing rigidity to the steel shapes, the concrete acts as a fire-proofing material to the steel shapes. It also serves as a corrosion barrier shielding the steel from corroding under adverse environmental conditions. For concrete filled members (Figure 1.37b), structural steel tubes are filled with concrete. In both concrete encased and concrete filled sections, the rigidity of the concrete often eliminates the problem of local buckling experienced by some slender elements of the steel sections.

Some disadvantages associated with composite sections are that concrete creeps and shrinks. Furthermore, uncertainties with regard to the mechanical bond developed between the steel shape and the concrete often complicate the design of beam–column joints.

1.13.1   Composite Columns

According to the LRFD Specification (AISC 1999), a compression member is regarded as a composite column if

- The cross-sectional area of the steel section is at least 4% of the total composite area. If this condition is not satisfied, the member should be designed as a reinforced concrete column.
- Longitudinal reinforcements and lateral ties are provided for concrete encased members. The cross-sectional area of the reinforcing bars shall be 0.007 in.^{2}/in. (180 mm^2/m) of bar spacing. To avoid spalling, lateral ties shall be placed at a spacing not greater than $\frac{2}{3}$ the least dimension of

(a) 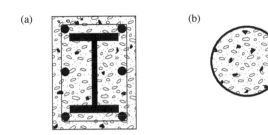 (b)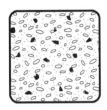

FIGURE 1.37 Composite columns: (a) concrete encased composite section and (b) concrete filled composite sections.

the composite cross-section. For fire and corrosion resistance, a minimum clear cover of 1.5 in. (38 mm) shall be provided.

- The compressive strength of concrete f_c' used for the composite section falls within the range 3 ksi (21 MPa) to 8 ksi (55 MPa) for normal weight concrete and not less than 4 ksi (28 MPa) for light weight concrete. These limits are set because they represent the range of test data available for the development of the design equations.
- The specified minimum yield stress for the steel sections and reinforcing bars used in calculating the strength of the composite columns does not exceed 60 ksi (415 MPa). This limit is set because this stress corresponds to a strain below which the concrete remains unspalled and stable.
- The minimum wall thickness of the steel sections for concrete filled members is equal to $b\sqrt{(F_Y/3E)}$ for rectangular sections of width b and $D\sqrt{(F_Y/8E)}$ for circular sections of outside diameter D.

1.13.1.1 Design Compressive Strength

The design compressive strength, $\phi_c P_n$, shall exceed the factored compressive force, P_u. The design compressive strength is given as follows:

For $\lambda_c \leq 1.5$:

$$\phi_c P_n = \begin{cases} 0.85\left[(0.658^{\lambda_c^2})A_s F_{my}\right], & \text{if } \lambda_c \leq 1.5 \\ 0.85\left[\left(\dfrac{0.877}{\lambda_c^2}\right)A_s F_{my}\right], & \text{if } \lambda_c > 1.5 \end{cases} \tag{1.132}$$

where

$$\lambda_c = \frac{KL}{r_m \pi}\sqrt{\frac{F_{my}}{E_m}} \tag{1.133}$$

$$F_{my} = F_y + c_1 F_{yr}\left(\frac{A_r}{A_s}\right) + c_2 f_c'\left(\frac{A_c}{A_s}\right) \tag{1.134}$$

$$E_m = E + c_3 E_c\left(\frac{A_c}{A_s}\right) \tag{1.135}$$

where r_m is the radius of gyration of steel section and shall not be less than 0.3 times the overall thickness of the composite cross-section in the plane of buckling, A_c is the area of concrete, A_r is the area of longitudinal reinforcing bars, A_s is the area of steel shape, E is the modulus of elasticity of steel, E_c is the modulus of elasticity of concrete, F_y is the specified minimum yield stress of steel shape, F_{yr} is the specified minimum yield stress of longitudinal reinforcing bars, f_c' is the specified compressive strength of concrete, and c_1, c_2, c_3 are the coefficients given in the table below.

Type of composite section	c_1	c_2	c_3
Concrete encased shapes	0.7	0.6	0.2
Concrete-filled pipes and tubings	1.0	0.85	0.4

In addition to satisfying the condition $\phi_c P_n \geq P_u$, shear connectors spaced no more than 16 in. (405 mm) apart on at least two faces of the steel section in a symmetric pattern about the axes of the steel section shall be provided for concrete encased composite columns to transfer the interface shear force V_u' between steel and concrete. V_u' is given by

$$V_u' = \begin{cases} V_u\left(1 - \dfrac{A_s F_y}{P_n}\right) & \text{when the force is applied to the steel section} \\[4mm] V_u\left(\dfrac{A_s F_y}{P_n}\right) & \text{when the force is applied to the concrete encasement} \end{cases} \tag{1.136}$$

where V_u is the axial force in the column, A_s is the area of the steel section, F_y is the yield strength of the steel section, and P_n is the nominal compressive strength of the composite column without consideration of slenderness effect.

If the supporting concrete area in direct bearing is larger than the loaded area, the bearing condition for concrete must also be satisfied. Denoting $\phi_c P_{nc}$ $(= \phi_c P_{n,\text{composite section}} - \phi_c P_{n,\text{steel shape alone}})$ as the portion of compressive strength resisted by the concrete and A_B as the loaded area, the condition that needs to be satisfied is

$$\phi_c P_{nc} \leq 0.65[1.7 f_c' A_B] \tag{1.137}$$

1.13.2 Composite Beams

Composite beams used in construction can often be found in two forms: steel beams connected to a concrete slab by shear connectors and concrete encased steel beams.

1.13.2.1 Steel Beams with Shear Connectors

The design flexure strength for steel beams with shear connectors is $\phi_b M_n$. The resistance factor ϕ_b and nominal moment M_n are determined as follows:

Condition	ϕ_b	M_n
Positive moment region and $h/t_w \leq 3.76\sqrt{(E/F_{yf})}$	0.85	Determined from plastic stress distribution on the composite section
Positive moment region and $h/t_w > 3.76\sqrt{(E/F_{yf})}$	0.90	Determined from elastic stress superposition considering the effects of shoring
Negative moment region	0.90	Determined for the steel section alone using equations presented in Section 1.5

1.13.2.2 Concrete Encased Steel Beams

For steel beams fully encased in concrete, no additional anchorage for shear transfer is required if (a) at least $1\frac{1}{2}$ in. (38 mm) concrete cover is provided on top of the beam and at least 2 in. (51 mm) cover is provided over the sides and at the bottom of the beam and (b) spalling of concrete is prevented by adequate mesh or other reinforcing steel. The design flexural strength $\phi_b M_n$ can be computed using either an elastic analysis or a plastic analysis.

If an elastic analysis is used, ϕ_b shall be taken as 0.90. A linear strain distribution is assumed for the cross-section with zero strain at the neutral axis and maximum strains at the extreme fibers. The stresses are then computed by multiplying the strains by E (for steel) or E_c (for concrete). Maximum stress in steel shall be limited to F_y and maximum stress in concrete shall be limited to $0.85f_c'$. The tensile strength of concrete shall be neglected. M_n is to be calculated by integrating the resulting stress block about the neutral axis.

If a plastic analysis is used, ϕ_b shall be taken as 0.90 and M_n shall be assumed to be equal to M_p, the plastic moment capacity of the steel section alone.

1.13.3 Composite Beam–Columns

Composite beam–columns shall be designed to satisfy the interaction equation of Equation 1.73 or 1.74 whichever is applicable, with $\phi_c P_n$ calculated based on Equations 1.132 to 1.135, P_e calculated using the equation $P_e = A_s F_{my}/\lambda_c^2$ and $\phi_b M_n$ calculated using the following equation (Galambos and Chapuis 1980):

$$\phi_b M_n = 0.90\left[ZF_y + \frac{1}{3}(h_2 - 2c_r)A_r F_{yr} + \left(\frac{h_2}{2} - \frac{A_w F_y}{1.7f_c' h_1}\right)A_w F_y\right] \tag{1.138}$$

where Z is the plastic section modulus of the steel section, c_r is the average of the distance measured from the compression face to the longitudinal reinforcement in that face and the distance measured from the tension face to the longitudinal reinforcement in that face, h_1 is the width of the composite section perpendicular to the plane of bending, h_2 is the width of the composite section parallel to the plane of bending, A_r is the cross-sectional area of longitudinal reinforcing bars, A_w is the web area of the encased steel shape ($=0$ for concrete-filled tubes), F_y is the yield stress of the steel section, and F_{yr} is the yield stress of reinforcing bars.

If $0 < (P_u/\phi_c P_n) \leq 0.3$, a linear interpolation of $\phi_b M_n$ calculated using the above equation assuming $P_u/\phi_c P_n = 0.3$ and that calculated for beams with $P_u/\phi_c P_n = 0$ (see Section 1.13.2) should be used.

1.13.4 Composite Floor Slabs

Composite floor slabs (Figure 1.38) can be designed as shored or unshored. In shored construction, temporary shores are used during construction to support the dead and accidental live loads until the concrete cures. The supporting beams are designed on the basis of their ability to develop composite action to support all factored loads after the concrete cures. In unshored construction, temporary shores are not used. As a result, the steel beams alone must be designed to support the dead and accidental live loads before the concrete has attained 75% of its specified strength. After the concrete is cured, the composite section should have adequate strength to support all factored loads.

Composite action for the composite floor slabs shown in Figure 1.38 is developed as a result of the presence of shear connectors. If sufficient shear connectors are provided so that the maximum flexural strength of the composite section can be developed, the section is referred to as fully composite. Otherwise, the section is referred to as partially composite. The flexural strength of a partially composite section is governed by the shear strength of the shear connectors. The horizontal shear force V_h that should be designed for at the interface of the steel beam and the concrete slab is given by

In regions of positive moment:

$$V_h = \min\left(0.85f_c' A_c, A_s F_y, \sum Q_n\right) \tag{1.139}$$

In regions of negative moment:

$$V_h = \min\left(A_r F_{yr}, \sum Q_n\right) \tag{1.140}$$

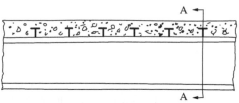

Composite floor slab with stud shear connectors

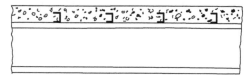

Composite floor slab with channel shear connectors

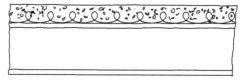

Composite floor slab with spiral shear connectors

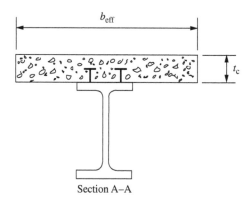

Section A–A

FIGURE 1.38   Composite floor slabs.

where

f_c' = compressive strength of concrete

A_c = effective area of the concrete slab $= t_c b_{eff}$

t_c = thickness of the concrete slab

b_{eff} = effective width of the concrete slab

   = min $(L/4, s)$ for an interior beam

   = min $(L/8 +$ distance from beam centerline to edge of slab, $s/2 +$ distance from beam centerline to edge of slab) for an exterior beam

L = beam span measured from center-to-center of supports

s = spacing between centerline of adjacent beams

A_s = cross-sectional area of the steel beam

F_y = yield stress of the steel beam

A_r = area of reinforcing steel within the effective area of the concrete slab

F_{yr} = yield stress of the reinforcing steel
$\sum Q_n$ = sum of nominal shear strengths of the shear connectors

The nominal shear strength of a shear connector (used without a formed steel deck) is given by

For a stud shear connector:

$$Q_n = 0.5 A_{sc} \sqrt{f_c' E_c} \leq A_{sc} F_u \tag{1.141}$$

For a channel shear connector:

$$Q_n = 0.3(t_f + 0.5t_w) L_c \sqrt{f_c' E_c} \tag{1.142}$$

where A_{sc} is the cross-sectional area of the shear stud (in.^{2}), f_c' is the compressive strength of concrete (ksi), E_c is the modulus of elasticity of concrete (ksi), F_u is the minimum specified tensile strength of the stud shear connector (ksi), t_f is the flange thickness of the channel shear connector (in.), t_w is the web thickness of the channel shear connector (in.), and L_c is the length of the channel shear connector (in.).

If a formed steel deck is used, Q_n must be reduced by a reduction factor. The reduction factor depends on whether the deck ribs are perpendicular or parallel to the steel beam.

For deck ribs perpendicular to steel beam:

$$\frac{0.85}{\sqrt{N_r}} \left(\frac{w_r}{h_r}\right) \left[\left(\frac{H_s}{h_r}\right) - 1.0\right] \leq 1.0 \tag{1.143}$$

When only a single stud is present in a rib perpendicular to the steel beam, the reduction factor expressed in Equation 1.143 shall not exceed 0.75.

For deck ribs parallel to steel beam:

$$0.6 \left(\frac{w_r}{h_r}\right) \left[\left(\frac{H_s}{h_r}\right) - 1.0\right] \leq 1.0 \tag{1.144}$$

The reduction factor expressed in Equation 1.144 is applicable only if $(w_r/h_r) < 1.5$. In the above equations, N_r is the number of stud connectors in one rib at a beam intersection, not to exceed three in computations regardless of the actual number of studs installed, w_r is the average width of the concrete rib or haunch, h_r is the nominal rib height, and H_s is the length of stud connector after welding, not to exceed the value $h_r + 3$ in. (75 mm) in computations regardless of the actual length.

For full composite action, the number of connectors required between the *maximum* moment point and the *zero* moment point of the beam is given by

$$N = \frac{V_h}{Q_n} \tag{1.145}$$

For partial composite action, the number of connectors required is governed by the condition $\phi_b M_n \geq M_u$, where $\phi_b M_n$ is governed by the shear strength of the connectors.

The placement and spacing of the shear connectors should comply with the following guidelines:

- The shear connectors shall be uniformly spaced between the points of maximum moment and zero moment. However, the number of shear connectors placed between a concentrated load point and the nearest zero moment point must be sufficient to resist the factored moment M_u.
- Except for connectors installed in the ribs of formed steel decks, shear connectors shall have at least 1 in. (25.4 mm) of lateral concrete cover. The slab thickness above the formed steel deck shall not be less than 2 in. (50 mm).
- Unless located over the web, the diameter of shear studs must not exceed 2.5 times the thickness of the beam flange. For the formed steel deck, the diameter of stud shear connectors shall not exceed $\frac{3}{4}$ in. (19 mm), and shall extend not less than $1\frac{1}{2}$ in. (38 mm) above the top of the steel deck.

- The longitudinal spacing of the studs should fall in the range six times the stud diameter to eight times the slab thickness if a solid slab is used or four times the stud diameter to eight times the slab thickness or 36 in. (915 mm), whichever is smaller, if a formed steel deck is used. Also, to resist uplift, the steel deck shall be anchored to all supporting members at a spacing not to exceed 18 in. (460 mm).

The design flexural strength $\phi_b M_n$ of the composite beam with shear connectors is determined as follows:

In regions of positive moments: For $h_c/t_w \leq 3.76/\sqrt{(E/F_{yf})}$, $\phi_b = 0.85$, M_n is the moment capacity determined using a plastic stress distribution assuming concrete crushes at a stress of $0.85f_c'$ and steel yields at a stress of F_y. If a portion of the concrete slab is in tension, the strength contribution of that portion of concrete is ignored. The determination of M_n using this method is very similar to the technique used for computing moment capacity of a reinforced concrete beam according to the ultimate strength method.

For $h_c/t_w > 3.76/\sqrt{(E/F_{yf})}$, $\phi_b = 0.90$, M_n is the moment capacity determined using superposition of elastic stress, considering the effect of shoring. The determination of M_n using this method is quite similar to the technique used for computing the moment capacity of a reinforced concrete beam according to the working stress method.

In regions of negative moment: $\phi_b M_n$ is to be determined for the steel section alone in accordance with the requirements discussed in Section 1.5.

To facilitate design, numerical values of $\phi_b M_n$ for composite beams with shear studs in solid slabs are given in tabulated form in the AISC-LRFD Manual. Values of $\phi_b M_n$ for composite beams with formed steel decks are given in a publication by the Steel Deck Institute (2001).

1.14 Plastic Design

Plastic analysis and design is permitted only for steels with yield stress not exceeding 65 ksi. The reason for this is that steels with high yield stress lack the ductility required for inelastic deformation at hinge locations. Without adequate inelastic deformation, moment redistribution, which is an important characteristic for PD, cannot take place.

In PD, the predominant limit state is the formation of plastic hinges. Failure occurs when sufficient plastic hinges have formed for a collapse mechanism to develop. To ensure that plastic hinges can form and can undergo large inelastic rotation, the following conditions must be satisfied:

- Sections must be compact. That is, the width–thickness ratios of flanges in compression and webs must not exceed λ_p in Table 1.8.
- For columns, the slenderness parameter λ_c (see Section 1.4) shall not exceed $1.5K$ where K is the effective length factor and P_u from gravity and horizontal loads shall not exceed $0.75A_g F_y$.
- For beams, the lateral unbraced length L_b shall not exceed L_{pd} where

For doubly and singly symmetric I-shaped members loaded in the plane of the web

$$L_{pd} = \frac{3600 + 2200(M_l/M_p)}{F_y} r_y \tag{1.146}$$

and *for solid rectangular bars and symmetric box beams*

$$L_{pd} = \frac{5000 + 3000(M_l/M_p)}{F_y} r_y \geq \frac{3000 \, r_y}{F_y} \tag{1.147}$$

In the above equations, M_l is the smaller end moment within the unbraced length of the beam, M_p is the plastic moment $(= Z_x F_y)$ of the cross-section, r_y is the radius of gyration about the minor axis, in inches, and F_y is the specified minimum yield stress, in ksi.

L_{pd} is not defined for beams bent about their minor axes, nor for beams with circular and square cross-sections because these beams do not experience lateral torsional bucking when loaded.

1.14.1 Plastic Design of Columns and Beams

Provided that the above limitations are satisfied, the design of columns shall meet the condition $1.7F_aA \geq P_u$ where F_a is the allowable compressive stress given in Equation 1.16, A is the gross cross-sectional area and P_u is the factored axial load.

The design of beams shall satisfy the conditions $M_p \geq M_u$, and $0.55F_yt_wd \geq V_u$ where M_u and V_u are the factored moment and shear, respectively. M_p is the plastic moment capacity, F_y is the minimum specified yield stress, t_w is the beam web thickness, and d is the beam depth. For beams subjected to concentrated loads, all failure modes associated with concentrated loads (see Sections 1.5.1.1.3 and 1.5.1.2.3) should also be prevented.

Except at the location where the last hinge forms, a beam bending about its major axis must be braced to resist lateral and torsional displacements at plastic hinge locations. The distance between adjacent braced points should not exceed l_{cr} given by

$$l_{cr} = \begin{cases} \left(\dfrac{1375}{F_y} + 25\right)r_y, & \text{if } -0.5 < \dfrac{M}{M_p} < 1.0 \\[3mm] \left(\dfrac{1375}{F_y}\right)r_y, & \text{if } -1.0 < \dfrac{M}{M_p} \leq -0.5 \end{cases} \tag{1.148}$$

where r_y is the radius of gyration about the weak axis, M is the smaller of the two end moments of the unbraced segment, and M_p is the plastic moment capacity.

M/M_p is taken as positive if the unbraced segment bends in reverse curvature and is taken as negative if the unbraced segment bends in single curvature.

1.14.2 Plastic Design of Beam–Columns

Beam–columns designed on the basis of plastic analysis shall satisfy the following interaction equations for stability (Equation 1.149) and for strength (Equation 1.150):

$$\frac{P_u}{P_{cr}} + \frac{C_mM_u}{\left(1 - \dfrac{P_u}{P_e}\right)M_m} \leq 1.0 \tag{1.149}$$

$$\frac{P_u}{P_y} + \frac{M_u}{1.18M_p} \leq 1.0 \tag{1.150}$$

where
P_u = factored axial load
P_{cr} = $1.7F_aA$, F_a is defined in Equation 1.16 and A is the cross-sectional area
P_y = yield load = AF_y
P_e = Euler buckling load = $\pi^2EI/(Kl)^2$
C_m = coefficient defined in Section 1.4
M_u = factored moment
M_p = plastic moment = ZF_y
M_m = maximum moment that can be resisted by the member in the absence of axial load
  = M_{px} if the member is braced in the weak direction
  = $\{1.07-[(l/r_y)\sqrt{F_y}]/3160\}M_{px} \leq M_{px}$ if the member is unbraced in the weak direction
l = unbraced length of the member
r_y = radius of gyration about the minor axis
M_{px} = plastic moment about the major axis = Z_xF_y
F_y = minimum specified yield stress, ksi

1.15   Reduced Beam Section

Reduced beam section (RBS) or dogbone connection is a type of connection in welded steel moment frames in which portions of the bottom beam flange or both top and bottom flanges are cut near the beam-to-column connection thereby reducing the flexural strength of the beam at the RBS region and thus force a plastic hinge to form in a region away from the connection (Engelhardt et al. 1996; Iwankiw and Carter 1996; Plumier 1997). The presence of this reduced section in the beam also tends to decrease the force demand on the beam flange welds and so mitigate the distress that may cause fracture in the connection. RBS can be bottom flange cut only, or both top and bottom flange cuts. Bottom flange RBS is used if it is difficult or impossible to cut the top flange of an existing beam (e.g., if the beam is attached to a concrete floor slab). Figure 1.39 shows some typical cut geometries for RBS. The constant cut offers the advantage of ease of fabrication. The tapered cut has the advantage of matching the beam's flexural strength to the flexural demand on the beam under a gravity load. The radius cut is relatively easy to fabricate and because the change in geometry of the cross-section is rather gradual, it also has the advantage of minimizing stress concentration. Based on experimental investigations (Engelhardt et al. 1998; Moore et al. 1999), the radius cut RBS has been shown to be a reliable connection for welded steel moment frames.

The key dimensions of a radius cut RBS is shown in Figure 1.40. The distance from the face of the column to the start of the cut is designated as a, the length and depth of the cut are denoted as b and c, respectively. Values of a, b, c are given as follows (Engelhardt et al. 1998; Gross et al. 1999):

$$a \approx (0.5 \text{ to } 0.75)b_f \tag{1.151}$$

$$b \approx (0.65 \text{ to } 0.85)d \tag{1.152}$$

$$c \approx 0.25b_f \quad \text{for a bottom flange RBS} \tag{1.153}$$

where b_f is the beam flange width and d is the beam depth. Using geometry, the cut radius R can be calculated as

$$R = \frac{4c^2 + b^2}{8c} \tag{1.154}$$

and the distance from the face of the column to the critical plastic section s_c is given by

$$s_c = a + \frac{b}{2} \tag{1.155}$$

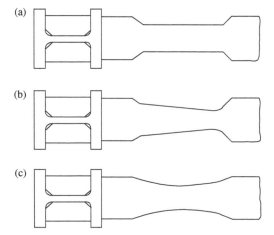

FIGURE 1.39   Reduced beam section cut geometries: (a) constant cut; (b) tapered cut; and (c) radius cut.

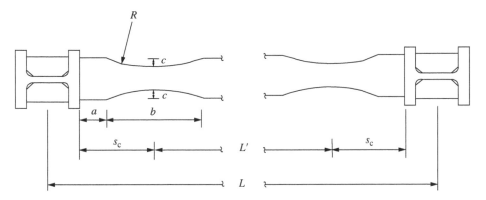

FIGURE 1.40 Key dimensions of a radius cut reduced beam section.

An optimal RBS is one in which the moment at the face of the column will be minimized. To achieve this condition, the following procedure is recommended (Gross et al. 1999):

- Set $c = 0.25b_f$
- Compute the RBS plastic section modulus using the equation

$$Z_{RBS} = Z_b - \frac{(ct_f)^2}{t_w} - ct_f(d - t_f) \tag{1.156}$$

where Z_b is the plastic section modulus of the full beam cross-section, c is the depth of cut as shown in Figure 1.40, and d, t_f, t_w are the beam depth, beam flange thickness, and web thickness, respectively.

- Compute η, the ratio of moment at the face of the column to plastic moment of the connecting beam, from the equation

$$\eta = 1.1\left(1 + \frac{2s_c}{L'}\right)\frac{Z_{RBS}}{Z_b} + \frac{wL's_c}{2Z_bF_{yf}} \tag{1.157}$$

where s_c is given in Equation 1.155 and shown in Figure 1.40, L' is the beam span between critical plastic sections (see Figure 1.40), Z_{RBS} and Z_b are defined in Equation 1.156, w is the magnitude of the uniformly distributed load on the beam, and F_{yf} is the beam flange yield strength.

- If $\eta < 1.05$, then the RBS dimensions are satisfactory. Otherwise, use RBS cutouts in both the top and bottom flanges, or consider using other types of moment connections (Gross et al. 1999).

Experimental studies (Uang and Fan 1999; Engelhardt et al. 2000; Gilton et al. 2000; Yu et al. 2000) of a number of radius cut RBS with or without the presence of a concrete slab have shown that the connections perform satisfactorily and exhibit sufficient ductility under cyclic loading. However, the use of RBS beams in a moment resistant frame tends to cause an overall reduction in frame stiffness of around 4 to 7% (Grubbs 1997). If the increase in frame drift due to this reduction in frame stiffness is appreciable, proper allowances must be made in the analysis and design of the frame.

1.16 Seismic Design

Special provisions (AISC 2002) apply for the design of steel structures to withstand earthquake loading. To ensure sufficient ductility, more stringent limiting width–thickness ratios than those shown in Table 1.8 for compression elements are required. These seismic limiting width thickness ratios are shown in Table 1.21.

TABLE 1.21   Seismic Limiting Width–Thickness Ratios for Compression Elements

Component element	Width–thickness ratio[a]	Limiting value[a], λ_p
Flanges of I-shaped rolled, hybrid, or welded beams, columns (in SMF system) and braces, flanges of channels and angles in flexure, legs of single angle, legs of double-angle members with separators, flanges of tees	b/t	$0.30\sqrt{(E/F_y)}$
Webs of tees	d/t	$0.30\sqrt{(E/F_y)}$
Flanges of I-shaped rolled, hybrid, or welded columns (in other framing systems)	b/t	$0.38\sqrt{(E/F_y)}$
Flanges of H-piles	b/t	$0.45\sqrt{(E/F_y)}$
Flat bars	b/t	2.5
Webs in flexural compression for beams in SMF	h_c/t_w	$2.45\sqrt{(E/F_y)}$
Other webs in flexural compression	h_c/t_w	$3.14\sqrt{(E/F_y)}$
Webs in combined flexural and axial compression	h_c/t_w	For $P_u/\phi_b P_y \leq 0.125$: $3.14(1 - 1.54 P_u/\phi_b P_y)\sqrt{(E/F_y)}$ For $P_u/\phi_b P_y > 0.125$: $1.12(2.33 - P_u/\phi_b P_y)\sqrt{(E/F_y)}$
Round HSS in axial compression or flexure	D/t	$0.044 E/F_y$
Rectangular HSS in axial compression or flexure	b/t or h_c/t	$0.64\sqrt{(E/F_y)}$
Webs of H-pile sections	h/t_w	$0.94\sqrt{(E/F_y)}$

[a] See Table 1.8 for definitions of the terms used, and replace F_y by F_{yf} for hybrid sections.

Moreover, the structure needs to be designed for loads and load combinations that include the effect of earthquakes, and if the applicable building code (e.g., UBC 1997; IBC 2000) requires that amplified seismic loads be used, the horizontal component of the earthquake load shall be multiplied by an overstrength factor Ω_0, as prescribed by the applicable building code. In lieu of a specific definition of Ω_0, the ones shown in Table 1.22 can be used.

The seismic provisions of AISC (2002) identify seven types of frames for seismic resistant design: special moment frames (SMF), intermediate moment frames (IMF), ordinary moment frames (OMF), special truss moment frames (STMF), special concentrically braced frames (SCBF), ordinary concentrically braced frames (OCBF), and eccentrically braced frames (EBF). Examples of these framing systems are shown in Figure 1.41. Specific requirements for the design of these frames are given by AISC (2002). Note that the word "special" is used if the frame is expected to withstand significant inelastic deformation, the word "intermediate" is used if the frame is expected to withstand moderate inelastic deformations, and the word "ordinary" is used if the frame is expected to withstand only limited inelastic deformations. For instance, all connections in an SMF must be designed to sustain an interstory drift angle of at least 0.04 radians at a moment (determined at the column face) equal to at least 80% of the nominal plastic moment of the connecting beam, while in an IMF the ductility value is reduced to 0.02 radians, and in an OMF a value of 0.01 radians is expected at a column face moment no less than 50% of the nominal connecting beam plastic moment. Also, connections used in SMF and IMF must be FR moment connections that are prequalified or tested for conformance with the aforementioned ductility requirement, whereas both FR and PR moment connections are permitted for OMF. To avoid lateral torsional instability, beams in SMF must have both flanges laterally supported at a distance not to exceed $0.086 r_y E/F_y$. This limit is more stringent than those (L_p) discussed in Section 1.5. In addition, lateral supports are needed at locations near concentrated forces, changes in cross-section, and regions of plastic hinges.

TABLE 1.22 Value of Ω_0

Seismic load resisting system	Ω_0
Moment frame systems	3
Eccentrically braced frames	2.5
All other systems meeting Part I (AISC 2002) requirements	2

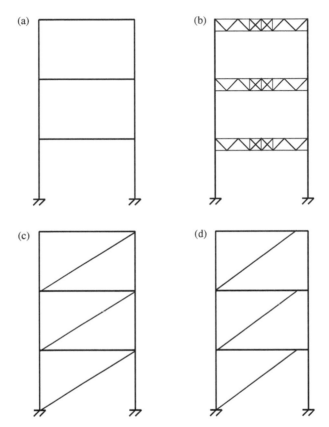

FIGURE 1.41 Types of frames: (a) moment frame; (b) truss moment frame; (c) concentrically braced frame; and (d) eccentrically braced frame.

If bolts are used, they should all be pretensioned high-strength bolts designed for standard or short-slotted (with direction of slot perpendicular to applied force) holes with hole deformation considered in the design (i.e., nominal bearing strength $\leq 2.4dtF_u$). Although design shear strength for the bolts is permitted to be calculated as that of bearing-type joints, all faying surfaces must be prepared as required for Class A or better slip-critical joints. Bolted connections for members that form part of the seismic load resisting system shall be designed to ensure that a ductile limit state (either in the connection or in the member) controls the design. If welds are used, they shall conform to procedures and standards outlined in AWS D1.1 (AWS 2000). If both bolts and welds are used, bolts shall not be designed to share load with welds on the same faying surface.

To avoid brittle fracture of welds, a minimum Charpy V-Notch (CVN) toughness of 20 ft lb (27 J) at $-20°$F ($-29°$C) is required for all welds in members and connections that are part of the seismic lateral load resisting system. For structures with enclosed SMF or IMF maintained at a temperature of 50°F (10°C) or higher, complete joint penetration welds of beam flanges to columns, column splices, and groove welds of shear tabs and beam webs to columns, shall be made with a filler material with minimum CVN toughness of 20 ft lb (27 J) at $-20°F$ $(-29°C)$ and 40 ft lb (54 J) at $70°F$ $(21°C)$. If the service temperature is lower than $50°F$ $(10°C)$, these minimum CVN toughness values shall be reduced accordingly.

Regardless of the type of framing system used, the required strength of the connections or related components must be designed for an expected yield strength F_{ye} of the connection members given by

$$F_{ye} = R_y F_y \qquad (1.158)$$

where R_y is an adjustment factor to account for the increase in demand on the connections when yielding occurs in the connecting members at the true yield stress (which is often higher than the minimum specified yield stress used in the design) during a severe earthquake. Values for R_y are given in Table 1.23.

The current design philosophy for moment frames is based on the strong column–weak beam concept. This condition is enforced in the design of SMF at all beam–column joints by the following equation

$$\frac{\sum M_{pc}^*}{\sum M_{pb}^*} > 1.0 \qquad (1.159)$$

where $\sum M_{pc}^*$ is the sum of column moments acting at the joint in question and $\sum M_{pb}^*$ is the sum of beam moments acting at the joint in question. They can be calculated as follows:

$$\sum M_{pc}^* = \sum Z_c \left(F_{yc} - \frac{P_{uc}}{A_g} \right) \qquad (1.160)$$

$$\sum M_{pb}^* = \sum (1.1 R_y M_p + M_v) \qquad (1.161)$$

where Z_c is the column plastic section modulus, F_{yc} is the specified minimum yield stress of the column, P_{uc} is the required axial strength of the column, A_g is the gross sectional area, and R_y are the values shown in Table 1.23, M_p is the plastic moment capacity of beam (for a reduced beam section, M_p can be calculated based on the minimum plastic section modulus of the reduced section), and M_v is the additional moment due to shear amplification from the location of plastic hinge to the column centerline.

While the use of the strong column–weak beam design tends to force plastic hinges to develop in the beams, there is a region in the column commonly referred to as the panel zone where high stresses are often developed. The panel zone is the region of the column to which the beams are attached. For the moment frame to be able to withstand the seismic loading, the panel zone must be strong enough to allow the frame to sustain inelastic deformation and dissipate energy before failure. For SMF, the design criterion for a panel zone (when the beam web is parallel to the column web) is

$$\phi_v R_v \geq R_u \qquad (1.162)$$

TABLE 1.23 Value of R_y

Application	R_y
Hot rolled structural steel shapes	
ASTM A36/A36M	1.5
ASTM A572/A572M Grade 42 (290)	1.3
ASTM A992/A992M	1.1
All others	1.1
HSS shapes	
ASTM A500, A501, A618, A847	1.3
Steel pipes	
ASTM A53, A53M	1.4
Plates	1.1
All other products	1.1

where $\phi_v = 0.75$ and R_v is the design shear strength of the panel zone and is given by

$$R_v = \begin{cases} 0.60 F_y d_c t_p \left[1 + \dfrac{3 b_{cf} t_{cf}^2}{d_b d_c t_p} \right], & \text{if } \dfrac{P_u}{P_y} \leq 0.75 \\[4mm] 0.60 F_y d_c t_p \left[1 + \dfrac{3 b_{cf} t_{cf}^2}{d_b d_c t_p} \right] \left[1.9 - \dfrac{1.2 P_u}{P_y} \right], & \text{if } \dfrac{P_u}{P_y} > 0.75 \end{cases} \qquad (1.163)$$

where b_{cf} is the column flange width, d_b is the overall beam depth, d_c is the overall column depth, t_{cf} is the column flange thickness, t_p is the thickness of the panel zone including the thickness of any doubler plates (if used), F_y is the specified minimum yield strength of the panel zone, P_u is the required strength of the column containing the panel zone, P_y is the yield strength of the column containing the panel zone, and R_u is the required shear strength of the panel zone.

In addition, the individual thickness t of the column web and doubler plates (if used) should satisfy the condition

$$t \geq (d_z + w_z)/90 \qquad (1.164)$$

where d_z is the panel zone depth between continuity plates and w_z is the panel zone width between column flanges.

If doubler plates are used, and are placed against the column web, they should be welded across the top and bottom edges to develop the proportion of the total force that is transmitted to the plates. When doubler plates are placed away from the column web, they should be placed symmetrically in pairs and welded to the continuity plates to develop the proportion of total force that is transmitted to the plates. In either case, the doubler plates should be welded to the column flanges using either complete joint penetration groove welds or fillet welds that can develop the design shear strength of the full plate thickness.

Glossary

ASD — Acronym for allowable stress design.
Beam–columns — Structural members whose primary function are to carry loads both along and transverse to their longitudinal axes.
Biaxial bending — Simultaneous bending of a member about two orthogonal axes of the cross-section.
Built-up members — Structural members made of structural elements jointed together by bolts, welds, or rivets.
Composite members — Structural members made of both steel and concrete.
Compression members — Structural members whose primary function is to carry loads along their longitudinal axes.
Design strength — Resistance provided by the structural member obtained by multiplying the nominal strength of the member by a resistance factor.
Drift — Lateral deflection of a building.
Factored load — The product of the nominal load and a load factor.
Flexural members — Structural members whose primary function are to carry loads transverse to their longitudinal axes.
Limit state — A condition in which a structural or structural component becomes unsafe (strength limit state) or unfit for its intended function (serviceability limit state).
Load factor — A factor to account for the unavoidable deviations of the actual load from its nominal value and uncertainties in structural analysis in transforming the applied load into a load effect (axial force, shear, moment, etc.).
LRFD — Acronym for load and resistance factor design.
PD — Acronym for plastic design.

Plastic hinge — A yielded zone of a structural member in which the internal moment is equal to the plastic moment of the cross-section.

Reduced beam section — A beam section with portions of flanges cut out to reduce the section moment capacity.

Resistance factor — A factor to account for the unavoidable deviations of the actual resistance of a member from its nominal value.

Service load — Nominal load expected to be supported by the structure or structural component under normal usage.

Sideways inhibited frames — Frames in which lateral deflections are prevented by a system of bracing.

Sideways uninhibited frames — Frames in which lateral deflections are not prevented by a system of bracing.

Shear lag — The phenomenon in which the stiffer (or more rigid) regions of a structure or structural component attract more stresses than the more flexible regions of the structure or structural component. Shear lag causes stresses to be unevenly distributed over the cross-section of the structure or structural component.

Tension field action — Postbuckling shear strength developed in the web of a plate girder. Tension field action can develop only if sufficient transverse stiffeners are provided to allow the girder to carry the applied load using truss-type action after the web has buckled.

References

AASHTO. 1997. *Standard Specification for Highway Bridges*, 16th ed. American Association of State Highway and Transportation Officials, Washington D.C.

AASHTO. 1998. *LRFD Bridge Design Specification*, 2nd ed. American Association of State Highway and Transportation Officials, Washington D.C.

AISC. 1989. *Manual of Steel Construction — Allowable Stress Design*, 9th ed. American Institute of Steel Construction, Chicago, IL.

AISC. 1992. *Manual of Steel Construction — Volume II Connections*. ASD 1st ed./LRFD 1st ed. American Institute of Steel Construction, Chicago, IL.

AISC. 1999. *Load and Resistance Factor Design Specification for Structural Steel Buildings*. American Institute of Steel Construction, Chicago, IL.

AISC. 2001. *Manual of Steel Construction — Load and Resistance Factor Design*, 3rd ed. American Institute of Steel Construction, Chicago, IL.

AISC. 2002. *Seismic Provisions for Structural Steel Buildings*. American Institute of Steel Construction, Chicago, IL.

Aminmansour, A. 2000. A New Approach for Design of Steel Beam-Columns. *AISC Eng. J.*, **37**(2):41–72.

ASCE. 1997. *Effective Length and Notional Load Approaches for Assessing Frame Stability: Implications for American Steel Design*. American Society of Civil Engineers, New York, NY.

ASCE. 2002. *Minimum Design Loads for Buildings and Other Structures*. SEI/ASCE 7-02. American Society of Civil Engineers, Reston, VA.

ASTM. 2001a. *Standard Specification for Carbon Steel Bolts and Studs, 60000 psi Tensile Strength (A307–00)*. American Society for Testing and Materials, West Conshohocken, PA.

ASTM. 2001b. *Standard Specification for Structural Bolts, Steel, Heat-Treated 120/105 ksi Minimum Tensile Strength (A325-00)*. American Society for Testing and Materials, West Conshohocken, PA.

ASTM. 2001c. *Standard Specification for Heat-Treated Steel Structural Bolts, 150 ksi Minimum Tensile Strength (A490-00)*. American Society for Testing and Materials, West Conshohocken, PA.

ASTM. 2001d. *Standard Specification for Quenched and Tempered Steel Bolts and Studs (A449-00)*. American Society for Testing and Materials, West Conshohocken, PA.

AWS. 1987. *Welding Handbook*, 8th ed., **1**, *Welding Technology*. American Welding Society, Miami, FL.

AWS. 2000. *Structural Welding Code-Steel*. ANSI/AWS D1.1:2002, American Welding Society, Miami, FL.

Blodgett, O.W. Undated. Distortion. . . How to Minimize It with Sound Design Practices and Controlled Welding Procedures Plus Proven Methods for Straightening Distorted Members. *Bulletin G261*, The Lincoln Electric Company, Cleveland, OH.

Chen, W.F. (editor). 1987. *Joint Flexibility in Steel Frames*. Elsevier, London.

Chen, W.F. (editor). 2000. *Practical Analysis for Semi-Rigid Frame Design*. World Scientific, Singapore.

Chen, W.F. and Lui, E.M. 1991. *Stability Design of Steel Frames*. CRC Press, Boca Raton, FL.

Chen, W.F., Goto, Y., and Liew, J.Y.R. 1996. *Stability Design of Semi-Rigid Frames*. John Wiley & Sons, New York, NY.

CTBUH. 1993. *Semi-Rigid Connections in Steel Frames*. Council on Tall Buildings and Urban Habitat, Committee 43, McGraw-Hill, New York, NY.

Dewolf, J.T. and Ricker, D.T. 1990. *Column Base Plates*. Steel Design Guide Series 1, American Institute of Steel Construction, Chicago, IL.

Disque, R.O. 1973. Inelastic K-Factor in Column Design. *AISC Eng. J.*, **10**(2): 33–35.

Engelhardt, M.D., Winneberger, T., Zekany, A.J., and Potyraj, T.J. 1996. The Dogbone Connection: Part II. *Modern Steel Construction*, **36**(8): 46–55.

Engelhardt, M.D., Winneberger, T., Zekany, A.J., and Potyraj, T.J. 1998. Experimental Investigation of Dogbone Moment Connections. *AISC Eng. J.*, **35**(4): 128–139.

Engelhardt, M.D., Fry, G.T., Jones, S.L., Venti, M., and Holliday, S.D. 2000. Experimental Investigation of Reduced Beam Section Connections with Composite Slabs. Paper presented at the Fourth US–Japan Workshop on Steel Fracture Issues, San Francisco, 11 pp.

Faella, C., Piluso, V., and Rizzano, G. 2000. *Structural Steel Semirigid Connections*. CRC Press, Boca Raton, FL.

Galambos, T.V. (editor). 1998. *Guide to Stability Design Criteria for Metal Structures*, 5th ed., John Wiley & Sons, New York, NY.

Galambos, T.V. and Chapuis, J. 1980. LRFD Criteria for Composite Columns and Beam Columns. Washington University, Department of Civil Engineering, St. Louis, MO.

Gaylord, E.H., Gaylord, C.N., and Stallmeyer, J.E. 1992. *Design of Steel Structures*, 3rd ed. McGraw-Hill, New York, NY.

Gilton, C., Chi, B., and Uang, C.-M. 2000. Cyclic Response of RBS Moment Connections: Weak-Axis Configuration and Deep Column Effects. *Structural Systems Research Project Report No. SSRP-2000/03*, Department of Structural Engineering, University of California, San Diego, La Jolla, CA, 197 pp.

Gross, J.L., Engelhardt, M.D., Uang, C.-M., Kasai, K., and Iwankiw, N.R. 1999. *Modification of Existing Welded Steel Moment Frame Connections for Seismic Resistance*. Steel Design Guide Series 12, American Institute of Steel Construction, Chicago, IL.

Grubbs, K.V. 1997. The Effect of Dogbone Connection on the Elastic Stiffness of Steel Moment Frames. M.S. Thesis, University of Texas at Austin, 54 pp.

International Building Code (IBC). 2000. International Code Council. Falls Church, VA.

Iwankiw, N.R. and Carter, C. 1996. The Dogbone: A New Idea to Chew On. *Modern Steel Construction*, **36**(4): 18–23.

Kulak, G.L., Fisher, J.W., and Struik, J.H.A. 1987. *Guide to Design Criteria for Bolted and Riveted Joints*, 2nd ed. John Wiley & Sons, New York, NY.

Lee, G.C., Morrel, M.L., and Ketter, R.L. 1972. *Design of Tapered Members*. WRC Bulletin No. 173.

Marsh, M.L. and Burdette, E.G. 1985a. Multiple Bolt Anchorages: Method for Determining the Effective Projected Area of Overlapping Stress Cones. *AISC Eng. J.*, **22**(1): 29–32.

Marsh, M.L. and Burdette, E.G. 1985b. Anchorage of Steel Building Components to Concrete. *AISC Eng. J.*, **22**(1): 33–39.

Moore, K.S., Malley, J.O., and Engelhardt, M.D. 1999. *Design of Reduced Beam Section (RBS) Moment Frame Connections*. Steel Tips. Structural Steel Educational Council Technical Information & Product Service, 36 pp.

Munse, W.H. and Chesson, E., Jr. 1963. Riveted and Bolted Joints: Net Section Design. *ASCE J. Struct. Div.*, **89**(1): 107–126.

Plumier, A. 1997. The Dogbone: Back to the Future. *AISC Eng. J.*, **34**(2): 61–67.

Rains, W.A. 1976. A New Era in Fire Protective Coatings for Steel. *Civil Eng.*, ASCE, September: 80–83.

RCSC. 2000. *Load and Resistance Factor Design Specification for Structural Joints Using ASTM A325 or A490 Bolts*. American Institute of Steel Construction, Chicago, IL.

Rossow, E.C. 1996. *Analysis and Behavior of Structures*. Prentice Hall, Upper Saddle River, NJ.

Shipp, J.G. and Haninger, E.R. 1983. Design of Headed Anchor Bolts. *AISC Eng. J.*, **20**(2): 58–69.

SSRC 1993. *Is Your Structure Suitably Braced?* Structural Stability Research Council, Bethlehem, PA.

Steel Deck Institute. 2001. *Design Manual for Composite Decks, Form Decks and Roof Decks, Publication No. 30*. Steel Deck Institute, Fox River Grove, IL.

Uang, C.-M. and Fan, C.-C. 1999. Cyclic Instability of Steel Moment Connections with Reduced Beams Sections. *Structural Systems Research Project Report No. SSRP-99/21*, Department of Structural Engineering, University of California, San Diego, La Jolla, CA, 51 pp.

Uniform Building Code (UBC). 1997. Volume 2 — Structural Engineering Design Provisions, *International Conference of Building Officials*, Whittier, CA.

Yu, Q.S., Gilton, C., and Uang, C.-M. 2000. Cyclic Response of RBS Moment Connections: Loading Sequence and Lateral Bracing Effects. *Structural Systems Research Project Report No. SSRP-99/13*, Department of Structural Engineering, University of California, San Diego, La Jolla, CA, 119 pp.

Yura, J.A. 2001. Fundamentals of Beam Bracing. *AISC Eng. J.*, **38**(1): 11–26.

Further Reading

The following publications provide additional sources of information for the design of steel structures:

General Information

AISC Design Guide Series: Design Guide 1: *Column Base Plates*, Dewolf and Ricker; Design Guide 2: *Design of Steel and Composite Beams with Web Openings*, Darwin; Design Guide 3: *Considerations for Low-Rise Buildings*, Fisher and West; Design Guide 4: *Extended End-Plate Moment Connections*, Murray; Design Guide 5: *Design of Low- and Medium-Rise Steel Buildings*, Allison; Design Guide 6: *Load and Resistance Factor Design of W-Shapes Encased in Concrete*, Griffes; Design Guide 7: *Industrial Buildings — Roofs to Column Anchorage*, Fisher; Design Guide 8: *Partially Restrained Composite Connections*, Leon; Design Guide 9: *Torsional Analysis of Structural Steel Members*, Seaburg and Carter; Design Guide 10: *Erection Bracing of Low-Rise Structural Steel Frames*, Fisher and West; Design Guide 11: *Floor Vibration Due to Human Activity*, Murray, Allen and Ungar; Design Guide 12: *Modification of Existing Steel Welded Moment Frame Connections for Seismic Resistance*, Gross, Engelhardt, Uang, Kasai and Iwankiw (1999); Design Guide 13: *Wide-Flange Column Stiffening at Moment Connections*, Carter. American Institute of Steel Construction, Chicago, IL.

Chen, W.F. and Lui, E.M. 1987. *Structural Stability — Theory and Implementation*. Elsevier, New York, NY.

Chen, W.F. and Kim, S.-E. 1997. *LRFD Steel Design Using Advanced Analysis*. CRC Press, Boca Raton, FL.

Englekirk, R. 1994. *Steel Structures — Controlling Behavior through Design*. John Wiley & Sons, New York, NY.

Fukumoto, Y. and Lee, G. 1992. *Stability and Ductility of Steel Structures under Cyclic Loading*. CRC Press, Boca Raton, FL.

Stability of Metal Structures — A World View. 1991. 2nd ed. Lynn S. Beedle (editor-in-chief), Structural Stability Research Council, Lehigh University, Bethlehem, PA.

Trahair, N.S. 1993. *Flexural–Torsional Buckling of Structures*. CRC Press, Boca Raton, FL.

Allowable Stress Design

Adeli, H. 1988. *Interactive Microcomputer-Aided Structural Steel Design*. Prentice Hall, Englewood Cliffs, NJ.

Cooper, S.E. and Chen, A.C. 1985. *Designing Steel Structures — Methods and Cases*. Prentice Hall, Englewood Cliffs, NJ.

Crawley, S.W. and Dillon, R.M. 1984. *Steel Buildings Analysis and Design*, 3rd ed. John Wiley & Sons, New York, NY.

Fanella, D.A., Amon, R., Knobloch, B., and Mazumder, A. 1992. *Steel Design for Engineers and Architects*, 2nd ed. Van Nostrand Reinhold, New York, NY.

Kuzmanovic, B.O. and Willems, N. 1983. *Steel Design for Structural Engineers*, 2nd ed. Prentice Hall, Englewood Cliffs, NJ.

McCormac, J.C. 1981. *Structural Steel Design*. 3rd ed. Harper & Row, New York, NY.

Segui, W.T. 1989. *Fundamentals of Structural Steel Design*. PWS-KENT, Boston, MA.

Spiegel, L. and Limbrunner, G.F. 2002. *Applied Structural Steel Design*, 4th ed. Prentice Hall, Upper Saddle River, NJ.

Plastic Design

Horne, M.R. and Morris, L.J. 1981. *Plastic Design of Low-Rise Frames*, Constrado Monographs. Collins, London.

Plastic Design in Steel—A Guide and Commentary. 1971. 2nd ed. ASCE Manual No. 41, ASCE-WRC, New York, NY.

Load and Resistance Factor Design

Geschwindner, L.F., Disque, R.O., and Bjorhovde, R. 1994. *Load and Resistance Factor Design of Steel Structures*, Prentice Hall, Englewood Cliffs, NJ.

McCormac, J.C. 1995. *Structural Steel Design — LRFD Method*, 2nd ed. Harper & Row, New York, NY.

Salmon, C.G. and Johnson, J.E. 1996. *Steel Structures — Design and Behavior*, 4th ed. Harper & Row, New York, NY.

Segui, W.T. 1999. *LRFD Steel Design*, 2nd ed. Brooks/Cole, Pacific Grove, CA.

Smith, J.C. 1996. *Structural Steel Design — LRFD Approach*, 2nd ed. John Wiley & Sons, New York, NY.

Tamboli, A.R. 1997. *Steel Design Handbook — LRFD Method*, McGraw-Hill, New York, NY.

Relevant Websites

www.aisc.org
www.aws.org
www.boltcouncil.org
www.icbo.org
www.iccsafe.org
www.sbcci.org
www.steel.org

2

Steel Frame Design Using Advanced Analysis

S. E. Kim
Department of Civil Engineering,
Sejong University,
Seoul, South Korea

Wai-Fah Chen
College of Engineering,
University of Hawaii at Manoa,
Honolulu, HI

2.1 Introduction

The steel design methods used in the United States are allowable stress design (ASD), plastic design (PD), and load and resistance factor design (LRFD). In ASD, the stress computation is based on a first-order elastic analysis, and the geometric nonlinear effects are implicitly accounted for in the member design equations. In PD, a first-order plastic-hinge analysis is used in the structural analysis. PD allows inelastic force redistribution throughout the structural system. Since geometric nonlinearity and gradual yielding effects are not accounted for in the analysis of PD, they are approximated in member design equations. In LRFD, a first-order elastic analysis with amplification factors or a direct second-order elastic analysis is used to account for geometric nonlinearity, and the ultimate strength of beam–column members is implicitly reflected in the design interaction equations. All three design methods require separate member capacity checks including the calculation of the K-factor. In the following, the characteristics of the LRFD method are briefly described.

The strength and stability of a structural system and its members are related, but the interaction is treated separately in the current American Institute of Steel Construction (AISC)-LRFD Specification [1]. In current practice, the interaction between the structural system and its members is represented by the effective length factor. This aspect is described in the following excerpt from SSRC Technical Memorandum No. 5 [2]:

> Although the maximum strength of frames and the maximum strength of component members are interdependent (but not necessarily coexistent), it is recognized that in many structures it is not practical to take this interdependence into account rigorously. At the same time, it is known that difficulties are encountered in complex frameworks when attempting to compensate automatically in column design for the instability of the entire frame (for example, by adjustment of column effective length). Therefore, SSRC recommends that, in design practice, the two aspects, stability of separate members and elements of the structure and stability of the structure as a whole, be considered separately.

This design approach is marked in Figure 2.1 as the indirect analysis and design method.

In the current AISC-LRFD Specification [1], first- or second-order elastic analysis is used to analyze a structural system. In using first-order elastic analysis, the first-order moment is amplified by B_1 and B_2 factors to account for second-order effects. In the Specification, the members are isolated from a structural system, and they are then designed by the member strength curves and interaction equations as given in the Specifications, which implicitly account for the second-order effects, inelasticity, residual stresses, and geometric imperfections [3]. The column curve and the beam curve were developed by a curve-fit to both theoretical solutions and experimental data, while the beam–column interaction equations were determined by a curve-fit to the so-called "exact" plastic-zone solutions generated by Kanchanalai [4].

In order to account for the influence of a structural system on the strength of individual members, the effective length factor is used as illustrated in Figure 2.2. The effective length method generally provides a good design of framed structures. However, several difficulties are associated with the use of the effective length method, which are as follows:

1. The effective length approach cannot accurately account for the interaction between the structural system and its members. This is because the interaction in a large structural system is too complex to be represented by the simple effective length factor K. As a result, this method cannot accurately predict the actual strengths required of its framed members.

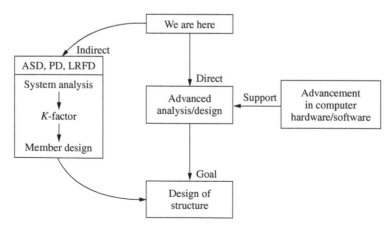

FIGURE 2.1 Analysis and design methods.

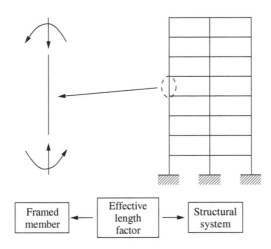

FIGURE 2.2 Interaction between a structural system and its component members.

2. The effective length method cannot capture the inelastic redistributions of internal forces in a structural system, since the first-order elastic analysis with B_1 and B_2 factors accounts only for second-order effects and not the inelastic redistribution of internal forces. The effective length method provides a conservative estimation of the ultimate load-carrying capacity of a large structural system.
3. The effective length method cannot predict the failure modes of a structural system subject to a given load. This is because the LRFD interaction equation does not provide any information about failure modes of a structural system at the factored loads.
4. The effective length method is not user friendly for a computer-based design.
5. The effective length method requires a time-consuming process of separate member capacity checks involving the calculation of K-factors.

With the development of computer technology, two aspects, the stability of separate members and the stability of the structure as a whole, can be treated rigorously for the determination of the maximum strength of the structures. This design approach is shown in Figure 2.1 as the direct analysis and design method. The development of the direct approach to design is called "advanced analysis" or, more specifically, "second-order inelastic analysis for frame design." In this direct approach, there is no need to compute the effective length factor, since separate member capacity checks encompassed by the specification equations are not required. With the current available computing technology, it is feasible to employ advanced analysis techniques for direct frame design. This method was considered impractical for design office use in the past. The purpose of this chapter is to present a practical, direct method of steel frame design, using advanced analysis, that will produce almost identical member sizes as those of the LRFD method.

The advantages of advanced analysis in design use are outlined as follows:

1. Advanced analysis is another tool used by structural engineers in steel design, and its adoption is not mandatory but will provide a flexibility of options to the designer.
2. Advanced analysis captures the limit state strength and stability of a structural system and its individual members directly, so separate member capacity checks encompassed by specification equations are not required.
3. Compared to the LRFD and ASD, advanced analysis provides more information of structural behavior by direct inelastic second-order analysis.
4. Advanced analysis overcomes the difficulties due to incompatibility between the elastic global analysis and the limit state member design in the conventional LRFD method.

5. Advanced analysis is user friendly for a computer-based design, but the LRFD and ASD methods are not, since they require the calculation of the K-factor along with the analysis to separate member capacity checks.
6. Advanced analysis captures the inelastic redistribution of internal forces throughout a structural system, and allows an economic use of material for highly indeterminate steel frames.
7. It is now feasible to employ advanced analysis techniques that were considered impractical for design office use in the past, since the power of personal computers and engineering workstations is rapidly increasing.
8. Member sizes determined by advanced analysis are close to those determined by the LRFD method, since the advanced analysis method is calibrated against the LRFD column curve and beam–column interaction equations. As a result, advanced analysis provides an alternative to the LRFD.
9. Since advanced analysis is a structure-based analysis and design approach, it is more appropriate for performance-based fire and seismic design than conventional member-based design approaches [5].

Among various advanced analyses, including plastic zone, quasi-plastic hinge, elastic–plastic hinge, notional-load plastic hinge, and refined plastic-hinge methods, the refined plastic-hinge method is recommended since it retains the efficiency and simplicity of computation and accuracy for practical use. The method is developed by imposing simple modifications on the conventional elastic–plastic hinge method. These include a simple modification to account for the gradual sectional stiffness degradation at the plastic-hinge locations and to include the gradual member stiffness degradation between two plastic hinges.

The key considerations of the conventional LRFD method and the practical advanced analysis method are compared in Table 2.1. While the LRFD method does account for key behavioral effects implicitly in its column strength and beam–column interaction equations, the advanced analysis method accounts for these effects explicitly through stability functions, stiffness degradation functions, and geometric imperfections, which are discussed in detail in Section 2.2.

Advanced analysis holds many answers about the real behavior of steel structures and, as such, the authors recommend the proposed design method to engineers seeking to perform frame design with efficiency and rationality, yet consistent with the present LRFD Specification. In the following sections, we will present a practical advanced analysis method for the design of steel frame structures with LRFD. The validity of the approach will be demonstrated by comparing case studies of actual members and frames with the results of analysis/design based on exact plastic-zone solutions and

TABLE 2.1   Key Considerations of Load and Resistance Factor Design (LRFD) and Proposed Methods

Key considerations	LRFD	Proposed methods
Second-order effects	Column curve B_1, B_2 factors	Stability function
Geometric imperfection	Column curve	Explicit imperfection modeling method $\psi = 1/500$ for unbraced frame $\delta_c = L_c/1000$ for braced frame Equivalent notional load method $\alpha = 0.002$ for unbraced frame $\alpha = 0.004$ for braced frame Further reduced tangent modulus method $E_t' = 0.85E_t$
Stiffness degradation associated with residual stresses	Column curve	CRC tangent modulus
Stiffness degradation associated with flexure	Column curve Interaction equations	Parabolic degradation function
Connection nonlinearity	No procedure	Power model/rotational spring

LRFD designs. The wide range of case studies and comparisons should confirm the validity of this advanced method.

2.2 Practical Advanced Analysis

This section presents a practical advanced analysis method for the direct design of steel frames by eliminating separate member capacity checks by the specification. The refined plastic-hinge method was developed and refined by simply modifying the conventional elastic–plastic hinge method to achieve both simplicity and a realistic representation of the actual behavior [6,7]. Verification of the method will be given in the next section to provide the final confirmation of the validity of the method.

Connection flexibility can be accounted for in advanced analysis. Conventional analysis and design of steel structures are usually carried out under the assumption that beam-to-column connections are either fully rigid or ideally pinned. However, most connections in practice are "semirigid" and their behavior lies between these two extreme cases. In the AISC-LRFD Specification [1], two types of construction are designated: Type FR (fully restrained) construction and Type PR (partially restrained) construction. The LRFD Specification permits the evaluation of the flexibility of connections by "rational means."

Connection behavior is represented by its moment–rotation relationship. Extensive experimental work on connections has been performed, and a large body of moment–rotation data collected. With this database, researchers have developed several connection models including: linear, polynomial, *B*-spline, power, and exponential. Herein, the three-parameter power model proposed by Kishi and Chen [8] is adopted.

Geometric imperfections should be modeled in frame members when using advanced analysis. Geometric imperfections result from unavoidable error during fabrication or erection. For structural members in building frames, the types of geometric imperfections are out-of-straightness and out-of-plumbness. Explicit modeling and equivalent notional loads were used in the past to account for geometric imperfections. In this section, a new method based on further reduction of the tangent stiffness of members is developed [6,9]. This method provides a simple means to account for the effect of imperfection without inputting notional loads or explicit geometric imperfections.

The practical advanced analysis method described in this section is limited to two-dimensional braced, unbraced, and semirigid frames subjected to static loads. The spatial behavior of frames is not considered, and lateral torsional buckling is assumed to be prevented by adequate lateral bracing. A compact W-section is assumed so that sections can develop full plastic moment capacity without local buckling. Both strong-axis and weak-axis bending of wide flange sections have been studied using the practical advanced analysis method [6]. In recent developments, several studies have used advanced analysis of structures by including effects such as spatial behavior [10–12], member local buckling [13,14], and lateral torsional buckling [15,16]. The present method may be considered an interim analysis/design procedure between the conventional LRFD method widely used now and a more rigorous advanced analysis/design method such as the plastic-zone method to be developed in the future for practical use.

2.2.1 Second-Order Refined Plastic-Hinge Analysis

In this section, a method called the *refined plastic-hinge* approach is presented. This method is comparable to the elastic–plastic hinge analysis in efficiency and simplicity, but without its limitations. In this analysis, stability functions are used to predict second-order effects. The benefit of stability functions is that they make the analysis method practical by using only one element per beam column. The refined plastic-hinge analysis uses a two-surface yield model and an effective tangent modulus to account for stiffness degradation due to distributed plasticity in framed members. The member stiffness is assumed to degrade gradually as the second-order forces at critical locations approach the cross-section plastic strength. Column tangent modulus is used to represent the effective stiffness of the member when it is loaded with a high axial load. Thus, the refined plastic-hinge model approximates the effect of distributed plasticity along the element length caused by initial imperfections and large bending and axial force actions. In fact, researches by Liew et al. [7,17], Kim and Chen [9], and Kim [6] have shown that refined plastic-hinge analysis captures the interaction of strength and stability of structural systems and that of their component elements. This type of analysis method may, therefore, be classified as an *advanced analysis* and separate specification member capacity checks are not required.

2.2.1.1 Stability Function

To capture second-order effects, stability functions are recommended since they lead to large savings in modeling and solution efforts by using one or two elements per member. The simplified stability functions reported by Chen and Lui [18] or an alternative may be used. Considering the prismatic beam–column element, the incremental force–displacement relationship of this element may be written as

$$\begin{bmatrix} \dot{M}_A \\ \dot{M}_B \\ \dot{P} \end{bmatrix} = \frac{EI}{L} \begin{bmatrix} S_1 & S_2 & 0 \\ S_2 & S_1 & 0 \\ 0 & 0 & A/I \end{bmatrix} \begin{bmatrix} \dot{\theta}_A \\ \dot{\theta}_B \\ \dot{e} \end{bmatrix} \tag{2.1}$$

where S_1, S_2 are stability functions, $\dot{M}_A$, $\dot{M}_B$ are incremental end moments, $\dot{P}$ is the incremental axial force, $\dot{\theta}_A$, $\dot{\theta}_B$ are incremental joint rotations, $\dot{e}$ is the incremental axial displacement, A, I, L are area, moment of inertia, and length of beam–column element, respectively, and E is the modulus of elasticity.

In this formulation, all members are assumed to be adequately braced to prevent out-of-plane buckling and their cross-sections are compact to avoid local buckling.

2.2.1.2 Cross-Section Plastic Strength

Based on the AISC-LRFD bilinear interaction equations [1], the cross-section plastic strength may be expressed as Equation 2.2. These AISC-LRFD cross-section plastic strength curves may be adopted for both strong-axis and weak-axis bendings (Figure 2.3):

$$\frac{P}{P_y} + \frac{8}{9}\frac{M}{M_p} = 1.0 \quad \text{for} \quad \frac{P}{P_y} \geq 0.2 \tag{2.2a}$$

$$\frac{1}{2}\frac{P}{P_y} + \frac{M}{M_p} = 1.0 \quad \text{for} \quad \frac{P}{P_y} \leq 0.2 \tag{2.2b}$$

where P, M are second-order axial force and bending moment, P_y is the squash load, and M_p is the plastic moment capacity.

2.2.1.3 CRC Tangent Modulus

The CRC tangent modulus concept is employed to account for the gradual yielding effect due to residual stresses along the length of members under axial loads between two plastic hinges. In this concept, the elastic modulus E, instead of moment of inertia I, is reduced to account for the reduction of the elastic portion of the cross-section since the reduction of elastic modulus is easier to implement than that of moment of inertia for different sections. The reduction rate in stiffness between weak and strong axes is different, but this is not considered here because rapid degradation in stiffness in the weak-axis strength is compensated well by the stronger weak-axis plastic strength. As a result, this simplicity will make the present methods practical. From Chen and Lui [18], the CRC E_t is written as (Figure 2.4):

$$E_t = 1.0E \quad \text{for} \quad P \leq 0.5P_y \tag{2.3a}$$

$$E_t = 4\frac{P}{P_y}E\left(1 - \frac{P}{P_y}\right) \quad \text{for} \quad P > 0.5P_y \tag{2.3b}$$

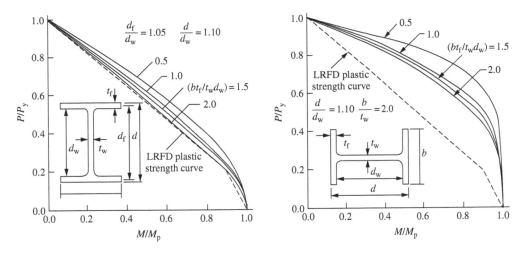

FIGURE 2.3 Strength interaction curves for wide-flange sections.

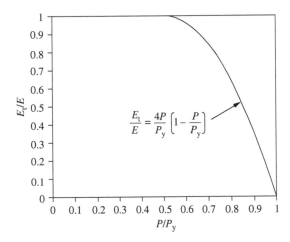

FIGURE 2.4 Member tangent stiffness degradation derived from the CRC column curve.

2.2.1.4 Parabolic Function

The tangent modulus model in Equation 2.3 is suitable for $P/P_y > 0.5$, but is not sufficient to represent the stiffness degradation for cases with small axial forces and large bending moments. A gradual stiffness degradation of the plastic hinge is required to represent the distributed plasticity effects associated with bending actions. We shall introduce the hardening plastic-hinge model to represent the gradual transition from elastic stiffness to zero stiffness associated with a fully developed plastic hinge. When the hardening plastic hinges are present at both ends of an element, the incremental force–displacement relationship may be expressed as [19]:

$$
\begin{bmatrix} \dot{M}_A \\ \dot{M}_B \\ \dot{P} \end{bmatrix} = \frac{E_t I}{L} \begin{bmatrix} \eta_A \left[S_1 - \dfrac{S_2^2}{S_1}(1 - \eta_B) \right] & \eta_A \eta_B S_2 & 0 \\ \eta_A \eta_B S_2 & \eta_B \left[S_1 - \dfrac{S_2^2}{S_1}(1 - \eta_A) \right] & 0 \\ 0 & 0 & A/I \end{bmatrix} \begin{bmatrix} \dot{\theta}_A \\ \dot{\theta}_B \\ \dot{e} \end{bmatrix} \quad (2.4)
$$

where $\dot{M}_A$, $\dot{M}_B$, $\dot{P}$ are incremental end moments and axial force, respectively, S_1, S_2 are stability functions, E_t is the tangent modulus, and η_A, η_B are the element stiffness parameters.

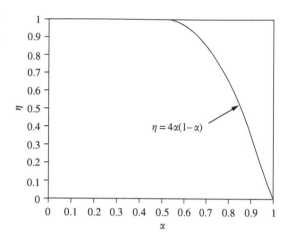

FIGURE 2.5 Parabolic plastic-hinge stiffness degradation function with $\alpha_0 = 0.5$ based on the load and resistance factor design sectional strength equation.

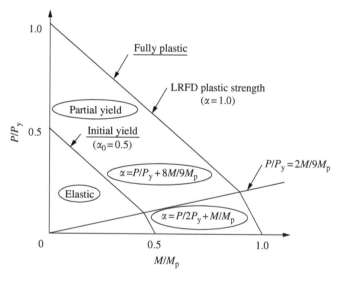

FIGURE 2.6 Smooth stiffness degradation for a work-hardening plastic hinge based on the load and resistance factor design sectional strength curve.

The parameter η represents a gradual stiffness reduction associated with flexure at sections. The partial plastification at cross-sections in the end of elements is denoted by $0 < \eta < 1$. η may be assumed to vary according to the parabolic expression (Figure 2.5):

$$\eta = 4\alpha(1 - \alpha) \quad \text{for } \alpha > 0.5 \tag{2.5}$$

where α is the force-state parameter obtained from the limit state surface corresponding to the element end (Figure 2.6):

$$\alpha = \frac{P}{P_y} + \frac{8}{9}\frac{M}{M_p} \quad \text{for } \frac{P}{P_y} \geq \frac{2}{9}\frac{M}{M_p} \tag{2.6a}$$

$$\alpha = \frac{1}{2}\frac{P}{P_y} + \frac{M}{M_p} \quad \text{for } \frac{P}{P_y} < \frac{2}{9}\frac{M}{M_p} \tag{2.6b}$$

where P, M are second-order axial force and bending moment at the cross-section, respectively and M_p is the plastic moment capacity.

2.2.2 Analysis of Semirigid Frames

2.2.2.1 Practical Connection Modeling

The three-parameter power model contains three parameters: initial connection stiffness R_{ki}, ultimate connection moment capacity M_u, and shape parameter n. The power model may be written as (Figure 2.7):

$$m = \frac{\theta}{(1 + \theta^n)^{1/n}} \quad \text{for } \theta > 0, \ m > 0 \tag{2.7}$$

where $m = M/M_u$, $\theta = \theta_r/\theta_0$, θ_0 is the reference plastic rotation, M_u/R_{ki}, M_u is the ultimate moment capacity of the connection, R_{ki} is the initial connection stiffness, and n is the shape parameter. When the connection is loaded, the connection tangent stiffness, R_{kt}, at an arbitrary rotation, θ_r, can be derived by simply differentiating Equation 2.7 as

$$R_{kt} = \frac{dM}{d|\theta_r|} = \frac{M_u}{\theta_0(1 + \theta^n)^{1+1/n}} \tag{2.8}$$

When the connection is unloaded, the tangent stiffness is equal to the initial stiffness:

$$R_{kt} = \frac{dM}{d|\theta_r|} = \frac{M_u}{\theta_0} = R_{ki} \tag{2.9}$$

It is observed that a small value of the power index, n, makes a smooth transition curve from the initial stiffness, R_{kt}, to the ultimate moment, M_u. On the contrary, a large value of the index, n, makes the transition more abruptly. In the extreme case, when n is infinity, the curve becomes a bilinear line consisting of the initial stiffness, R_{ki} and the ultimate moment capacity, M_u.

2.2.2.2 Practical Estimation of Three Parameters Using a Computer Program

An important task for the practical use of the power model is to determine the three parameters for a given connection configuration. One difficulty in determining the three parameters is the need for numerical iteration, especially to estimate the ultimate moment, M_u. A set of nomographs was

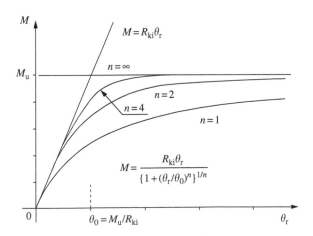

FIGURE 2.7 Moment–rotation behavior of the three-parameter model.

proposed by Kishi et al. [20] to overcome the difficulty. Even though the purpose of these nomographs is to allow the engineer to rapidly determine the three parameters for a given connection configuration, the nomographs require other efforts for engineers to know how to use those, and the values of the nomographs are approximate.

Herein, one simple way to avoid difficulties described above is presented. A direct and easy estimation of the three parameters may be achieved by use of a simple computer program 3PARA.f. The operating procedure of the program is shown in Figure 2.8. The input data CONN.DAT may be easily generated corresponding to the input format listed in Table 2.2.

As for the shape parameter n, the equations developed by Kishi et al. [20] are implemented here. Using a statistical technique for n values, empirical equations of n are determined as a linear function of $\log_{10}\theta_0$ shown in Table 2.3. This n value may be calculated using 3PARA.f.

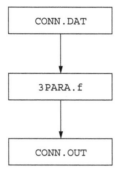

FIGURE 2.8 Operating procedure of computer program estimating the three parameters.

TABLE 2.2 Input Format

Line	Input data	Remark
1	ITYPE F_y E	Connection type and material properties
2	l_t t_t k_t g_t W d	Top/seat-angle data
3	l_a t_a k_a g_a	Web-angle data

ITYPE = Connection type (1 = top- and seat-angle connection, 2 = with web-angle connection).
F_y = yield strength of angle
E = Young's modulus (= 29,000 ksi)
l_t = length of top angle
t_t = thickness of top angle
k_t = k value of top angle
g_t = gauge of top angle (= 2.5 in., typical)
W = width of nut (W = 1.25 in. for 3/4D bolt, W = 1.4375 in. for 7/8D bolt)
d = depth of beam
l_a = length of web angle
t_a = thickness of web angle
k_a = k value of web angle
g_a = gauge of web angle

Notes:
1. Top- and seat-angle connections need lines 1 and 2 for input data, and top and seat angles with web-angle connections need line 1, 2, and 3.
2. All input data are in free format.
3. Top- and seat-angle sizes are assumed to be the same.
4. Bolt sizes of top angle, seat angle, and web angle are assumed to be the same.

2.2.2.3 Load–Displacement Relationship Accounting for Semirigid Connection

The connection may be modeled as a rotational spring in the moment–rotation relationship represented by Equation 2.10. Figure 2.9 shows a beam–column element with semirigid connections at both ends. If the effect of connection flexibility is incorporated into the member stiffness, the incremental element force–displacement relationship of Equation 2.1 is modified as [18,19]

$$
\begin{bmatrix} \dot{M}_A \\ \dot{M}_B \\ \dot{P} \end{bmatrix} = \frac{E_t I}{L} \begin{bmatrix} S_{ii}^* & S_{ij}^* & 0 \\ S_{ij}^* & S_{jj}^* & 0 \\ 0 & 0 & A/I \end{bmatrix} \begin{bmatrix} \dot{\theta}_A \\ \dot{\theta}_B \\ \dot{e} \end{bmatrix}
\tag{2.10}
$$

where

$$
S_{ii}^* = \left(S_{ii} + \frac{E_t I S_{ii} S_{jj}}{L R_{ktB}} - \frac{E_t I S_{ij}^2}{L R_{ktB}} \right) \bigg/ R^*
\tag{2.11a}
$$

$$
S_{jj}^* = \left(S_{jj} + \frac{E_t I S_{ii} S_{jj}}{L R_{ktA}} - \frac{E_t I S_{ij}^2}{L R_{ktA}} \right) \bigg/ R^*
\tag{2.11b}
$$

$$
S_{ij}^* = S_{ij} / R^*
\tag{2.11c}
$$

$$
R^* = \left(1 + \frac{E_t I S_{ii}}{L R_{ktA}} \right) \left(1 + \frac{E_t I S_{jj}}{L R_{ktB}} \right) - \left(\frac{E_t I}{L} \right)^2 \frac{S_{ij}^2}{R_{ktA} R_{ktB}}
\tag{2.11d}
$$

where R_{ktA}, R_{ktB} are tangent stiffnesses of connections A and B, respectively, S_{ii}, S_{ij} are generalized stability functions, and S_{ii}^*, S_{ij}^* are modified stability functions that account for the presence of end connections. The tangent stiffness (R_{ktA}, R_{ktB}) accounts for the different types of semirigid connections (see Equation 2.8).

TABLE 2.3 Empirical Equations for Shape Parameter n

Connection type	n
Single web-angle connection	$0.520 \log_{10} \theta_0 + 2.291$ for $\log_{10} \theta_0 > -3.073$ 0.695 for $\log_{10} \theta_0 < -3.073$
Double web-angle connection	$1.322 \log_{10} \theta_0 + 3.952$ for $\log_{10} \theta_0 > -2.582$ 0.573 for $\log_{10} \theta_0 < -2.582$
Top- and seat-angle connection	$2.003 \log_{10} \theta_0 + 6.070$ for $\log_{10} \theta_0 > -2.880$ 0.302 for $\log_{10} \theta_0 < -2.880$
Top- and seat-angle connection with double web angle	$1.398 \log_{10} \theta_0 + 4.631$ for $\log_{10} \theta_0 > -2.721$ 0.827 for $\log_{10} \theta_0 < -2.721$

Source: From Kishi, N., Goto, Y., Chen, W.F., and Matsuoka, K.G. 1993. *Eng. J.*, AISC, pp. 90–107. With permission.

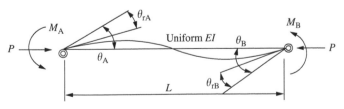

FIGURE 2.9 Beam–column element with semirigid connections.

2.2.3 Geometric Imperfection Methods

Geometric imperfection modeling combined with the Column Research Council (CRC) tangent modulus model is discussed in what follows. There are three methods: the explicit imperfection modeling method, the equivalent notional load method, and the further reduced tangent modulus method.

2.2.3.1 Explicit Imperfection Modeling Method

2.2.3.1.1 Braced Frames

The refined plastic-hinge analysis implicitly accounts for the effects of both residual stresses and spread of yielded zones. To this end, refined plastic-hinge analysis may be regarded as equivalent to the plastic-zone analysis. As a result, geometric imperfections are necessary only to consider fabrication error. For braced frames, member out-of-straightness, rather than frame out-of-plumbness, needs to be used for geometric imperfections. This is because the P–Δ effect due to the frame out-of-plumbness is diminished by braces. The ECCS [21,22], AS [23], and Canadian Standard Association (CSA) [24,25] specifications recommend an initial crookedness of column equal to 1/1000 times the column length. The AISC Code recommends the same maximum fabrication tolerance of $L_c/1000$ for member out-of-straightness. In this study, a geometric imperfection of $L_c/1000$ is adopted.

The ECCS [21,22], AS [23], and CSA [24,25] specifications recommend the out-of-straightness varying parabolically with a maximum in-plane deflection at the mid-height. They do not, however, describe how the parabolic imperfection should be modeled in analysis. Ideally, many elements are needed to model the parabolic out-of-straightness of a beam–column member, but it is not practical. In this study, two elements with a maximum initial deflection at the mid-height of a member are found adequate for capturing the imperfection. Figure 2.10 shows the out-of-straightness modeling for a braced beam–column member. It may be observed that the out-of-plumbness is equal to 1/500 when the half-segment of the member is considered. This value is identical to that of sway frames as discussed in recent papers by Kim and Chen [9,26,27]. Thus, it may be stated that the imperfection values are essentially identical for both sway and braced frames. It is noted that this explicit modeling method in braced frames requires the inconvenient imperfection modeling at the center of columns although the inconvenience is much lesser than that of the conventional LRFD method for frame design.

2.2.3.1.2 Unbraced Frames

The CSA [23,24] and the AISC Code of Standard Practice [1] set the limit of erection out-of-plumbness at $L_c/500$. The maximum erection tolerances in the AISC are limited to 1 in. toward the exterior

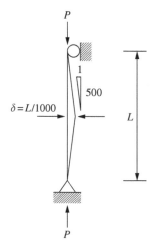

FIGURE 2.10 Explicit imperfection modeling of a braced member.

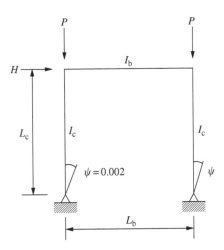

FIGURE 2.11 Explicit imperfection modeling of an unbraced frame.

of buildings and 2 in. toward the interior of buildings less than 20 stories. Considering the maximum permitted average lean of 1.5 in. in the same direction of a story, the geometric imperfection of $L_c/500$ can be used for buildings up to six stories with each story approximately 10 ft high. For taller buildings, this imperfection value of $L_c/500$ is conservative since the accumulated geometric imperfection calculated by 1/500 times building height is greater than the maximum permitted erection tolerance.

In this study, we shall use $L_c/500$ for the out-of-plumbness without any modification because the system strength is often governed by a weak story that has an out-of-plumbness equal to $L_c/500$ [28] and a constant imperfection has the benefit of simplicity in practical design. The explicit geometric imperfection modeling for an unbraced frame is illustrated in Figure 2.11.

2.2.3.2 Equivalent Notional Load Method

2.2.3.2.1 Braced Frames
The ECCS [21,22] and the CSA [23,24] introduced the equivalent load concept, which accounted for the geometric imperfections in an unbraced frame, but not in braced frames. The notional load approach for braced frames is also necessary to use the proposed methods for braced frames.

For braced frames, an equivalent notional load may be applied at mid-height of a column since the ends of the column are braced. An equivalent notional load factor equal to 0.004 is proposed here, and it is equivalent to the out-of-straightness of $L_c/1000$. When the free body of the column shown in Figure 2.12 is considered, the notional load factor, α, results in 0.002 with respect to one-half of the member length. Here, as in explicit imperfection modeling, the equivalent notional load factor is the same in concept for both sway and braced frames.

One drawback of this method for braced frames is that it requires tedious input of notional loads at the center of each column. Another is the axial force in the columns must be known in advance to determine the notional loads before analysis, but these are often difficult to calculate for large structures subject to lateral wind loads. To avoid this difficulty, it is recommended that either the explicit imperfection modeling method or the further reduced tangent modulus method be used.

2.2.3.2.2 Unbraced Frames
The geometric imperfections of a frame may be replaced by the equivalent notional lateral loads expressed as a fraction of the gravity loads acting on the story. Herein, the equivalent notional load factor of 0.002 is used. The notional load should be applied laterally at the top of each story. For sway frames subject to combined gravity and lateral loads, the notional loads should be added to the lateral loads. Figure 2.13 shows an illustration of the equivalent notional load for a portal frame.

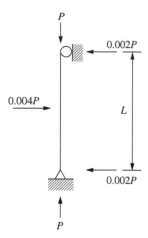

FIGURE 2.12 Equivalent notional load modeling for geometric imperfection of a braced member.

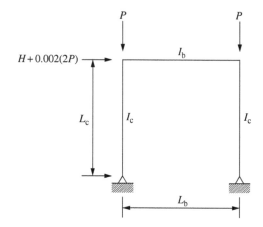

FIGURE 2.13 Equivalent notional load modeling for geometric imperfection of an unbraced frame.

2.2.3.3 Further Reduced Tangent Modulus Method

2.2.3.3.1 Braced Frames

The idea of using the reduced tangent modulus concept is to further reduce the tangent modulus, E_t, to account for further stiffness degradation due to geometrical imperfections. The degradation of member stiffness due to geometric imperfections may be simulated by an equivalent reduction of member stiffness. This may be achieved by a further reduction of tangent modulus [6,9]:

$$E'_t = 4\frac{P}{P_y}\left(1 - \frac{P}{P_y}\right)E\xi_i \quad \text{for } P > 0.5P_y \tag{2.12a}$$

$$E'_t = E\xi_i \quad \text{for } P \le 0.5P_y \tag{2.12b}$$

where E'_t is the reduced E_t and ξ_i is the reduction factor for geometric imperfection.

Herein, a reduction factor of 0.85 is used; the further reduced tangent modulus curves for the CRC E_t with geometric imperfections are shown in Figure 2.14. The further reduced tangent modulus concept satisfies one of the requirements for advanced analysis recommended by the SSRC task force report [29], that is, "The geometric imperfections should be accommodated implicitly within the element model.

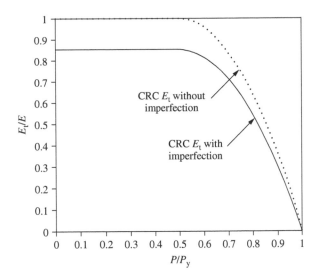

FIGURE 2.14 Further reduced CRC tangent modulus for members with geometric imperfections.

This would parallel the philosophy behind the development of most modern column strength expressions. That is, the column strength expressions in specifications such as the AISC-LRFD implicitly include the effects of residual stresses and out-of-straightness."

The advantage of this method over the other two methods is its convenience for design use, because it eliminates the inconvenience of explicit imperfection modeling or equivalent notional loads. Another benefit of this method is that it does not require the determination of the direction of geometric imperfections, often difficult to determine in a large system. On the other hand, in the other two methods, the direction of geometric imperfections must be taken correctly in coincidence with the deflection direction caused by bending moments, otherwise the wrong direction of geometric imperfection in braced frames may help the bending stiffness of columns rather than reduce it.

2.2.3.3.2 Unbraced Frames

The idea of the further reduced tangent modulus concept may also be used in the analysis of unbraced frames. Herein, as in the braced frame case, an appropriate reduction factor of 0.85 to E_t can be used [27,30,31]. The advantage of this approach over the other two methods is its convenience and simplicity because it completely eliminates the inconvenience of explicit imperfection modeling or the notional load input.

2.2.4 Numerical Implementation

The nonlinear global solution methods may be subdivided into two subgroups: (1) iterative methods and (2) simple incremental method. Iterative methods such as Newton–Raphson method, modified Newton–Raphson method, and quasi-Newton method satisfy equilibrium equations at specific external loads. In these methods, the equilibrium out-of-balance present following linear load step is eliminated (within tolerance) by taking corrective steps. The iterative methods possess the advantage of providing the exact load–displacement frame; however, they are inefficient, especially for practical purposes, in the trace of the hinge-by-hinge formation due to the requirement of the numerical iteration process.

The simple incremental method is a direct nonlinear solution technique. This numerical procedure is straightforward in concept and implementation. The advantage of this method is its computational efficiency. This is especially true when the structure is loaded into the inelastic region since tracing the hinge-by-hinge formation is required in the element stiffness formulation. For a finite increment size, this approach approximates only the nonlinear structural response, and equilibrium between the external applied loads and the internal element forces is not satisfied. To avoid this, an improved incremental method is used in this program. The applied load increment is automatically reduced to minimize the error when the change in the element stiffness parameter ($\Delta\eta$) exceeds a defined tolerance. To prevent plastic hinges from forming within a constant-stiffness load increment, load step sizes less than or equal to the specified increment magnitude are internally computed so that plastic hinges form only after the load increment. Subsequent element stiffness formations account for the stiffness reduction due to the presence of the plastic hinges. For elements partially yielded at their ends, a limit is placed on the magnitude of the increment in the element end forces.

The applied load increment in the above solution procedure may be reduced for any of the following reasons:

1. Formation of new plastic hinge(s) prior to the full application of incremental loads.
2. The increment in the element nodal forces at plastic hinges is excessive.
3. Nonpositive definiteness of the structural stiffness matrix.

As the stability limit point is approached in the analysis, large step increments may overstep a limit point. Therefore, a smaller step size is used near the limit point to obtain accurate collapse displacements and second-order forces.

2.3 Verifications

In the previous section, a practical advanced analysis method was presented for a direct two-dimensional frame design. The practical approach of geometric imperfections and semirigid connections was also discussed together with the advanced analysis method. The practical advanced analysis method was developed using simple modifications to the conventional elastic–plastic hinge analysis.

In this section, the practical advanced analysis method will be verified by the use of several benchmark problems available in the literature. Verification studies are carried out by comparing with the plastic-zone solutions as well as the conventional LRFD solutions. The strength predictions and the load–displacement relationships are checked for a wide range of steel frames including axially loaded columns, portal frame, six-story frame, and semirigid frames [6]. The three imperfection modelings, including explicit imperfection modeling, equivalent notional load modeling, and further reduced tangent modulus modeling, are also verified for a wide range of steel frames [6].

2.3.1 Axially Loaded Columns

The AISC-LRFD column strength curve is used for the calibration since it properly accounts for second-order effects, residual stresses, and geometric imperfections in a practical manner. In this study, the column strength of the proposed methods is evaluated for columns with slenderness parameters, $\left[\lambda_c = (KL/r)\sqrt{F_y/(\pi^2 E)}\right]$, varying from 0 to 2, which is equivalent to slenderness ratios (L/r) from 0 to 180 when the yield stress is equal to 36 ksi.

In explicit imperfection modeling, the two-element column is assumed to have an initial geometric imperfection equal to $L_c/1000$ at column mid-height. The predicted column strengths are compared with the LRFD curve in Figure 2.15. The errors are found to be less than 5% for slenderness ratios up to 140 (or λ_c up to 1.57). This range includes most columns used in engineering practice.

In the equivalent notional load method, notional loads equal to 0.004 times the gravity loads are applied mid-height to the column. The strength predictions are the same as those of the explicit imperfection model (Figure 2.16).

In the further reduced tangent modulus method, the reduced tangent modulus factor equal to 0.85 results in an excellent fit to the LRFD column strengths. The errors are less than 5% for columns of all slenderness ratios. These comparisons are shown in Figure 2.17.

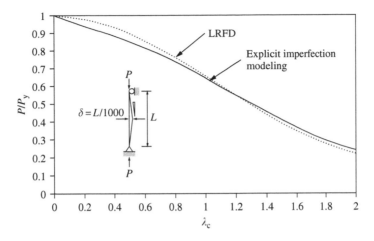

FIGURE 2.15 Comparison of strength curves for an axially loaded pin-ended column (explicit imperfection modeling method).

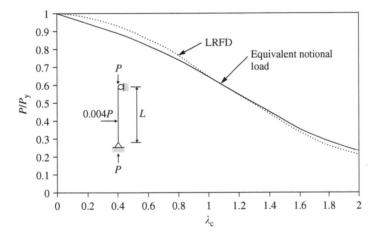

FIGURE 2.16 Comparison of strength curves for an axially loaded pin-ended column (equivalent notional load method).

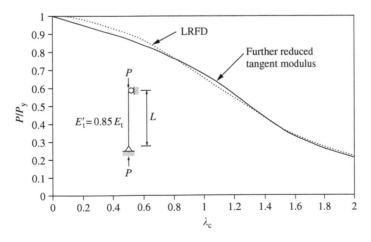

FIGURE 2.17 Comparison of strength curves for an axially loaded pin-ended column (further reduced tangent modulus method).

2.3.2 Portal Frame

Kanchanalai [4] performed extensive analyses of portal and leaning column frames, and developed exact interaction curves based on plastic-zone analyses of simple sway frames. Note that the simple frames are more sensitive in their behavior than the highly redundant frames. His studies formed the basis of the interaction equations in AISC-LRFD design specifications [1,32,33]. In his studies, the stress–strain relationship was assumed elastic–perfectly plastic with a 36-ksi yield stress and a 29,000-ksi elastic modulus. The members were assumed to have a maximum compressive residual stress of $0.3F_y$. Initial geometric imperfections were not considered, and thus an adjustment of his interaction curves is made to account for this. Kanchanalai further performed experimental work to verify his analyses, which covered a wide range of portal and leaning column frames with the slenderness ratios of 20, 30, 40, 50, 60, 70, and 80, and relative stiffness ratios (G) of 0, 3, and 4. The ultimate strength of each frame was presented in the form of interaction curves consisting of the nondimensional first-order moment ($HL_c/2M_p$ in portal frames or HL_c/M_p in leaning column frames in the x-axis) and the nondimensional axial load (P/P_y in the y-axis).

In this study, the AISC-LRFD interaction curves are used for strength comparisons. The strength calculations are based on the LeMessurier K factor method [34] since it accounts for story buckling and results in more accurate predictions. The inelastic stiffness reduction factor, τ [1], is used to calculate K in the LeMessurier's procedure. The resistance factors ϕ_b and ϕ_c in the LRFD equations are taken as 1.0 to obtain the nominal strength. The interaction curves are obtained by the accumulation of a set of moments and axial forces that result in unity on the value of the interaction equation.

When a geometric imperfection of $L_c/500$ is used for unbraced frames, including leaning column frames, most of the strength curves fall within an area bounded by the plastic-zone curves and the LRFD curves. In portal frames, the conservative errors are less than 5%, an improvement on the LRFD error of 11%, and the maximum unconservative error is not more than 1%, shown in Figure 2.18. In leaning column frames, the conservative errors are less than 12%, as opposed to the 17% error of the LRFD, and the maximum unconservative error is not more than 5%, as shown in Figure 2.19.

When a notional load factor of 0.002 is used, the strengths predicted by this method are close to those given by explicit imperfection modeling method (Figure 2.20 and Figure 2.21).

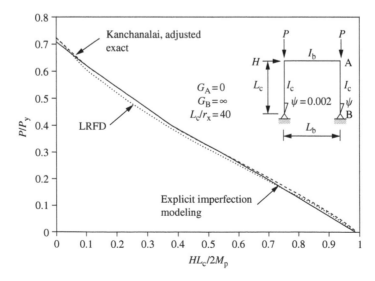

FIGURE 2.18 Comparison of strength curves for a portal frame subject to strong-axis bending with $L_c/r_x = 40$, $G_A = 0$ (explicit imperfection modeling method).

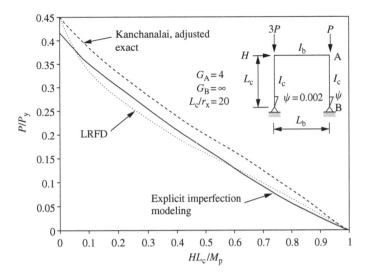

FIGURE 2.19 Comparison of strength curves for a leaning column frame subject to strong-axis bending with $L_c/r_x = 20$, $G_A = 4$ (explicit imperfection modeling method).

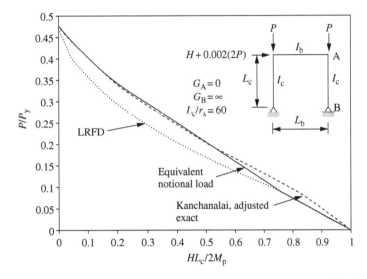

FIGURE 2.20 Comparison of strength curves for a portal frame subject to strong-axis bending with $L_c/r_x = 60$, $G_A = 0$ (equivalent notional load method).

When the reduced tangent modulus factor of 0.85 is used for portal and leaning column frames, the interaction curves generally fall between the plastic zone and LRFD curves. In portal frames, the conservative error is less than 8% (better than 11% error of the LRFD) and the maximum unconservative error is not more than 5% (Figure 2.22). In leaning column frames, the conservative error is less than 7% (better than 17% error of the LRFD), and the maximum unconservative error is not more than 5% (Figure 2.23).

2.3.3 Six-Story Frame

Vogel [35] presented the load–displacement relationships of a six-story frame using plastic-zone analysis. The frame is shown in Figure 2.24. Based on ECCS recommendations, the maximum

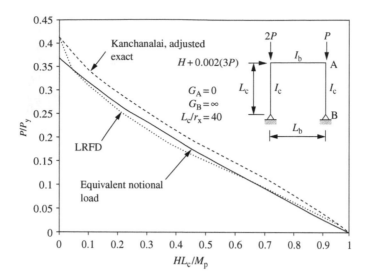

FIGURE 2.21 Comparison of strength curves for a leaning column frame subject to strong-axis bending with $L_c/r_x = 40$, $G_A = 0$ (equivalent notional load method).

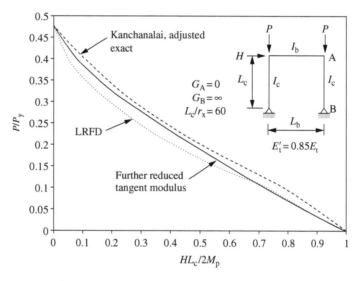

FIGURE 2.22 Comparison of strength curves for a portal frame subject to strong-axis bending with $L_c/r_x = 60$, $G_A = 0$ (further reduced tangent modulus method).

compressive residual stress is $0.3F_y$ when the ratio of depth to width (d/b) is greater than 1.2, and is $0.5F_y$ when the d/b ratio is less than 1.2 (Figure 2.25). The stress–strain relationship is elastic–plastic with strain hardening as shown in Figure 2.26. The geometric imperfections are $L_c/450$.

For comparison, the out-of-plumbness of $L_c/450$ is used in the explicit modeling method. The notional load factor of 1/450 and the reduced tangent modulus factor of 0.85 are used. The further reduced tangent modulus is equivalent to the geometric imperfection of $L_c/500$. Thus, the geometric imperfection of $L_c/4500$ is additionally modeled in the further reduced tangent modulus method, where $L_c/4500$ is the difference between the Vogel's geometric imperfection of $L_c/450$ and the proposed geometric imperfection of $L_c/500$.

The load–displacement curves in the proposed methods together with Vogel's plastic-zone analysis are compared in Figure 2.27. The errors in strength prediction by the proposed methods

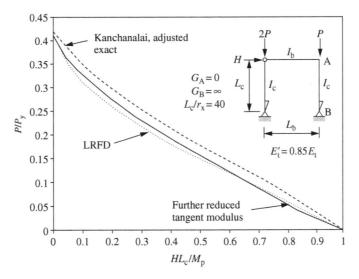

FIGURE 2.23 Comparison of strength curves for a leaning column frame subject to strong-axis bending with $L_c/r_x = 40$, $G_A = 0$ (further reduced tangent modulus method).

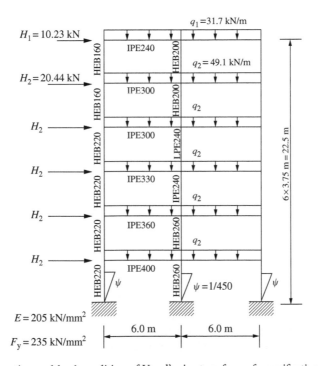

FIGURE 2.24 Configuration and load condition of Vogel's six-story frame for verification study.

are less than 1%. Explicit imperfection modeling and the equivalent notional load method under-predict lateral displacements by 3%, and the further reduced tangent modulus method shows a good agreement in displacement with the Vogel's exact solution. Vogel's frame is a good example of how the reduced tangent modulus method predicts lateral displacement well under reasonable load combinations.

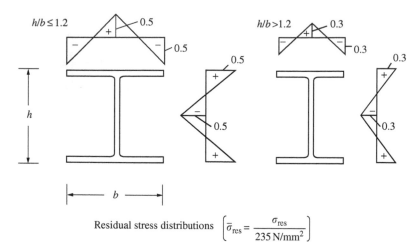

FIGURE 2.25   Residual stresses of cross-section for Vogel's frame.

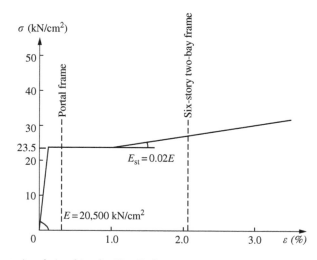

FIGURE 2.26   Stress–strain relationships for Vogel's frame.

2.3.4 Semirigid Frame

In the open literature, no benchmark problems solving semirigid frames with geometric imperfections are available for a verification study. An alternative is to separate the effects of semirigid connections and geometric imperfections. In the previous sections, the geometric imperfections were studied and comparisons between proposed methods, plastic-zone analyses, and conventional LRFD methods were made. Herein, the effect of semirigid connections will be verified by comparing analytical and experimental results.

Stelmack [36] studied the experimental response of two flexibly connected steel frames. A two-story, one-bay frame in his study is selected as a benchmark for the present study. The frame was fabricated from A36 W5 × 16 sections, with pinned base supports (Figure 2.28). The connections were bolted top and seat angles (L4 × 4 × $\frac{1}{2}$) made of A36 steel and A325 $\frac{3}{4}$in.-diameter bolts (Figure 2.29). The experimental moment–rotation relationship is shown in Figure 2.30. A gravity load of 2.4 kip was applied at third points along the beam at the first level, followed by a lateral load application. The lateral load–displacement relationship was provided in Stelmack.

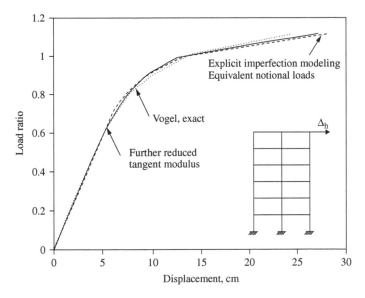

FIGURE 2.27   Comparison of displacements for Vogel's six-story frame.

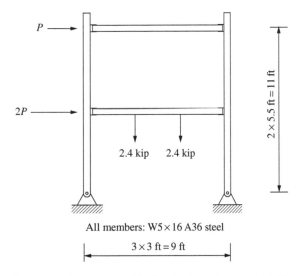

FIGURE 2.28   Configuration and load condition of Stelmack's two-story semirigid frame.

   Herein, the three parameters of the power model are determined by curve fitting and the program 3PARA.f presented in Section 2.2. The three parameters obtained by the curve-fit are $R_{ki} = 40{,}000$ k-in./rad, $M_u = 220$ k-in., and $n = 0.91$. We obtain three parameters of $R_{ki} = 29{,}855$ kip/rad, $M_u = 185$ k-in, and $n = 1.646$ with 3PARA.f.

   The moment–rotation curves given by experiment and curve fitting show good agreement (Figure 2.30). The parameters given by the Kishi–Chen equations and by experiment show some deviation (Figure 2.30). In spite of this difference, the Kishi–Chen equations, using the computer program 3PARA.f, are a more practical alternative in design since experimental moment–rotation curves are not usually available [30]. In the analysis, the gravity load is first applied and then the lateral load. The lateral displacements given by the proposed methods and by the experimental method compare well (Figure 2.31). The proposed method adequately predicts the behavior and strength of semirigid connections.

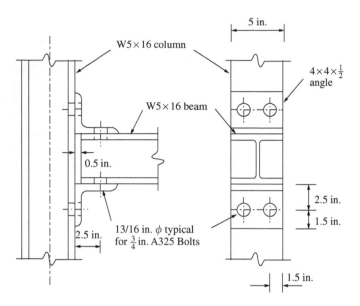

FIGURE 2.29 Top- and seat-angle connection details.

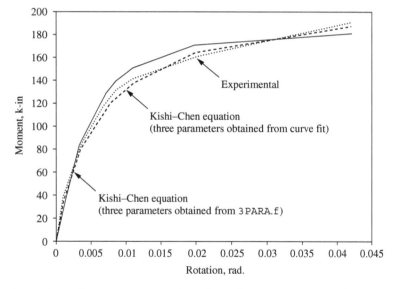

FIGURE 2.30 Comparison of moment–rotation relationships of semirigid connection by experiment and Kishi–Chen equation.

2.4 Analysis and Design Principles

In the preceding section, the proposed advanced analysis method was verified using several benchmark problems available in the literature. Verification studies were carried out by comparing it to the plastic-zone and conventional LRFD solutions. It was shown that practical advanced analysis predicted the behavior and failure mode of a structural system with reliable accuracy.

In this section, analysis and design principles are summarized for the practical application of the advanced analysis method. Step-by-step analysis and design procedures for the method are presented.

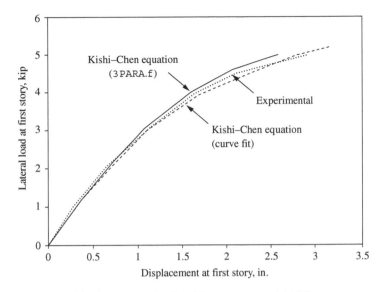

FIGURE 2.31 Comparison of displacements of Stelmack's two-story semirigid frame.

2.4.1 Design Format

Advanced analysis follows the format of LRFD. In LRFD, the factored load effect does not exceed the factored nominal resistance of the structure. Two safety factors are used: one is applied to loads and the other to resistances. This approach is an improvement on other models (e.g., ASD and PD) because both the loads and the resistances have unique factors for unique uncertainties. LRFD has the format

$$\phi R_n \geq \sum_{i=1}^{m} \gamma_i Q_{ni} \qquad (2.13)$$

where R_n is the nominal resistance of the structural member, Q_n is the nominal load effect (e.g., axial force, shear force, bending moment), ϕ is the resistance factor (≤ 1.0) (e.g., 0.9 for beams, 0.85 for columns), γ_i is the load factor (usually > 1.0) corresponding to Q_{ni} (e.g., $1.4D$ and $1.2D + 1.6L + 0.5S$), i is the type of load (e.g., D=dead load, L=live load, S=snow load), and m is the number of load type.

Note that the LRFD [32] uses separate factors for each load and therefore reflects the uncertainty of different loads and combinations of loads. As a result, a relatively uniform reliability is achieved.

The main difference between the conventional LRFD method and advanced analysis methods is that the left-hand side of Equation 2.13 (ϕR_n) in the LRFD method is the resistance or strength of the component of a structural system, but in the advanced analysis method it represents the resistance or the load-carrying capacity of the whole structural system.

2.4.2 Loads

Structures are subjected to various loads including dead, live, impact, snow, rain, wind, and earthquake loads. Structures must be designed to prevent failure and limit excessive deformation; thus, an engineer must anticipate the loads a structure may experience over its service life with reliability.

Loads may be classified as static or dynamic. Dead loads are typical of static loads, and wind or earthquake loads are dynamic. Dynamic loads are usually converted to equivalent static loads in conventional design procedures, and it may be adopted in advanced analysis as well [37].

2.4.3 Load Combinations

The load combinations in advanced analysis methods are based on the LRFD combinations [1]. Six factored combinations are provided by the LRFD Specification. One must be used to determine member sizes. Probability methods were used to determine the load combinations listed in the LRFD Specification (LRFD-A4). Each factored load combination is based on the load corresponding to the 50-year recurrence as follows:

$$(1)\ 1.4D \tag{2.14a}$$

$$(2)\ 1.2D + 1.6L + 0.5(L_r \text{ or } S \text{ or } R) \tag{2.14b}$$

$$(3)\ 1.2D + 1.6(L_r \text{ or } S \text{ or } R) + (0.5L \text{ or } 0.8W) \tag{2.14c}$$

$$(4)\ 1.2D + 1.6W + 0.5L + 0.5(L_r \text{ or } S \text{ or } R) \tag{2.14d}$$

$$(5)\ 1.2D \pm 1.0E + 0.5L + 0.2S \tag{2.14e}$$

$$(6)\ 0.9D \pm (1.6W \text{ or } 1.0E) \tag{2.14f}$$

where D is the dead load (the weight of the structural elements and the permanent features on the structure), L is the live load (occupancy and moveable equipment), L_r is the roof live load, W is the wind load, S is the snow load, E is the earthquake load, and R is the rainwater or ice load.

The LRFD Specification specifies an exception that the load factor on live load, L, in combination (3)–(5) must be 1.0 for garages, areas designated for public assembly, and all areas where the live load is greater than 100 psf.

2.4.4 Resistance Factors

The AISC-LRFD cross-section strength equations may be written as

$$\frac{P}{\phi_c P_y} + \frac{8}{9}\frac{M}{\phi_b M_p} = 1.0 \quad \text{for} \quad \frac{P}{\phi_c P_y} \geq 0.2 \tag{2.15a}$$

$$\frac{1}{2}\frac{P}{\phi_c P_y} + \frac{M}{\phi_b M_p} = 1.0 \quad \text{for} \quad \frac{P}{\phi_c P_y} < 0.2 \tag{2.15b}$$

where P, M are second-order axial force and bending moment, respectively, P_y is the squash load, M_p is the plastic moment capacity, and ϕ_c, ϕ_b are the resistance factors for axial strength and flexural strength, respectively.

Figure 2.32 shows the cross-section strength including the resistance factors ϕ_c and ϕ_b. The reduction factors ϕ_c and ϕ_b are built into the analysis program and are thus automatically included in the calculation of the load-carrying capacity. The reduction factors are 0.85 for axial strength and 0.9 for flexural strength, corresponding to the AISC-LRFD Specification [1]. For connections, the ultimate moment, M_u, is reduced by the reduction factor 0.9.

2.4.5 Section Application

The AISC-LRFD Specification uses only one column curve for rolled and welded sections of W, WT, and HP shapes, pipe, and structural tubing. The specification also uses some interaction equations for doubly and singly symmetric members including W, WT, and HP shapes, pipe, and structural tubing, even though the interaction equations were developed on the basis of W shapes by Kanchanalai [4].

The present advanced analysis method was developed by calibration with the LRFD column curve and interaction equations described in Section 2.3. To this end, it is concluded that the proposed method can be used for various rolled and welded sections, including W, WT, and HP shapes, pipe, and structural tubing without further modifications.

2.4.6 Modeling of Structural Members

Different types of advanced analysis are (1) plastic-zone method, (2) quasi-plastic hinge method, (3) elastic–plastic hinge method, and (4) refined plastic-hinge method. An important consideration in making these advanced analyses practical is the required number of elements for a member in order to predict realistically the behavior of frames.

A sensitivity study of advanced analysis is performed on the required number of elements for a beam member subject to distributed transverse loads. A two-element model adequately predicts the strength of a member. To model a parabolic out-of-straightness in a beam column, a two-element model with a maximum initial deflection at the mid-height of a member adequately captures imperfection effects. The required number of elements in modeling each member to provide accurate predictions of the strengths is summarized in Table 2.4. It is concluded that practical advanced analysis is computationally efficient.

2.4.7 Modeling of Geometric Imperfection

Geometric imperfection modeling is required to account for fabrication and erection tolerances. The imperfection modeling methods used here are the explicit imperfection, the equivalent notional load, and the further reduced tangent modulus models. Users may choose one of these three models in an advanced analysis. The magnitude of geometric imperfections is listed in Table 2.5.

Geometric imperfection modeling is required for a frame but not a truss element, since the program computes the axial strength of a truss member using the LRFD column strength equations, which account for geometric imperfections.

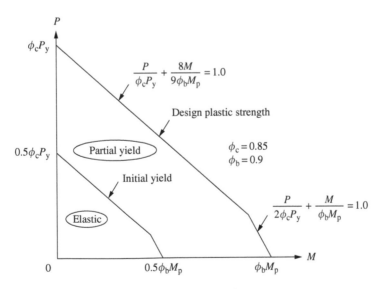

FIGURE 2.32 Stiffness degradation model including reduction factors.

TABLE 2.4 Necessary Number of Elements

Member	Number of elements
Beam member subject to uniform loads	2
Column member of braced frame	2
Column member of unbraced frame	1

TABLE 2.5   Magnitude of Geometric Imperfection

Geometric imperfection method	Magnitude
1. Explicit imperfection modeling method	$\psi = 2/1000$ for unbraced frames
	$\psi = 1/1000$ for braced frames
2. Equivalent notional load method	$\alpha = 2/1000$ for unbraced frames
	$\alpha = 4/1000$ for braced frames
3. Further reduced tangent modulus method	$E_t' = 0.85E_t$

2.4.8 Load Application

It is necessary, in an advanced analysis, to input proportional increment load (not the total loads) to trace nonlinear load–displacement behavior. The incremental loading process can be achieved by scaling down the combined factored loads by a number between 10 and 50. For a highly redundant structure (such as one greater than six stories), dividing by about 10 is recommended, and for a nearly statically determinate structure (such as a portal frame), the incremental load may be factored down by 50. One may choose a number between 10 and 50 to reflect the redundancy of a particular structure. Since a highly redundant structure has the potential to form many plastic hinges and the applied load increment is automatically reduced as new plastic hinges form, the larger incremental load (i.e., the smaller scaling number) may be used.

2.4.9 Analysis

Analysis is important in the proposed design procedures, since the advanced analysis method captures key behaviors including second-order and inelasticity in its analysis program. Advanced analysis does not require separate member capacity checks by the specification equations. On the other hand, the conventional LRFD method accounts for inelastic second-order effects in its design equations (not in analysis). The LRFD method requires tedious separate member capacity checks. Input data used for advanced analysis is easily accessible to users, and the input format is similar to the conventional linear elastic analysis. The format will be described in detail in Section 2.5. Analyses can be simply carried out by executing the program described in Section 2.5. This program continues to analyze with increased loads and stops when a structural system reaches its ultimate state.

2.4.10 Load-Carrying Capacity

Because consideration at moment redistribution may not always be desirable, two approaches (including and excluding inelastic moment redistribution) are presented. First, the load-carrying capacity, including the effect of inelastic moment redistribution, is obtained from the final loading step (limit state) given by the computer program. Second, the load-carrying capacity without the inelastic moment redistribution is obtained by extracting the force sustained when the first plastic hinge is formed. Generally, advanced analysis predicts the same member size as the LRFD method when moment redistribution is not considered. Further illustrations on these two choices will be presented in Section 2.6.

2.4.11 Serviceability Limits

The serviceability conditions specified by the LRFD consist of five limit states: (1) deflection, vibration, and drift; (2) thermal expansion and contraction; (3) connection slip; (4) camber; and (5) corrosion. The most common parameter affecting the design serviceability of steel frames is the deflections.

TABLE 2.6   Deflection Limitations of Frame

Item	Deflection ratio
Floor girder deflection for service live load	$L/360$
Roof girder deflection	$L/240$
Lateral drift for service wind load	$H/400$
Interstory drift for service wind load	$H/300$

Based on the studies by the Ad Hoc Committee [38] and Ellingwood [39], the deflection limits recommended (Table 2.6) were proposed for general use. At service load levels, no plastic hinges are permitted anywhere in the structure to avoid permanent deformation under service loads.

2.4.12 Ductility Requirements

Adequate inelastic rotation capacity is required for members in order to develop their full plastic moment capacity. The required rotation capacity may be achieved when members are adequately braced and their cross-sections are compact. The limitations of compact sections and lateral unbraced length in what follows lead to an inelastic rotation capacity of at least three and seven times the elastic rotation corresponding to the onset of the plastic moment for nonseismic and seismic regions, respectively.

Compact sections are capable of developing the full plastic moment capacity, M_p, and sustaining large hinge rotation before the onset of local buckling. The compact section in the LRFD Specification is defined as:

1. Flange
   * For nonseismic region

$$\frac{b_f}{2t_f} \le 0.38\sqrt{\frac{E_s}{F_y}} \tag{2.16}$$

   * For seismic region

$$\frac{b_f}{2t_f} \le 0.31\sqrt{\frac{E_s}{F_y}} \tag{2.17}$$

where E_s is the modulus of elasticity, b_f is the width of flange, t_f is the thickness of flange, and F_y is the yield stress.

2. Web
   * For nonseismic region

$$\frac{h}{t_w} \le 3.76\sqrt{\frac{E_s}{F_y}}\left(1 - \frac{2.75P_u}{\phi_b P_y}\right) \quad \text{for} \quad \frac{P_u}{\phi_b P_y} \le 0.125 \tag{2.18a}$$

$$\frac{h}{t_w} \le 1.12\sqrt{\frac{E_s}{F_y}}\left(2.33 - \frac{P_u}{\phi_b P_y}\right) \ge 1.49\sqrt{\frac{E_s}{F_y}} \quad \text{for} \quad \frac{P_u}{\phi_b P_y} > 0.125 \tag{2.18b}$$

   * For seismic region

$$\frac{h}{t_w} \le 3.05\sqrt{\frac{E_s}{F_y}}\left(1 - \frac{1.54P_u}{\phi_b P_y}\right) \quad \text{for} \quad \frac{P_u}{\phi_b P_y} \le 0.125 \tag{2.19a}$$

$$\frac{h}{t_w} \le 1.12\sqrt{\frac{E_s}{F_y}}\left(2.33 - \frac{P_u}{\phi_b P_y}\right) \quad \text{for} \quad \frac{P_u}{\phi_b P_y} > 0.125 \tag{2.19b}$$

where h is the clear distance between flanges, t_w is the thickness of the web, and F_y is the yield strength.

In addition to the compactness of the section, the lateral unbraced length of beam members is also a limiting factor for the development of the full plastic moment capacity of members. The LRFD provisions provide the limit on spacing of braces for beam as:

- For nonseismic region

$$L_{pd} \leq \left[0.12 + 0.076 \left(\frac{M_1}{M_2} \right) \right] \left(\frac{E_s}{F_y} \right) r_y \tag{2.20a}$$

- For seismic region

$$L_{pd} \leq 0.086 \left(\frac{E_s}{F_y} \right) r_y \tag{2.20b}$$

where L_{pd} is the unbraced length, r_y is the radius of gyration about y-axis, F_y is the yield strength, M_1, M_2 are smaller and larger end moments, and M_1/M_2 is the positive in double curvature bending.

The AISC-LRFD Specification explicitly specifies the limitations for beam members as described above, but not for beam–column members. More studies are necessary to determine the reasonable limits leading to adequate rotation capacity of beam–column members. Based on White's study [40], the limitations for beam members seem to be used for beam–column members until the specification provides the specific values for beam–column members.

2.4.13 Adjustment of Member Sizes

If one of the following three conditions — strength, serviceability, or ductility — is not satisfied, appropriate adjustments of the member sizes should be made. This can be done by referring to the sequence of plastic hinge formation shown in the P.OUT. For example, if the load-carrying capacity of a structural system is less than the factored load effect, the member with the first plastic hinge should be replaced with a stronger member. On the other hand, if the load-carrying capacity exceeds the factored load effect significantly, members without plastic hinges may be replaced with lighter members. If lateral drift exceeds drift requirements, columns or beams should be sized up, or a braced structural system should be considered instead to meet this serviceability limit.

In semirigid frames, behavior is influenced by the combined effects of members and connections. As an illustration, if an excessive lateral drift occurs in a structural system, the drift may be reduced by increasing member sizes or using more rigid connections. If the strength of a beam exceeds the required strength, it may be adjusted by reducing the beam size or using more flexible connections. Once the member and connection sizes are adjusted, the iteration leads to an optimum design. Figure 2.33 shows a flow chart of analysis and design procedure in the use of advanced analysis.

2.5 Computer Program

This section describes the practical advanced analysis program (PAAP) for two-dimensional steel frame design [6,19]. The program integrates the methods and techniques developed in Sections 2.2 and 2.3. The names of variables and arrays correspond as closely as possible to those used in theoretical derivations. The main objective of this section is to present an educational version of software to enable engineers and graduate students to perform planar frame analysis for a more realistic prediction of the strength and behavior of structural systems.

The instructions necessary for user input into PAAP are presented in Section 2.5.4. Except for the requirement to input geometric imperfections and incremental loads, the input data format

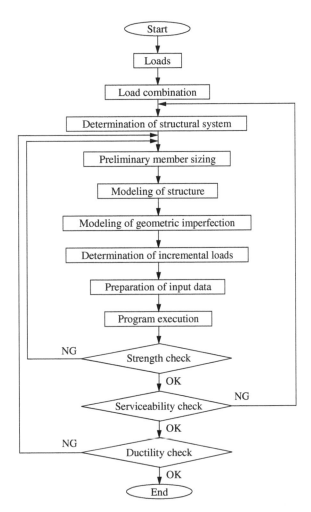

FIGURE 2.33 Analysis and design procedure.

of the program is basically the same as that of the usual linear elastic analysis program. The user is advised to read all the instructions, paying particular attention to the appended notes, to achieve an overall view of the data required for a specific PAAP analysis. The reader should recognize that no system of units is assumed in the program, and take the responsibility to make all units consistent. Mistaken unit conversion and input are a common source of erroneous results.

2.5.1 Program Overview

This FORTRAN program is divided into three parts: DATAGEN, INPUT, and PAAP. The first program, DATAGEN, reads an input data file, P.DAT, and generates a modified data file, INFILE. The second program, INPUT, rearranges INFILE into three working data files, DATA0, DATA1, and DATA2. The third program, PAAP, reads the working data files and provides two output files named P.OUT1 and P.OUT2. P.OUT1 contains an echo of the information from the input data file, P.DAT. This file may be used to check for numerical and incompatibility errors in input data. P.OUT2 contains the load and displacement information for various joints in the structure as well as the element joint forces for all types of elements at every load step. The load–displacement results are presented at the end of every load increment. The sign conventions for loads and displacements should follow the frame degrees of freedom, as shown in Figure 2.34 and Figure 2.35.

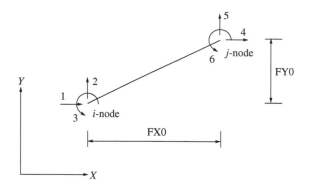

FIGURE 2.34 Degrees of freedom numbering for the frame element.

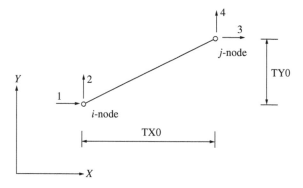

FIGURE 2.35 Degrees of freedom numbering for the truss element.

The element joint forces are obtained by summing the product of the element incremental displacements at every load step. The element joint forces act in the global coordinate system and must be in equilibrium with applied forces. After the output files are generated, the user can view these files on the screen or print them with the MS-DOS PRINT command. The schematic diagram in Figure 2.36 sets out the operation procedure used by PAAP and its supporting programs [6].

2.5.2 Hardware Requirements

This program has been tested in two computer processors. It was first tested on an IBM 486 or equivalent personal computer system using Microsoft's FORTRAN 77 compiler v1.00 and Lahey's FORTRAN 77 compiler v5.01. Then, its performance in the workstation environment was tested on a Sun 5 using a Sun FORTRAN 77 compiler. The program sizes of DATAGEN, INPUT, and PAAP are 8, 9, and 94 kB, respectively. The total size of the three programs is small, 111 kB (= 0.111 MB), and so a 3.5-in. high-density diskette (1.44 MB) can accommodate the three programs and several example problems.

The memory required to run the program depends on the size of the problem. A computer with minimum 640 K of memory and a 30 MB hard disk is generally required. For the PC applications, the array sizes are restricted as follows:

1. Maximum total degrees of freedom, MAXDOF = 300
2. Maximum translational degrees of freedom, MAXTOF = 300
3. Maximum rotational degrees of freedom, MAXROF = 100
4. Maximum number of truss elements, MAXTRS = 150
5. Maximum number of connections, MAXCNT = 150

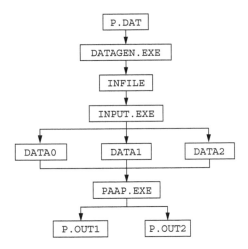

FIGURE 2.36   Operating procedures of the proposed program.

It is possible to run bigger jobs in UNIX workstations by modifying the above values in the PARAMETER and COMMON statements in the source code.

2.5.3 Execution of Program

A computer diskette is provided in *LRFD Steel Design Using Advanced Analysis*, by Chen and Kim [37], containing four directories with following files:

1. Directory PSOURCE
   - DATAGEN.FOR
   - INPUT.FOR
   - PAAP.FOR
2. Directory PTEST
   - DATAGEN.EXE
   - INPUT.EXE
   - PAAP.EXE
   - RUN.BAT (batch file)
   - P.DAT (input data for a test run)
   - P.OUT1 (output for a test run)
   - P.OUT2 (output for a test run)
3. Directory PEXAMPLE
   - All input data for the example problems presented in Section 6 of *Handbook of Structural Engineering, Second Edition*, by Chen and Lui [44]
4. Directory CONNECT
   - 3PARA.FOR (program for semirigid connection parameters)
   - 3PARA.EXE
   - CONN.DAT (input data)
   - CONN.OUT (output for three parameters)

To execute the programs, one must first copy them onto the hard disk (i.e., copy DATAGEN.EXE, INPUT.EXE, PAAP.EXE, RUN.BAT, and P.DAT from the directory PTEST on the diskette to the hard disk). Before launching the program, the user should test the system by running the sample example provided in the directory. The programs are executed by issuing the command RUN. The batch file, RUN.BAT executes DATAGEN, INPUT, and PAAP in sequence. The output files produced are P.OUT1 and P.OUT2. When the authors' and the user's compilers are different, the program (PAAP.EXE) may not be executed. This problem may be easily solved by recompiling the source programs in the directory PSOURCE.

The input data for all the problems in Section 2.6 are provided in the directory PEXAMPLE. The user may use the input data for his or her reference and confirmation of the results presented in Section 2.6. It should be noted that RUN is a batch command to facilitate the execution of PAAP. Entering the command RUN will write the new files including DATA0, DATA1, DATA2, P.OUT1, and P.OUT2 over the old ones. Therefore, output files should be renamed before running a new problem.

The program can generate the output files P.OUT1 and P.OUT2 in a reasonable time period, as described in the following. The run time on an IBM 486 PC with memory of 640 K to get the output files for the eight-story frame shown in Figure 2.37 is taken as 4 min 10 s and 2 min 30 s in real time rather than CPU time by using Microsoft FORTRAN and Lahey FORTRAN, respectively. In the Sun 5, the run time varies approximately 2 to 3 min depending on the degree of occupancy by users.

The directory CONNECT contains the program that computes the three parameters needed for semirigid connections. The operation procedure of the program, the input data format, and two examples were presented in Section 2.2.2.

2.5.4 User Manual

2.5.4.1 Analysis Options

PAAP was developed on the basis of the theory presented in Section 2.2. While the purpose of the program is basically for advanced analysis using second-order inelastic concept, the program can also be used for first- and second-order elastic analyses. For a first-order elastic analysis, the total factored load should be applied in one load increment to suppress numerical iteration in the nonlinear analysis algorithm. For a second-order elastic analysis, a yield strength of an arbitrarily large value should be assumed for all members to prevent yielding.

2.5.4.2 Coordinate System

A two-dimensional (x, y) global coordinate system is used for the generation of all the input and output data associated with the joints. The following input and output data are prepared with respect to the global coordinate system:

1. Input data
   - joint coordinates
   - joint restraints
   - joint load

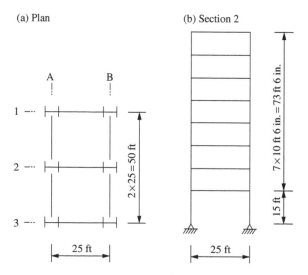

FIGURE 2.37   Configuration of the unbraced eight-story frame.

2. Output data
   - joint displacement
   - member forces

2.5.4.3 Type of Elements

The analysis library consists of three elements: a plane frame, a plane truss, and a connection. The connection is represented by a zero-length rotational spring element with a user-specified nonlinear moment–rotation curve. Loading is allowed only at nodal points. Geometric and material nonlinearities can be accounted for by using an iterative load-increment scheme. Zero-length plastic hinges are lumped at the element ends.

2.5.4.4 Locations of Nodal Points

The geometric dimensions of the structures are established by placing joints (or nodal points) on the structures. Each joint is given an identification number and is located in a plane associated with a global two-dimensional coordinate system. The structural geometry is completed by connecting the predefined joints with structural elements, which may be a frame, a truss, or a connection. Each element also has an identification number.

The following are some of the factors that need to be considered in placing joints in a structure:

1. The number of joints should be sufficient to describe the initial geometry and the response behavior of the structures.
2. Joints need to be located at points and lines of discontinuity (e.g., at changes in material properties or section properties).
3. Joints should be located at points on the structure where forces and displacements need to be evaluated.
4. Joints should be located at points where concentrated loads will be applied. The applied loads should be concentrated and act on the joints.
5. Joints should be located at all support points. Support conditions are represented in the structural model by restricting the movement of the specific joints in specific directions.
6. Second-order inelastic behavior can be captured by the use of one or two elements per member corresponding to the following guidelines:
   - Beam member subjected to uniform loads: two elements
   - Column member of braced frames: two elements
   - Column member of unbraced frames: one element.

2.5.4.5 Degrees of Freedom

A two-joint frame element has six displacement components as shown in Figure 2.34. Each joint can translate in the global x- and y-directions, and rotate about the global z-axis. The directions associated with these displacement components are known as degrees of freedom of the joint. A two-joint truss element has four degrees of freedom as shown in Figure 2.35. Each joint has two translational degrees of freedom and no rotational component.

If the displacement of a joint corresponding to any one of its degrees of freedom is known to be zero (such as at a support), then it is labeled an inactive degree of freedom. Degrees of freedom where the displacements are not known are termed active degree of freedoms. In general, the displacement of an inactive degree of freedom is usually known, and the purpose of the analysis is to find the reaction in that direction. For an active degree of freedom, the applied load is known (it could be zero), and the purpose of the analysis is to find the corresponding displacement.

2.5.4.6 Units

There are no "built-in" units in PAAP. The user must prepare the input in a consistent set of units. The output produced by the program will conform to the same set of units. Therefore, if the user chooses to use kips and inches as the input units, all the dimensions of the structure must be entered in inches and all the loads in kip. The material properties should also conform to these units. The output units will then be in kips and inches, so that the frame member axial force will be in kips, bending moments will be in kip-inches, and displacements will be in inches. Joint rotations, however, are in radians, irrespective of units.

2.5.4.7 Input Instructions

In this section, the input sequence and data structure used to create an input file called `P.DAT` are described. The analysis program, `PAAP`, can analyze any structures with up to 300 degrees of freedom, but it is possible to recompile the source code to accommodate more degrees of freedom by changing the size of the arrays in the `PARAMETER` and `COMMON` statements. The limitation of degree of freedom can be solved by using dynamic storage allocation. This procedure is common in finite element programs [41,42], and will be used in the next release of the program.

The input data file is prepared in a specific format. The input data consists of 13 data sets, including five control data, three section property data, three element data, one boundary condition, and one load data set:

1. Title
2. Analysis and design control
3. Job control
4. Total number of element types
5. Total number of elements
6. Connection properties
7. Frame element properties
8. Truss element properties
9. Connection element data
10. Frame element data
11. Truss element data
12. Boundary conditions
13. Incremental loads

Input of all data sets are mandatory, but some of the data associates with elements (data sets 6–11) may be skipped depending on whether the use of the element. The order of data sets in the input file must be strictly maintained. Instructions for inputting data are summarized in Table 2.7.

2.6 Design Examples

In previous sections, the concept, verifications, and computer program of the practical advanced analysis method for steel frame design have been presented. The present advanced analysis method has been developed and refined to achieve both simplicity in use and, as far as possible, a realistic representation of behavior and strength. The advanced analysis method captures the limit state strength and stability of a structural system and its individual members. As a result, the method can be used for practical frame design without the tedious separate member capacity checks, including the calculation of *K* factor.

The aim of this section is to provide further confirmation of the validity of the LRFD-based advanced analysis methods for practical frame design. The comparative design examples in this section show the detailed design procedure for advanced and LRFD design procedures [6]. The design procedures conform to those described in Section 2.4 and may be grouped into four basic steps: (1) load condition, (2) structural modeling, (3) analysis, and (4) limit state check. The design examples cover simple structures, truss structures, braced frames, unbraced frames, and semirigid frames. The three practical models — explicit imperfection, equivalent notional load, and further reduced tangent modulus — are used for the design examples. Member sizes determined by advanced procedures are compared with those determined by the LRFD, and good agreement is generally observed.

TABLE 2.7 Input Data Format for the Program PAAP

Data set	Column	Variable	Description
Title	A70	—	Job title and general comments
Analysis and design control		IGEOIM	Geometric imperfection method
	1–5		0: No geometric imperfection (default)
			1: Explicit imperfection modeling
			2: Equivalent notional load
			3: Further reduced tangent modulus
		ILRFD	Strength reduction factor $\phi_c = 0.85$, $\phi_b = 0.9$
	6–10		0: No reduction factors considered (default)
			1: Reduction factors considered
Job control	1–5	NNODE	Total number of nodal points of the structure
	6–10	NBOUND	Total number of supports
	11–15	NINCRE	Allowable number of load increments (default = 100); at least two or three times larger than the scaling number
Total number of element types	1–5	NCTYPE	Number of connection types (1–30)
	6–10	NFTYPE	Number of frame types (1–30)
	11–15	NTTYPE	Number of truss types (1–30)
Total number of elements	1–5	NUMCNT	Number of connection elements (1–150)
	6–10	NUMFRM	Number of frame elements (1–100)
	11–15	NUMTRS	Number of truss elements (1–150)
Connection property	1–5	ICTYPE	Connection type number
	6–15[a]	M_u	Ultimate moment capacity of connection
	16–25[a]	R_{ki}	Initial stiffness of connection
	26–35[a]	N	Shape parameter of connection
Frame element property	1–5	IFTYPE	Frame type number
	6–15[a]	A	Cross-section area
	16–25[a]	I	Moment of inertia
	26–35[a]	Z	Plastic section modulus
	36–45[a]	E	Modulus of elasticity
	46–55[a]	FY	Yield stress
	55–60	IFCOL	Identification of column member, IFCOL = 1 for column (default = 0)
Truss element property	1–5	ITYPE	Truss type number
	6–15[a]	A	Cross-section area
	15–25[a]	I	Moment of inertia
	25–35[a]	E	Modulus of elasticity
	36–45[a]	FY	Yield stress
	46–50	ITCOL	Identification of column member, ITCOL = 1 for column (default = 0)
Connection element data	1–5	LCNT	Connection element number
	6–10	IFMCNT	Frame element number containing the connection
	11–15	IEND	Identification of element ends containing the connection
	16–20	JDCNT	1: Connection attached at element end i
	21–25	NOSMCN	2: Connection attached at element end j
	26–30	NELINC	Connection type number
			Number of same elements for automatic generation (default = 1)
			Element number (IFMCNT) increment of automatically generated elements (default = 1)
Frame element data	1–5	LFRM	Frame element number
	6–15[a]	FXO	Horizontal projected length; positive for i–j direction in global x direction
	16–25[a]	FYO	Vertical projected length; positive for i–j direction in global y direction

TABLE 2.7 Continued

Data set	Column	Variable	Description
	26–30	JDFRM	Frame type number
	31–35	IFNODE	Number of node i
	36–40	JFNODE	Number of node j
	41–45	NOSMFE	Number of same elements for automatic generation (default = 1)
	46–50	NODINC	Node number increment of automatically generated elements (default = 1)
Truss element data	1–5	LTRS	Truss element number
	6–15[a]	TXO	Horizontal projected length; positive for i–j direction in global x direction
	16–25[a]	TYO	Vertical projected length; positive for i–j direction in global y direction
	26–30	JDTRS	Truss type number
	31–35	ITNODE	Number of node i
	36–40	JTNODE	Number of node j
	51–55	NOSMTE	Number of same elements for automatic generation (default = 1)
	56–60	NODINC	Node number increment of automatically generated elements (default = 1)
Boundary condition	1–5	NODE	Node number of support
	6–10	XFIX	XFIX = 1 for restrained in global x-direction (default = 0)
	11–15	YFIX	YFIX = 1 for restrained in global y-direction (default = 0)
	16–20	RFIX	RFIX = 1 for restrained in rotation (default = 0)
	21–25	NOSMBD	Number of same boundary conditions for automatic generation (default = 1)
	26–30	NODINC	Node number increment of automatically generated supports (default = 1)
Incremental loads	1–5	NODE	Node number where a load applied
	6–15[a]	XLOAD	Incremental load in global x direction (default = 0)
	16–25[a]	YLOAD	Incremental load in global y direction (default = 0)
	26–35[a]	RLOAD	Incremental moment in global θ-direction (default = 0)
	36–40	NOSMLD	Number of same loads for automatic generation (default = 1)
	41–45	NODINC	Node number increment of automatically generated loads (default = 1)

[a] Indicates that real value (F or E format) should be entered; otherwise input the integer value (I format).

The design examples are limited to two-dimensional steel frames, so that the spatial behavior is not considered. Lateral torsional buckling is assumed to be prevented by adequate lateral braces. Compact W sections are assumed so that sections can develop their full plastic moment capacity without buckling locally. All loads are statically applied.

2.6.1 Roof Truss

Figure 2.38 shows a hinged-jointed roof truss subject to gravity loads of 201 kip at the joints. A36 steel pipe is used. All member sizes are assumed identical.

2.6.1.1 Design by Advanced Analysis

Step 1: Load condition and preliminary member sizing. The critical factored load condition is shown in Figure 2.38. The member forces of the truss may be obtained (Figure 2.39) using equilibrium conditions. The maximum compressive force is 67.1 kip. The effective length is the same as the actual length (22.4 ft) since K is 1.0. The preliminary member size of steel pipe is 6 in. diameter with 0.28 in. thickness ($\phi P_n = 81$ kip), obtained using the column design table in the LRFD Specification.

Step 2: Structural modeling. Each member is modeled with one truss element without geometric imperfection since the program computes the axial strength of the truss member with the LRFD column strength equations, which indirectly account for geometric imperfections. An incremental load of 0.51 kip is determined by dividing the factored load of 201 kip by a scaling factor of 40 as shown in Figure 2.40.

Step 3: Analysis. Referring to the input instructions described in Section 2.5.4, the input data may be easily generated, as listed in Table 2.8. Note that the total number of supports (NBOUND) in the hinged-jointed truss must be equal to the total number of nodal points, since the nodes of a truss element are restrained against rotation. Programs DATAGEN, INPUT, and PAAP are executed in sequence by entering the batch file command RUN on the screen.

Step 4: Check of load-carrying capacity. Truss elements 10 and 13 fail at load step 48, with loads at nodes 6, 7, and 8 being 241 kip. Since this truss is statically determinant, failure of one member leads to failure of the whole system. Load step 49 shows a sharp increase in displacement and indicates a system failure. The member force of element 10 is 80.41 kip ($F_x = 72.0$ kip, $F_y = 35.7$ kip). Since the load-carrying capacity of 241 kip at nodes 6, 7, and 8 is greater than the applied load of 20 kip, the member size is adequate.

Step 5: Check of serviceability. Referring to P.OUT2, the deflection at node 3 corresponding to load step 1 is equal to 0.02 in. This deflection may be considered elastic since the behavior of the beam

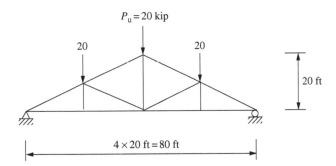

FIGURE 2.38 Configuration and load condition of the hinged-jointed roof truss.

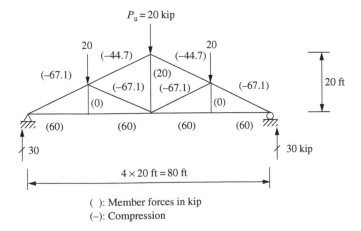

FIGURE 2.39 Member forces of the hinged-jointed roof truss.

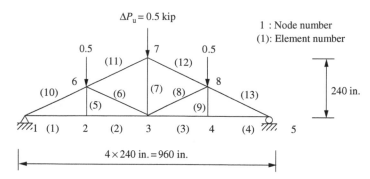

FIGURE 2.40 Modeling of the hinged-jointed roof truss.

TABLE 2.8 Input Data, P.DAT, of the
Hinged-Jointed Roof Truss

```
Roof truss
    0    1
    8    8    100
    0    0    1
    0    0    13
    1       5.58        28.1     29000.0        36.0
    1     240.0          0.0   1    1    2    4
    5       0.0        120.0    1    2    6
    6    -240.0        120.0    2    3    6
    7       0.0        240.0    1    3    7
    8     240.0        120.0    2    3    8
    9       0.0        120.0    1    4    8
   10     240.0        120.0    2    1    6
   11     240.0        120.0    2    6    7
   12     240.0       -120.0    2    7    8
   13     240.0       -120.0    2    8    5
    1    1    1    1
    2              1    3
    5         1    1
    6              1    3
    6             -0.5                   3
```

is linear and elastic under small loads. The total deflection of 0.8 in. is obtained by multiplying the deflection of 0.02 in. by the scaling factor of 40. The deflection ratio over the span length is 1/1200, which meets the limitation 1/360. The deflection at the service load will be smaller than that above since the factored load is used for the calculation of deflection above.

2.6.1.2 Comparison of Results

The advanced analysis and LRFD methods predict the same member size of steel pipe with 6 in. diameter and 0.28 in. thickness. The load-carrying capacities of element 10 predicted by these two methods are the same, 80.5 kip. This is because the truss system is statically determinant, rendering inelastic moment redistribution of little or no benefit.

2.6.2 Unbraced Eight-Story Frame

Figure 2.37 shows an unbraced eight-story, one-bay frame with hinged supports. All beams are rigidly connected to the columns. The column and beam sizes are the same. All beams are continuously braced about their weak axis. Bending is primarily about the strong axis at the column. A36 steel is used for all members.

2.6.2.1 Design by Advanced Analysis

Step 1: Load condition and preliminary member sizing. The uniform gravity loads are converted to equivalent concentrated loads, as shown in Figure 2.41. The preliminary column and beam sizes are selected as W33 × 130 and W21 × 50.

Step 2: Structural modeling. Each column is modeled with one element since the frame is unbraced and the maximum moment in the member occurs at the ends. Each beam is modeled with two elements.

The explicit imperfection and the further reduced tangent modulus models are used in this example, since they are easier in preparing the input data compared to equivalent notional load models. Figure 2.42 shows the model for the eight-story frame. The explicit imperfection model uses an out-of-plumbness of 0.2%, and in the further reduced tangent modulus model, $0.85E_t$ is used.

Herein, a scaling factor of 10 is used due to the high indeterminacy. The load increment is automatically reduced if the element stiffness parameter, η, exceeds the predefined value 0.1. The 54 load steps required to converge on the solution are given in P.OUT2.

Step 3: Analysis. The input data may be easily generated, as listed in Table 2.9. Programs are executed in sequence by typing the batch file command RUN.

Step 4: Check of load-carrying capacity. From the output file P.OUT, the ultimate load-carrying capacity of the structure is obtained as 5.24 and 5.18 kip with respect to the lateral load at roof in load combination 2 by the imperfection method and the reduced tangent modulus method, respectively. This load-carrying capacity is 3 and 2% greater, respectively, than the applied factored load of 5.12 kip. As a result, the preliminary member sizes are satisfactory.

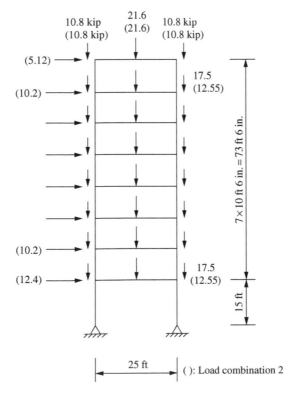

FIGURE 2.41 Concentrated load condition converted from the distributed load for the two-story frame.

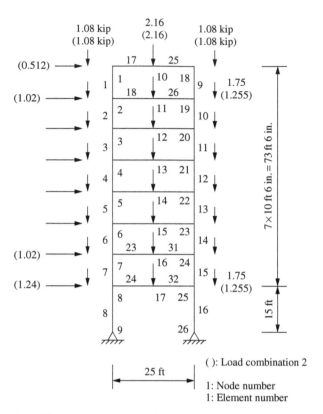

FIGURE 2.42   Structural modeling of the eight-story frame.

Step 5: Check of serviceability. The lateral drift at the roof level by the wind load (1.0W) is 5.37 in. and the drift ratio is 1/198, which does not satisfy the drift limit of 1/400. When W40 × 174 and W24 × 76 are used for column and beam members, respectively, the lateral drift is reduced to 2.64 in. and the drift ratio is 1/402, which satisfies the limit 1/400. The design of this frame is thus governed by serviceability rather than strength.

2.6.2.2   Comparison of Results

The sizes predicted by the proposed methods are W33 × 130 columns and W21 × 50 beams. They do not, however, meet serviceability conditions, and must therefore be increased to W40 × 174 and W24 × 76 members. The LRFD method results in the same (W40 × 174) column but a larger (W27 × 84) beam (Figure 2.43).

2.6.3   Two-Story, Four-Bay Semirigid Frame

Figure 2.44 shows a two-story, four-bay semirigid frame. The height of each story is 12 ft and it is 25 ft wide. The spacing of the frames is 25 ft. The frame is subjected to a distributed gravity and concentrated lateral loads. The roof beams connections are the top and seat L6 × 4.0 × 3/8 × 7 angle with double web angles of L4 × 3.5 × 1/4 × 5.5 made of A36 steel. The floor beam connections are the top and seat angles of L6 × 4 × 9/16 × 7 with double web angles of L4 × 3.5 × 5/16 × 8.5. All fasteners are A325 $\frac{3}{4}$in.-diameter bolts. All members are assumed to be continuously braced laterally.

2.6.3.1   Design by Advanced Analysis

Step 1: Load condition and preliminary member size. The load conditions are shown in Figure 2.44. The initial member sizes are selected as W8 × 21, W12 × 22, and W16 × 40 for the columns, the roof beams, and the floor beams, respectively.

TABLE 2.9 Input Data, P.DAT, of the Explicit Imperfection Modeling for the Unbraced Eight-Story Frame: (a) Explicit Imperfection Modeling and (b) Further Reduced Tangent Modulus

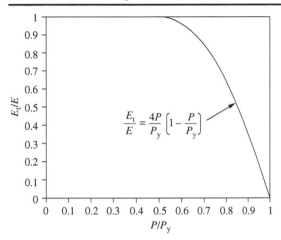

$$\frac{E_t}{E} = \frac{4P}{P_y}\left[1 - \frac{P}{P_y}\right]$$

(b) Unbraced eight-story frame, further reduced tangent modulus

```
      3     1
     26     2
      0     2    0
      0    32    0
      1    38.30    6710.00    467.00   29000.00    36.00    1
      2    14.70     984.00    110.00   29000.00    36.00
      1     0.        126.00    1     2    1     7
      8     0.        180.00    1     9    8
      9     0.        126.00    1    19   18     7
     16     0.        180.00    1    26   25
     17   150.00        0.00    2     1   10     8
     25   150.00        0.00    2    10   18     8
      9     1     1
     26     1     1
      1     0.512    -1.080
      2     1.020    -1.255               6
      8     1.240    -1.255
     10              -2.160
     11              -2.510               7
     18              -1.080
     19              -1.255               7
```

Step 2: Structural modeling. Each column is modeled with one element and beam with two elements. The distributed gravity loads are converted to equivalent concentrated loads on the beam, as shown in Figure 2.45. In explicit imperfection modeling, the geometric imperfection is obtained by multiplying the column height by 0.002. In the equivalent notional load method, the notional load is 0.002 times the total gravity load plus the lateral load. In the further reduced tangent modulus method, the program automatically accounts for geometric imperfection effects. Although users can choose any of these three models, the further reduced tangent modulus model is the only one presented herein. The incremental loads are computed by dividing the concentrated load by the scaling factor of 20.

Step 3: Analysis. The three parameters of the connections can be computed by the use of the computer program 3PARA. Corresponding to the input format in Table 2.2, the input data, CONN.DAT, may be generated, as shown in Table 2.10.

Referring to the input instructions (Section 2.5.4), the input data are written in the form shown in Table 2.11. Programs DATAGEN, INPUT, and PAAP are executed sequentially by typing "RUN." The program will continue to analyze with increasing load steps up to the ultimate state.

Step 4: Check of load-carrying capacity. As shown in output file, P.OUT2, the ultimate load-carrying capacities of the load combinations 1 and 2 are 46.2 and 42.9 kip, respectively, at nodes 7–13 (Figure 2.45). Compared to the applied loads, 45.5 and 31.75 kip, the initial member sizes are adequate.

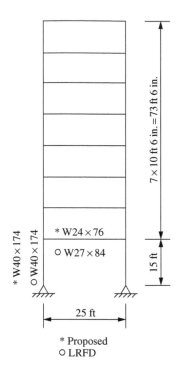

FIGURE 2.43 Comparison of member sizes of the eight-story frame.

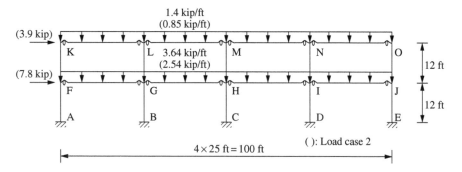

FIGURE 2.44 Configuration and load condition of two-story, four-bay semirigid frame.

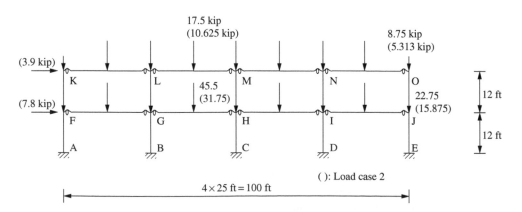

FIGURE 2.45 Concentrated load condition converted form the distributed load for the two-story, one-bay semirigid frame.

Step 5: Check of serviceability. The lateral displacement at roof level corresponding to 1.0W is computed as 0.51 in. from the computer output P.OUT2. The drift ratio is 1/565, which satisfies the limitation 1/400. The preliminary member sizes are satisfactory.

2.6.3.2 Comparison of Results

The member sizes by the advanced analysis and the LRFD method are compared in Figure 2.46. The beam members are one size larger in the advanced analysis method, and the interior columns are one size smaller.

TABLE 2.10 Input Data, CONN.DAT, of Connection for (a) Roof Beam and (b) Floor Beam

		(a) Roof beam			
2	36.0	29,000	2.5	1.25	12
7	0.375	0.875	2.5		
5.5	0.25	0.6875			

		(b) Floor beam			
2	36.0	29.000	2.5	1.25	16
7	0.5625	1.0625	2.5		
8.5	0.3125	0.75			

TABLE 2.11 Input Data, P.DAT, of the Four-Bay, Two-Story Semirigid Frame: (a) Load Case 1 and (b) Load Case 2

```
(a) Four-bay two-story semi-rigid frame (Load case 1)
    3    1
   23    5  100
    2    3    0
   16   26    0
    1   1361.0   607384.0     0.927
    2    446.0    90887.0     1.403
    1      6.16       75.3    20.4   29000.0   36.0   1
    2      6.48      156.0    29.3   29000.0   36.0
    3     11.8       518.0    72.9   29000.0   36.0
    1   11    1    1
    2   12    2    1
    3   13    1    1
    4   14    2    1
    5   15    1    1
    6   16    2    1
    7   17    1    1
    8   18    2    1
    9   19    1    2
   10   20    2    2
   11   21    1    2
   12   22    2    2
   13   23    1    2
   14   24    2    2
   15   25    1    2
   16   26    2    2
    1        0.0     144.0   1    1    6
    2        0.0     144.0   1    2    8
    3        0.0     144.0   1    3   10
    4        0.0     144.0   1    4   12
    5        0.0     144.0   1    5   14
    6        0.0     144.0   1    6   15
    7        0.0     144.0   1    8   17    3    2
   10        0.0     144.0   1   14   23
   11      150.0       0.0   3    6    7    8
   19      150.0       0.0   2   15   16    8
    1    1    1    1    5
    6              -1.1375
    7              -2.2750              7
   14              -1.1375
   15              -0.4375
   16              -0.8750              7
   23              -0.4375
```

Principles of Structural Design

TABLE 2.11 Continued

```
(b) Four-bay two-story semi-rigid frame (Load case 2)
        3     1
       23     5   100
        2     3     0
       16    26     0
        1   1361.0    607384.0      0.927
        2    446.0     90887.0      1.403
        1      6.16        75.3     20.4    29000.0      36.0    1
        2      6.48       156.0     29.3    29000.0      36.0
        3     11.8        518.0     72.9    29000.0      36.0
        1    11     1     1
        2    12     2     1
        3    13     1     1
        4    14     2     1
        5    15     1     1
        6    16     2     1
        7    15     1     1
        8    16     2     1
        9    19     1     2
       10    20     2     2
       11    21     1     2
       12    22     2     2
       13    23     1     2
       14    24     2     2
       15    25     1     2
       16    26     2     2
        1      0.0        144.0     1     1     6
        2      0.0        144.0     1     2     8
        3      0.0        144.0     1     3    10
        4      0.0        144.0     1     4    12
        5      0.0        144.0     1     5    14
        6      0.0        144.0     1     6    15
        7      0.0        144.0     1     8    17     3     2
       10      0.0        144.0     1    14    23
       11    150.0          0.0     3     6     7     8
       19    150.0          0.0     2    15    16     8
        1      1     1     1     5
        6      0.39      -0.7938
        7                -1.5880                      7
       14                -0.7938
       15      0.195     -0.2657
       16                -0.5313                      7
       23                -0.2657
```

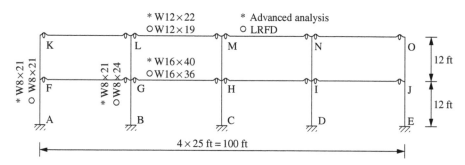

FIGURE 2.46   Comparison of member sizes of the two-story, four-bay semirigid frame.

Glossary

Advanced analysis — Analysis predicting directly the stability of a structural system and its component members and not needing separate member capacity checks.

ASD — Acronym for allowable stress design.

Beam columns — Structural members whose primary function is to carry axial force and bending moment.

Braced frame — Frame in which lateral deflection is prevented by braces or shear walls.

Column — Structural member whose primary function is to carry axial force.

CRC — Acronym for Column Research Council.

Drift — Lateral deflection of a building.

Ductility — Ability of a material to undergo a large deformation without a significant loss in strength.

Factored load — The product of the nominal load and a load factor.

Flexural member — Structural member whose primary function is to carry bending moment.

Geometric imperfection — Unavoidable geometric error during fabrication and erection.

Limit state — A condition in which a structural or structural component becomes unsafe (strength limit state) or unfit for its intended function (serviceability limit state).

Load factor — A factor to account for the unavoidable deviations of the actual load from its nominal value and uncertainties in structural analysis.

LRFD — Acronym for load resistance factor design.

Notional load — Load equivalent to geometric imperfection.

PD — Acronym for plastic design.

Plastic hinge — A yield section of a structural member in which the internal moment is equal to the plastic moment of the cross-section.

Plastic zone — A yield zone of a structural member in which the stress of a fiber is equal to the yield stress.

Refined plastic hinge analysis — Modified plastic hinge analysis accounting for gradual yielding of a structural member.

Resistance factors — A factor to account for the unavoidable deviations of the actual resistance of a member or a structural system from its nominal value.

Second-order analysis — Analysis to use equilibrium equations based on the deformed geometry of a structure under load.

Semirigid connection — Beam-to-column connection whose behavior lies between fully rigid and ideally pinned connection.

Service load — Nominal load under normal usage.

Stability function — Function to account for the bending stiffness reduction due to axial force.

Stiffness — Force required to produce unit displacement.

Unbraced frame — Frame in which lateral deflections are not prevented by braces or shear walls.

References

[1] American Institute of Steel Construction. 2001. *Load and Resistance Factor Design Specification*, 3rd ed., Chicago.

[2] SSRC. 1981. General principles for the stability design of metal structures, Technical Memorandum No. 5, Civil Engineering, ASCE, February, pp. 53–54.

[3] Chen, W.F. and Lui, E.M. 1986. *Structural Stability — Theory and Implementation*, Elsevier, New York.

[4] Kanchanalai, T. 1977. The Design and Behavior of Beam–Columns in Unbraced Steel Frames, AISI Project No. 189, Report No. 2, Civil Engineering/Structures Research Lab., University of Texas at Austin.

[5] Hwa, K. 2003. *Toward Advanced Analysis in Steel Frame Design*, PhD dissertation, Department of Civil and Environmental Engineering, University of Hawaii at Manoa, Honolulu, HI.

[6] Kim, S.E. 1996. *Practical Advanced Analysis for Steel Frame Design*, PhD thesis, School of Civil Engineering, Purdue University, West Lafayette, IN.

[7] Liew, J.Y.R., White, D.W., and Chen, W.F. 1993. Second-order refined plastic hinge analysis of frame design: Part I, *J. Struct. Eng.*, ASCE, 119(11), 3196–3216.

[8] Kishi, N. and Chen, W.F. 1990. Moment–rotation relations of semi-rigid connections with angles, *J. Struct. Eng.*, ASCE, 116(7), 1813–1834.

[9] Kim, S.E. and Chen, W.F. 1996. Practical advanced analysis for steel frame design, *ASCE Structural Congress XIV*, Chicago, Special Proceeding Volume on Analysis and Computation, April, pp. 19–30.

[10] Chen, W.F., Kim, S.E., and Choi, S.H. 2001. Practical second-order inelastic analysis for three-dimensional steel frames, *Steel Struct.*, 1, 213–223.

[11] Kim, S.E. and Lee, D.H. 2002. Second-order distributed plasticity analysis of space steel frames, *Eng. Struct.*, 24, 735–744.

[12] Kim, S.E., Park, M.H., and Choi, S.H. 2001. Direct design of three-dimensional frames using practical advanced analysis, *Eng. Struct.*, 23, 1491–1502.

[13] Avery, P. 1998. *Advanced Analysis of Steel Frames Comprising Non-compact Sections*, PhD thesis, School of Civil Engineering, Queensland University of Technology, Brisbane, Australia.

[14] Kim, S.E. and Lee, J.H. 2001. Improved refined plastic-hinge analysis accounting for local buckling, *Eng. Struct.*, 23, 1031–1042.

[15] Kim, S.E., Lee, J.H., and Park, J.S. 2002. 3D second-order plastic hinge analysis accounting for lateral torsional buckling, *Int. J. Solids Struct.*, 39, 2109–2128.

[16] Wongkeaw, K. and Chen, W.F. 2002. Consideration of out-of-plane buckling in advanced analysis for planar steel frame design, *J. Constr. Steel Res.*, 58, 943–965.

[17] Liew, J.Y.R., White, D.W., and Chen, W.F. 1993. Second-order refined plastic-hinge analysis for frame design: Part 2, *J. Struct. Eng.*, ASCE, 119(11), 3217–3237.

[18] Chen, W.F. and Lui, E.M. 1992. *Stability Design of Steel Frames*, CRC Press, Boca Raton, FL.

[19] Liew, J.Y.R. 1992. *Advanced Analysis for Frame Design*, PhD thesis, School of Civil Engineering, Purdue University, West Lafayette, IN.

[20] Kishi, N., Goto, Y., Chen, W.F., and Matsuoka, K.G. 1993. Design aid of semi-rigid connections for frame analysis, *Eng. J.*, AISC, 4th quarter, 90–107.

[21] ECCS. 1991. *Essentials of Eurocode 3 Design Manual for Steel Structures in Building*, ECCS-Advisory Committee 5, No. 65.

[22] ECCS. 1984. *Ultimate Limit State Calculation of Sway Frames with Rigid Joints*, Technical Committee 8 — Structural stability technical working group 8.2-system, Publication No. 33.

[23] Standards Australia. 1990. *AS4100–1990, Steel Structures*, Sydney, Australia.

[24] Canadian Standard Association. 1994. *Limit States Design of Steel Structures*, CAN/CSA-S16.1-M94.

[25] Canadian Standard Association. 1989. *Limit States Design of Steel Structures*, CAN/CSA-S16.1-M89.

[26] Kim, S.E. and Chen, W.F. 1996. Practical advanced analysis for braced steel frame design, *J. Struct. Eng.*, ASCE, 122(11), 1266–1274.

[27] Kim, S.E. and Chen, W.F. 1996. Practical advanced analysis for unbraced steel frame design, *J. Struct. Eng.*, ASCE, 122(11), 1259–1265.

[28] Maleck, A.E., White, D.W., and Chen, W.F. 1995. Practical application of advanced analysis in steel design, *Proc. 4th Pacific Structural Steel Conf.*, Vol. 1, Steel Structures, pp. 119–126.

[29] White, D.W. and Chen, W.F., Eds. 1993. *Plastic Hinge Based Methods for Advanced Analysis and Design of Steel Frames: An Assessment of the State-of-the-art*, SSRC, Lehigh University, Bethlehem, PA.

[30] Kim, S.E. and Chen, W.F. 1996. Practical advanced analysis for semi-rigid frame design, *Eng. J.*, AISC, 33(4), 129–141.

[31] Kim, S.E. and Chen, W.F. 1996. Practical advanced analysis for frame design — Case study, *SSSS J.*, 6(1), 61–73.

[32] American Institute of Steel Construction. 1994. *Load and Resistance Factor Design Specification*, 2nd ed., Chicago.

[33] American Institute of Steel Construction. 1986. *Load and Resistance Factor Design Specification for Structural Steel Buildings*, Chicago.

[34] LeMessurier, W.J. 1977. A practical method of second order analysis, Part 2 — Rigid Frames. *Eng. J.*, AISC, 2nd quarter, 14(2), 49–67.

[35] Vogel, U. 1985. Calibrating frames, *Stahlbau*, 10, 1–7.

[36] Stelmack, T.W. 1982. *Analytical and Experimental Response of Flexibly-Connected Steel Frames,* MS dissertation, Department of Civil, Environmental, and Architectural Engineering, University of Colorado.

[37] Chen, W.F. and Kim, S.E. 1997. *LRFD Steel Design Using Advanced Analysis,* CRC Press, Boca Raton, FL.

[38] Ad Hoc Committee on Serviceability. 1986. Structural serviceability: a critical appraisal and research needs, *J. Struct. Eng.,* ASCE, 112(12), 2646–2664.

[39] Ellingwood, B.R. 1989. Serviceability Guidelines for Steel Structures, *Eng. J.,* AISC, 26, 1st Quarter, 1–8.

[40] White, D.W. 1993. Plastic hinge methods for advanced analysis of steel frames, *J. Constr. Steel Res.,* 24(2), 121–152.

[41] Cook, R.D., Malkus, D.S., and Plesha, M.E. 1989. *Concepts and Applications of Finite Element Analysis,* 3rd ed., John Wiley & Sons, New York.

[42] Hughes, T.J.R. 1987. *The Finite Element Method: Linear Static and Dynamic Finite Element Analysis,* Prentice Hall, Englewood Cliffs, NJ.

[43] American Institute of Steel Construction. 2002. *Seismic Provisions for Structural Steel Buildings,* Chicago.

[44] Chen, W.F. and Lui, E.M. 2005. *Handbook of Structural Engineering, Second Edition,* CRC Press, Boca Raton, FL.

3

Cold-Formed Steel Structures

Wei-Wen Yu
*Department of Civil Engineering,
University of Missouri,
Rolla, MO*

3.1 Introduction

Cold-formed steel members as shown in Figure 3.1 are widely used in building construction, bridge construction, storage racks, highway products, drainage facilities, grain bins, transmission towers, car bodies, railway coaches, and various types of equipment. These sections are cold-formed from carbon,

FIGURE 3.1   Various shapes of cold-formed steel sections (courtesy of Yu, W.W. 1991).

low-alloy steel, or stainless steel sheet, strip, plate, or flat bar in cold-rolling machines or by press brake or bending brake operations. The thicknesses of such members usually range from 0.0149 in. (0.378 mm) to about 0.25 in. (6.35 mm) even though steel plates and bars as thick as 1 in. (25.4 mm) can be cold-formed into structural shapes.

The use of cold-formed steel members in building construction began around the 1850s in both the United States and Great Britain. However, such steel members were not widely used in buildings in the United States until the 1940s. At present, cold-formed steel members are widely used as construction materials worldwide.

Compared with other materials such as timber and concrete, cold-formed steel members can offer the following advantages: (1) lightness, (2) high strength and stiffness, (3) ease of prefabrication and mass production, (4) fast and easy erection and installation, and (5) economy in transportation and handling, to name a few.

From the structural design point of view, cold-formed steel members can be classified into two major types: (1) individual structural framing members (Figure 3.2) and (2) panels and decks (Figure 3.3).

In view of the fact that the major function of the individual framing members is to carry load, structural strength and stiffness are the main considerations in design. The sections shown in Figure 3.2 can be used as primary framing members in buildings up to four or five stories in height. In tall multistory buildings, the main framing is typically of heavy hot-rolled shapes and the secondary elements such as wall studs, joists, decks, or panels may be of cold-formed steel members. In this case, the heavy hot-rolled steel shapes and the cold-formed steel sections supplement each other.

The cold-formed steel sections shown in Figure 3.3 are generally used for roof decks, floor decks, wall panels, and siding material in buildings. Steel decks not only provide structural strength to carry loads, but they also provide a surface on which flooring, roofing, or concrete fill can be applied as shown in Figure 3.4. They can also provide space for electrical conduits. The cells of cellular panels can also be used

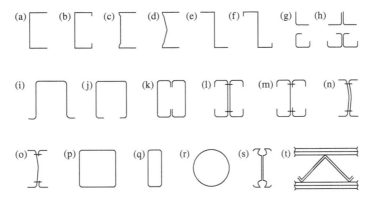

FIGURE 3.2 Cold-formed steel sections used for structural framing (courtesy of Yu, W.W. 1991).

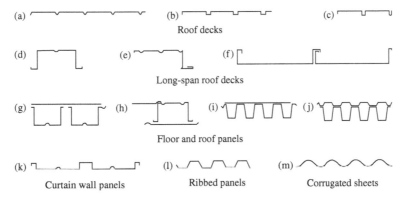

FIGURE 3.3 Decks, panels, and corrugated sheets (courtesy of Yu, W.W. 1991).

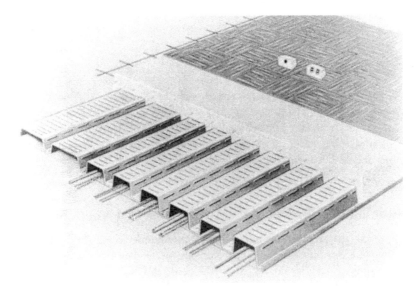

FIGURE 3.4 Cellular floor decks (courtesy of Yu, W.W. 1991).

as ducts for heating and air conditioning. For *composite slabs*, steel decks are used not only as formwork during construction, but also as reinforcement of the composite system after the concrete hardens. In addition, load-carrying panels and decks not only withstand loads normal to their surface, but they can also act as shear diaphragms to resist forces in their own planes if they are adequately interconnected to each other and to supporting members.

During recent years, cold-formed steel sections have been widely used in residential construction and pre-engineered metal buildings for industrial, commercial, and agricultural applications. Metal building systems are also used for community facilities such as recreation buildings, schools, and churches. For additional information on cold-formed steel structures, see Yu [1], Rhodes [2], Hancock et al. [3], and Ghersi et al. [4].

3.2 Design Standards

Design standards and recommendations are now available in Australia [5], Austria [6], Canada [7], Czech Republic [8], Finland [9], France [10], Germany [11], India [12], Italy [13], Japan [14], Mexico [7], The Netherlands [15], New Zealand [16], The People's Republic of China [17], The Republic of South Africa [18], Sweden [19], Romania [20], the United Kingdom [21], the United States [7], Russia [22], and elsewhere. Since 1975, the European Convention for Constructional Steelwork has prepared several documents for the design and testing of cold-formed sheet steel used in buildings. In 1996, Eurocode 3 [23] provided design information for cold-formed steel members.

This chapter presents discussions on the design of cold-formed steel structural members for use in buildings. It is mainly based on the current American Iron and Steel Institute (AISI) North American Specification [7] for allowable strength design (ASD), load and resistance factor design (LRFD), and limit states design (LSD). It should be noted that in addition to the AISI specification, many trade associations and professional organizations have issued special design and construction requirements for using cold-formed steel members as floor and roof decks [24], roof trusses [25], open web steel joists [26], transmission poles and towers [27], storage racks [28], shear diaphragms [7,29], composite slabs [30], metal buildings [31], light framing systems [32–34], guardrails, structural supports for highway signs, luminaries, and traffic signals [36], automotive structural components [37], and others. For the design of cold-formed stainless steel structural members, see SEI/ASCE Standard 8-02 [35].

3.3 Design Bases

For cold-formed steel design, three design approaches are being used in the North American Specification [7]. They are (1) ASD, (2) LRFD, and (3) LSD. The ASD and LRFD methods are used in the United States and Mexico, while the LSD method is used in Canada. The unit systems used in the North American Specification are (1) U.S. customary units (force in kip and length in inches), (2) SI units (force in newtons and length in millimeters), and (3) MKS units (force in kilograms and length in centimeters).

3.3.1 Allowable Strength Design (United States and Mexico)

In the ASD approach, the *required allowable strengths* (moments, axial forces, and shear forces) in structural members are computed by accepted methods of structural analysis for all applicable load combinations of nominal loads as stipulated by the applicable building code. In the absence of an applicable building code, the nominal loads and load combinations should be those stipulated in the American Society of Civil Engineers (ASCE) Standard ASCE 7 [38].

The required allowable strengths should not exceed the allowable *design strengths* permitted by the applicable design standard. The allowable design strength is determined by dividing the *nominal strength* by a factor of safety as follows:

$$R_a = R_n/\Omega \qquad (3.1)$$

where R_a is the allowable design strength, R_n is the nominal strength, and Ω is the *factor of safety*. For the design of cold-formed steel structural members using the AISI ASD method, the factors of safety are given in Table 3.1. For details, see the AISI Specification [7].

3.3.2 Load and Resistance Factor Design (United States and Mexico)

Two types of *limit states* are considered in the LRFD method. They are (1) the limit state of strength required to resist the extreme loads during its life and (2) the limit state of serviceability for a structure to perform its intended function.

For the limit state of strength, the general format of the LRFD method is expressed by the following equation:

$$R_u \leq \phi R_n \qquad (3.2)$$

where $R_u = \sum \gamma_i Q_i$ is the required strength, ϕR_n is the design strength, γ_i is the load factor, Q_i is the load effect, ϕ is the resistance factor, and R_n is the nominal strength.

The structure and its components should be designed so that the design strengths, ϕR_n, are equal to or greater than the required strengths, $\sum \gamma_i Q_i$, which are computed on the basis of the factored nominal loads and load combinations as stipulated by the applicable building code or, in the absence of an applicable code, as stipulated in ASCE 7 [38].

In addition, the following LRFD criteria can be used for roof and floor composite construction using cold-formed steel:

$$1.2D_s + 1.6C_w + 1.4C \qquad (3.3)$$

where D_s is the weight of the steel deck, C_w is the weight of wet concrete during construction, and C is the construction load, including equipment, workmen, and formwork, but excluding the weight of the wet concrete. Table 3.1 lists the ϕ factors, which are used for the LRFD and LSD methods for the design of cold-formed steel members and connections. For details, see the AISI Specification [7].

3.3.3 Limit States Design (Canada)

The methodology for the LSD method is the same as for the LRFD method, except that the load factors, load combinations, target reliability indices, and assumed live-to-dead ratio used in the development of the design criteria are different. In addition, a few different terms are used in the LSD method.

Like the LRFD method, the LSD method specifies that the structural members and connections should be designed to have factored resistance equal to or greater than the effect of factored loads as follows:

$$\phi R_n \geq R_f \qquad (3.4)$$

where R_f is the effect of factored loads, R_n is the nominal resistance, ϕ is the resistance factor, and ϕR_n is the factored resistance.

For the LSD method, the load factors and load combinations used for the design of cold-formed steel structures are based on the National Building Code of Canada. For details, see appendix B of the North American Specification [7]. The resistance factors are also listed in Table 3.1.

TABLE 3.1 Factors of Safety, Ω, and Resistance Factors, ϕ, used in the North American Specification [7]

Type of strength	ASD factor of safety, Ω	LRFD resistance factor, ϕ	LSD resistance factor, ϕ
Tension members			
For yielding	1.67	0.90	0.90
For fracture away from the connection	2.00	0.75	0.75
For fracture at the connection (see connections)			
Flexural members			
Bending strength			
For sections with stiffened or partially stiffened compression flanges	1.67	0.95	0.90
For sections with unstiffened compression flanges	1.67	0.90	0.90
Laterally unbraced beams	1.67	0.90	0.90
Beams having one-flange through fastened-to-deck or sheathing (C- or Z-sections)	1.67	0.90	0.90
Beams having one-flange fastened to a standing seam roof system	1.67	0.90	a
Web design			
Shear strength	1.60	0.99	0.80
Web crippling			
Built-up sections	1.65–2.00	0.75–0.90	0.60–0.80
Single-web channel and C-sections	1.65–2.00	0.75–0.90	0.65–0.80
Single-web Z-sections	1.65–2.00	0.75–0.90	0.65–0.80
Single-hat sections	1.70–2.00	0.75–0.90	0.65–0.75
Multiweb deck sections	1.65–2.25	0.65–0.90	0.55–0.80
Stiffeners			
Transverse stiffeners	2.00	0.85	0.80
Concentrically loaded compression members	1.80	0.85	0.80
Combined axial load and bending			
For tension	1.67	0.95	0.90
For compression	1.80	0.85	0.80
For bending	1.67	0.90–0.95	0.90
Closed cylindrical tubular members			
Bending strength	1.67	0.95	0.90
Axial compression	1.80	0.85	0.80
Wall studs and wall assemblies			
Wall studs in compression	1.80	0.85	0.80
Wall studs in bending	1.67	0.90–0.95	0.90
Diaphragm construction	2.00–3.00	0.50–0.65	0.50
Welded connections			
Groove welds			
Tension or compression	1.70	0.90	0.80
Shear (welds)	1.90	0.80	0.70
Shear (base metal)	1.70	0.90	0.80
Arc spot welds			
Welds	2.55	0.60	0.50
Connected part	2.20–3.05	0.50–0.70	0.40–0.60
Minimum edge distance	2.20–2.55	0.60–0.70	0.50–0.60
Tension	2.50–3.00	0.50–0.60	0.40–0.50
Arc seam welds			
Welds	2.55	0.60	0.50
Connected part	2.55	0.60	0.50
Fillet welds			
Longitudinal loading (connected part)	2.55–3.05	0.50–0.60	0.40–0.50
Transverse loading (connected part)	2.35	0.65	0.60
Welds	2.55	0.60	0.50
Flare groove welds			
Transverse loading (connected part)	2.55	0.60	0.50
Longitudinal loading (connected part)	2.80	0.55	0.45
Welds	2.55	0.60	0.50

TABLE 3.1   Continued

Type of strength	ASD factor of safety, Ω	LRFD resistance factor, ϕ	LSD resistance factor, ϕ
Resistance welds	2.35	0.65	0.55
Shear lag effect	2.50	0.60	0.50
Bolted connections			
Shear, spacing, and edge distance	2.00–2.22	0.60–0.70	a
Fracture in net section (shear lag)	2.00–2.22	0.55–0.65	0.55
Bearing strength	2.22–2.50	0.60–0.65	0.50–0.55
Shear strength of bolts	2.40	0.65	0.55
Tensile strength of bolts	2.00–2.25	0.75	0.65
Screw connections	3.00	0.50	0.40
Rupture			
Shear rupture	2.00	0.75	a
Block shear rupture			
For bolted connection	2.22	0.65	a
For welded connections	2.50	0.60	a

[a] See appendix B of the North American Specification for the provisions applicable to Canada [7].

3.4   Materials and Mechanical Properties

In the AISI Specification [7], 15 different steels are presently listed for the design of cold-formed steel members. Table 3.2 lists steel designations, American Society for Testing Materials (ASTM) designations, and *yield points*, tensile strengths, and elongations for these steels. From a structural standpoint, the most important properties of steel are as follows:

1. Yield point or yield strength, F_y
2. Tensile strength, F_u
3. Stress–strain relationship
4. Modulus of elasticity, tangent modulus, and shear modulus
5. Ductility
6. Weldability
7. Fatigue strength

In addition, formability, durability, and toughness are also important properties of cold-formed steel.

3.4.1   Yield Point, Tensile Strength, and Stress–Strain Relationship

As listed in Table 3.2, the yield points or yield strengths of all 15 different steels range from 24 to 80 ksi (166 to 552 MPa or 1687 to 5624 kg/cm^2). The tensile strengths of the same steels range from 42 to 100 ksi (290 to 690 MPa or 2953 to 7030 kg/cm^2). The ratios of the tensile strength to yield point vary from 1.08 to 1.88. As far as the stress–strain relationship is concerned, the stress–strain curve can be either the sharp-yielding type (Figure 3.5a) or the gradual-yielding type (Figure 3.5b).

3.4.2   Strength Increase from Cold Work of Forming

The mechanical properties (yield point, tensile strength, and ductility) of cold-formed steel sections, particularly at the corners, are sometimes substantially different from those of the flat steel sheet, strip, plate, or bar before forming. This is because the cold-forming operation increases the yield point and tensile strength and at the same time decreases the ductility. The effects of cold work on the mechanical properties of corners usually depend on several parameters. The ratios of tensile strength to yield point, F_u/F_y, and inside bend radius to thickness, R/t, are considered to be the most important factors to affect the change in mechanical properties of cold-formed steel sections. Design equations are given in the AISI

TABLE 3.2 Mechanical Properties of Steels Referred to in the AISI North American Specification[a,b]

Steel designation	ASTM designation[c]	Yield point, F_y (ksi)	Tensile strength, F_u (ksi)	Elongation (%) In 2-in. gage length	In 8-in. gage length
Structural steel	A36	36	58–80	23	—
High-strength low-alloy structural steel	A242				
	$\frac{3}{4}$ in. and below	50	70	—	18
	$\frac{3}{4}$ in. to $1\frac{1}{2}$ in.	46	67	21	18
Low and intermediate tensile strength carbon plates, shapes, and bars	A283 Grade A	24	45–60	30	27
	B	27	50–65	28	25
	C	30	55–75	25	22
	D	33	60–80	23	20
Cold-formed welded and seamless carbon steel structural tubing in rounds and shapes	A500				
	Round tubing				
	A	33	45	25	—
	B	42	58	23	—
	C	46	62	21	—
	D	36	58	23	—
	Shaped tubing				—
	A	39	45	25	—
	B	46	58	23	—
	C	50	62	21	—
	D	36	58	23	—
Structural steel with 50-ksi minimum yield point	A529 Grade 50	50	70–100	21	—
	55	55	70–100	20	—
High-strength low-alloy columbium–vanadium steels of structural quality	A572 Grade 42	42	60	24	20
	50	50	65	21	18
	60	60	75	18	16
	65	65	80	17	15
High-strength low-alloy structural steel with 50-ksi minimum yield point	A588	50	70	21	18
Hot-rolled and cold-rolled high-strength low-alloy steel sheet and strip with improved corrosion resistance	A606				
	Hot-rolled as rolled coils; annealed, or normalized; cold rolled	45	65	22	—
	Hot-rolled as rolled cut lengths	50	70	22	—
Zinc-coated steel sheets of structural quality	A653				
	SS Grade 33	33	45	20	—
	37	37	52	18	—
	40	40	55	16	—
	50 Class 1	50	65	12	—
	50 Class 3	50	70	12	—
	HSLAS Grade 40	40	50	22	—
	50	50	60	20	—
	60	60	70	16	—
	70	70	80	12	—
	80	80	90	10	—

TABLE 3.2 Continued

Steel designation	ASTM designation[c]	Yield point, F_y (ksi)	Tensile strength, F_u (ksi)	Elongation (%)	
				In 2-in. gage length	In 8-in. gage length
Aluminum–zinc alloy	A792 Grade 33	33	45	20	—
coated by the hot-dip	37	37	52	18	—
process (general	40	40	55	16	—
requirements)	50	50	65	12	—
Cold-formed welded and	A847	50	70	19	—
seamless high-strength,					
low-alloy structural					
tubing with improved					
atmospheric corrosion					
resistance					
Zinc–5% aluminum	A875				
alloy-coated steel sheet	SS Grade 33	33	45	20	—
by the hot-dip process	37	37	52	18	—
	40	40	55	16	—
	40	40	55	16	—
	50 Class 3	50	70	12	—
	HSLAS Type A Grade 50	50	60	20	—
	60	60	70	16	—
	70	70	80	12	—
	80	80	90	10	—
	HSLAS Type B Grade 50	50	60	22	—
	60	60	70	18	—
	70	70	80	14	—
	80	80	90	12	—
Metal- and nonmetal-	A1003				
coated carbon steel sheet	ST Grade 33H	33	d	10	
	37 H	37	d	10	—
	40 H	40	d	10	—
	50 H	50	d	10	—
Cold-rolled steel sheet,	A1008				
carbon structural, high-	SS Grade 25	25	42	26	—
strength low-alloy	30	30	45	24	—
with improved	33 Types 1 and 2	33	48	22	—
formability	40 Types 1 and 2	40	52	20	—
	HSLAS Grade 45 Class 1	45	60	22	—
	45 Class 2	45	55	22	—
	50 Class 1	50	65	20	—
	50 Class 2	50	60	20	—
	55 Class 1	55	70	18	—
	55 Class 2	55	65	18	—
	60 Class 1	60	75	16	—
	60 Class 2	60	70	16	—
	65 Class 1	65	80	15	—
	65 Class 2	65	75	15	—
	70 Class 1	70	85	14	—
	70 Class 2	70	80	14	—

TABLE 3.2   Continued

Steel designation	ASTM designation[c]	Yield point, F_y (ksi)	Tensile strength, F_u (ksi)	Elongation (%) In 2-in. gage length	Elongation (%) In 8-in. gage length
	HSLAS-F Grade 50	50	60	22	—
	60	60	70	18	—
	70	70	80	16	—
	80	80	90	14	—
Hot-rolled steel sheet and strip, carbon, structural, high-strength low-alloy with improved formability	A1011				
	SS Grade 30	30	49	21–25	19[e]
	33	33	52	18–23	18[e]
	36 Type 1	36	53	17–22	17[e]
	36 Type 2	36	58–80	16–21	16[e]
	40	40	55	15–21	16[e]
	45	45	60	13–19	14[e]
	50	50	65	11–17	12[e]
	55	55	70	9–15	10[e]
	HSLAS Grade 45 Class 1	45	60	23–25	—
	45 Class 2	45	55	23–25	—
	50 Class 1	50	65	20–22	—
	50 Class 2	50	60	20–22	—
	55 Class 1	55	70	18–20	—
	55 Class 2	55	65	18–20	—
	60 Class 1	60	75	16–18	—
	60 Class 2	60	70	16–18	—
	65 Class 1	65	80	14–16	—
	65 Class 2	65	75	14–16	—
	70 Class 1	70	85	12–14	—
	70 Class 2	70	80	12–14	—
	HSLAS-F Grade 50	50	60	22–24	
	60	60	70	20–22	—
	70	70	80	18–20	—
	80	80	90	16–18	—

[a] The tabulated values are based on ASTM standards.
[b] 1 in. = 25.4 mm; 1 ksi = 6.9 MPa = 70.3 kg/cm^2.
[c] Structural Grade 80 of A653, A875, and A1008 steel and Grade 80 of A792 are allowed in AISI Specification under special conditions. For these grades, F_y = 80 ksi, F_u = 82 ksi, and elongations are unspecified. See AISI Specification for reduction of yield point and tensile strength.
[d] For type H of A1003 steel, the minimum tensile strength is not specified. The ratio of tensile strength to yield strength should not be less than 1.08.
[e] For A1011 steel, the specified minimum elongation in 2-in. gage length varies with the thickness of the steel sheet and strip. The elongation in 8-in. gage length is for thickness below 0.23 in.

Specification [7] for computing the tensile yield point of corners and the average full-section tensile yield point for design purposes.

3.4.3 Modulus of Elasticity, Tangent Modulus, and Shear Modulus

The strength of cold-formed steel members that are governed by buckling depends not only on the yield point but also on the modulus of elasticity, E, and the tangent modulus, E_t. A value of E = 29,500 ksi (203 GPa or 2.07 × 10^6 kg/cm^2) is used in the AISI Specification for the design of cold-formed steel structural members. This E value is slightly larger than the value of 29,000 ksi (200 GPa or 2.04 × 10^6 kg/cm^2), which is used in the American Institute of Steel Construction (AISC) Specification for the design of hot-rolled shapes. The tangent modulus is defined by the slope of the stress–strain curve at any given stress level as shown in Figure 3.5b. For sharp-yielding steels, $E_t = E$ up to the yield, but with gradual-yielding steels, $E_t = E$ only up to the proportional limit, f_{pr} (Figure 3.5b). Once the stress exceeds the proportional

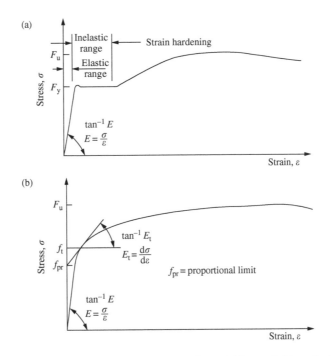

FIGURE 3.5 Stress–strain curves of steel sheet or strip: (a) sharp-yielding and (b) gradual-yielding (courtesy of Yu, W.W. 1991).

limit, the tangent modulus E_t becomes progressively smaller than the initial modulus of elasticity. For cold-formed steel design, the shear modulus is taken as $G = 11,300$ ksi (77.9 GPa or 0.794×10^6 kg/cm^2) according to the AISI Specification.

3.4.4 Ductility

According to the AISI Specification, the ratio F_u/F_y for the steels used for structural framing members should not be less than 1.08, and the total elongation should not be less than 10% for a 2-in. (50.8 mm) gage length. If these requirements cannot be met, the following limitations should be satisfied when such a material is used for purlins and girts: (1) local elongation in a $\frac{1}{2}$-in. (12.7 mm) gage length across the fracture should not be less than 20% and (2) uniform elongation outside the fracture should not be less than 3%. It should be noted that the required ductility for cold-formed steel structural members depends mainly on the type of application and the suitability of the material. The same amount of ductility that is considered necessary for individual framing members may not be needed for roof panels, siding, and similar applications. For this reason, even though Structural Grade 80 of ASTM A653, A875, and 1008 steel, and Grade 80 of A792 steel do not meet the AISI requirements for the F_u/F_y ratio and the elongation, these steels can be used for roofing, siding, and similar applications provided that (1) the yield strength, F_y, used for design is taken as 75% of the specified minimum yield point or 60 ksi (414 MPa or 4218 kg/cm^2), whichever is less, and (2) the tensile strength, F_u, used for design is taken as 75% of the specified minimum tensile stress or 62 ksi (427 MPa or 4359 kg/cm^2). For multiple web configurations a reduced yield point is permitted by the AISI Specification [7] for determining the nominal flexural strength of the section on the basis of the initiation of yielding.

3.5 Element Strength

For cold-formed steel members, the width to thickness ratios of individual elements are usually large. These thin elements may buckle locally at a stress level lower than the yield point of steel when they are

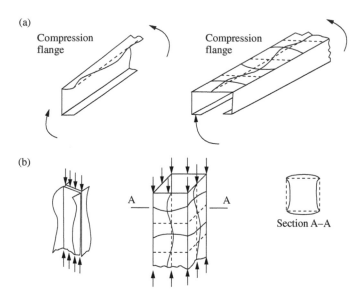

FIGURE 3.6 Local buckling of compression elements: (a) beams and (b) columns (courtesy of Yu, W.W. 1991).

subjected to compression in flexural bending and axial compression as shown in Figure 3.6. Therefore, for the design of such thin-walled sections, *local buckling* and postbuckling strength of thin elements have often been the major design considerations. In addition, shear buckling and web crippling should also be considered in the design of beams.

3.5.1 Maximum Flat Width to Thickness Ratios

In cold-formed steel design, the maximum *flat width to thickness ratio, w/t,* for flanges is limited to the following values in the AISI Specification [7]:

1. *Stiffened compression element* having one longitudinal edge connected to a web or flange element, the other stiffened by
   - Simple lip — 60
   - Any other kind of adequate stiffener — 90
2. *Stiffened compression element* with both longitudinal edges connected to other stiffened element — 500
3. *Unstiffened compression element* — 60

For the design of beams, the maximum depth to thickness ratio, h/t, for webs is:

1. Unreinforced webs, $(h/t)_{max} = 200$
2. Webs that are provided with transverse stiffeners:
   - Using bearing stiffeners only, $(h/t)_{max} = 260$
   - Using bearing stiffeners and intermediate stiffeners, $(h/t)_{max} = 300$

3.5.2 Stiffened Elements under Uniform Compression

The strength of a stiffened compression element such as the compression flange of a hat section is governed by yielding if its w/t ratio is relatively small. It may be governed by local buckling as shown in Figure 3.7 at a *stress* level less than the yield point if its w/t ratio is relatively large.

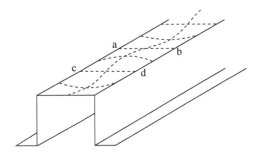

FIGURE 3.7   Local buckling of stiffened compression flange of hat-shaped beam.

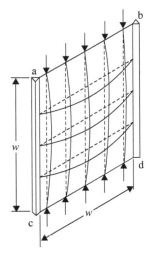

FIGURE 3.8   Postbuckling strength model (courtesy of Yu, W.W. 1991).

The elastic local buckling stress, F_{cr}, of simply supported square plates and long plates can be determined as follows:

$$F_{cr} = \frac{k\pi^2 E}{12(1 - \mu^2)(w/t)^2} \tag{3.5}$$

where k is the local buckling coefficient, E is the modulus of elasticity of steel $= 29.5 \times 10^3$ ksi (203 GPa or 2.07×10^6 kg/cm^2), w is the width of the plate, t is the thickness of the plate, and μ is the Poisson's ratio $= 0.3$.

It is well known that stiffened compression elements will not collapse when the local buckling stress is reached. An additional load can be carried by the element after buckling by means of a redistribution of stress. This phenomenon is known as postbuckling strength and is most pronounced for elements with large w/t ratios.

The mechanism of the postbuckling action can easily be visualized from a square plate model as shown in Figure 3.8 [39]. It represents the portion abcd of the compression flange of the hat section illustrated in Figure 3.7. As soon as the plate starts to buckle, the horizontal bars in the grid of the model will act as tie rods to counteract the increasing deflection of the longitudinal struts.

In the plate, the stress distribution is uniform prior to its buckling. After buckling, a portion of the prebuckling load of the center strip transfers to the edge portion of the plate. As a result, a nonuniform stress distribution is developed, as shown in Figure 3.9. The redistribution of stress continues until the stress at the edge reaches the yield point of steel and then the plate begins to fail.

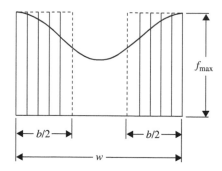

FIGURE 3.9 Stress distribution in stiffened compression elements.

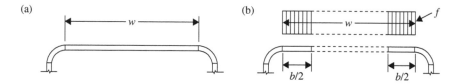

FIGURE 3.10 Effective design width of stiffened compression elements: (a) actual element and (b) effective element, b and stress, f, on effective elements.

For cold-formed steel members, a concept of "effective width" has long been used for practical design. In this approach, instead of considering the nonuniform distribution of stress over the entire width of the plate w, it is assumed that the total load is carried by a fictitious effective width b, subjected to a uniformly distributed stress equal to the edge stress f_{max}, as shown in Figure 3.9. The width b is selected so that the area under the curve of the actual nonuniform stress distribution is equal to the sum of the two parts of the equivalent rectangular shaded area with a total width b and an intensity of stress equal to the edge stress f_{max}. Based on the research findings of von Karman, Sechler, Donnell, and Winter [40], the following equations have been developed in the AISI Specification for computing the *effective design width*, b, for stiffened elements under uniform compression [7].

3.5.2.1 Strength Determination

$$\text{When } \lambda \leq 0.673, \quad b = w \tag{3.6}$$

$$\text{When } \lambda > 0.673, \quad b = \rho w \tag{3.7}$$

where b is the effective design width of uniformly compressed element for strength determination (Figure 3.10), w is the flat width of compression element, and ρ is the reduction factor determined from Equation 3.8:

$$\rho = (1 - 0.22/\lambda)/\lambda \leq 1 \tag{3.8}$$

where λ is the plate slenderness factor determined from Equation 3.9:

$$\lambda = \sqrt{f/F_{cr}} = (1.052/\sqrt{k})(w/t)(\sqrt{f/E}) \tag{3.9}$$

where k is equal to 4.0 is the plate buckling coefficient for stiffened elements supported by a web on each longitudinal edge as shown in Figure 3.10, t is the thickness of compression element, E is the modulus of elasticity, and f is the maximum compressive edge stress in the element without considering the factor of safety.

3.5.2.2 Serviceability Determination

For serviceability determination, Equations 3.6 through 3.9 can also be used for computing the effective design width of compression elements, except that the compressive stress should be computed on the basis of the effective section at the load for which serviceability is determined.

The relationship between ρ and λ according to Equation 3.8 is shown in Figure 3.11.

EXAMPLE 3.1

Calculate the effective width of the compression flange of the box section (Figure 3.12) to be used as a beam bending about the *x*-axis. Use $F_y = 33$ ksi. Assume that the beam webs are fully effective and that the bending moment is based on initiation of yielding.

Solution

Because the compression flange of the given section is a uniformly compressed stiffened element, which is supported by a web on each longitudinal edge, the effective width of the flange for strength determination can be computed by using Equations 3.6 through 3.9 with $k = 4.0$.

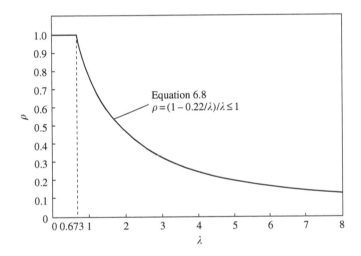

FIGURE 3.11 Reduction factor, ρ, versus slenderness factor, λ (courtesy of Yu, W.W. 1991).

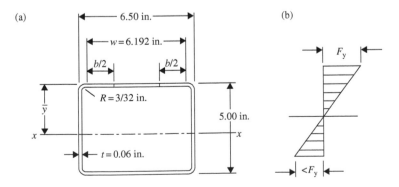

FIGURE 3.12 Example 3.1: (a) tubular section and (b) stress distribution for yield moment (courtesy of Yu, W.W. 1991).

Assume that the bending strength of the section is based on initiation of yielding, $\bar{y} \geq 2.50$ in. Therefore, the slenderness factor λ for $f = F_y$ can be computed from Equation 3.9, that is,

$k \quad = 4.0$

$w \quad = 6.50 - 2\,(R + t) = 6.192$ in.

$w/t = 103.2$

$f \quad = 33$ ksi

$\lambda \quad = \sqrt{f/F_{cr}} = (1.052/\sqrt{k})(w/t)\sqrt{f/E}$

$\quad = (1.052/\sqrt{4.0})(103.2)\sqrt{33/29{,}500} = 1.816$

Since $\lambda > 0.673$, use Equations 3.7 and 3.8 to compute the effective width, b, as follows:

$$b = \rho w = [(1 - 0.22/\lambda)/\lambda]w$$
$$= [(1 - 0.22/1.816)/1.816](6.192) = 3.00 \text{ in.}$$

3.5.3 Stiffened Elements with Stress Gradient

When a flexural member is subjected to bending moment, the beam web is under the stress gradient condition (Figure 3.13), in which the compression portion of the web may buckle due to the compressive stress caused by bending. The effective width of the beam web can be determined from the following AISI provisions.

3.5.3.1 Strength Determination

The effective widths, b_1 and b_2, as shown in Figure 3.13, should be determined from the following procedure:

1. Calculate the effective width b_e on the basis of Equations 3.6 through 3.9 with f_1 substituted for f and with k determined as follows:

$$k = 4 + 2(1 + \psi)^3 + 2(1 + \psi) \tag{3.10}$$

where

$$\psi = |f_2/f_1| \quad \text{(absolute value)} \tag{3.11}$$

2. For $h_o/b_o \leq 4$

$$b_1 = b_e/(3 + \psi) \tag{3.12}$$

$$b_2 = b_e/2 \quad \text{when } \psi > 0.236 \tag{3.13a}$$

$$b_2 = b_e - b_1 \quad \text{when } \psi \leq 0.236 \tag{3.13b}$$

where h_o is the out-to-out depth of web and b_o is the out-to-out width of the compression flange.

3. For $h_o/b_o > 4$

$$b_1 = b_e/(3 + \psi) \tag{3.12}$$

$$b_2 = b_e/(1 + \psi) - b_1 \tag{3.14}$$

The value of $(b_1 + b_2)$ should not exceed the compression portion of the web calculated on the basis of effective section.

For the effective width of b_1 and b_2 for C-section webs with holes under stress gradient, see the AISI Specification [7].

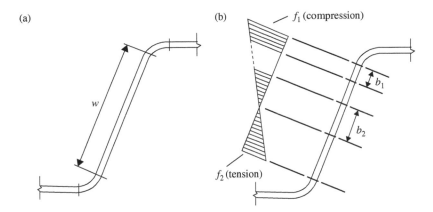

FIGURE 3.13 Stiffened elements with stress gradient: (a) actual element and (b) effective element and stress on effective elements.

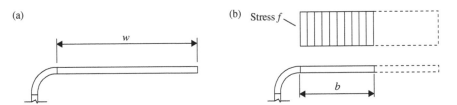

FIGURE 3.14 Effective design width of unstiffened compression elements: (a) actual element and (b) effective element and stress on effective elements.

3.5.3.2 Serviceability Determination

The effective widths used in determining serviceability should be determined as above, except that f_{d1} and f_{d2} are substituted for f_1 and f_2, where f_{d1} and f_{d2} are the computed stresses f_1 and f_2 as shown in Figure 3.13 based on the effective section at the load for which serviceability is determined.

3.5.4 Unstiffened Elements under Uniform Compression

The effective width of unstiffened elements under uniform compression as shown in Figure 3.14 can also be computed by using Equations 3.6 through 3.9, except that the value of k should be taken as 0.43 and the flat width w is measured as shown in Figure 3.14.

3.5.5 Uniformly Compressed Elements with an Edge Stiffener

The following equations can be used to determine the effective width of the uniformly compressed elements with an edge stiffener as shown in Figure 3.15.

3.5.5.1 Strength Determination

For $w/t \leq 0.328S$:

$$I_a = 0 \quad \text{(no edge stiffener needed)} \tag{3.15}$$

$$b = w \tag{3.16}$$

$$b_1 = b_2 = w/2 \tag{3.17}$$

$$d_s = d_s' \quad \text{for simple lip stiffener} \tag{3.18}$$

$$A_s = A_s' \quad \text{for other stiffener shapes} \tag{3.19}$$

(a)

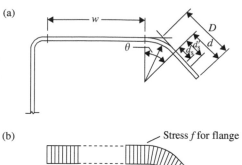

(b)

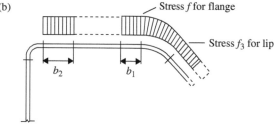

FIGURE 3.15 Compression elements with an edge stiffener.

For $w/t > 0.328S$:

$$b_1 = b/2 \, (R_1) \tag{3.20}$$
$$b_2 = b - b_1 \tag{3.21}$$
$$d_s = d_s' R_I \quad \text{for simple lip stiffeners} \tag{3.22}$$
$$A_s = A_s' R_I \quad \text{for other stiffener shapes} \tag{3.23}$$

where

$$(R_1) = I_s/I_a \leq 1 \tag{3.24}$$
$$I_a = 399t^4 \left[\frac{w/t}{S} - 0.328\right]^3 \leq t^4 \left[115\frac{w/t}{S} + 5\right] \tag{3.25}$$
$$S = 1.28\sqrt{E/f} \tag{3.26}$$

The effective width, b, is calculated in accordance with Equations 3.6 through 3.9 with k determined as follows.

For simple lip edge stiffener ($140° \geq \theta \geq 40°$)

$D/w \leq 0.25$:

$$k = 3.57(R_I)^n + 0.43 \leq 4 \tag{3.27}$$

$0.25 < D/w \leq 0.8$:

$$k = (4.82 - 5D/w)(R_I)^n + 0.43 \leq 4 \tag{3.28}$$

For other edge stiffener shapes

$$k = 3.57(R_I)^n + 0.43 \leq 4 \tag{3.29}$$

where

$$n = \left[0.582 - \frac{w/t}{4S}\right] \geq \frac{1}{3} \tag{3.30}$$

In the above equations, k is the plate buckling coefficient, d, w, and D are the dimensions shown in Figure 3.15, d_s is the reduced effective width of the stiffener, d_s' is the effective width of the stiffener calculated as unstiffened element under uniform compression, and b_1 and b_2 are the effective widths shown in Figure 3.15, A_s is the reduced area of the stiffener, I_a is the adequate moment of inertia of the stiffener, so that each component element will behave as a stiffened element, and I_s and A'_s are the moments of inertia of the full-section of the stiffener, about its own centroidal axis parallel to the element to be stiffened, and the effective area of the stiffener, respectively.

For the stiffener shown in Figure 3.15

$$I_s = (d^3 t \sin^2\theta)/12 \tag{3.31}$$
$$A'_s = d'_s t \tag{3.32}$$

3.5.5.2 Serviceability Determination

The effective width, b_d, used in determining serviceability is calculated as in Section 3.5.5.1, except that f_d is substituted for f.

3.5.6 Uniformly Compressed Elements with Intermediate Stiffeners

The effective width of uniformly compressed elements with intermediate stiffeners can also be determined from the AISI Specification, which includes separate design rules for compression elements with only one intermediate stiffener and compression elements with more than one intermediate stiffener.

3.5.6.1 Uniformly Compressed Elements with One Intermediate Stiffener

The following equations can be used to determine the effective width of the uniformly compressed elements with one intermediate stiffener as shown in Figure 3.16.

3.5.6.1.1 Strength Determination
For $b_o/t \le S$

$$I_a = 0 \quad \text{(no intermediate stiffener required)} \tag{3.33}$$
$$b = w \tag{3.34}$$
$$A_s = A'_s \tag{3.35}$$

For $b_o/t > S$

$$A_s = A'_s(R_I) \tag{3.36}$$

where

$$R_I = I_s/I_a \le 1 \tag{3.37}$$

where for $S < b_o/t < 3S$

$$I_a = t^4 \left[50 \frac{b_o/t}{S} - 50 \right] \tag{3.38}$$

and for $b_o/t \ge 3S$

$$I_a = t^4 \left[128 \frac{b_o/t}{S} - 285 \right] \tag{3.39}$$

The effective width, b, is calculated in accordance with Equations 3.6 through 3.9 with k determined as follows:

$$k = 3(R_I)^n + 1 \tag{3.40}$$

where

$$n = \left[0.583 - \frac{b_o/t}{12S} \right] \ge \frac{1}{3} \tag{3.41}$$

All symbols in the above equations have been defined previously and are shown in Figure 3.16.

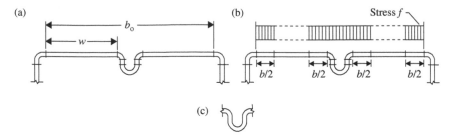

FIGURE 3.16 Compression elements with one intermediate stiffener: (a) actual elements, (b) effective elements and stress on effective elements, and (c) stiffener section.

3.5.6.1.2 Serviceability Determination

The effective width, b_d, used in determining serviceability is calculated as in Section 3.5.6.1.1, except that f_d is substituted for f.

3.5.6.2 Uniformly Compressed Elements with More Than One Intermediate Stiffener

In the AISI North American Specification [7], new design equations are provided for determining the effective width of uniformly compressed stiffened elements with multiple intermediate stiffeners.

3.6 Member Design

This chapter deals with the design of the following cold-formed steel structural members: (1) tension members, (2) *flexural members*, (3) concentrically loaded compression members, (4) combined axial load and bending, and (5) closed cylindrical tubular members. The nominal strength equations with factors of safety (Ω) and resistance factors (ϕ) are provided in the Specification [7] for the given limit states.

3.6.1 Sectional Properties

The sectional properties of a member such as area, moment of inertia, section modulus, and radius of gyration are calculated by using the conventional methods of structural design. These properties are based on full cross-section dimensions, effective widths, or net section, as applicable.

For the design of tension members, the nominal tensile strength is presently based on the gross and net sections. However, for flexural members and axially loaded compression members, the full dimensions are used when calculating the critical moment or load, while the effective dimensions, evaluated at the stress corresponding to the critical moment or load, are used to calculate the nominal strength.

3.6.2 Linear Method for Computing Sectional Properties

Because the thickness of cold-formed steel members is usually uniform, the computation of sectional properties can be simplified by using a "linear" or "midline" method. In this method, the material of each element is considered to be concentrated along the centerline or midline of the steel sheet and the area elements are replaced by straight or curved "line elements." The thickness dimension t is introduced after the linear computations have been completed. Thus, the total area $A = Lt$, and the moment of inertia of the section $I = I't$, where L is the total length of all line elements and I' is the moment of inertia

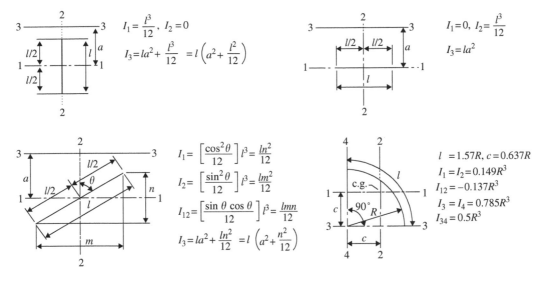

FIGURE 3.17   Properties of line elements.

of the centerline of the steel sheet. The moments of inertia of straight line elements and circular line elements are shown in Figure 3.17.

3.6.3 Tension Members

In the United States and Mexico, the nominal tensile strength of axially loaded cold-formed steel tension members is determined by the following equations:

1. *For yielding:*

$$T_n = A_g F_y \tag{3.42}$$

2. *For fracture away from connection:*

$$T_n = A_n F_u \tag{3.43}$$

where T_n is the nominal strength of the member when loaded in tension, A_g is the gross area of cross-section, A_n is the net area of cross-section, F_y is the design yield point of steel, and F_u is the tensile strength of steel.

3. *For fracture at connection:* The nominal tensile strength is also limited by sections E2.7, E3, and E4 of the North American Specification [7] for tension members using welded connections, bolted connections, and screw connections. For details, see chapter E of the Specification under the title "Connections and Joints."

In Canada, the design of tension members is based on appendix B of the AISI North American Specification [7].

3.6.4 Flexural Members

For the design of flexural members, consideration should be given to several design features: (1) bending strength and deflection, (2) shear strength of webs and combined bending and shear, (3) web crippling strength and combined bending and web crippling, and (4) bracing requirements. For some cases, special considerations should also be given to shear lag and flange curling due to the use of thin materials.

3.6.4.1 Bending Strength

Bending strengths of flexural members are differentiated according to whether or not the member is laterally braced. If such members are fully supported laterally, they are designed according to the nominal section strength. Otherwise, if they are laterally unbraced, then the bending strength may be governed by the lateral–torsional buckling strength. Section 3.9 discusses the flange distortional buckling. For C- or Z-sections with tension flange attached to deck or sheathing and with compression flange laterally unbraced, the nominal bending strength may be reduced according to the AISI Specification.

3.6.4.1.1 Nominal Section Strength

Two design procedures are now used in the AISI Specification for determining the nominal section strength. They are (1) initiation of yielding and (2) inelastic reserve capacity.

According to procedure I on the basis of initiation of yielding, the nominal moment, M_n, of the cross-section is the effective yield moment, M_y, determined for the effective areas of flanges and the beam web. The effective width of the compression flange and the effective depth of the web can be computed from the design equations given in Section 3.5 on Element Strength. The yield moment of a cold-formed steel flexural member is defined by the moment at which an outer fiber (tension, compression, or both) first attains the yield point of the steel. Figure 3.18 shows three types of stress distribution for yield moment based on different locations of the neutral axis. Accordingly, the nominal section strength for initiation of yielding can be computed as follows:

$$M_n = M_y = S_e F_y \qquad (3.44)$$

where S_e is the elastic section modulus of the effective section calculated with the extreme compression or tension fiber at F_y and F_y is the design yield stress.

For cold-formed steel design, S_e is usually computed by using one of the following two cases:

1. If the neutral axis is closer to the tension than to the compression flange (case c of Figure 3.18), the maximum stress occurs in the compression flange, and therefore the plate slenderness factor λ (Equation 3.9) and the effective width of the compression flange are determined by the w/t ratio and $f = F_y$. This procedure is also applicable to those beams for which the neutral axis is located at the mid-depth of the section (case a).
2. If the neutral axis is closer to the compression than to the tension flange (case b), the maximum stress of F_y occurs in the tension flange. The stress in the compression flange depends on the location of the neutral axis, which is determined by the effective area of the section. The latter cannot be determined unless the compressive stress is known. The closed-form solution of this type of design is possible but would be a very tedious and complex procedure. It is, therefore, customary to determine the sectional properties of the section by successive approximation.

See Examples 3.2 and 3.3 for the calculation of nominal bending strengths.

EXAMPLE 3.2

Use the ASD and LRFD methods to check the adequacy of the I-section with unstiffened flanges as shown in Figure 3.19. The nominal moment is based on the initiation of yielding using $F_y = 50$ ksi.

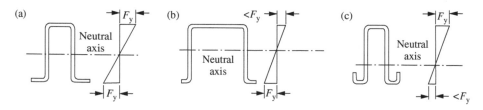

FIGURE 3.18 Stress distribution for yield moment (based on initiation of yielding).

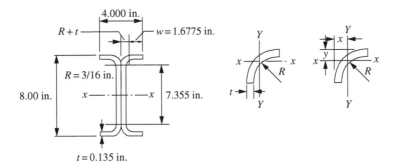

FIGURE 3.19   Example 3.2 (courtesy of Yu, W.W. 1991).

Assume that lateral bracing is adequately provided. The dead load moment $M_D = 30$ in. kip and the live load moment $M_L = 150$ in. kip.

Solution

ASD method

1. Location of neutral axis. For $R = \frac{3}{16}$ in. and $t = 0.135$ in., the sectional properties of the corner element are as follows:

$$I_x = I_y = 0.0003889 \text{ in.}^4$$
$$A = 0.05407 \text{ in.}^2$$
$$x = y = 0.1564 \text{ in.}$$

For the unstiffened compression flange

$$w = 1.6775 \text{ in.,} \quad w/t = 12.426$$

Using $k = 0.43$ and $f = F_y = 50$ ksi,

$$\lambda = \sqrt{f/F_{cr}} = (1.052/\sqrt{k})(w/t)\sqrt{f/E} = 0.821 > 0.673$$
$$b = [(1 - 0.22/0.821)/0.821](1.6775) = 1.496 \text{ in.}$$

Assuming that the web is fully effective, the neutral axis is located at $y_{cg} = 4.063$ in. Since $y_{cg} > d/2$, initial yield occurs in the compression flange. Therefore, $f = F_y$.

2. Check the web for full effectiveness as follows (Figure 3.20):

$$f_1 = 46.03 \text{ ksi (compression)}$$
$$f_2 = 44.48 \text{ ksi (tension)}$$
$$\psi = |f_2/f_1| = 0.966$$

Using Equation 3.10,

$$k = 4 + 2(1 + \psi)^3 + 2(1 + \psi) = 23.13$$
$$h = 7.355 \text{ in.}$$
$$h/t = 54.48$$
$$\lambda = \sqrt{f/F_{cr}} = (1.052/\sqrt{k})\,(54.48)\sqrt{46.03/29,500}$$
$$= 0.471 < 0.673$$
$$b_e = h = 7.355 \text{ in.}$$

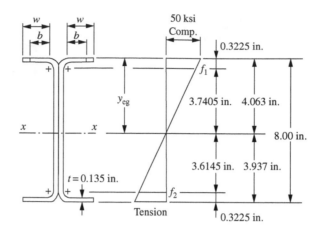

FIGURE 3.20 Stress distribution in webs (courtesy of Yu, W.W. 1991).

Since $h_o = 8.00$ in., $b_o = 2.00$ in., $h_o/b_o = 4$. Using Equations 3.12 and 3.13a to compute b_1 and b_2 as follows:

$$b_1 = b_e/(3+\psi) = 1.855 \text{ in.}$$
$$b_2 = b_e/2 = 3.6775 \text{ in.}$$

Since $b_1 + b_2 = 5.5325$ in. > 3.7405 in., the web is fully effective.

3. The moment of inertia I_x is

$$I_x = \sum(Ay^2) + 2I_{web} - \left(\sum A\right)(y_{cg})^2$$
$$= 25.382 \text{ in.}^4$$

The section modulus for the top fiber is

$$S_e = I_x/y_{cg} = 6.247 \text{ in.}^3$$

4. Based on initiation of yielding, the nominal moment for section strength is

$$M_n = S_e F_y = 312.35 \text{ in. kip}$$

5. The allowable design moment is

$$M_a = M_n/\Omega = 312.35/1.67 = 187.04 \text{ in. kip}$$

Based on the given data, the required allowable moment is

$$M = M_D + M_L = 30 + 150 = 180 \text{ in. kip}$$

Since $M < M_a$, the I-section is adequate for the ASD method.

LRFD method

1. Based on the nominal moment M_n computed above, the design moment is

$$\phi_b M_n = 0.90(312.35) = 281.12 \text{ in. kip}$$

2. According to the ASCE Standard 7 [38] the required moment for combined dead and live moments is

$$M_u = 1.2M_D + 1.6M_L$$
$$= (1.2 \times 30) + (1.6 \times 150)$$
$$= 276.00 \text{ in. kip}$$

Since $\phi_b M_n > M_u$, the I-section is adequate for bending strength according to the LRFD approach.

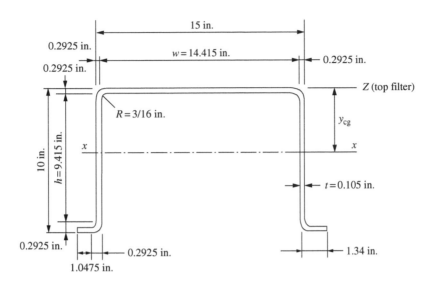

FIGURE 3.21   Example 3.3 (courtesy of Yu, W.W. 1991).

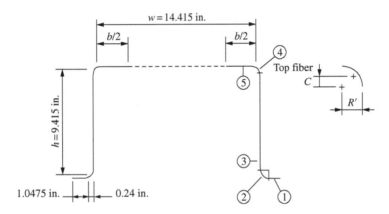

FIGURE 3.22   Line elements (courtesy of Yu, W.W. 1991).

EXAMPLE 3.3

Determine the nominal moment about the x-axis for the hat section with stiffened compression flange as shown in Figure 3.21. Assume that the yield point of steel is 50 ksi. Use the *linear method*. The nominal moment is determined by initiation of yielding.

Solution

Calculation of sectional properties. For the linear method, midline dimensions are as shown in Figure 3.22:

1. Corner element (Figure 3.17 and Figure 3.22)

$$R' = R + t/2 = 0.240 \text{ in.}$$

Arc length

$$L = 1.57R' = 0.3768 \text{ in.}$$
$$c = 0.637R' = 0.1529 \text{ in.}$$

2. Location of neutral axis
   - First approximation. For the compression flange,

$$w = 15 - 2(R + t) = 14.415 \text{ in.}$$
$$w/t = 137.29$$

Using Equations 3.6 through 3.9 and assuming $f = F_y = 50$ ksi,

$$\lambda = \sqrt{\frac{f}{F_{cr}}} = \frac{1.052}{\sqrt{4}}(137.29)\sqrt{\frac{50}{29,500}} = 2.973 > 0.673$$

$$\rho = \left(1 - \frac{0.22}{2.973}\right) \Big/ 2.973 = 0.311$$

$$b = \rho w = 0.311(14.415) = 4.483 \text{ in.}$$

By using the effective width of the compression flange and assuming that the web is fully effective, the neutral axis can be located as follows:

Element	Effective length, L (in.)	Distance from top fiber, y (in.)	L_y (in.^{2})
1	$2 \times 1.0475 = 2.0950$	9.9475	20.8400
2	$2 \times 0.3768 = 0.7536$	9.8604	7.4308
3	$2 \times 9.4150 = 18.8300$	5.0000	94.1500
4	$2 \times 0.3768 = 0.7536$	0.1396	0.1052
5	$2 \times 0.3768 = 4.4830$	0.0525	0.2354
Total	$2 \times 0.3768 = 26.9152$		122.7614

$$y_{cg} = \frac{\sum(Ly)}{\sum L} = \frac{122.7614}{26.9152} = 4.561 \text{ in.}$$

Because the distance y_{cg} is less than the half-depth of 5.0 in., the neutral axis is closer to the compression flange and, therefore, the maximum stress occurs in the tension flange. The maximum compressive stress can be computed as follows:

$$f = 50\left(\frac{4.561}{10 - 4.561}\right) = 41.93 \text{ ksi}$$

Since the above computed stress is less than the assumed value, another trial is required.
   - Second approximation. Following several trials, assume that

$$f = 40.70 \text{ ksi}$$
$$\lambda = 2.682 > 0.673$$
$$b = 4.934 \text{ in.}$$

Element	Effective length, L (in.)	Distance from top fiber, y (in.)	L_y (in.^{2})	L_y^2 (in.^{3})
1	22.0950	9.9475	20.8400	207.3059
2	20.7536	9.8604	7.4308	73.2707
3	18.8300	5.0000	94.1500	470.7500
4	20.7536	0.1396	0.1052	0.0147
5	24.9340	0.0525	0.2590	0.0136
Total	27.3662		122.7850	751.3549

$$y_{cg} = \frac{122.7850}{27.3662} = 4.487 \text{ in.}$$

$$f = \left(\frac{4.487}{10 - 4.487}\right) = 40.69 \text{ ksi}$$

Since the above computed stress is close to the assumed value, it is alright.

3. Check the effectiveness of the web. Use the AISI Specification to check the effectiveness of the web element. From Figure 3.23,

$$f_1 = 50(4.1945/5.513) = 38.04 \text{ ksi} \quad \text{(compression)}$$
$$f_2 = 50(5.2205/5.513) = 47.35 \text{ ksi} \quad \text{(tension)}$$
$$\psi = |f_2/f_1| = 1.245.$$

Using Equation 3.10,

$$k = 4 + 2(1 + \psi)^3 + 2(1 + \psi)$$
$$= 4 + 2(2.245)^3 + 2(2.245) = 31.12$$
$$h/t = 9.415/0.105 = 89.67 < 200 \quad \text{OK.}$$
$$\lambda = \sqrt{\frac{f}{F_{cr}}} = \frac{1.052}{\sqrt{31.12}}(89.67)\sqrt{\frac{38.04}{29,500}} = 0.607 < 0.673$$
$$b_e = \quad h = 9.415 \text{ in.}$$

Since $h_o = 10.0$ in., $b_o = 15.0$ in., $h_o/b_o = 0.67 < 4$, use Equation 3.12 to compute b_1:

$$b_1 = b_e/(3 + \psi) = 2.218 \text{ in.}$$

Since $\psi > 0.236$,

$$b_2 = b_e/2 = 4.7075 \text{ in.}$$
$$b_1 + b_2 = 6.9255 \text{ in.}$$

Because the computed value of $(b_1 + b_2)$ is greater than the compression portion of the web (4.1945 in.), the web element is fully effective.

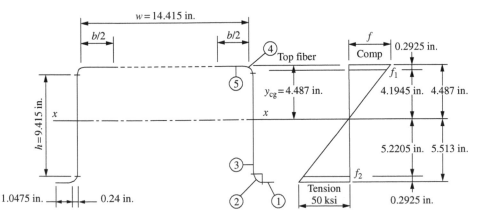

FIGURE 3.23 Effective lengths and stress distribution using fully effective webs (courtesy of Yu, W.W. 1991).

4. Moment of inertia and section modulus. The moment of inertia based on line elements is

$$\sum(Ly^2) = 751.3549$$

$$2I'_3 = 2\left(\tfrac{1}{12}\right)(9.415)^3 = 139.0944$$

$$I'_z = 2I'_3 + \sum(Ly^2) = 890.4493 \text{ in.}^3$$

$$\left(\sum L\right)(y_{cg})^2 = 27.3662(4.487)^2 = 550.9683 \text{ in.}^3$$

$$I'_x = I'_z - \left(\sum L\right)(y_{cg})^2 = 339.4810 \text{ in.}^3$$

The actual moment of inertia is

$$I_x = I'_x t = (339.4810)(0.105) = 35.646 \text{ in.}^4$$

The section modulus relative to the extreme tension fiber is

$$S_x = 35.646/5.513 = 6.466 \text{ in.}^3$$

Nominal moment. The nominal moment for section strength is

$$M_n = S_e F_y = S_x F_y = (6.466)(50) = 323.30 \text{ in.-kip}$$

Once the nominal moment is computed, the design moments for the ASD and LRFD methods can be determined as illustrated in Example 3.2.

According to procedure II of the AISI Specification, the nominal moment, M_n, is the maximum bending capacity of the beam by considering the inelastic reserve strength through partial plastification of the cross-section as shown in Figure 3.24. The inelastic stress distribution in the cross-section depends on the maximum strain in the compression flange, which is limited by the Specification for the given width to thickness ratio of the compression flange. On the basis of the maximum compression strain allowed in the Specification, the neutral axis can be located by Equation 3.45 and the nominal moment, M_n, can be determined by using Equation 3.46:

$$\int \sigma\, dA = 0 \tag{3.45}$$

$$\int \sigma y\, dA = M \tag{3.46}$$

where σ is the stress in the cross-section. For additional information, see Yu [1].

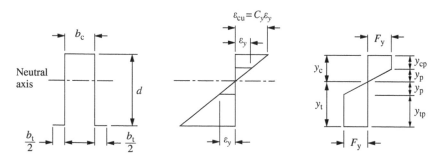

FIGURE 3.24 Stress distribution for maximum moment (inelastic reserve strength) (courtesy of Yu, W.W. 1991).

3.6.4.1.2 Lateral–Torsional Buckling Strength

The nominal lateral–torsional buckling strength of unbraced segments of singly, doubly, and point-symmetric sections subjected to lateral–torsional buckling, M_n, can be determined as follows:

$$M_n = S_c F_c \tag{3.47}$$

where S_c is the elastic section modulus of the effective section calculated at a stress F_c in the extreme compression fiber and F_c is the critical lateral–torsional buckling stress determined as follows:

1. For $F_e \geq 2.78 F_y$

$$F_c = F_y \tag{3.48}$$

2. For $2.78 F_y > F_e > 0.56 F_y$

$$F_c = \frac{10}{9} F_y \left(1 - \frac{10 F_y}{36 F_e} \right) \tag{3.49}$$

3. For $F_e \leq 0.56 F_y$

$$F_c = F_e \tag{3.50}$$

where F_e is the elastic critical lateral–torsional buckling stress.

1. For singly, doubly, and point-symmetric sections

   $F_e = C_b r_0 A \sqrt{\sigma_{ey} \sigma_t} / S_f$ for bending about the symmetry axis. For singly symmetric sections, x-axis is the axis of symmetry oriented such that the shear center has a negative x-coordinate. For point-symmetric sections, use $0.5 F_e$. Alternatively, F_e can be calculated using the equation for doubly symmetric I-sections, singly symmetric C-sections, or point-symmetric sections given in (2)

   $F_e = C_s A \sigma_{ex} [j + C_s \sqrt{j^2 + r_0^2 (\sigma_t / \sigma_{ex})}] / (C_{TF} S_f)$ for bending about the centroidal axis perpendicular to the symmetry axis for singly symmetric sections

   $C_s = +1$ for moment causing compression on shear center side of centroid
   $C_s = -1$ for moment causing tension on shear center side of centroid
   $\sigma_{ex} = \pi^2 E / (K_x L_x / r_x)^2$
   $\sigma_{ey} = \pi^2 E / (K_y L_y / r_y)^2$
   $\sigma_t = [GJ + \pi^2 E C_w / (K_t L_t)^2] / (A r_0^2)$
   A = full unreduced cross-sectional area
   S_f = elastic section modulus of full unreduced section relative to extreme compression fiber

   $$C_b = 12.5 M_{max} / (2.5 M_{max} + 3 M_A + 4 M_B + 3 M_C) \tag{3.51}$$

   In Equation 3.51

   M_{max} = absolute value of maximum moment in unbraced segment
   M_A = absolute value of moment at quarter point of unbraced segment
   M_B = absolute value of moment at centerline of unbraced segment
   M_C = absolute value of moment at three-quarter point at unbraced segment

   C_b is permitted to be conservatively taken as unity for all cases. For cantilevers or overhangs where the free end is unbraced, C_b is taken as unity.
   E = modulus of elasticity
   $C_{TF} = 0.6 - 0.4 \ (M_1 / M_2)$

where M_1 is the smaller and M_2 the larger bending moment at the ends of the unbraced length in the plane of bending, and where M_1 / M_2, the ratio of end moments, is positive when M_1 and M_2 have the same sign (reverse curvature bending) and negative when they are of opposite sign (single-curvature bending). When the bending moment at any point within an unbraced length is larger than that at both ends of this length, C_{TF} is taken as unity.

r_0 = Polar radius of gyration of cross-section about the shear center

$$= \sqrt{r_x^2 + r_y^2 + x_0^2} \tag{3.52}$$

r_x, r_y = radii of gyration of cross-section about centroidal principal axes
G = shear modulus
K_x, K_y, K_t = effective length factors for bending about the x- and y-axes and for twisting
L_x, L_y, L_t = unbraced length of compression member for bending about x- and y-axes, and for twisting
x_0 = distance from shear center to centroid along the principal x-axis, taken as negative
J = St. Venant torsion constant of cross-section
C_w = torsional warping constant of cross-section

$$j = \left[\int_A x^3 \, dA + \int_A xy^2 \, dA \right] \Big/ (2I_y) - x_0 \tag{3.53}$$

2. For I-sections, singly symmetric C-sections, or Z-sections bent about the centroidal axis perpendicular to the web (x-axis), the following equations are permitted to be used in lieu of (1) to calculate F_e:

$$F_e = C_b \pi^2 E d I_{yc} / [S_f (K_y L_y)^2] \tag{3.54}$$

for doubly symmetric I-sections and singly symmetric C-sections and

$$F_e = C_b \pi^2 E d I_{yc} / [2S_f (K_y L_y)^2] \tag{3.55}$$

for point-symmetric Z-sections. In Equations 3.54 and 3.55 d is the depth of the section and I_{yc} is the moment of inertia of compression portion of section about centroidal axis of the entire section parallel to web, using full unreduced section.

EXAMPLE 3.4

Determine the nominal moment for lateral–torsional buckling strength for the I-beam used in Example 3.2. Assume that the beam is braced laterally at both ends and midspan. Use $F_y = 50$ ksi.

Solution

Calculation of sectional properties
Based on the dimensions given in Example 3.2 (Figure 3.19 and Figure 3.20), the moment of inertia, I_x, and the section modulus, S_f, of the full-section can be computed as shown in the following table:

Element	Area, A (in.^{2})	Distance from mid-depth, y (in.)	Ay^2 (in.^{4})
Flanges	4(1.6775)(0.135) = 0.9059	3.9325	14.0093
Corners	4(0.05407) = 0.2163	3.8436	3.1955
Webs	2(7.355)(0.135) = 1.9859	0	0
Total	3.1081		17.2048

$$2I_{\text{web}} = 2(1/12)(0.135)(7.355)^3 = 8.9522$$

$$I_x = 26.1570 \text{ in.}^4$$

$$S_f = I_x/(8/2) = 6.54 \text{ in.}^3$$

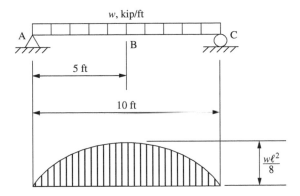

FIGURE 3.25 Example 3.4 (courtesy of Yu, W.W. 1991).

The value of I_{yc} can be computed as shown below:

Element	Area, A (in.²)	Distance from y-axis, x-axis (in.)	Ax^2 (in.⁴)
Flanges	$4(1.6775)(0.135) = 0.9059$	1.1613	1.2217
Corners	$4(0.05407) = 0.2163$	0.1564	0.0053
Webs	$2(7.355)(0.135) = 1.9859$	0.0675	0.0090
Total	3.1081	$I_{\text{flanges}} = 4(1/12)0.135(1.6775)^3 = 1.2360$	
			0.2124
		$I_y = 1.4484$ in.⁴	

$$I_{yc} = I_y/2 = 0.724 \text{ in.}^4$$

Considering the lateral supports at both ends and midspan, and the moment diagram shown in Figure 3.25, the value of C_b for the segment AB or BC is 1.30 according to Equation 3.51.

Using Equation 3.54,

$$F_e = C_b \pi^2 E \frac{dI_{yc}}{S_f(K_y L_y)^2}$$

$$= (1.30)\pi^2(29,500)\frac{(8)(0.724)}{(6.54)(1 \times 5 \times 12)^2} = 93.11 \text{ ksi}$$

$$0.56F_y = 28 \text{ ksi}$$
$$2.78F_y = 139 \text{ ksi}$$

Since $2.78F_y > F_e > 0.56F_y$, from Equation 3.49

$$F_c = \frac{10}{9}F_y\left(1 - \frac{10F_y}{36F_e}\right)$$

$$= \frac{10}{9}(50)\left[1 - \frac{10(50)}{36(93.11)}\right]$$

$$= 47.27 \text{ ksi}$$

Based on Equation 3.47, the nominal moment for lateral–torsional buckling strength is

$$M_n = S_c F_c$$

where S_c is the elastic section modulus of the effective section calculated at a compressive stress of $f = 47.27$ ksi. By using the same procedure illustrated in Example 3.2, $S_c = 6.295$ in.^{3}. Therefore, the nominal moment for lateral–torsional buckling strength is

$$M_n = (6.295)(47.27) = 297.6 \text{ in.-kip.}$$

For C- or Z-sections having the tension flange through-fastened to deck or sheathing with the compression flange laterally unbraced and loaded in a plane parallel to the web, the nominal flexural strength is determined by $M_n = RS_eF_y$, where R is a reduction factor [7]. A similar approach is used for beams having one-flange fastened to a standing seam roof system.

For closed box members, the nominal lateral–torsional buckling strength can be determined on the basis of the following elastic critical lateral buckling stress:

$$F_e = \frac{C_b \pi}{K_y L_y S_f} \sqrt{EGJI_y} \tag{3.56}$$

where I_y is the moment of inertia of full unreduced section about centroidal axis parallel to web and J is the torsional constant of box section.

3.6.4.1.3 Unusually Wide Beam Flanges and Short Span Beams

When beam flanges are unusually wide, special consideration should be given to the possible effects of shear lag and flange curling. Shear lag depends on the type of loading and the span to width ratio and is independent of the thickness. Flange curling is independent of span length but depends on the thickness and width of the flange, the depth of the section, and the bending stresses in both tension and compression flanges.

To consider the shear lag effects, the effective widths of both tension and compression flanges should be used according to the AISI Specification.

When a beam with unusually wide and thin flanges is subjected to bending, the portion of the flange most remote from the web tends to deflect toward the neutral axis due to the effect of longitudinal curvature of the beam and the applied bending stresses in both flanges. For the purpose of controlling the excessive flange curling, AISI Specification provides an equation to limit the flange width.

3.6.4.2 Shear Strength

The shear strength of beam webs is governed by either yielding or buckling of the web element, depending on the depth to thickness ratio, h/t, and the mechanical properties of steel. For beam webs having small h/t ratios, the nominal shear strength is governed by shear yielding. When the h/t ratio is large, the nominal shear strength is controlled by elastic shear buckling. For beam webs having moderate h/t ratios, the shear strength is based on inelastic shear buckling.

For the design of beam webs without holes, the AISI Specification provides the following equations for determining the nominal shear strength:

1. For $h/t \le \sqrt{Ek_v/F_y}$

$$V_n = 0.60F_y ht \tag{3.57}$$

2. For $\sqrt{Ek_v/F_y} < h/t \le 1.51\sqrt{Ek_v/F_y}$

$$V_n = 0.60t^2 \sqrt{k_v F_y E} \tag{3.58}$$

3. For $h/t > 1.51\sqrt{Ek_v/F_y}$

$$V_n = \pi^2 E k_v t^3 / [12(1 - \mu^2)h] = 0.904 E k_v t^3 / h \tag{3.59}$$

where V_n is the nominal shear strength of the beam, h is the depth of the flat portion of the web measured along the plane of the web, t is the web thickness, and k_v is the shear buckling coefficient determined as follows:

1. For unreinforced webs, $k_v = 5.34$.
2. For beam webs with transverse stiffeners satisfying the AISI requirements

when $a/h \leq 1.0$

$$k_v = 4.00 + \frac{5.34}{(a/h)^2}$$

when $a/h > 1.0$

$$k_v = 5.34 + \frac{4.00}{(a/h)^2}$$

where a is the shear panel length for unreinforced web element, h is the clear distance between transverse stiffeners for reinforced web elements.

For a web consisting of two or more sheets, each sheet should be considered as a separate element carrying its share of the shear force.

For the design of C-section webs with holes, the above nominal shear strength should be multiplied by a factor q_s specified in section C3.2.2 of the AISI Specification [7].

3.6.4.3 Combined Bending and Shear

For continuous beams and cantilever beams, high bending stresses often combine with high shear stresses at the supports. Such beam webs must be safeguarded against buckling due to the combination of bending and shear stresses. Based on the AISI Specification, the moment and shear should satisfy the interaction equations listed in Table 3.3.

3.6.4.4 Web Crippling

For cold-formed steel beams, transverse stiffeners are not frequently used for beam webs. The webs may cripple due to the high local intensity of the load or reaction as shown in Figure 3.26. Because the theoretical analysis of web crippling is rather complex due to the involvement of many factors, the present AISI design equations are based on the extensive experimental investigations conducted at Cornell University, University of Missouri-Rolla, University of Waterloo, and University of Sydney under four loading conditions: (1) end one-flange (EOF) loading, (2) interior one-flange (IOF) loading, (3) end two-flange (ETF) loading, and (4) interior two-flange (ITF) loading [7]. The loading conditions used for the tests are illustrated in Figure 3.27.

TABLE 3.3 Interaction Equations Used for Combined Bending and Shear

	ASD	LRFD and LSD
Beams with unreinforced webs	$M \leq M_n/\Omega_b$ and $V \leq V_n/\Omega_v$	$\bar{M} \leq \phi_b M_n$ and $\bar{V} \leq \phi_v V_n$
	$\left(\dfrac{\Omega_b M}{M_{nxo}}\right)^2 + \left(\dfrac{\Omega_v V}{V_n}\right)^2 \leq 1.0$ (3.60)	$\left(\dfrac{\bar{M}}{\phi_b M_{nxo}}\right)^2 + \left(\dfrac{\bar{V}}{\phi_v V_n}\right)^2 \leq 1.0$ (3.61)
Beams with transverse web stiffeners	$0.6\left(\dfrac{\Omega_b M}{M_{nxo}}\right) + \left(\dfrac{\Omega_v V}{V_n}\right) \leq 1.3$ (3.62)	$0.6\left(\dfrac{\bar{M}}{\phi_b M_{nxo}}\right) + \left(\dfrac{\bar{V}}{\phi_v V_n}\right) \leq 1.3$ (3.63)

Note: M is the bending moment, V is the unfactored shear force, Ω_b and Ω_v are the factors of safety for bending and shearing, respectively, ϕ_b is the resistance factor for bending, ϕ_v is the resistance factor for shear, M_n is the nominal flexural strength when bending alone exists, M_{nxo} is the nominal flexural strength about the centroidal x-axis determined in accordance with the specification excluding the consideration of lateral–torsional buckling, $\bar{M}$ is the required flexural strength (M_u for LRFD and M_f for LSD), V_n is the nominal shear strength when shear alone exists, and $\bar{V}$ is the required shear strength (V_u for LRFD and V_f for LSD).

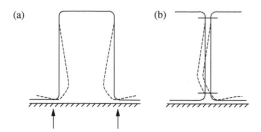

FIGURE 3.26   Web crippling of cold-formed steel beams.

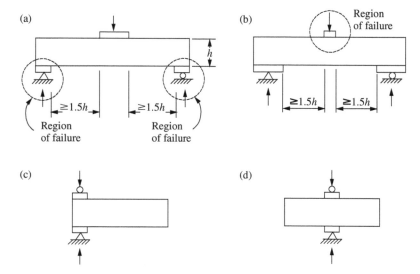

FIGURE 3.27   Loading conditions for web crippling tests: (a) EOF loading, (b) IOF loading, (c) ETF loading, and (d) ITF loading (courtesy of Yu, W.W. 1991).

The nominal web crippling strength of webs without holes for a given loading condition can be determined from the following AISI equation [7] on the basis of the thickness of web element, design yield point, the bend radius to thickness ratio, the depth to thickness ratio, the bearing length to thickness ratio, and the angle between the plane of the web and the plane of the bearing surface:

$$P_n = Ct^2 F_y \sin\theta \left[1 - C_R\sqrt{\frac{R}{t}}\right]\left[1 + C_N\sqrt{\frac{N}{t}}\right]\left[1 - C_h\sqrt{\frac{h}{t}}\right] \tag{3.64}$$

where P_n is the nominal web crippling strength, C is the coefficient, C_h is the web slenderness coefficient, C_N is the bearing length coefficient, C_R is the inside bend radius coefficient, F_y is the design yield point, h is the flat portion of web measured in the plane of the web, N is the bearing length ($\frac{3}{4}$ in. [19 mm] minimum), R is the inside bend radius, t is the web thickness, and θ is the angle between the plane of the web and the plane of the bearing surface, $45° \leq \theta \leq 90°$. Values of C, C_R, C_N, C_h, factor of safety, and resistance factor are listed in separate tables of the AISI Specification [7] for built-up sections, single web channel and C-sections, single web Z-sections, single hat sections, and multiweb deck sections.

For C-section webs with holes, the web crippling strength determined in accordance with the above equation should be reduced by using a reduction factor as given in the Specification [7].

3.6.4.5 Combined Bending and Web Crippling

For combined bending and web crippling, the design of beam webs should be based on the interaction equations provided in the AISI Specification [7]. These equations are presented in Table 3.4.

3.6.4.6 Bracing Requirements

In cold-formed steel design, braces should be designed to restrain lateral bending or twisting of a loaded beam and to avoid local crippling at the points of attachment. When C-sections and Z-sections are used as beams and loaded in the plane of the web, the AISI Specification [7] provides design requirements to restrain twisting of the beam under the following two conditions: (1) the top flange is connected to deck or sheathing material in such a manner as to effectively restrain lateral deflection of the connected flange and (2) neither flange is connected to sheathing. In general, braces should be designed to satisfy the strength and stiffness requirements. For beams using symmetrical cross-sections such as I-beams, the AISI Specification does not provide specific requirements for braces. However, the braces may be designed for a capacity of 2% of the force resisted by the compression portion of the beam. This is a frequently used rule of thumb but is a conservative approach, as proven by a rigorous analysis.

3.6.5 Concentrically Loaded Compression Members

Axially loaded cold-formed steel *compression members* should be designed for the following limit states: (1) yielding, (2) overall column buckling (flexural buckling, torsional buckling, or *torsional–flexural buckling*), and (3) local buckling of individual elements. The governing failure mode depends on the configuration of the cross-section, thickness of material, unbraced length, and end restraint. For distorsional buckling of compression members, see Section 3.9 on Direct Strength Method.

TABLE 3.4 Interaction Equations for Combined Bending and Web Crippling

	ASD		LRFD and LSD	
Shapes having single unreinforced webs	$1.2\left(\dfrac{\Omega_w P}{P_n}\right) + \left(\dfrac{\Omega_b M}{M_{nxo}}\right) \leq 1.5$	(3.65)	$1.07\left(\dfrac{\bar{P}}{\phi_w P_n}\right) + \left(\dfrac{\bar{M}}{\phi_b M_{nxo}}\right) \leq 1.42$	(3.66)
Shapes having multiple unreinforced webs such as I-sections	$1.1\left(\dfrac{\Omega_w P}{P_n}\right) + \left(\dfrac{\Omega_b M}{M_{nxo}}\right) \leq 1.5$	(3.67)	$0.82\left(\dfrac{\bar{P}}{\phi_w P_n}\right) + \left(\dfrac{\bar{M}}{\phi_b M_{nxo}}\right) \leq 1.32$	(3.68)
Support point of two nested Z-shapes	$\dfrac{M}{M_{no}} + 0.85\dfrac{P}{P_n} \leq \dfrac{1.65}{\Omega}$	(3.69)	$\dfrac{\bar{M}}{M_{no}} + 0.85\dfrac{\bar{P}}{P_n} \leq 1.65\phi$	(3.70)
	$M \leq M_{no}/\Omega_b$ and $P \leq P_n/\Omega_w$		$\bar{M} \leq \phi_b M_{no}$ and $\bar{P} \leq \phi_w P_n$	

Note: The AISI Specification includes some exception clauses, under which the effect of combined bending and web crippling need not be checked. P is the concentrated load or reaction in presence of bending moment, P_n is the nominal web crippling strength for concentrated load or reaction in the absence of bending moment (for Equations 3.65, 3.66, 3.67, and 3.68), P_n is the nominal web crippling strength assuming single web interior one-flange loading for the nested Z-sections, that is, the sum of the two webs evaluated individually (for Equations 3.69 and 3.70), $\bar{P}$ is the required strength for concentrated load or reaction in the presence of bending moment (P_u for LRFD and P_f for LSD), M is the applied bending moment at, or immediately adjacent to, the point of application of the concentrated load or reaction, M_{no} is the nominal yield moment for nested Z-sections, that is, the sum of two sections evaluated individually, M_{nxo} is the nominal flexural strength about the centroidal x-axis determined in accordance with the specification excluding the consideration of lateral–torsional buckling, $\bar{M}$ is the required flexural strength at, or immediately adjacent to, the point of application of the concentrated load or reaction $\bar{P}$ (M_u for LRFD and M_f for LSD), Ω is the factor of safety $= 1.75$, Ω_b and Ω_w are the factors of safety for bending and web crippling, respectively, ϕ is the resistance factor (0.9 for LRFD and 0.80 for LSD), ϕ_b is the resistance factor for bending, and ϕ_w is the resistance factor for web crippling.

3.6.5.1 Yielding

A very short, compact column under axial load may fail by yielding. For this case, the nominal axial strength is the yield load, that is,

$$P_n = P_y = A_g F_y \tag{3.71}$$

where A_g is the gross area of the column and F_y is the yield point of steel.

3.6.5.2 Overall Column Buckling

Overall column buckling may be one of the following three types:

1. *Flexural buckling — bending about a principal axis.* The elastic flexural buckling stress is

$$F_e = \frac{\pi^2 E}{(KL/r)^2} \tag{3.72}$$

   where E is the modulus of elasticity, K is the effective length factor for flexural buckling (Figure 3.28), L is the unbraced length of member for flexural buckling, and r is the radius of gyration of the full-section.

2. *Torsional buckling — twisting about shear center.* The elastic torsional buckling stress is

$$F_e = \frac{1}{A r_0^2}\left[GJ + \frac{\pi^2 E C_w}{(K_t L_t)^2} \right] \tag{3.73}$$

   where A is the full cross-sectional area, C_w is the torsional warping constant of cross-section, G is the shear modulus, J is the St. Venant torsion constant of cross-section, K_t is the effective length factor for twisting, L_t is the unbraced length of member for twisting, and r_0 is the polar radius of gyration of cross-section about shear center.

3. *Torsional–flexural buckling — bending and twisting simultaneously.* The elastic torsional–flexural buckling stress is

$$F_e = \left[(\sigma_{ex} + \sigma_t) - \sqrt{(\sigma_{ex} + \sigma_t)^2 - 4\beta \sigma_{ex}\sigma_t} \right] \bigg/ (2\beta) \tag{3.74}$$

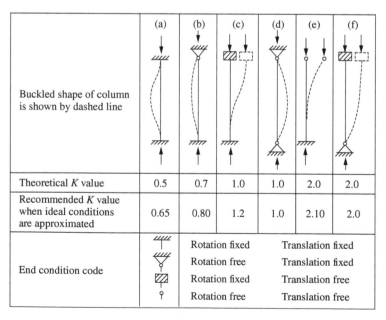

FIGURE 3.28 Effective length factor K for concentrically loaded compression members.

where $\beta = 1 - (x_0/r_0)^2$, $\sigma_{ex} = \pi^2 E/(K_x L_x/r_x)^2$, $\sigma_t =$ Equation 3.73, and x_0 is the distance from shear center to centroid along the principal x-axis.

For doubly symmetric and *point-symmetric shapes* (Figure 3.29), the overall column buckling can be either flexural type or torsional type. However, for singly symmetric shapes (Figure 3.30), the overall column buckling can be either flexural buckling or torsional–flexural buckling.

For overall column buckling, the nominal axial strength is determined by Equation 3.75:

$$P_n = A_e F_n \tag{3.75}$$

where A_e is the effective area determined for the stress F_n and F_n is the nominal buckling stress determined as follows:

For $\lambda_c \leq 1.5$

$$F_n = (0.658^{\lambda_c^2}) F_y \tag{3.76}$$

For $\lambda_c > 1.5$

$$F_n = \left[\frac{0.877}{\lambda_c^2}\right] F_y \tag{3.77}$$

The use of the effective area A_e in Equation 3.75 is to reflect the effect of local buckling on the reduction of column strength. In Equations 3.76 and 3.77,

$$\lambda_c = \sqrt{F_y/F_e} \tag{3.78}$$

where F_e is the least of elastic flexural buckling stress (Equation 3.72), torsional buckling stress (Equation 3.73), and torsional–flexural buckling stress (Equation 3.74), whichever is applicable.

Concentrically loaded angle sections should be designed for an additional bending moment in accordance with section C5.2 of the Specification [7].

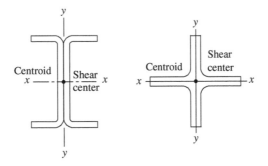

FIGURE 3.29 Doubly symmetric shapes.

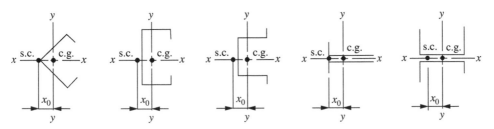

FIGURE 3.30 Singly symmetric shapes (courtesy of Yu, W.W. 1991).

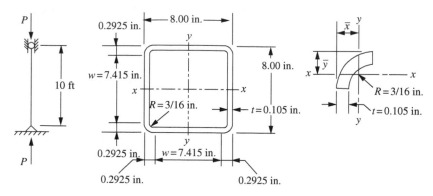

FIGURE 3.31 Example 7.5 (courtesy of Yu, W.W. 1991).

For nonsymmetric shapes whose cross-sections do not have any symmetry, either about an axis or about a point, the elastic torsional–flexural buckling stress should be determined by rational analysis or by tests. See AISI Design Manual [41].

In addition to the above design provisions for the design of axially loaded columns, the AISI Specification also provides design criteria for compression members having one-flange through-fastened to deck or sheathing.

EXAMPLE 3.5

Determine the allowable axial load for the square tubular column shown in Figure 3.31. Assume that $F_y = 40$ ksi, $K_x L_x = K_y L_y = 10$ ft, and the dead to live load ratio is $\frac{1}{5}$. Use the ASD and LRFD methods.

Solution

ASD method

Since the square tube is a doubly symmetric closed section, it will not be subject to torsional–flexural buckling. It can be designed by flexural buckling.

1. *Sectional properties of full-section.*

$w = 8.00 - 2(R + t) = 7.415$ in.

$A = 4(7.415 \times 0.105 + 0.0396) = 3.273$ in.^{2}

$I_x = I_y = 2(0.105)[(1/12)(7.415)^3 + 7.415(4 - 0.105/2)^2] + 4(0.0396)(4.0 - 0.1373)^2 = 33.763$ in.^{4}

$r_x = r_y = \sqrt{I_x/A} = \sqrt{33.763/3.273} = 3.212$ in.

2. *Nominal buckling stress, F_n.* According to Equation 3.72, the elastic flexural buckling stress, F_e, is computed as follows:

$$\frac{KL}{r} = \frac{10 \times 12}{3.212} = 37.36 < 200, \quad \text{OK.}$$

$$F_e = \frac{\pi^2 E}{(KL/r)^2} = \frac{\pi^2 (29,500)}{(37.36)^2} = 208.597 \text{ ksi}$$

$$\lambda_c = \sqrt{\frac{F_y}{F_e}} = \sqrt{\frac{40}{208.597}} = 0.438 < 1.5$$

$$F_n = (0.658^{\lambda_c^2})F_y = (0.658^{0.438^2})40 = 36.914 \text{ ksi}$$

3. *Effective area, A_e.* Because the given square tube is composed of four stiffened elements, the effective width of stiffened elements subjected to uniform compression can be computed from Equations 3.6 through 3.9 by using $k = 4.0$:

$$w/t = 7.415/0.105 = 70.619$$

$$\lambda = \sqrt{\frac{f}{F_{cr}}} = \frac{1.052}{\sqrt{k}} \left(\frac{w}{t}\right) \sqrt{\frac{F_n}{E}}$$

$$= 1.052/\sqrt{4}(70.619)\sqrt{36.914/29,500} = 1.314$$

Since $\lambda > 0.673$, from Equation 3.7

$$b = \rho w$$

where

$$\rho = (1 - 0.22/\lambda)/\lambda = (1 - 0.22/1.314)/1.314 = 0.634$$

Therefore,

$$b = (0.634)(7.415) = 4.701 \text{ in.}$$

The effective area is

$$A_e = 3.273 - 4(7.415 - 4.701)(0.105) = 2.133 \text{ in.}^2$$

4. *Nominal and allowable loads.* Using Equation 3.75, the nominal load is

$$P_n = A_e F_n = (2.133)(36.914) = 78.738 \text{ kip}$$

The allowable load is

$$P_a = P_n/\Omega_c = 78.738/1.80 = 43.74 \text{ kip}$$

LRFD method

In the ASD method, the nominal axial load, P_n, was computed to be 78.738 kip. The design axial load for the LRFD method is

$$\phi_c P_n = 0.85(78.738) = 66.93 \text{ kip}$$

Based on the load combination of dead and live loads, the required axial load is

$$P_u = 1.2P_D + 1.6P_L = 1.2P_D + 1.6(5P_D) = 9.2P_D$$

where P_D is the axial load due to dead load and P_L is the axial load due to live load.

By using $P_u = \phi_c P_n$, the values of P_D and P_L are computed as follows:

$$P_D = 66.93/9.2 = 7.28 \text{ kip}$$
$$P_L = 5P_D = 36.40 \text{ kip}$$

Therefore, the allowable axial load is

$$P_a = P_D + P_L = 43.68 \text{ kip}$$

It can be seen that the allowable axial loads determined by the ASD and LRFD methods are practically the same.

3.6.6 Combined Axial Load and Bending

The AISI Specification provides interaction equations for combined axial load and bending.

3.6.6.1 Combined Tensile Axial Load and Bending

For combined tensile axial load and bending, the required strengths should satisfy the interaction equations presented in Table 3.5. These equations are to prevent yielding of the tension flange and to prevent failure of the compression flange of the member.

3.6.6.2 Combined Compressive Axial Load and Bending

Cold-formed steel members under combined compressive axial load and bending are usually referred to as *beam–columns*. Such members are often found in framed structures, trusses, and exterior wall studs. For the design of these members, the required strengths should satisfy the AISI interaction equations presented in Table 3.6.

TABLE 3.5 Interaction Equations for Combined Tensile Axial Load and Bending

	ASD		LRFD and LSD	
Check tension flange	$\dfrac{\Omega_b M_x}{M_{nxt}} + \dfrac{\Omega_b M_y}{M_{nyt}} + \dfrac{\Omega_t T}{T_n} \leq 1.0$	(3.79)	$\dfrac{\bar{M}_x}{\phi_b M_{nxt}} + \dfrac{\bar{M}_y}{\phi_b M_{nyt}} + \dfrac{\bar{T}}{\phi_t T_n} \leq 1.0$	(3.80)
Check compression flange	$\dfrac{\Omega_b M_x}{M_{nx}} + \dfrac{\Omega_b M_y}{M_{ny}} - \dfrac{\Omega_t T}{T_n} \leq 1.0$	(3.81)	$\dfrac{\bar{M}_x}{\phi_b M_{nx}} + \dfrac{\bar{M}_y}{\phi_b M_{ny}} - \dfrac{\bar{T}}{\phi_t T_n} \leq 1.0$	(3.82)

Note: M_{nx} and M_{ny} are the nominal flexural strengths about the centroidal *x*- and *y*-axes, respectively, M_{nxt}, $M_{nyt} = S_{ft}F_y$, $\bar{M}_x$ and $\bar{M}_y$ are the required flexural strengths with respect to the centroidal axes (M_{ux} and M_{uy} for LRFD, M_{fx} and M_{fy} for LSD), M_x and M_y are the required moments with respect to the centroidal axes of the section, S_{ft} is the section modulus of the full-section for the extreme tension fiber about the appropriate axis, T is the required tensile axial load, T_n is the nominal tensile axial strength, $\bar{T}$ is the required tensile axial strength (T_u for LRFD and T_f for LSD), ϕ_b is the resistance factor for bending, ϕ_t is the resistance factor for tension (0.95 for LRFD and 0.90 for LSD), Ω_b is the safety factor for bending, and Ω_t is the safety factor for tension.

TABLE 3.6 Interaction Equations for Combined Compressive Axial Load and Bending

ASD		LRFD and LSD	
When $\Omega_c P/P_n \leq 0.15$, $\dfrac{\Omega_c P}{P_n} + \dfrac{\Omega_b M_x}{M_{nx}} + \dfrac{\Omega_b M_y}{M_{ny}} \leq 1.0$	(3.83)	When $\bar{P}/\phi_c P_n \leq 0.15$, $\dfrac{\bar{P}}{\phi_c P_n} + \dfrac{\bar{M}_x}{\phi_b M_{nx}} + \dfrac{\bar{M}_y}{\phi_b M_{ny}} \leq 1.0$	(3.84)
When $\Omega_c P/P_n > 0.15$, $\dfrac{\Omega_c P}{P_n} + \dfrac{\Omega_b C_{mx} M_x}{M_{nx}\alpha_x} + \dfrac{\Omega_b C_{my} M_y}{M_{ny}\alpha_y} \leq 1.0$ (3.85)		When $\bar{P}/\phi_c P_n > 0.15$, $\dfrac{\bar{P}}{\phi_c P_n} + \dfrac{C_{mx}\bar{M}_x}{\phi_b M_{nx}\alpha_x} + \dfrac{C_{my}\bar{M}_y}{\phi_b M_{ny}\alpha_y} \leq 1.0$	(3.86)
$\dfrac{\Omega_c P}{P_{no}} + \dfrac{\Omega_b M_x}{M_{nx}} + \dfrac{\Omega_b M_y}{M_{ny}} \leq 1.0$	(3.87)	$\dfrac{\bar{P}}{\phi_c P_{no}} + \dfrac{\bar{M}_x}{\phi_b M_{nx}} + \dfrac{\bar{M}_y}{\phi_b M_{ny}} \leq 1.0$	(3.88)

Note: M_x and M_y are the required moments with respect to the centroidal axes of the effective section determined for the required axial strength alone, M_{nx} and M_{ny} are the nominal flexural strengths about the centroidal axes, $\bar{M}_x$ and $\bar{M}_y$ are the required flexural strengths with respect to the centroidal axes of the effective section determined for the required axial strength alone (M_{ux} and M_{uy} for LRFD, M_{fx} and M_{fy} for LSD), P is the required axial load, P_n is the nominal axial strength determined in accordance with Equation 3.75, P_{no} is the nominal axial strength determined in accordance with Equation 3.75, for $F_n = F_y$, $\bar{P}$ is the required compressive axial strength (P_u for LRFD and P_f for LSD), $\alpha_x = 1 - \Omega_c P/P_{EX}$ (for Equation 3.85), $\alpha_y = 1 - \Omega_c P/P_{EY}$ (for Equation 3.85), $\alpha_x = 1 - \bar{P}/P_{EX}$ (for Equation 3.86), $\alpha_y = 1 - \bar{P}/P_{EY}$ (for Equation 3.86), $P_{EX} = \pi^2 EI_x/(K_x L_x)^2$, $P_{EY} = \pi^2 EI_y/(K_y L_y)^2$, Ω_b is the factor of safety for bending, Ω_c is the factor of safety for concentrically loaded compression, and C_{mx} and C_{my} are the coefficients whose value is taken as follows:

1. For compression members in frames subject to joint translation (side sway), $C_m = 0.85$.
2. For restrained compression members in frames braced against joint translation and not subjected to transverse loading between their supports in the plane of bending, $C_m = 0.6 - 0.4(M_1/M_2)$, where M_1/M_2 is the ratio of the smaller to the larger moment at the ends of that portion of the member under consideration which is unbraced in the plane of bending. M_1/M_2 is positive when the member is bent in reverse curvature and negative when it is bent in single curvature.
3. For compression members in frames braced against joint translation in the plane of loading and subjected to transverse loading between their supports, the value of C_m may be determined by rational analysis. However, in lieu of such analysis, the following values may be used: (1) for members whose ends are restrained, $C_m = 0.85$ and (2) for members whose ends are unrestrained, $C_m = 1.0$.

I_x, I_y, L_x, L_y, K_x, and K_y have been defined previously.

3.6.7 Closed Cylindrical Tubular Members

Thin-walled closed cylindrical tubular members are economical sections for compression and torsional members because of their large ratio of radius of gyration to area, the same radius of gyration in all directions, and the large torsional rigidity. The AISI design provisions are limited to the ratio of outside diameter to wall thickness, D/t, not being greater than $0.441E/F_y$.

3.6.7.1 Bending Strength

For cylindrical tubular members subjected to bending, the nominal flexural strengths are as follows according to the D/t ratio:

$$M_n = F_c S_f$$

1. For $D/t \le 0.0714 \ E/F_y$

$$F_c = 1.25 F_y \tag{3.89}$$

2. For $0.0714 E/F_y < D/t \le 0.318 E/F_y$

$$F_c = [0.970 + 0.020(E/F_y)/(D/t)]F_y \tag{3.90}$$

3. For $0.318 E/F_y < D/t \le 0.441 E/F_y$

$$F_c = 0.328 E/(D/t) \tag{3.91}$$

where D is the outside diameter of the cylindrical tube, t is the wall thickness, F_c is the critical buckling stress, and S_f is the elastic section modulus of full, unreduced cross-section. Other symbols have been defined previously.

3.6.7.2 Compressive Strength

When cylindrical tubes are used as concentrically loaded compression members, the nominal axial strength is determined by Equation 3.75, except that (1) the elastic buckling stress, F_e, is determined for flexural buckling by using Equation 3.72 and (2) the effective area, A_e, is calculated by Equation 3.92 given below:

$$A_e = A_0 + R(A - A_0) \tag{3.92}$$

where

$$R = F_y/(2F_e) \le 1.0$$
$$A_0 = \{0.037/[(DF_y)/(tE)] + 0.667\}A \le A \quad \text{for } D/t \le 0.441E/F_y \tag{3.93}$$
$$A = \text{area of unreduced cross-section}$$

In the above equations, the value A_0 is the reduced area due to the effect of local buckling [1,7].

3.7 Connections and Joints

Welds, bolts, screws, rivets, and other special devices such as metal stitching and adhesives are generally used for cold-formed steel connections. The AISI Specification contains only the design provisions for welded connections, bolted connections, and screw connections. These design equations are based primarily on the experimental data obtained from extensive test programs.

3.7.1 Welded Connections

Welds used for cold-formed steel constructions may be classified as arc welds (or fusion welds) and resistance welds. Arc welding is usually used for connecting cold-formed steel members to each other as well as connecting such thin members to heavy, hot-rolled steel framing members. It is used for groove welds, arc spot welds, arc seam welds, fillet welds, and flare groove welds (Figures 3.32 to 3.36). The AISI design provisions for welded connections are applicable only for cold-formed steel structural members, in which the thickness of the thinnest connected part is 0.18 in. (4.57 mm) or less. Otherwise, when the thickness of connected parts is thicker than 0.18 in. (4.57 mm), the welded connection should be designed according to the AISC Specifications [42,43]. Additional design information on structural welding of sheet steels can also be found in the AWS Code [44].

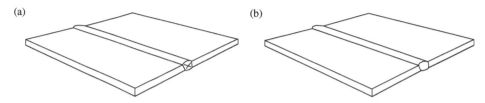

FIGURE 3.32 Groove welds.

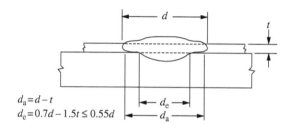

$$d_a = d - t$$
$$d_e = 0.7d - 1.5t \leq 0.55d$$

FIGURE 3.33 Arc spot weld — single thickness of sheet.

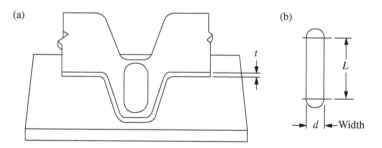

FIGURE 3.34 Arc seam weld.

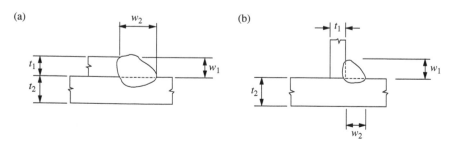

FIGURE 3.35 Fillet welds.

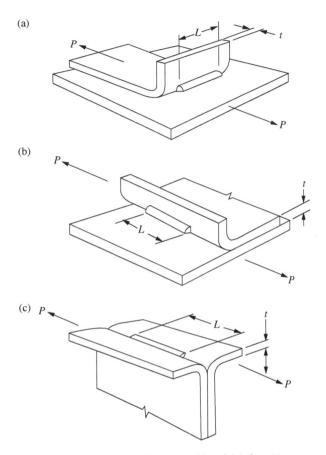

FIGURE 3.36 Flare groove welds: (a, b) flare bevel groove weld and (c) flare V-groove weld.

3.7.1.1 Arc Welds

According to the AISI Specification, the nominal strengths of arc welds can be determined from the equations given in Table 3.7. The design strengths can then be computed by using the factor of safety or resistance factor provided in Table 3.1.

3.7.1.2 Resistance Welds

The nominal shear strengths of resistance spot welds are provided in the AISI Specification [7] according to the thickness of the thinnest outside sheet and the unit used for the nominal shear strength. They are applicable for all structural grades of low-carbon steel, uncoated or galvanized with 0.9 oz/ft^2 of sheet or less, and medium carbon and low-alloy steels.

3.7.2 Bolted Connections

Due to the thinness of the connected parts, the design of bolted connections in cold-formed steel construction is somewhat different from that in hot-rolled heavy construction. The AISI design provisions are applicable only to cold-formed members or elements less than $\frac{3}{16}$ in. (4.76 mm) in thickness. For materials not less than $\frac{3}{16}$ in. (4.76 mm), the bolted connection should be designed in accordance with the AISC Specifications (42,43).

In the AISI Specification, five types of bolts (A307, A325, A354, A449, and A490) are used for connections in cold-formed steel construction, in which A449 and A354 bolts should be used as

TABLE 3.7 Nominal Strength Equations for Arc Welds

Type of weld	Type of strength	Nominal strength, P_n
Groove welds (Figure 3.32)	Tension or compression	$Lt_e F_y$
	Shear strength of weld	$Lt_e(0.6F_{xx})$
	Shear strength of connected part	$Lt_e(F_y/\sqrt{3})$
Arc spot welds (Figure 3.33)	Shear strength	
	Strength of weld	$0.589d_e^2 F_{xx}$
	Strength of connected part	
	1. $d_a/t \le 0.815\sqrt{E/F_u}$	$2.20td_a F_u$
	2. $0.815\sqrt{E/F_u} < (d_a/t) < 1.397\sqrt{E/F_u}$	$0.28[1 + (5.59\sqrt{E/F_u})/(d_a/t)](td_a F_u)$
	3. $d_a/t \ge 1.397\sqrt{E/F_u}$	$1.40td_a F_u$
	Shear strength of connected part based on end distance	$eF_u t$
	Tensile strength	
	Strength of weld	$0.785d_e^2 F_{xx}$
	Strength of connected part	$0.8\,(F_u/F_y)^2 td_a F_u$
Arc seam welds (Figure 3.34)	Shear strength	$[\pi d_e^2/4 + Ld_e]0.75F_{xx}$
	Strength of connected part	$2.5tF_u(0.25L + 0.96d_a)$
Fillet welds (Figure 3.35)	Shear strength of weld (for $t > 0.10$ in.)	$0.75t_w LF_{xx}$
	Strength of connected part	
	1. Longitudinal loading	
	$L/t < 25$	$[1 - (0.01L/t)]tLF_u$
	$L/t \ge 25$	$0.75tLF_u$
	2. Transverse loading	tLF_u
Flare groove welds (Figure 3.36)	Shear strength of weld (for $t > 0.10$ in.)	$0.75t_w LF_{xx}$
	Strength of connected part	
	1. Transverse loading	$0.833tLF_u$
	2. Longitudinal loading	
	For $t \le t_w < 2t$ or if lip height $< L$	$0.75tLF_u$
	For $t_w \ge 2t$ and lip height $\ge L$	$1.50tLF_u$

Note: d is the visible diameter of outer surface of arc spot weld, d_a is the average diameter of the arc spot weld at midthickness of t, $d_a = (d - t)$ for single sheet or multiple sheets not more than four sheets, d_e is the effective diameter of fused area at the plane of maximum shear transfer, $d_e = 0.7d - 1.5t \le 0.55d$, e is the distance measured in the line of force from the centerline of a weld to the nearest edge of an adjacent weld or to the end of the connected part toward which the force is directed, F_u is the tensile strength of the connected part, F_y is the yield point of steel, F_{xx} is the filler metal strength designation in AWS electrode classification, L is the length of the weld, P_n is the nominal strength of the weld, t is the thickness of the connected sheet, t_e is the effective throat dimension for groove weld, t_w is the effective throat for fillet welds or flare groove weld not filled flush to surface, $t_w = 0.707w_1$ or $0.707w_2$, whichever is smaller, and w_1 and w_2 are the leg of the weld, respectively.

equivalents of A325 and A490 bolts, respectively, whenever a diameter smaller than $\frac{1}{2}$ in. (12.7 mm) is required.

On the basis of the failure modes occurring in the tests of bolted connections, the AISI criteria deal with three major design considerations for the connected parts: (1) longitudinal shear failure, (2) tensile failure, and (3) bearing failure. The nominal strength equations are given in Table 3.8.

In addition, design strength equations are provided for shear and tension in bolts. Accordingly, the AISI nominal strength for shear and tension in bolts can be determined as follows:

$$P_n = A_b F_n \qquad (3.94)$$

where A_b is the gross cross-sectional area of bolt and F_n is the nominal shear or tensile stress given in Table 3.9 for ASD and LRFD. For the LSD method, see appendix B of the North American Specification [7].

TABLE 3.8 Nominal Strength Equations for Bolted Connections (ASD and LRFD)[a]

Type of strength	Nominal strength, P_n
Shear strength based on spacing and edge distance	teF_u
Tensile strength in net section (shear lag)	
1. Flat sheet connections not having staggered holes	
With washers under bolt head and nut	
Single bolt in the line of force	$(0.1 + 3d/s)F_u A_n \leq F_u A_n$
Multiple bolts in the line of force	$F_u A_n$
Without washers under bolt head or nut	
Single bolt in the line of force	$(2.5d/s)F_u A_n \leq F_u A_n$
Multiple bolt in the line of force	$F_u A_n$
2. Flat sheet connections having staggered hole (see AISI Specification for the modified A_n)	Same as item (1), except for A_n
3. Other than flat sheet (see AISI Specification for the effective net area A_e)	$F_u A_e$
Bearing strength	
1. Without consideration of bolt hole deformation (see AISI Specification for bearing factor, C, and modification factor, m_f)	$m_f C dt F_u$
2. With consideration of bolt hole deformation (see AISI Specification for coefficient α)	$(4.64\alpha t + 1.53)dt F_u$

[a] For the LSD method, see appendix B of the AISI North American Specification [7] for shear strength based on spacing and edge distance and for tensile strength in net section.

Note: A_n is the net area of the connected part, d is the nominal diameter of bolt, F_u is the tensile strength of the connected part, s is the sheet width divided by the number of bolt holes in cross-section being analyzed, and t is the thickness of the thinnest connected part.

TABLE 3.9 Nominal Tensile and Shear Stresses for Bolts (ASD and LRFD)[a]

Description of bolts	Nominal tensile stress, F_{nt} (ksi)	Nominal shear stress, F_{nv} (ksi)
A307 bolts, Grade A, $\frac{1}{4}$ in. $\leq d < \frac{1}{2}$ in.	40.5	24.0
A307 bolts, Grade A, $d \geq \frac{1}{2}$ in.	45.0	27.0
A325 bolts, when threads are not excluded from shear planes	90.0	54.0
A325 bolts, when threads are excluded from shear planes	90.0	72.0
A354 Grade BD bolts, $\frac{1}{4}$ in. $\leq d < \frac{1}{2}$ in., when threads are not excluded from shear planes	101.0	59.0
A354 Grade BD bolts, $\frac{1}{4}$ in. $\leq d < \frac{1}{2}$ in., when threads are excluded from shear planes	101.0	90.0
A449 bolts, $\frac{1}{4}$ in. $\leq d < \frac{1}{2}$ in., when threads are not excluded from shear planes	81.0	47.0
A449 bolts, $\frac{1}{4}$ in. $\leq d < \frac{1}{2}$ in., when threads are excluded from shear planes	81.0	72.0
A490 bolts, when threads are not excluded from shear planes	112.5	67.5
A490 bolts, when threads are excluded from shear planes	112.5	90.0

[a] For the LSD method, see appendix B of the AISI North American Specification [7].

Note: 1 in. $= 25.4$ mm, 1 ksi $= 6.9$ MPa $= 70.3$ kg/cm^2.

For bolts subjected to the combination of shear and tension, the reduced nominal tension stress for the ASD and LRFD is given in Table 3.10. For the LSD method, refer to appendix B of the North American Specification.

3.7.3 Screw Connections

Screws can provide a rapid and effective means to fasten sheet metal siding and roofing to framing members and to connect individual siding and roofing panels. Design equations are presently given in the AISI Specification for determining the nominal shear strength and the nominal tensile strength of connected parts and screws. These design requirements should be used for self-tapping screws with diameters greater than or equal to 0.08 in. (2.03 mm) but not exceeding $\frac{1}{4}$ in. (6.35 mm). The screw can be thread-forming or thread-cutting, with or without drilling point. The spacing between the centers of screws should not be less than $3d$ and the distance from the center of a screw to the edge of any part should not be less than $1.5d$, where d is the diameter of screw. In the direction of applied force, the end distance is also limited by the shear strength of the connected part.

According to the AISI North American Specification, the nominal strength per screw is determined from Table 3.11. See Figure 3.37 and Figure 3.38 for t_1, t_2, F_{u1}, and F_{u2}.

For the convenience of designers, the following table gives the correlation between the common number designation and the nominal diameter for screws.

Number designation	Nominal diameter, d (in.)[a]
0	0.060
1	0.073
2	0.086
3	0.099
4	0.112
5	0.125
6	0.138
7	0.151
8	0.164
10	0.190
12	0.216
1/4	0.250

[a] 1 in. = 25.4 mm.

In addition to the design requirements discussed above, the AISI North American Specification also includes some provisions for built-up compression members composed of two sections in contact or for compression elements joined to other parts of built-up members by intermittent connections.

3.7.4 Rupture

In connection design, due consideration should be given to shear rupture, tension rupture, and block shear rupture. For details, see the AISI Specification [7].

3.8 Structural Systems and Assemblies

In the past, cold-formed steel components have been used in different structural systems and assemblies such as metal buildings, shear diaphragms, shell roof structures, wall stud assemblies, residential construction, and composite construction.

TABLE 3.10 Nominal Tension Stresses, F'_{nt}, for Bolts Subjected to the Combination of Shear and Tension (ASD and LRFD)[a]

Description of bolts	Threads not excluded from shear planes	Threads excluded from shear planes
	(A) ASD method	
A325 bolts	$110 - 3.6f_v \leq 90$	$110 - 2.8f_v \leq 90$
A354 Grade BD bolts	$122 - 3.6f_v \leq 101$	$122 - 2.8f_v \leq 101$
A449 bolts	$100 - 3.6f_v \leq 81$	$100 - 2.8f_v \leq 81$
A490 bolts	$136 - 3.6f_v \leq 112.5$	$136 - 2.8f_v \leq 112.5$
A307 bolts, Grade A		
When $\frac{1}{4}$ in. $\leq d < \frac{1}{2}$ in.	$52 - 4f_v \leq 40.5$	$52 - 4f_v \leq 40.5$
When $d \geq \frac{1}{2}$ in.	$58.5 - 4f_v \leq 45$	$58.5 - 4f_v \leq 45$
	(B) LRFD method	
A325 bolts	$113 - 2.4f_v \leq 90$	$113 - 1.9f_v \leq 90$
A354 Grade BD bolts	$127 - 2.4f_v \leq 101$	$127 - 1.9f_v \leq 101$
A449 bolts	$101 - 2.4f_v \leq 81$	$101 - 1.9f_v \leq 81$
A490 bolts	$141 - 2.4f_v \leq 112.5$	$141 - 1.9f_v \leq 112.5$
A307 bolts, Grade A		
When $\frac{1}{4}$ in. $\leq d < \frac{1}{2}$ in.	$47 - 2.4f_v \leq 40.5$	$47 - 2.4f_v \leq 40.5$
When $d \geq \frac{1}{2}$ in.	$52 - 2.4f_v \leq 45$	$52 - 2.4f_v \leq 45$

[a] For the LSD method, see appendix B of the AISI North American Specification [7].
Note: d is the nominal diameter of bolt and f_v is the shear stress based on gross cross-sectional area of bolt; 1 in. = 25.4 mm, 1 ksi = 6.9 MPa = 70.3 kg/cm^2.

TABLE 3.11 Nominal Strength Equations for Screws

Type of strength	Nominal strength
Shear strength	
1. Connection shear limited by tilting and bearing	
For $t_2/t_1 < 1.0$, use smallest of three considerations	1. $P_{ns} = 4.2(t_2^3 d)^{1/2} F_{u2}$
	2. $P_{ns} = 2.7 t_1 d F_{u1}$
	3. $P_{ns} = 2.7 t_2 d F_{u2}$
For $t_2/t_1 \geq 2.5$, use smallest of two considerations	1. $P_{ns} = 2.7 t_1 d F_{u1}$
	2. $P_{ns} = 2.7 t_2 d F_{u2}$
For $1.0 < t_2/t_1 < 2.5$, use linear interpolation	
2. Connection shear limited by end distance	$P_{ns} = te F_u^a$
3. Shear in screws	$P_{ns} = 0.8 P_{ss}$
Tensile strength	
1. Connection tension	
Pull-out strength	$P_{not} = 0.85 t_c d F_{u2}$
Pull-over strength	$P_{nov} = 1.5 t_1 d_w F_{u1}$
2. Tension in screws	$P_{nt} = 0.8 P_{ts}$

[a] For the LSD method see appendix B of the AISI North American Specification [7].
Note: d is the nominal diameter of screw, d_w is the larger of screw head diameter and washer diameter, F_{u1} is the tensile strength of member in contact with the screw head (Figure 3.38), F_{u2} is the tensile strength of member not in contact with the screw head (Figure 3.37), P_{ns} is the nominal shear strength per screw, P_{ss} is the nominal shear strength per screw as reported by the manufacturer or determined by tests, P_{nt} is the nominal tension strength per screw, P_{not} is the nominal pull-out strength per screw, P_{nov} is the nominal pull-over strength per screw, P_{ts} is the nominal tension strength per screw as reported by the manufacturer or determined by tests, t_1 is the thickness of member in contact with the screw head (Figure 3.38), t_2 is the thickness of member not in contact with the screw head (Figure 3.37), and t_c is the lesser of depth penetration and thickness t_2.

3.8.1 Metal Buildings

Standardized metal buildings have been widely used in industrial, commercial, and agricultural applications. This type of metal building has also been used for community facilities because of

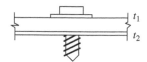

Tilting	$P_{ns} = 4.2\,(t_2^3 d)^{1/2} F_{u2}$ or
Bearing	$P_{ns} = 2.7\,t_1 d F_{u1}$ or
Bearing	$P_{ns} = 2.7\,t_2 d F_{u2}$

FIGURE 3.37 Screw connection for $t_2/t_1 \leq 1.0$.

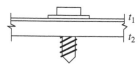

Tilting	N/A
Bearing	$P_{ns} = 2.7\,t_1 d F_{u1}$ or
Bearing	$P_{ns} = 2.7\,t_2 d F_{u2}$

FIGURE 3.38 Screw connection for $t_2/t_1 \geq 2.5$.

its attractive appearance, fast construction, low maintenance, easy extension, and lower long-term cost.

In general, metal buildings are made of welded rigid frames with cold-formed steel sections used for purlins, girts, roofs, and walls. In the United States, the design of standardized metal buildings is often based on the *Metal Building Systems Manual* published by the Metal Building Manufacturers' Association [31]. This manual contains load applications, crane loads, serviceability, common industry practices, guide specifications, AISC-MB certification, wind load commentary, fire protection, wind, snow, and rain data by US county, glossary, appendix and bibliography. In other countries, many design concepts and building systems have also been developed.

3.8.2 Shear Diaphragms

In building construction, it has been a common practice to provide a separate bracing system to resist horizontal loads due to wind load or earthquake. However, steel floor and roof panels, with or without concrete fill, are capable of resisting horizontal in-plane loads in addition to the beam strength for gravity loads if they are adequately interconnected to each other and to the supporting frame. For the same reason, wall panels not only provide enclosure surface and support normal loads, but they can also provide diaphragm action in their own planes.

The structural performance of a diaphragm construction can be evaluated by either calculations or tests. Several analytical procedures exist, and are summarized in the literature [29,45–47]. Tested performance can be measured by the procedures of the *Standard Method for Static Load Testing of Framed Floor, Roof and Wall Diaphragm Construction for Buildings*, ASTM E455 [41,48]. A general discussion of structural diaphragm behavior is given by Yu [1].

Shear diaphragms should be designed for both strength and stiffness. After the nominal shear strength is established by calculations or tests, the design strength can be determined on the basis of the factor of safety or resistance factor given in the Specification [7]. Six cases are currently classified in the AISI Specification for the design of shear diaphragms according to the type of failure mode, type of connections, and type of loading. Because the quality of mechanical connectors is easier to control than that of welded connections, a relatively smaller factor of safety or larger resistance factor is used for mechanical connections. As far as the loading is concerned, the factors of safety for earthquake are slightly larger than those for wind due to the ductility requirements of seismic loading.

3.8.3 Shell Roof Structures

Shell roof structures such as folded-plate and hyperbolic paraboloid roofs have been used in building construction for churches, auditoriums, gymnasiums, schools, restaurants, office buildings, and airplane hangars. This is because the effective use of steel panels in roof construction is not only to provide an economical structure but also to make the building architecturally attractive and flexible for future extension. The design methods used in engineering practice are mainly based on the successful investigation of shear diaphragms and the structural research on shell roof structures.

A folded-plate roof structure consists of three major components. They are (1) steel roof panels, (2) fold line members at ridges and valleys, and (3) end frame or end walls as shown in Figure 3.39. Steel roof panels can be designed as simply supported slabs in the transverse direction between fold lines. The reaction of the panels is then applied to fold lines as a line loading, which can be resolved into two components parallel to the two adjacent plates. These load components are carried by an inclined deep girder spanned between end frames or end walls. These deep girders consist of fold line members as flanges and steel panels as a web element. The longitudinal flange force in fold line members can be obtained by dividing the bending moment of the deep girder by its depth. The shear force is resisted by the diaphragm action of the steel roof panels. In addition to the strength, the deflection characteristics of the folded-plate roof should also be investigated, particularly for long-span structures. In the past, it has been found that a method similar to the Williot diaphragm for determining truss deflections can also be used for the prediction of the deflection of a steel folded-plate roof. The in-plane deflection of each plate should be computed as a sum of the deflections due to flexure, shear, and seam slip, considering the plate temporarily separated from the adjacent plates. The true displacement of the fold line can then be determined analytically or graphically by a Williot diagram. The above discussion deals with a simplified method. The finite-element method can provide a more detailed analysis for various types of loading, support, and material.

The hyperbolic paraboloid roof has also gained popularity due to the economical use of materials and its appearance. This type of roof can be built easily with either single-layer or double-layer standard steel roof deck panels because the hyperbolic paraboloid has straight line generators. Figure 3.40 shows four common types of hyperbolic paraboloid roofs, which may be modified or varied in other ways to achieve a striking appearance. The method of analysis depends on the curvature of the shell used for the roof. If the uniformly loaded shell is deep, the membrane theory may be used. For the case of a shallow shell or a deep shell subjected to unsymmetrical loading, the finite-element method will provide accurate results. Using the membrane theory, the panel shear for a uniformly loaded hyperbolic paraboloid roof can be determined by $wab/2h$, where w is the applied load per unit surface area, a and b are horizontal projections, and h is the amount of corner depression of the surface. This panel shear force should be carried by tension and compression framing members. For additional design information, see Yu [1].

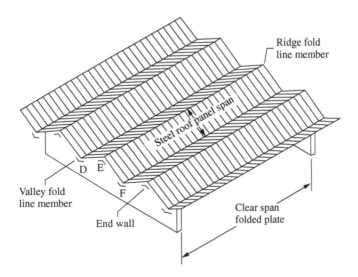

FIGURE 3.39 Folded-plate structure (courtesy of Yu, W.W. 1991).

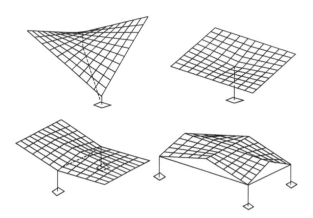

FIGURE 3.40   Types of hyperbolic paraboloid roofs (courtesy of Yu, W.W. 1991).

3.8.4 Wall Stud Assemblies

Cold-formed steel I-, C-, Z-, or box-type studs are widely used in walls with their webs placed perpendicular to the wall surface. The walls may be made of different materials, such as fiber board, lignocellulosic board, plywood, or gypsum board. If the wall material is strong enough and there is adequate attachment provided between wall material and studs for lateral support of the studs, then the wall material can contribute to the structural economy by increasing the usable strength of the studs substantially.

The AISI North American Specification provides the requirements for two types of stud design. The first type is "All Steel Design," in which the wall stud is designed as an individual compression member neglecting the structural contribution of the attached sheathing. The second type is "Sheathing Braced Design," in which consideration is given to the bracing action of the sheathing material due to the shear rigidity and the rotational restraint provided by the sheathing. Both solid and perforated webs are permitted. The subsequent discussion deals with the sheathing braced design of wall studs.

3.8.4.1 Wall Studs in Compression

The AISI design provisions are used to prevent three possible modes of failure. The first requirement is for column buckling between fasteners in the plane of the wall (Figure 3.41). For this case, the limit state may be (1) flexural buckling, (2) torsional buckling, or (3) torsional–flexural buckling depending on the geometric configuration of the cross-section and the spacing of the fasteners. The nominal compressive strength is based on the stud itself without considering any interaction with the sheathing material.

The second requirement is for overall column buckling of wall studs braced by shear diaphragms on both flanges (Figure 3.42). For this case, the AISI Specification provides equations for calculating the critical stresses to determine the nominal axial strength by considering the shear rigidity of the sheathing material. These lengthy equations can be found in Section D4 of the AISI North American Specification [7].

The third requirement is to prevent shear failure of the sheathing by limiting the shear strain within the permissible value for a given sheathing material.

3.8.4.2 Wall Studs in Bending

The nominal flexural strength of wall studs is determined by the nominal section strength by using the "All Steel Design" approach and neglecting the structural contribution of the attached sheathing material.

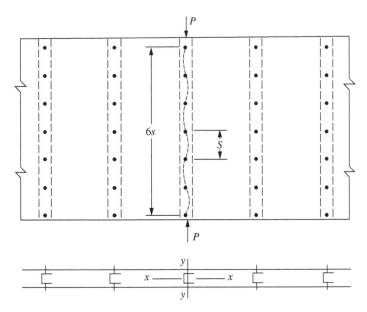

FIGURE 3.41 Buckling of studs between fasteners (courtesy of Yu, W.W. 1991).

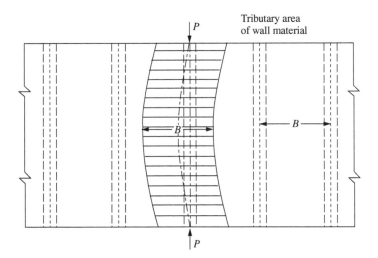

FIGURE 3.42 Overall column buckling of studs (courtesy of Yu, W.W. 1991).

3.8.4.3 Wall Studs with Combined Axial Load and Bending

The AISI interaction equations discussed in Table 3.6 are also applicable to wall studs subjected to combined axial load and bending with the exception that the nominal flexural strength is evaluated by excluding lateral–torsional buckling considerations.

3.8.5 Residential Construction

In recent years, cold-formed steel members have been increasingly used in residential construction as roof trusses, wall framing, and floor systems (Figure 3.43). Because of the lack of standard sections and

FIGURE 3.43 Steel house using cold-formed members for walls, joists, and trusses.

design tables, prescriptive standards have been developed by the National Association of Home Builders Research Center and the Housing and Urban Development. The sectional properties and load-span design tables for a selected group of C-sections have been calculated in accordance with the AISI Specification. For the design of cold-formed steel trusses and shear wall using steel studs, design guides have been published by the American Iron and Steel Institute.

To eliminate regulatory barriers and increase the reliability and cost competitiveness of cold-formed steel framing through improved design and installation standards, since 1998, the AISI Committee on Framing Standards has developed and published four new ANSI-accredited consensus standards. These publications include (1) General Provisions, (2) Truss Design, (3) Header Design, and (4) Prescriptive Method for One and Two Family Dwellings [25]. Two new standards for Wall Stud Design and Lateral Resistance Design are being developed by the AISI Committee on Framing Standards at present (2003).

3.8.6 Composite Construction

Cold-formed steel decks have been used successfully in composite roof and floor construction. For this type of application, the steel deck performs the dual role of serving as a form for the wet concrete during construction and as positive reinforcements for the slab during service.

As far as the design method for the composite slab is concerned, many designs have been based on the SDI Specification for composite steel floor deck [24]. This document contains requirements and recommendations on materials, design, connections, and construction practice. A design handbook is also available from Steel Deck Institute [49]. Since 1984, the American Society of Civil Engineers has published a standard specification for the design and construction of composite slabs [30].

When the composite construction is composed of steel beams or girders with cold-formed steel deck, the design should be based on the AISC Specifications [42,43].

3.9 Computer-Aided Design and Direct Strength Method

The design method discussed in Sections 3.5 and 3.6 deals with the limit states of yielding, local buckling, and overall buckling of structural members. According to the current edition of the AISI North American Specification [7] and previous editions of the AISI Specification, the *effective width design approach* has been and is being used to determine the structural strength of the member. Because hand calculations are excessively lengthy and difficult for solving complicated problems, a large number of computer programs have been prepared by many universities and companies for the analysis and design of cold-formed steel structures. Some of the computer programs can be found on the AISI website at www.steel.org and the website of Wei-Wen Yu Center for Cold-Formed Steel Structures at www.umr.edu/~ccfss.

In recent years, the direct strength method has been developed for determining the member strength [3,50,51]. Instead of using the effective width design approach of individual elements, this method uses the entire cross-section for elastic buckling determination and incorporates local, distortional, and overall buckling in the design process. The cross-section elastic buckling solution insures interelement compatibility and equilibrium at element junctures. Thus, the actual restraining effects of adjoining elements are taken into account. This new method can be used as a rational engineering analysis as permitted by section A1.1(b) of the 2001 edition of the AISI *North American Specification for the Design of Cold-Formed Steel Structural Members*. For details, see the forthcoming new appendix in the North American Specification [7,51]. The computer program for the Direct Strength Method can be obtained from the website at www.ce.jhu.edu/bschafer.

Glossary

ASD (allowable strength design) — A method of proportioning structural components such that the allowable stress, allowable force, or allowable moment is not exceeded by the required allowable strength of the component determined by the load effects of all appropriate combinations of nominal loads. This method is used in the United States and Mexico.

Beam–column — A structural member subjected to combined compressive axial load and bending.

Buckling load — The load at which a compressed element, member, or frame assumes a deflected position.

Cold-formed steel members — Shapes that are manufactured by press-braking blanks sheared from sheets, cut lengths of coils or plates, or by roll forming cold- or hot-rolled coils or sheets.

Composite slab — A slab in which the load-carrying capacity is provided by the composite action of concrete and steel deck (as reinforcement).

Compression members — Structural members whose primary function is to carry concentric loads along their longitudinal axes.

Design strength — R_n/Ω, for ASD or ϕR_n for LRFD and LSD (force, moment, as appropriate), provided by the structural component.

Distorsional buckling — A mode of buckling involving change in cross-sectional shape, excluding local buckling.

Effective design width — Reduced flat width of an element for design purposes. The reduced design width is termed the effective width or effective design width.

Effective length — The equivalent length KL used in design equations.

Factor of safety — A ratio of the stress (or strength) at incipient failure to the computed stress (or strength) at design load (or service load).

Flat width to thickness ratio — The flat width of an element measured along its plane, divided by its thickness.

Flexural members (beams) — Structural members whose primary function is to carry transverse loads or moments.

Limit state — A condition in which a structure or component becomes unsafe (strength limit state) or no longer useful for its intended function (serviceability limit state).

Load factor — A factor that accounts for unavoidable deviations of the actual load from the nominal load.

Local buckling — Buckling of elements only within a section, where the line junctions between elements remain straight and angles between elements do not change.

LRFD (load and resistance factor design) — A method of proportioning structural components such that no applicable limit state is exceeded when the structure is subjected to all appropriate load combinations of factored loads. This method is used in the United States and Mexico.

LSD (limit states design) — A method of proportioning structural components such that no applicable limit state is exceeded when the structure is subjected to all appropriate load combinations of factored loads. This method is used in Canada.

Multiple-stiffened elements — An element that is stiffened between webs, or between a web and a stiffened edge, by means of intermediate stiffeners that are parallel to the direction of stress. A subelement is the portion between adjacent stiffeners or between web and intermediate stiffener or between edge and intermediate stiffener.

Nominal loads — The loads specified by the applicable code not including load factors.

Nominal strength — The capacity of a structure or component to resist the effects of loads, as determined by computations using specified material strengths and dimensions with equations derived from accepted principles of structural mechanics or by tests of scaled models, allowing for modeling effects and differences between laboratory and field conditions.

Point-symmetric section — A point-symmetric section is a section symmetrical about a point (centroid) such as a Z-section having equal flanges.

Required strength — Load effect (force, moment, as appropriate) acting on the structural component determined by structural analysis from the factored loads for LRFD and LSD or nominal loads for ASD (using the most appropriate critical load combinations).

Resistance factor — A factor that accounts for unavoidable deviations of the actual strength from the nominal value.

Stiffened or partially stiffened compression elements — A stiffened or partially stiffened compression element is a flat compression element of which both edges parallel to the direction of stress are stiffened by a web, flange, stiffening lip, intermediate stiffener, or the like.

Stress — Stress as used in this chapter means force per unit area and is expressed in ksi (kips per square inch) for U.S. customary units, MPa for SI units, and kg/cm^2 for the MKS system.

Tensile strength — Maximum stress reached in a tension test.

Thickness — The thickness of any element or section should be the base steel thickness, exclusive of coatings.

Torsional–flexural buckling — A mode of buckling in which compression members can bend and twist simultaneously without change in cross-sectional shape.

Unstiffened compression elements — A flat compression element that is stiffened at only one edge parallel to the direction of stress.

Yield point — Yield point as used in this chapter means either yield point or yield strength of steel.

References

[1] Yu, W.W. 1991. *Cold-Formed Steel Design*, 2nd Edition, John Wiley & Sons, New York, NY (see also 2000, 3rd Edition).
[2] Rhodes, J. 1991. *Design of Cold-Formed Steel Structures*, Elsevier Publishing Co., New York, NY.
[3] Hancock, G.J., Murray, T.M., and Ellifritt, D.S. 2001. *Cold-Formed Steel Structures to the AISI Specification*, Marcel Dekker, Inc., New York, NY.
[4] Ghersi, A., Landolfo, R., and Mazzolani, F.M. 2002. *Design of Metallic Cold-Formed Thin-Walled Members*, Spon, London.
[5] Standards Australia. 1996. *Cold-Formed Steel Structures*, Australia, Suppl 1, 1998.
[6] Lagereinrichtungen. 1974. ONORM B 4901, Dec.
[7] American Iron and Steel Institute. 2001. *North American Specification for the Design of Cold-Formed Steel Structural Members*, Washington, DC.
[8] Czechoslovak State Standard. 1987. *Design of Light Gauge Cold-Formed Profiles in Steel Structures*, CSN 73 1402.

[9] Finnish Ministry of Environment. 1988. *Building Code Series of Finland: Specification for Cold-Formed Steel Structures.*

[10] Centre Technique Industriel de la Construction Metallique. 1978. *Recommendations pourle Calcul des Constructions a Elements Minces en Acier.*

[11] DIN 18807, 1987. *Trapezprofile im Hochbau* (Trapezoidal Profiled Sheeting in Building). Deutsche Norm (German Standard).

[12] Indian Standards Institution. 1975. *Indian Standard Code of Practice for Use of Cold-Formed Light-Gauge Steel Structural Members in General Building Construction*, IS:801-1975.

[13] CNR, 10022. 1984. *Profilati di acciaio formati a freddo. Istruzioni per l' impiego nelle construzioni.*

[14] Architectural Institute of Japan. 1985. *Recommendations for the Design and Fabrication of Light Weight Steel Structures*, Japan.

[15] Groep Stelling Fabrikanten — GSF. 1977. *Richtlijnen Voor de Berekening van Stalen Industriele Magezijnstellingen*, RSM.

[16] Standards New Zealand. 1996. *Cold-Formed Steel Structures*, New Zealand, Suppl 1, 1998.

[17] People's Republic of China National Standard. 2002. *Technical Code for Cold-formed/Thin-Walled Steel Structures*, GB50018-2002, Beijing, China.

[18] South African Institute of Steel Construction. 1995. *Code of Practice for the Design of Structural Steelwork.*

[19] Swedish Institute of Steel Construction. 1982. *Swedish Code for Light Gauge Metal Structures*, Publication 76.

[20] *Romanian Specification for Calculation of Thin-Walled Cold-Formed Steel Members*, STAS 10108/2-83.

[21] British Standards Institution. 1991. *British Standard: Structural Use of Steelwork in Building. Part 5. Code of Practice for Design of Cold-Formed Sections*, BS 5950: Part 5: 1991.

[22] State Building Construction of USSR. 1988. *Building Standards and Rules: Design Standards — Steel Construction*, Part II, Moscow.

[23] Eurocode 3. 1996. *Design of Steel Structures, Part 1–3: General Rules–Supplementary Rules for Cold-Formed Thin Gauge Members and Sheeting.*

[24] Steel Deck Institute. 1995. *Design Manual for Composite Decks, Form Decks, Roof Decks, and Cellular Floor Deck with Electrical Distribution*, Publication No. 29. Fox River Grove, IL.

[25] American Iron and Steel Institute. 2001. *Standards for Cold-Formed Steel Framing: (a) General Provisions, (b) Truss Design, (c) Header Design, and (d) Prescriptive Method for One and Two Family Dwellings*, Washington, DC.

[26] Steel Joist Institute. 1995. *Standard Specification and Load Tables for Open Web Steel Joists*, 40th Edition, Myrtle Beach, SC.

[27] American Society of Civil Engineers. 1988. *Guide for Design of Steel Transmission Towers, Manual 52*, New York, NY.

[28] Rack Manufacturers Institute. 2002. *Specification for the Design, Testing, and Utilization of Industrial Steel Storage Racks*, Charlotte, NC.

[29] Luttrell, L.D. 1987. *Steel Deck Institute Diaphragm Design Manual*, 2nd Edition, Steel Deck Institute, Canton, OH.

[30] American Society of Civil Engineers. 1984. *Specification for the Design and Construction of Composite Steel Deck Slabs*, ASCE Standard, New York, NY.

[31] Metal Building Manufacturers Association. 2002. *Metal Building Systems Manual*, Cleveland, OH.

[32] American Iron and Steel Institute. 1996. *Residential Steel Framing Manual*, including the Cold-Formed Steel Framing Design Guide prepared by T.W.J. Trestain, Washington, DC.

[33] Association of the Wall and Ceiling Industries–International and Metal Lath/Steel Framing Association. 1979. *Steel Framing Systems Manual*, Chicago, IL.

[34] Steel Stud Manufacturers Association. 2001. *Product Technical Information*, Chicago, IL.

[35] American Society of Civil Engineers. 2002. *Specification for the Design of Cold-Formed Stainless Steel Structural Members*, SEI/ASCE-8-02, Reston, VA.

[36] American Iron and Steel Institute. 1983. *Handbook of Steel Drainage and Highway Construction Products*, Washington, DC.

[37] American Iron and Steel Institute. 2000. *Automotive Steel Design Manual*, Washington, DC.

[38] American Society of Civil Engineers. 2002. *Minimum Design Loads for Buildings and Other Structures*, ASCE 7, Washington, DC.

[39] Winter, G. 1970. *Commentary on the Specification for the Design of Cold-Formed Steel Structural Members*, American Iron and Steel Institute, New York, NY.

[40] Winter, G. 1947. Strength of Thin Steel Compression Flanges, *Trans.* ASCE, 112.

[41] American Iron and Steel Institute. 2002. *Cold-Formed Steel Design Manual*, Washington, DC.

[42] American Institute of Steel Construction. 1989. *Specification for Structural Steel Buildings — Allowable Stress Design and Plastic Design*, Chicago, IL.

[43] American Institute of Steel Construction. 2001. *Load and Resistance Factor Design Specification for Structural Steel Buildings*, Chicago, IL.

[44] American Welding Society. 1998. *Structural Welding Code — Sheet Steel*, AWS D1.3-98, Miami, FL.

[45] American Iron and Steel Institute. 1967. *Design of Light-Gage Steel Diaphragms*, New York, NY.

[46] Bryan, E.R. and Davies, J.M. 1981. *Steel Diaphragm Roof Decks — A Design Guide with Tables for Engineers and Architects*, Granada Publishing, New York, NY.

[47] Department of Army. 1992. *Seismic Design for Buildings*, US Army Technical Manual 5-809-10, Washington, DC.

[48] American Society for Testing and Materials. 1993. *Standard Method for Static Load Testing of Framed Floor, Roof and Wall Diaphragm Construction for Buildings*, ASTM E455, Philadelphia, PA.

[49] Heagler, R.B., Luttrell, L.D., and Easterling, W.S. 1997. *Composite Deck Design Handbook*, Steel Deck Institute, Fox River Grove, IL.

[50] Schafer, B.W. and Pekoz, T. 1998. Direct Strength Prediction of Cold-Formed Steel Members using Numerical Elastic Buckling Solutions, *Thin-Walled Structures: Research and Development*, Elsevier, New York, NY.

[51] Schafer, B.W. 2002. Progress on the Direct Strength Method, *Proceedings of the 16th International Specialty Conference on Cold-Formed Steel Structures*, University of Missouri-Rolla, Rolla, MO.

Further Reading

Guide to Stability Design Criteria for Metal Structures, edited by T.V. Galambos, presents general information, interpretation, new ideas, and research results on a full range of structural stability concerns. It was published by John Wiley & Sons in 1998.

Cold-Formed Steel in Tall Buildings, edited by W.W. Yu, R. Baehre, and T. Toma, provides readers with information needed for the design and construction of tall buildings, using cold-formed steel for structural members and architectural components. It was published by McGraw-Hill in 1993.

Thin-Walled Structures, edited by J. Rhodes, J. Loughlan, and K.P. Chong, is an international journal that publishes papers on theory, experiment, design, etc. related to cold-formed steel sections, plate and shell structures, and others. It is published by Elsevier Applied Science. A special issue of the journal on cold-formed steel structures was edited by J. Rhodes and W.W. Yu, guest editors, and published in 1993.

Proceedings of the International Specialty Conference on Cold-Formed Steel Structures, edited by W.W. Yu, J.H. Senne, and R.A. LaBoube, have been published by the University of Missouri–Rolla since 1971. These publications contain technical papers presented at the International Specialty Conferences on Cold-Formed Steel Structures.

"Cold-Formed Steel Structures," by J. Rhodes and N.E. Shanmugan, *The Civil Engineering Handbook* (W.F. Chen and J.Y.R. Liew, Editors-in-Chief), presents discussions on cold-formed steel sections, local buckling of plate elements, and the design of cold-formed steel members and connections. It was published by CRC Press in 2003.

4

Reinforced Concrete Structures

Austin Pan
*T.Y. Lin International,
San Francisco, CA*

4.1 Introduction

Reinforced concrete is a composite material. A lattice or cage of steel bars is embedded in a matrix of Portland cement concrete (see Figure 4.1). The specified compressive strength of the concrete typically ranges from 3,000 to 10,000 psi. The specified yield strength of the reinforcing steel is normally 60,000 psi. Reinforcement bar sizes range from $\frac{3}{8}$ to $2\frac{1}{4}$ in. in diameter (see Table 4.1). The steel reinforcement bars are manufactured with lugs or protrusion to ensure a strong bond between the steel and concrete for composite action. The placement location of the steel reinforcement within the concrete is specified by the concrete *cover*, which is the clear distance between the surface of the concrete and the reinforcement. Steel bars may be bent or hooked.

The construction of a reinforced concrete structural element requires molds or forms usually made of wood or steel supported on temporary shores or falsework (see Photo 4.1). The reinforcement bars are typically cut, bent, and wired together into a mat or cage before they are positioned into the forms. To maintain the specified clear cover, devices such as bar chairs or small blocks are used to support the rebars. Concrete placed into the forms must be vibrated well to remove air pockets. After placement, exposed concrete surfaces are toweled and finished, and sufficient time must be allowed for the concrete to set and cure to reach the desired strength.

The key structural design concept of reinforced concrete is the placement of steel in regions in the concrete where tension is expected. Although concrete is relatively strong in compression, it is weak in tension. Its tensile cracking strength is approximately 10% of its compressive strength. To overcome this weakness, steel reinforcement is used to resist tension; otherwise, the structure will crack excessively and may fail. This strategic combination of steel and concrete results in a composite material that has high strength and retains the versatility and economic advantages of concrete.

To construct concrete structures of even greater structural strength, very high-strength steel, such as Grade 270 strands, may be used instead of Grade 60 reinforcement bars. However, the high strength levels of Grade 270 steel is attained at high strain levels. Therefore, for this type of steel to work effectively with concrete, the high-strength strands must be prestrained or prestressed. This type of structure is

PHOTO 4.1 A 30-story reinforced concrete building under construction. The Pacific Park Plaza is one of the largest reinforced concrete structures in the San Francisco Bay area. It survived the October 17, 1989, Loma Prieta earthquake without damage. Instrumentation in the building recorded peak horizontal accelerations of 0.22g at the base and 0.39g at the top of the building (courtesy of Mr. James Tai, T.Y. International, San Francisco).

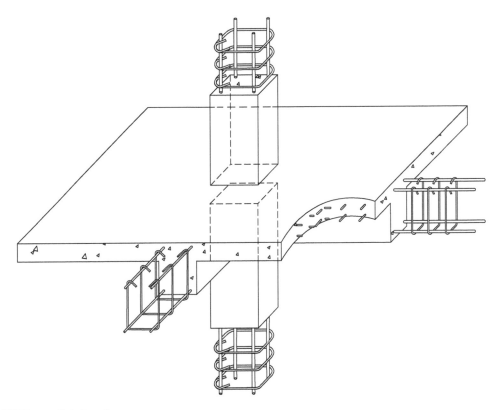

FIGURE 4.1 Reinforced concrete structure.

TABLE 4.1 Reinforcing Bar Properties

Bar size	Nominal properties		
	Diameter (in.)	Area (in.^{2})	Weight (lb/ft)
3	0.375	0.11	0.376
4	0.500	0.20	0.668
5	0.625	0.31	1.043
6	0.750	0.44	1.502
7	0.875	0.60	2.044
8	1.000	0.79	2.670
9	1.128	1.00	3.400
10	1.270	1.27	4.303
11	1.410	1.56	5.313
14	1.693	2.25	7.650
18	2.257	4.00	13.600

Note: Yield stress of ASTM 615 Grade 60 bar = 60,000 psi; modulus of elasticity of reinforcing steel = 29,000,000 psi.

referred to as *prestressed concrete*. Prestressed concrete is considered an extension of reinforced concrete, but it has many distinct features. It is not the subject of this chapter.

4.2 Design Codes

The primary design code for reinforced concrete structures in U.S. design practice is given by the American Concrete Institute (ACI) 318. The latest edition of this code is dated 2002 and is the main reference of this chapter. Most local and state jurisdictions, as well as many national organizations, have adopted ACI 318 for the coverage of reinforced concrete in their design codes. There may be minor changes or additions. The ACI code is incorporated into International Building Code (IBC), as well as the bridge design codes of the American Association of State Highway and Transportation Officials (AASHTO). The ACI Code is recognized internationally; design concepts and provision adopted by other countries are similar to those found in ACI 318.

4.3 Material Properties

With respect to structural design, the most important property of concrete that must be specified by the structural designer is the compressive strength. The typical compressive strength specified, f'_c, is one between 3000 and 8000 psi. For steel reinforcement, Grade 60 (American Society for Testing and Materials [ASTM] A615), with specified yield strength $f_y = 60,000$ psi, has become the industry standard in the United States and is widely available (see Photo 4.2). Material properties of concrete relevant for structural design practice are given in Table 4.2.

PHOTO 4.2 Installation of reinforcing bars in the Pacific Park Plaza building (courtesy of Mr. James Tai, T.Y. International, San Francisco).

TABLE 4.2 Concrete Properties

Concrete strength f'_c (psi)	Modulus of elasticity, $57,000\sqrt{f'_c}$ (psi)	Modulus of rupture, $7.5\sqrt{f'_c}$ (psi)	One-way, shear baseline, $2\sqrt{f'_c}$ (psi)	Two-way, shear baseline, $4\sqrt{f'_c}$ (psi)
3000	3,122,019	411	110	219
4000	3,604,997	474	126	253
5000	4,030,509	530	141	283
6000	4,415,201	581	155	310
7000	4,768,962	627	167	335
8000	5,098,235	671	179	358

Note: Typical range of normal-weight concrete = 145 to 155 pcf; typical range of lightweight concrete = 90 to 120 pcf.

4.4 Design Objectives

For reinforced concrete structures, the design objectives of the structural engineer typically consist of the following:

1. To configure a workable and economical structural system. This involves the selection of the appropriate structural types and laying out the locations and arrangement of structural elements such as columns and beams.
2. To select structural dimensions, depth and width, of individual members, and the concrete cover.
3. To determine the required reinforcement, both longitudinal and transverse.
4. Detailing of reinforcement such as development lengths, hooks, and bends.
5. To satisfy serviceability requirements such as deflections and crack widths.

4.5 Design Criteria

In achieving the design objectives, there are four general design criteria of SAFE that must be satisfied:

1. *Safety, strength, and stability.* Structural systems and member must be designed with sufficient margin of safety against failure.
2. *Aesthetics.* Aesthetics include such considerations as shape, geometrical proportions, symmetry, surface texture, and articulation. These are especially important for structures of high visibility such as signature buildings and bridges. The structural engineer must work in close coordination with planners, architects, other design professionals, and the affected community in guiding them on the structural and construction consequences of decisions derived from aesthetical considerations.
3. *Functional requirements.* A structure must always be designed to serve its intended function as specified by the project requirements. Constructability is a major part of the functional requirement. A structural design must be practical and economical to build.
4. *Economy.* Structures must be designed and built within the target budget of the project. For reinforced concrete structures, economical design is usually not achieved by minimizing the amount of concrete and reinforcement quantities. A large part of the construction cost are the costs of labor, formwork, and falsework. Therefore, designs that replicate member sizes and simplify reinforcement placement to result in easier and faster construction will usually result in being more economical than a design that achieves minimum material quantities.

4.6 Design Process

Reinforced concrete design is often an iterative trial-and-error process and involves the judgment of the designer. Every project is unique. The design process for reinforced concrete structures typically consists of the following steps:

1. Configure the structural system.
2. Determine design data: design loads, design criteria, and specifications. Specify material properties.
3. Make a first estimate of member sizes, for example, based on rule-of-thumb ratios for deflection control in addition to functional or aesthetic requirements.
4. Calculate member cross-sectional properties; perform structural analysis to obtain internal force demands: moment, axial force, shear force, and torsion. Review magnitudes of deflections.
5. Calculate the required longitudinal reinforcement based on moment and axial force demands. Calculate the required transverse reinforcement from the shear and torsional moment demands.

6. If members do not satisfy the SAFE criteria (see previous section), modify the design and make changes to steps 1 and 3.
7. Complete the detailed evaluation of member design to include additional load cases and combinations, and strength and serviceability requirements required by code and specifications.
8. Detail reinforcement. Develop design drawings, notes, and construction specifications.

4.7 Modeling of Reinforced Concrete for Structural Analysis

After a basic structural system is configured, member sizes selected, and loads determined, the structure is analyzed to obtain internal force demands. For simple structures, analysis by hand calculations or approximate methods would suffice (see Section 4.8); otherwise, structural analysis software may be used. For most reinforced concrete structures, a linear elastic analysis, assuming the gross moment of inertia of cross-sections and neglecting the steel reinforcement area, will provide results of sufficient accuracy for design purposes. The final design will generally be conservative even though the analysis does not reflect the actual nonlinear structural behavior because member design is based on ultimate strength design and the ductility of reinforced concrete enables force redistributions (see Sections 4.9 and 4.11). Refined modeling using nonlinear analysis is generally not necessary unless it is a special type of structure under severe loading situations like high seismic forces.

For structural modeling, the concrete modulus E_c given in Table 4.2 can be used for input. When the ends of beam and column members are cast together, the rigid end zone modeling option should be selected since its influence is often significant. Reinforced concrete floor systems should be modeled as rigid diaphragms by master slaving the nodes on a common floor. Tall walls or cores can be modeled as column elements. Squat walls should be modeled as plate or shear wall elements. If foundation conditions and soil conditions are exceptional, then the foundation system will need more refined modeling. Otherwise, the structural model can be assumed to be fixed to the ground. For large reinforced concrete systems or when geometrical control is important, the effects of creep and shrinkage and construction staging should be incorporated in the analysis.

If slender columns are present in the structure, a second-order analysis should be carried out that takes into account cracking by using reduced or effective cross-sectional properties (see Table 4.3 and Section 4.14). If a refined model and nonlinear analysis is called for, then the moment curvature analysis results will be needed for input into the computer analysis (see Section 4.10).

4.8 Approximate Analysis of Continuous Beams and One-Way Slabs

Under typical conditions, for continuous beams and one-way slabs with more than two spans the approximate moment and shear values given in Figure 4.2 may be used in lieu of more accurate analysis methods. These values are from ACI 8.3.3.

TABLE 4.3   Suggested Effective Member Properties for Analysis

Member	Effective moment of inertia for analysis
Beam	$0.35I_g$
Column	$0.70I_g$
Wall — uncracked	$0.70I_g$
Wall — cracked	$0.35I_g$
Flat plates and flat slabs	$0.25I_g$

Note: I_g is the gross uncracked moment of inertia. Use gross areas for input of cross-sectional areas.

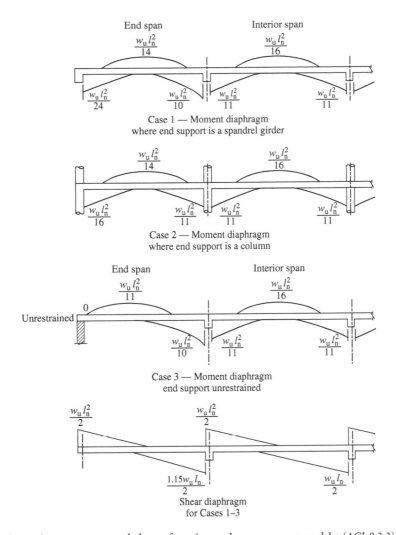

FIGURE 4.2 Approximate moment and shear of continuous beams or one-way slabs (ACI 8.3.3).

4.9 Moment Redistribution

The moment values of a continuous beam obtained from structural analysis may be adjusted or redistributed according to guidelines set by ACI 8.4. Negative moment can be adjusted down or up, but not more than $1000\varepsilon_t$ or 20% (see Notation section for ε_t). After the negative moments are adjusted in a span, the positive moment must also be adjusted to maintain the statical equilibrium of the span (see Section 4.13.12). Redistribution of moment is permitted to account for the ductile behavior of reinforcement concrete members.

4.10 Second-Order Analysis Guidelines

When a refined second-order analysis becomes necessary, as in the case where columns are slender, ACI 10.10.1 places a number of requirements on the analysis.

1. The analysis software should have been validated with test results of indeterminate structures and the predicted ultimate load within 15% of the test results.

2. The cross-section dimensions used in the analysis model must be within 10% of the dimensions shown in the design drawings.
3. The analysis should be based on factored loads.
4. The analysis must consider the material and geometrical nonlinearity of the structure, as well as the influence of cracking.
5. The effects of long-term effects, such as creep shrinkage and temperature effects, need to be assessed.
6. The effect of foundation settlement and soil–structure interaction needs to be evaluated.

A number of commercial software are available that meet the first requirement. If the second requirement is not met, the analysis must be carried out a second time. For the fourth requirement, the moment–curvature or moment–rotation curves need to be developed for the members to provide the accurate results. Alternatively, the code permits approximating the nonlinear effects by using the effective moment of inertias given in Table 4.3. Under the long-term influences of creep and shrinkage, and for stability checks, the effective moment of inertia needs to be further reduced by dividing it by $(1 + \beta_d)$.

4.11 Moment–Curvature Relationship of Reinforced Concrete Members

Member curvature ϕ can be defined as rotation per unit length. It is related to the applied moment M and the section stiffness by the relationship $EI = M/\phi$. A typical moment–curvature diagram of a reinforced concrete beam is shown in Figure 4.3. The reduction in slope of the curve (EI) is the result of concrete cracking and steel yielding. The moment–curvature relationship is a basic parameter of deformation. This information is needed for input if a nonlinear analysis is carried out. For an unconfined reinforced concrete beam section, the point of first cracking is usually

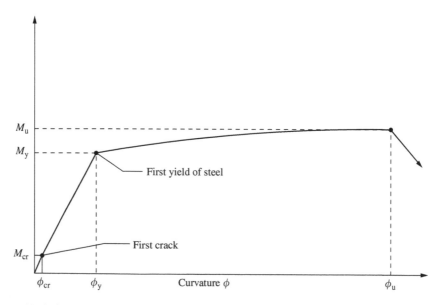

FIGURE 4.3   Typical moment–curvature diagram of a reinforced concrete beam.

neglected for input; the curvature points of first yield ϕ_y and ultimate ϕ_u are calculated from the following formulas:

$$\phi_y = \frac{f_y/E_s}{d(1-k)} \tag{4.1}$$

where

$$k = \left[(\rho + \rho')^2 n^2 + 2\left(\rho + \frac{\rho' d'}{d}\right)n\right]^{1/2} - (\rho + \rho')n \tag{4.2}$$

At ultimate

$$\phi_u = \frac{0.85\beta_1 E_s f_c'}{f_y^2(\rho - \rho')}\varepsilon_c\left\{1 + (\rho + \rho')n - \left[(\rho + \rho')^2 n^2 + 2\left(\rho + \frac{\rho' d'}{d}\right)n\right]^{1/2}\right\} \tag{4.3}$$

The concrete strain at ultimate ε_c is usually assumed to be a value between 0.003 and 0.004 for unconfined concrete. Software is available to obtain more refined moment–curvature relationships and to include other variables. If the concrete is considered confined, then an enhanced concrete stress–strain relationship may be adopted. For column members, the strain compatibility analysis must consider the axial load.

4.12 Member Design for Strength

4.12.1 Ultimate Strength Design

The main requirement of structural design is for the structural capacity, S_C, to be equal to or greater than the structural demand, S_D:

$$S_C \geq S_D$$

Modification factors are included in each side of the equation. The structural capacity S_C is equal to the nominal strength F_n multiplied by a capacity reduction safety factor ϕ:

$$S_C = \phi F_n$$

The nominal strength F_n is the internal ultimate strength at that section of the member. It is usually calculated by the designer according to formulas derived from the theory of mechanics and strength of materials. These strength formulas have been verified and calibrated with experimental testing. They are generally expressed as a function of the cross-section geometry and specified material strengths. There are four types of internal strengths: nominal moment M_n, shear V_n, axial P_n, and torsional moment T_n.

The capacity reduction safety factor ϕ accounts for uncertainties in the theoretical formulas, empirical data, and construction tolerances. The ϕ factor values specified by ACI are listed in Table 4.4.

The structural demand, S_D, is the internal force (moment, shear, axial, or torsion) at the section of the member resulting from the loads on the structure. The structural demand is usually obtained by carrying out a structural analysis of the structure using hand, approximate methods, or computer software. Loads to be input are specified by the design codes and the project specifications and normally include dead, live, wind, and earthquake loads. Design codes such and ACI, IBC, and AASHTO also specify the values of safety factors that should be multiplied with the specified loads and how different types of loads should be combined (i.e., $S_D = 1.2\text{Dead} + 1.6\text{Live}$). ACI load factors and combinations are listed in Table 4.5.

Combining the two equations above, a direct relationship between the nominal strength F_n and the structural demand S_D can be obtained

$$F_n \geq S_D/\phi \tag{4.4}$$

This relationship is convenient because the main design variables, such as reinforcement area, which are usually expressed in terms F_n, can be related directly to the results of the structural analysis.

TABLE 4.4   ACI Strength Reduction Factors ϕ

Nominal strength condition	Strength reduction factor ϕ
Flexure (tension-controlled)	0.90
Compression-controlled (columns)	
Spiral transverse reinforcement	0.70[a]
Other transverse reinforcement	0.65[a]
Shear and torsion	0.75
Bearing on concrete	0.65
Structural plain concrete	0.55

[a] ϕ is permitted to be linearly increased to 0.90 as the tensile strain in the extreme steel increases from the compression-controlled strain of 0.005.

Note: Under seismic conditions strength reduction factors may require modifications.

TABLE 4.5   ACI Load Factors

Load case	Structurals demand S_D or (required strength U)
1	$1.4(D+F)$
2	$1.2(D+F+T)+1.6(L+H)+0.5(L_r$ or S or $R)$
3	$1.2D+1.6(L_r$ or S or $R)+(1.0L$ or $0.8W)$
4	$1.2D+1.6W+1.0L+0.5(L_r$ or S or $R)$
5	$1.2D+1.0E+1.0L+0.2S$
6	$0.9D+1.6W+1.6H$
7	$0.9D+1.0E+1.6H$

Note: D is the dead load, or related internal moments and forces, E is the seismic load, F is the weight and pressure of well-defined fluids, H is the weight and pressure of soils, water in soil, or other materials, L is the live load, L_r is the roof live load, R is the rain load, S is the snow load, T is the time-dependent load (temperature, creep, shrinkage, differential settlement, etc.), and W is the wind load.

4.12.2 Beam Design

The main design steps for beam design and the formulas for determining beam capacity are outlined in the following.

4.12.2.1 Estimate Beam Size and Cover

Table 4.6 may be referenced for selecting a beam thickness. For practical construction, the minimum width of a beam is about 12 in. Economical designs are generally provided when the beam width to thickness ratio falls in the range of $\frac{1}{2}$ to 1. Minimum concrete covers are listed in Table 4.7 and typically should not be less than 1.5 in.

4.12.2.2 Moment Capacity

Taking a beam segment, flexural bending induces a force couple (see Figure 4.4). Internal tension N_T is carried by the reinforcement (the tensile strength of concrete is low and its tension carrying capacity is neglected). Reinforcement at the ultimate state is required to yield, hence

$$N_T = A_s f_y \tag{4.5}$$

At the opposite side of the beam, internal compression force N_C is carried by the concrete. Assuming a simplified rectangular stress block for concrete (uniform stress of $0.85f_c'$),

$$N_C = 0.85f_c' ab \tag{4.6}$$

To satisfy equilibrium, internal tension must be equal to internal compression, $N_C = N_T$. Hence, the depth of the rectangular concrete stress block a can be expressed as

$$a = \frac{A_s f_y}{0.85f_c' b} \tag{4.7}$$

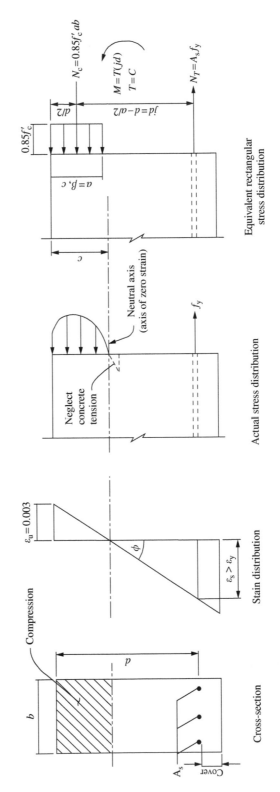

FIGURE 4.4 Mechanics of reinforced concrete beam under flexure.

TABLE 4.6 Minimum Depth of Beams

| Member | Minimum thickness, h | | | |
| | Support condition (L = span length) | | | |
	Simply supported	One end continuous	Both ends continuous	Cantilever
Beams or one-way joists	$L/16$	$L/18.5$	$L/21$	$L/8$
One-way slabs	$L/20$	$L/24$	$L/28$	$L/10$

Notes:
1. Applicable to normal-weight concrete members reinforced with Grade 60 steel and members not supported or attached to partitions or other construction likely to be damaged by large deflection.
2. For reinforcement f_y other than 60,000 psi, the h values above should be multiplied by $(0.4 + f_y/100,000)$.
3. For lightweight concrete of weight W_c (pcf), the h values above should be multiplied by $(1.65 - 0.005 W_c)$, but should not be less than 1.09.

TABLE 4.7 Minimum Concrete Cover

Exposure condition and member type	Minimum cover (in.)
Concrete not exposed to weather or in contact with ground	
Beams, columns	$1\frac{1}{2}$
Slabs, joist, walls	
No. 11 bar and smaller	$\frac{3}{4}$
No. 14 and No. 18 bars	$1\frac{1}{2}$
Concrete exposed to weather or earth	
No. 5 bar and smaller	$1\frac{1}{2}$
No. 6 through No. 18 bars	2
Concrete cast against and permanently exposed to earth	3

The moment capacity of the beam section ϕM_n may be expressed as the tension force multiplied by the moment arm of the force couple.

$$\phi M_n = \phi A_s f_y \left(d - \frac{a}{2} \right) \tag{4.8}$$

The strength reduction factor for flexure ϕ is 0.9.

4.12.2.3 Determination of Required Flexural Reinforcement Area

The maximum moment demand is determined from the structural analysis of the structure under the specified loads and load combinations, M_u. The nominal moment capacity M_n that the cross-section must supply is therefore

$$M_n = M_u/\phi \tag{4.9}$$

The beam cross-section dimensions, width b and thickness h, would be determined first or a first trial selected; the depth of the beam to the centroid of the tension reinforcement can be estimated by

$$d = h - \text{concrete cover} - \text{stirrup diameter} - \text{tension reinforcement bar radius} \tag{4.10}$$

A reasonable size of the stirrup and reinforcement bar can be assumed, if not known (a No. 4 or No. 5 bar size for stirrups is reasonable).

Rearranging the moment capacity equations presented in the previous section, the required flexural reinforcement is obtained by solving for A_s

$$A_s = \frac{M_n}{f_y\left(d - \frac{1}{2}\left(A_s f_y/0.85 f_c' b\right)\right)} \tag{4.11}$$

The required tension reinforcement area A_s is obtained from the quadratic expression

$$A_s = \frac{f_y d \pm \sqrt{(f_y d)^2 - 4 M_n K_m}}{2 K_m} \tag{4.12}$$

where K_m is a material constant:

$$K_m = \frac{f_y^2}{1.7 f_c' b} \tag{4.13}$$

Then, the sizes and quantity of bars are selected. Minimum requirements for reinforcement area and spacing must be satisfied (see the next two sections).

4.12.2.4 Limits on Flexural Reinforcement Area

1. *Minimum reinforcement area for beams:*

$$A_{s,min} = \frac{3\sqrt{f_c'}}{f_y} b_w d \geq 200 b_w d / f_y \tag{4.14}$$

2. *Maximum reinforcement for beams:* The maximum reinforcement A_s must satisfy the requirement that the net tensile strain ε_t (extreme fiber strain less effects of creep, shrinkage, and temperature) is not less than 0.004. The net tensile strain is solved from the compatibility of strain (see Figure 4.4).

$$\varepsilon_t = 0.003 \frac{d - c}{c} \tag{4.15}$$

The neutral axis location c is related to the depth of the compression stress block a by the relationship (ACI 10.2.7.3)

$$c = a/\beta_1 \tag{4.16}$$

The factor β_1 is dependent on the concrete strength as shown in Figure 4.5.

4.12.2.5 Detailing of Longitudinal Reinforcement

Clear spacing between parallel bars should be large enough to permit the coarse aggregate to pass through to avoid honeycombing. The minimum clear spacing should be d_b, but it should not be less than 1 in. For crack control, center-to-center spacing of bars should not exceed

$$\frac{540}{f_s} - 2.5 c_c \leq \frac{432}{f_s} \tag{4.17}$$

where f_s (in ksi) is the stress in the reinforcement at service load, which may be assumed to be 60% of the specified yields strength. Typically, the maximum spacing between bars is about 10 in. The maximum bar spacing rule ensures that crack widths fall below approximately 0.016 in. For very aggressive exposure environments, additional measures should be considered to guard against corrosion, such as reduced concrete permeability, increased cover, or application of sealants.

If the depth of the beam is large, greater than 36 in., additional reinforcement should be placed at the side faces of the tension zone to control cracking. The amount of skin reinforcement to add need not exceed one half of the flexural tensile reinforcement and it should be spread out for a distance $d/2$. The spacing of the skin reinforcement need not exceed $d/6$, 12 in., and $1000 A_b/(d - 30)$.

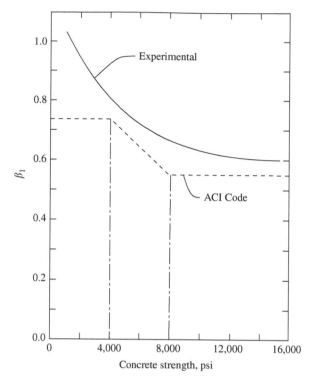

FIGURE 4.5 Relation between β_1 and concrete strength.

To ease reinforcement cage fabrication, a minimum of two top and two bottom bars should run continuously through the span of the beam. These bars hold up the transverse reinforcement (stirrups). At least one fourth of all bottom (positive) reinforcement should run continuously. If moment reversal is expected at the beam–column connection, that is, stress reversal from compression to tension, bottom bars must be adequately anchored into the column support to develop the yield strength.

The remaining top and bottom bars may be cut short. However, it is generally undesirable to cut bars within the tension zone (it causes loss of shear strength and ductility). It is good practice to run bars well into the compression zone, at least a distance d, $12d_b$ or $l_n/16$ beyond the point of inflection (PI) (see Figure 4.6). Cut bars must also be at least one development length l_d in length measured from each side of their critical sections, which are typically the point of peak moment where the yield strength must be developed. See Section 4.17 for development lengths.

To achieve structural integrity of the structural system, beams located at the perimeter of the structure should have minimum continuous reinforcement that ties the structure together to enhance stability, redundancy, and ductile behavior. Around the perimeter at least one sixth of the top (negative) longitudinal reinforcement at the support and one quarter of the bottom (positive) reinforcement should be made continuous and tied with closed stirrups (or open stirrups with minimum 135° hooks). Class A splices may be used to achieve continuity. Top bars should be spliced at the midspan, bottom bars at or near the support.

4.12.2.6 Beams with Compression Reinforcement

Reinforcement on the compression side of the cross-section (see Figure 4.4) usually does not increase in flexural capacity significantly, typically less than 5%, and for most design purposes its contribution to

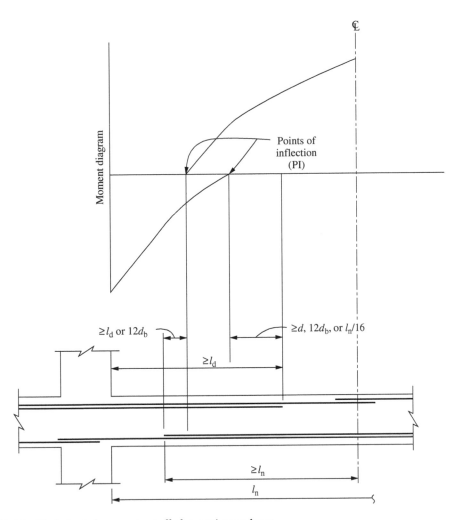

FIGURE 4.6 Typical reinforcement cutoffs for continuous beam.

strength can be neglected. The moment capacity equation considering the compression reinforcement area A_s' located at a distance d' from the compression fiber is

$$\phi M_n = A_s' f_y (d - d') + (A_s - A_s') f_y \left(d - \frac{a}{2} \right) \tag{4.18}$$

where

$$a = \frac{[A_s - A_s'(1 - 0.85 f_c'/f_y)] f_y}{0.85 f_c' b} \tag{4.19}$$

The above expressions assume the compression steel yield, which is typically the case (compression steel quantity is not high). For the nonyielding case, the stress in the steel needs to be determined by a stress–strain compatibility analysis.

Despite its small influence on strength, compression reinforcement serves a number of useful serviceability functions. It is needed for supporting the transverse shear reinforcement in the fabrication of the steel cage. It helps to reduce deflections and long-term creep, and it enhances ductile performance.

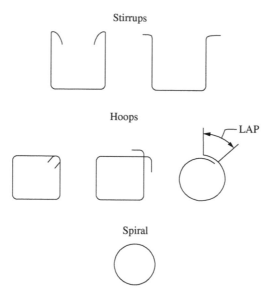

FIGURE 4.7   Typical types of transverse reinforcement.

4.12.2.7 Shear Capacity of Beams

Shear design generally follows after flexural design. The shear capacity ϕV_n of a beam consists of two parts: (1) the shear provided by the concrete itself V_c and (2) that provided by the transverse reinforcement V_s.

$$\phi V_n = \phi(V_c + V_s) \tag{4.20}$$

The strength reduction factor ϕ for shear is 0.85. The nominal shear capacity of the concrete may be taken as the simple expression

$$V_c = 2\sqrt{f_c'}\,b_w d \tag{4.21a}$$

which is in pound and inch units. An alternative empirical formula that allows a higher concrete shear capacity is

$$V_c = \left(1.9\sqrt{f_c'} + 2500\rho_w\frac{V_u d}{M_u}\right)b_w d \le 3.5\sqrt{f_c'}\,b_w d \tag{4.21b}$$

where M_u is the factored moment occurring simultaneously with V_u at the beam section being checked. The quantity $V_u d/M_u$ should not be taken greater than 1.0.

Transverse shear reinforcements are generally of the following types (see Figure 4.7): stirrups, closed hoops, spirals, or circular ties. In addition, welded wire fabric, inclined stirrups, or longitudinal bars bent at an angle may be used. For shear reinforcement aligned perpendicular to the longitudinal reinforcement, the shear capacity provided by transverse reinforcement is

$$V_s = \frac{A_v f_y d}{s} \le 8\sqrt{f_c'}\,b_w d \tag{4.22a}$$

When spirals or circular ties or hoops are used with this formula, d should be taken as 0.8 times the diameter of the concrete cross-section, and A_v should be taken as two times the bar area.

When transverse reinforcement is inclined at an angle α with respect to the longitudinal axis of the beam, the transverse reinforcement shear capacity becomes

$$V_s = \frac{A_v f_y(\sin\alpha_i + \cos\alpha_i)d}{s} \le 8\sqrt{f_c'}\,b_w d \tag{4.22b}$$

The shear formulas presented above were derived empirically, and their validity has also been tested by many years of design practice. A more rational design approach for shear is the strut-and-tie model, which is given as an alternative design method in ACI Appendix A. Shear designs following the strut-and-tie approach, however, often result in designs requiring more transverse reinforcement steel since the shear transfer ability of concrete is neglected.

4.12.2.8 Determination of Required Shear Reinforcement Quantities

The shear capacity must be greater than the shear demand V_u, which is based on the structural analysis results under the specified loads and governing load combination

$$\phi V_n \geq V_u \tag{4.23}$$

Since the beam cross-section dimensions b_w and d would usually have been selected by flexural design beforehand or governed by functional or architectural requirements, the shear capacity provided by the concrete V_c can be calculated by Equations 4.21a or 4.21b. From the above equations, the required shear capacity to be provided by shear reinforcement must satisfy the following:

$$V_s \geq \frac{V_u}{\phi} - V_c \tag{4.24}$$

Inserting V_s from this equation into Equation 4.22a, the required spacing and bar area of the shear reinforcement (aligned perpendicular to the longitudinal reinforcement) must satisfy the following:

$$\frac{s}{A_v} \leq \frac{f_y d}{V_s} \tag{4.25}$$

For ease of fabrication and bending, a bar size in the range of No. 4 to No. 6 is selected, then the required spacing s along the length of the beam is determined, usually rounded down to the nearest $\frac{1}{2}$ in.

In theory, the above shear design procedure can be carried out at every section along the beam. In practice, a conservative approach is taken and shear design is carried out at only one or two locations of maximum shear, typically at the ends of the beam, and the same reinforcement spacing s is adopted for the rest of the beam. Where the beam ends are cast integrally or supported by a column, beam, wall, or support element that introduces a region of concentrated compression, the maximum value of the shear demand need not be taken at the face of the support, but at a distance d away (see Figure 4.8).

Transverse reinforcement in the form of closed stirrups is preferred for better ductile performance and structural integrity. For beams located at the perimeter of the structure, ACI requires closed stirrups (or open stirrups within minimum 135° hooks). In interior beams, if closed stirrups are not provided, at least one quarter of the bottom (positive) longitudinal reinforcement at midspan should be made continuous over the support, or at the end support, detailed with a standard hook.

4.12.2.8.1 *Minimum Shear Reinforcement and Spacing Limits*

After the shear reinforcement and spacing are selected they should be checked against minimum requirements. The minimum shear reinforcement required is

$$A_{vmin} = 0.75\sqrt{f_c'}\,\frac{b_w s}{f} \geq \frac{50 b_w s}{f_y} \tag{4.26}$$

This minimum shear area applies in the beam where $V_u \geq \phi V_2 / 2$. It does not apply to slabs, footings, and concrete joists. The transverse reinforcement spacing s should not exceed $d/2$ nor 24 in. These spacing limits become $d/4$ and 12 in. when V_s exceeds $4\sqrt{f_c'}b_w d$.

When significant torsion exists, additional shear reinforcement may be needed to resist torsion. This is covered in Section 4.16.

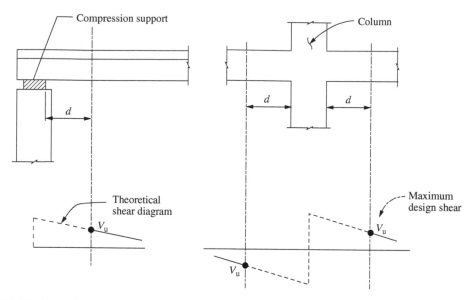

FIGURE 4.8 Typical support conditions for locating factored shear force V_u.

4.12.2.8.2 Modifications for High-Strength and Lightweight Concretes

For concretes with compressive strengths greater than 10,000 psi, the values of $\sqrt{f_c'}$ in all the shear capacity and design equations above should not exceed 100 psi. For lightweight concretes, $\sqrt{f_c'}$ should be multiplied by 0.75 for all-lightweight concrete, or 0.85 for sand-lightweight concrete. If the tensile strength f_{ct} of the concrete is specified, $\sqrt{f_c'}$ may be substituted by $f_{ct}/6.7$, but should not be greater than $\sqrt{f_c'}$.

4.12.2.9 Detailing of Transverse Reinforcement

Transverse reinforcement should extend close to the compression face of a member, as far as cover allows, because at ultimate state deep cracks may cause loss of anchorage. Stirrup should be hooked around a longitudinal bar by a standard stirrup hoop (see Figure 4.9). It is preferable to use transverse reinforcement size No. 5 or smaller. It is more difficult to bend a No. 6 or larger bar tightly around a longitudinal bar. For transverse reinforcement sizes No. 6, No. 7, and No. 8, a standard stirrup hook must be accompanied by a minimum embedment length of $0.014d_b f_y/\sqrt{f_c'}$ measured between the midheight of the member and the outside end of the hook.

4.12.3 One-Way Slab Design

When the load normal to the surface of a slab is transferred to the supports primarily in one major direction, the slab is referred to as a one-way slab. For a slab panel supported on all four edges, one-way action occurs when the aspect ratio, the ratio of its long-to-short span length, is greater than 2. Under one-way action, the moment diagram remains essentially constant across the width of the slab. Hence, the design procedure of a one-way slab can be approached by visualizing the slab as an assembly of the same beam strip of unit width. This beam strip can be designed using the same design steps and formulas presented in the previous section for regular rectangular beams.

   The required cover for one-way slab is less than for beams, typically $\frac{3}{4}$ in. The internal forces in one-way slabs are usually lower, so smaller bar sizes are used. The design may be controlled by the minimum temperature and shrinkage reinforcement. Shear is rarely a controlling factor for one-way slab design. Transverse reinforcement is difficult to install in one-way slabs. It is more economical to thicken or haunch the slab.

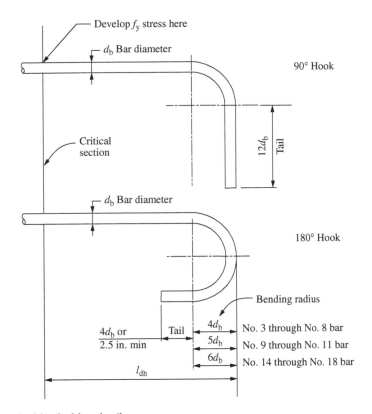

FIGURE 4.9   Standard hooked bar details.

4.12.3.1 Shrinkage and Temperature Reinforcement (ACI 4.12)

For Grade 60 reinforcement, the area of shrinkage and temperature reinforcement should be 0.0018 times the gross concrete area of the slab. Bars should not be spaced farther than five times the slab thickness or 18 in. The shrinkage and temperature requirements apply in both directions of the slab, and the reinforcement must be detailed with adequate development length where yielding is expected.

4.12.4 T-Beam Design

Where a slab is cast integrally with a beam, the combined cross-section acts compositely (see Figure 4.10). The design of T-beam differs from that of a rectangular beam only in the positive moment region, where part of the internal compression force occurs in the slab portion. The design procedures and formulas for T-beam design are the same as for rectangular beams, except for the substitution of b in the equations with an effective width b_{eff} at positive moment sections. The determination of b_{eff} is given in Figure 4.10. The effective width b_{eff} takes into account the participation of the slab in resisting compression. In the rare case where the depth of the compression stress block a exceeds the slab thickness, a general stress–strain compatibility analysis would be required. For shear design the cross-section width should be taken as the width of the web b_w.

4.12.4.1 Requirements for T-Beam Flanges

If the T-beam is an isolated beam and the flanges are used to provide additional compression area, the flange thickness should be not less than one half the width of the web and the effective flange width not more than four times the width of the web. For a slab that forms part of the T-beam flange and if the slab primary flexural reinforcement runs parallel to the T-beam, adequate transverse reinforcement needs

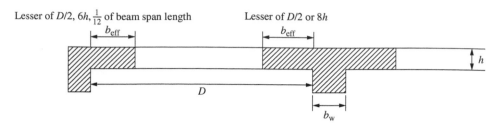

Lesser of $D/2$, $6h$, $\frac{1}{12}$ of beam span length   Lesser of $D/2$ or $8h$

FIGURE 4.10   T-beam section.

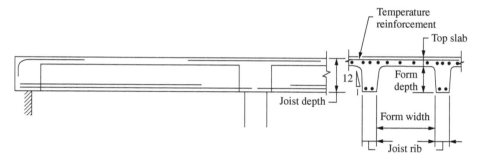

FIGURE 4.11   One-way joist.

to be provided in the slab by treating the flange as a cantilever. The full cantilevering length is taken for an isolated T-beam; otherwise, the effective flange length is taken.

4.12.5 One-Way Joist Design

A one-way joist floor system consists of a series of closely spaced T-beams (see Figure 4.11). The ribs of joists should not be less than 4 in. in width and should have a depth not more than 3.5 times the minimum width of the rib. Flexural reinforcement is determined by T-section design. The concrete ribs normally have sufficient shear capacity so that shear reinforcement is not necessary. A 10% increase is allowed in the concrete shear capacity calculation, V_c, if the clear spacing of the ribs does not exceed 30 in. Alternatively, higher shear capacity can be obtained by thickening the rib at the ends of the joist where the high shear demand occurs. If shear reinforcements are added, they are normally in the form of single-leg stirrups. The concrete forms or fillers that form the joists may be left in place; their vertical stems can be considered part of the permanent joist design if their compressive strength is at least equal to the joist. The slab thickness over the permanent forms should not be less than $\frac{1}{12}$ of the clear distance between ribs or less than 1.5 in. Minimum shrinkage and temperature reinforcement need to be provided in the slab over the joist stems. For structural integrity, at least one bottom bar in the joist should be continuous or spliced with a Class A tension splice (see Section 4.17) over continuous supports. At discontinuous end supports, bars should be terminated with a standard hook.

4.13 Two-Way Floor Systems

Design assuming one-way action is not applicable in many cases, such as when a floor panel is bounded by beams with a long to short aspect ratio of less than 2. Loads on the floor are distributed in both directions, and such a system is referred to as a two-way system (see Figure 4.12). The design approach of two-way floor systems remains in many ways similar to that of the one-way slab, except that the floor slab should now be visualized as being divided into a series of slab strips spanning *both* directions of the floor panel (see Figure 4.12). In the case of one-way slabs, each slab strip carries the same design moment diagram. In two-way systems, the design moment diaphragm varies from one strip to another. Slab strips

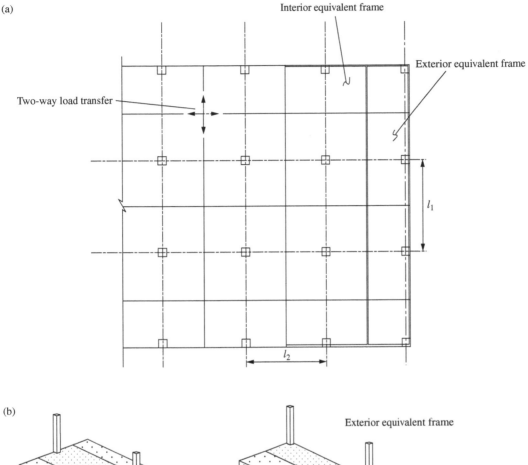

FIGURE 4.12 (a) Two-way floor system and (b) equivalent frames.

closer to the column support lines would generally carry a higher moment than strips at midspan. Hence, a key design issue for two-way floor design becomes one of analysis, on how to obtain an accurate estimate of internal force distribution among the slab strips. After this issue is resolved, and the moment diagrams of each strip are obtained, the flexural reinforcement design of each slab strip follows the same procedures and formulas as previously presented for one-way slabs and beams. Of course, the analysis of two-way floor systems can also be solved by computer software, using the finite element method, and a number of structural analysis software have customized floor slab analysis modules. The ACI Code contains an approximate manual analysis method, the Direct Design Method, for two-way floors, which is practical for design purposes. A more refined approximate method, the Equivalent Frame Method, is also available in the ACI.

If a floor system is regular in layout and stiffness (ACI 13.6.1), the Direct Design Method may be used to obtain the moment diagrams for the slab strips of two-way floor systems. The Direct Design Method is based on satisfying the global statical equilibrium of each floor panel. The relative stiffnesses of the panel components (e.g., slab, beam, drop panels) are then considered in distributing the statical moment. The subsequent sections present the application of the ACI Direct Design Method for different types of two-way systems.

General detailing of two-way slabs. The required slab reinforcement areas are taken at the critical sections, generally at the face of supports around the perimeter of the panel and at the midspans of the column and middle strips. The maximum spacing of reinforcement should not exceed two times the slab thickness or that required for temperature reinforcement (see Section 4.12.3.1). All bottom bars in slab panels that run perpendicular to the edge of the floor should be extended to the edge and anchored into the edge beam, column, or wall that exists there.

Opening in slabs of any size is permitted in the area common to intersecting middle strips (see Figure 4.12). But the original total reinforcement in the slab panel should be maintained by transferring bars to the sides of the opening. In intersecting column strips, not more than one eighth the width of the column strip should be interrupted by an opening. In the area common to one column strip and one middle strip, not more than one quarter of the reinforcement should be interrupted by an opening. If a larger opening is required, then edge beams or bands of reinforcement around the opening should be added.

4.13.1　Two-Way Slab with Beams

This system is shown in Figure 4.13. It consists of a slab panel bounded with beams supported on columns. Since the long to short aspect ratio of the panels is less than 2, a significant portion of the floor loading is transferred in the long direction. And the stiffness of the integral beams draws in load.

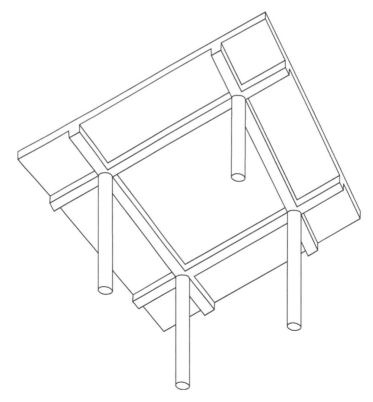

FIGURE 4.13　Two-way slab with beams.

TABLE 4.8 Minimum Thickness of Flat Plates (Two-Way Slabs without Interior Beams)

Yield strength, f_y (psi)	Without drop panels			With drop panels		
	Exterior panels		Interior panels	Exterior panels		Interior panels
	Without edge beams	With edge beams		Without edge beams	With edge beams	
40,000	$l_n/33$	$l_n/36$	$l_n/36$	$l_n/36$	$l_n/40$	$l_n/40$
60,000	$l_n/30$	$l_n/33$	$l_n/33$	$l_n/33$	$l_n/36$	$l_n/36$
75,000	$l_n/28$	$l_n/31$	$l_n/31$	$l_n/31$	$l_n/34$	$l_n/34$

Notes:
1. l_n is length of clear in long direction, face-to-face of support.
2. Minimum thickness if slabs without drop panels should not be less than 5 in.
3. Minimum thickness of slabs with drop panel should not be less than 4 in.

The minimum thickness of two-way slabs is dependent on the relative stiffness of the beams α_m. If $0.2 \leq \alpha_m \leq 2.0$, the slab thickness should not be less than 5 in. or

$$\frac{l_n\left(0.8 + (f_y/200,000)\right)}{35 + 5\beta(\alpha_m - 0.2)} \tag{4.27}$$

If $\alpha_m > 2.0$, the denominator in the above equation should be replaced with $(36 + 9\beta)$, but the thickness should not be less than 3.5 in. When $\alpha_m < 0.2$, the minimum thickness is given by Table 4.8.

4.13.1.1 Column Strips, Middle Strips, and Equivalent Frames

For the Direct Design Method, to take into account the change of the moment across the panel, the floor system is divided into column and middle strips in each direction. The column strip has a width on each side of a column centerline equal to $0.25l_2$ or $0.25l_1$, whichever is less (see Figure 4.12). A middle strip is bounded by two column strips. The moment diagram across each strip is assumed to be constant and the reinforcement is designed for each strip accordingly.

In the next step of the Direct Design Method, equivalent frames are set up. Each equivalent frame consists of the columns and beams that share a common column or grid line. Beams are attached to the slabs that extend to the half-panel division on each side of the grid line, so the width of each equivalent frame consists of one column strip and two half middle strips (see Figure 4.12). Equivalent frames are set up for all the grid lines in both directions of the floor system.

4.13.1.2 Total Factored Static Moment

The first analysis step of the Direct Design Method is determining the total static moment in each span of the equivalent frame

$$M_0 = \frac{w_u l_2 l_n^2}{8} \tag{4.28}$$

Note that w_u is the full, not half, factored floor load per unit area. The clear span l_n is measured from face of column to face of column. The static moment is the absolute sum of the positive midspan moment plus the average negative moment in each span (see Figure 4.14).

The next steps of the Direct Design Method involve procedures for distributing the static moment M_0 into the positive (midspan) and negative moment (end span) regions, and then on to the column and middle strips. The distribution procedures are approximate and reflect the relative stiffnesses of the frame components (Table 4.9).

4.13.1.3 Distribution of Static Moment to Positive and Negative Moment Regions

The assignment of the total factored static moment M_0 to the negative and positive moment regions is given in Figure 4.14. For interior spans, $0.65M_0$ is assigned to each negative moment region and $0.35M_0$

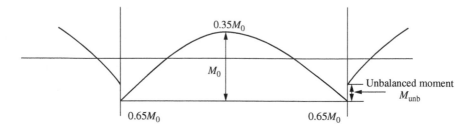

FIGURE 4.14 Static moment in floor panel.

TABLE 4.9 Distribution of Statical Moment for End Span Slab Panels

	Exterior edge unrestrained	Slab with beams between all supports	Slab without beams between interior supports		Exterior edge fully restrained
			Without edge beam	With edge beam	
Interior negative factored moment	0.75	0.70	0.70	0.70	0.65
Positive factored moment	0.63	0.57	0.52	0.50	0.35
Exterior negative factored moment	0	0.16	0.26	0.30	0.65

to the positive moment region. For the exterior span, the percentage of distribution is a function of the degree of restraint, as given in Table 4.9.

After the static moment is proportioned to the negative and positive regions, it is further apportioned on to the column and middle strips. For positive moment regions, the proportion of moment assigned to the column strips is given in Table 4.10. The parameter α_1 is a relative stiffness of the beam to slab, based on the full width of the equivalent frame:

$$\alpha_1 = \frac{E_{cb} I_b}{E_{cs} I_s} \tag{4.29}$$

For interior negative moment regions the proportion of moment assigned to the column strip is given by Table 4.11.

For negative moment regions of an exterior span, the moment assigned to the column follows Table 4.12, which takes into account the torsional stiffness of the edge beam. The parameter β_t is the ratio of torsional stiffness of edge beam section to flexural stiffness of a width of the slab equal to the center-to-center span length of the beam

$$\beta_t = \frac{E_{cb} C}{2 E_{cs} I_s} \tag{4.30}$$

The remaining moment, that was not proportioned to the column strips, is assigned to the middle strips.

Column strip moments need to be further divided into their slab and beam. The beam should be proportioned to take 85% of the column strip moment if $\alpha_1 l_1/l_2 \geq 1.0$. Linear interpolation is applied if this parameter is less than 1.0. If the beams are also part of a lateral force resisting system, then moments due to lateral forces should be added to the beams. After the assignment of moments, flexural reinforcement in the beams and slab strips can be determined following the same design procedures presented in Sections 4.12.2 and 4.12.3 for regular beams and one-way slabs.

TABLE 4.10 Distribution of Positive Moment in Column Strip

l_2/l_1	0.5	1.0	2.0
$(\alpha_1 l_2/l_1) = 0$	60	60	60
$(\alpha_1 l_2/l_1) \geq 1.0$	90	75	45

TABLE 4.11 Distribution of Interior Negative Moment in Column Strip

l_2/l_1	0.5	1.0	2.0
$(\alpha_1 l_2/l_1) = 0$	75	75	75
$(\alpha_1 l_2/l_1) \geq 1.0$	90	75	45

TABLE 4.12 Distribution of Negative Moment to Column of an Exterior Span

l_2/l_1		0.5	1.0	2.0
$(\alpha_1 l_2/l_1) = 0$	$\beta_t = 0$	100	100	100
	$\beta_t \geq 2.5$	75	75	75
$(\alpha_1 l_2/l_1) \geq 1.0$	$\beta_t = 0$	100	100	100
	$\beta_t \geq 2.5$	90	75	45

4.13.1.4 Shear Design

The shear in the beam may be obtained by assuming that floor loads act according to the 45° tributary areas of each respective beam. Additional shear from lateral loads and the direct loads on the beam should be added on. The shear design of the beam then follows the procedure presented in Section 4.12. Shear stresses in the floor slab are generally low, but they should be checked. The strip method, which approximates the slab shear by assuming a unit width of slab strip over the panel, may be used to estimate the shear force in the slab.

4.13.2 Flat Plates

Floor systems without beams are commonly referred to as flat plates, (see Figure 4.15). Flat plates are economical and functional because beams are eliminated and floor height clearances are reduced. Minimum thicknesses of flat plates are given in Table 4.8 and should not be less than 5 in. The structural design procedure is the same as for flat slab with beams, presented in the previous sections, except that for flat plates $\alpha_1 = 0$. Refer to Section 4.13.1.2 for the static moment calculation. For the exterior span the distribution of the static moment is given in Figure 4.14. Table 4.10 and Table 4.11 provide the application for moment assignments to column strips.

4.13.2.1 Transfer of Forces in slab–column connections

An important design requirement of the flat plate system is the transfer of forces between the slab and its supporting columns (see Figure 4.14 and Figure 4.16). This transfer mechanism is a complex one. The accepted design approach is to assume that a certain fraction of the unbalanced moment M_{unb} in the slab connection is transferred by direct bending into the column support. This γ_f fraction is estimated to be

$$\gamma_f = \frac{1}{1 + (2/3)\sqrt{b_1/b_2}} \tag{4.31}$$

The moment $\gamma_f M_{unb}$ is transferred over an effective slab width that extends 1.5 times the slab thickness outside each side face of the column or column capital support. The existing reinforcement in the column strip may be concentrated over this effective width or additional bars may be added.

The fraction of unbalanced moment not transferred by flexure γ_v ($\gamma_v = 1 - \gamma_f$) is transferred through eccentricity of shear that acts over an imaginary critical section perimeter located at a

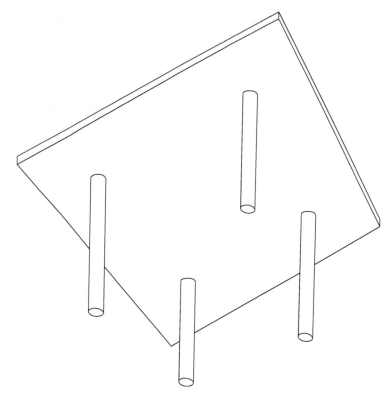

FIGURE 4.15 Flat plate.

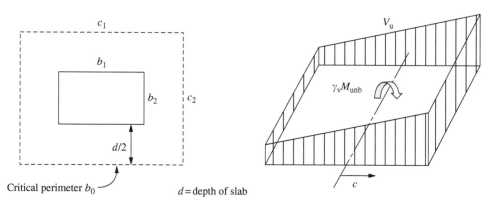

FIGURE 4.16 Transfer of shear in slab–column connections.

distance $d/2$ from the periphery of the column support (see Figure 4.16). Shear stress at the critical section is determined by combining the shear stress due to the direct shear demand V_u (which may be obtained from tributary loading) and that from the eccentricity of shear due to the unbalanced moment:

$$v_u = \frac{V_u}{A_c} \pm \frac{\gamma_v M_{unb} c}{J_c}$$ (4.32)

where the concrete area of the critical section $A_c = b_0 d = 2d(c_1 + c_2 + 2d)$, and J_c is the equivalent polar moment of inertia of the critical section

$$J_c = \frac{d(c_1 + d)^3}{6} + \frac{(c_1 + d)d^3}{6} + \frac{d(c_2 + d)(c_1 + d)^2}{2} \tag{4.33}$$

The maximum shear stress v_u on the critical section must not exceed the shear stress capacity defined by

$$\phi v_n = \phi V_c / b_0 d \tag{4.34}$$

The concrete shear capacity V_c for two-way action is taken to be the lowest of the following three quantities:

$$V_c = 4\sqrt{f_c'}b_0 d \tag{4.35}$$

$$V_c = \left(2 + \frac{4}{\beta_c}\right)4\sqrt{f_c'}b_0 d \tag{4.36}$$

$$V_c = \left(\frac{\alpha_s d}{b_0} + 2\right)4\sqrt{f_c'}b_0 d \tag{4.37}$$

where β_c is the ratio of long side to short side of the column. The factor α_s is 40 for interior columns, 30 for edge columns, or 20 for corner columns.

If the maximum shear stress demand exceeds the capacity, the designer should consider using a thicker slab or a larger column, or increasing the column support area with a column capital. Other options include insertion of shear reinforcement or shearhead steel brackets.

4.13.2.2 Detailing of Flat Plates

Refer to Figure 4.17 for minimum extensions for reinforcements. All bottom bars in the column strip should be continuous or spliced with a Class A splice. To prevent progressive collapse, at least two of the column strip bottom bars in each direction should pass within the column core or be anchored at the end supports. This provides catenary action to hold up the slab in the event of punching failure.

4.13.3 Flat Slabs with Drop Panels and/or Column Capitals

The capacity of flat plates may be increased with drop panels. Drop panels increase the slab thickness over the negative moment regions and enhance the force transfer in the slab–column connection. The minimum required configuration of drop panels is given in Figure 4.18. The minimum slab thickness is given in Table 4.8 and should not be less than 4 in.

Alternatively, or in combination with drop panels, column capitals may be provided to increase capacity (see Figure 4.19). The column capital geometry should follow a 45° projection. Column capitals increase the critical section of the slab–column force transfer and reduce the clear span lengths. The design procedure outlined for flat plates in the previous sections are applicable for flat slabs detailed with drop panels or column capitals.

4.13.4 Waffle Slabs

For very heavy floor loads or very long spans, waffle slab floor systems become viable (see Figure 4.20). A waffle slab can be visualized as being a very thick flat plate but with coffers to reduce weight and gain efficiency. The design procedure is therefore the same as for flat plates as presented in Section 4.13.2.

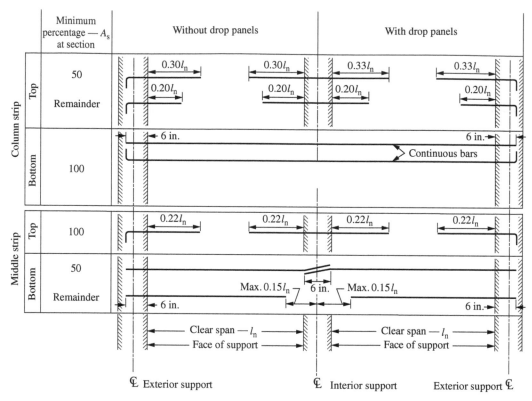

	Minimum percentage — A_s at section	Without drop panels		With drop panels	
Column strip — Top	50 Remainder	$0.30l_n$ $0.20l_n$	$0.30l_n$ $0.20l_n$	$0.33l_n$ $0.20l_n$	$0.33l_n$ $0.20l_n$
Column strip — Bottom	100	6 in.		Continuous bars 6 in.→	
Middle strip — Top	100	$0.22l_n$	$0.22l_n$	$0.22l_n$	$0.22l_n$
Middle strip — Bottom	50 Remainder	6 in.	Max. $0.15l_n$ 6 in. Max. $0.15l_n$		6 in.→

Clear span — l_n Clear span — l_n

Face of support   Face of support

℄ Exterior support   ℄ Interior support   Exterior support ℄

FIGURE 4.17   Detailing of flat plates.

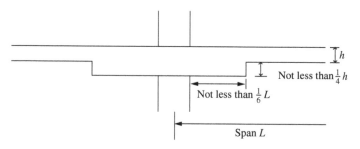

Not less than $\frac{1}{4}h$

Not less than $\frac{1}{6}L$

Span L

FIGURE 4.18   Drop panel dimensions.

The flexural reinforcement design is based on T-section strips instead of rectangular slab strips. Around column supports, the coffers may be filled in to act as column capitals.

4.14   Columns

Typical reinforcement concrete columns are shown in Figure 4.21. Longitudinal reinforcements in columns are generally distributed uniformly around the perimeter of the column section and run continuously through the height of the column. Transverse reinforcement may be in the form of rectangular hoops, ties, or spirals (Figure 4.21). Tall walls and core elements in buildings (Figure 4.22) are column-like in behavior and the design procedures presented in the following are applicable.

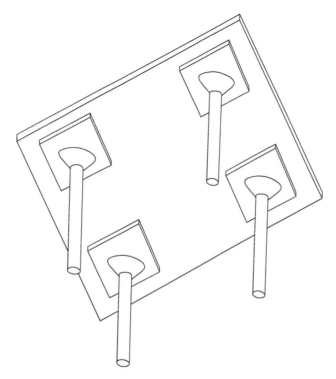

FIGURE 4.19   Flat slab with drop panels and column capital.

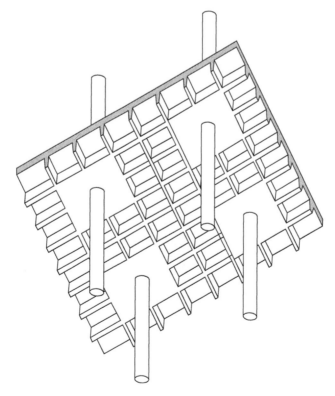

FIGURE 4.20   Waffle slab.

With rectangular hoops   With spirals or circular hoops

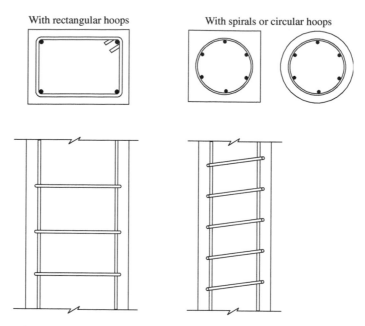

FIGURE 4.21 Typical reinforced concrete columns.

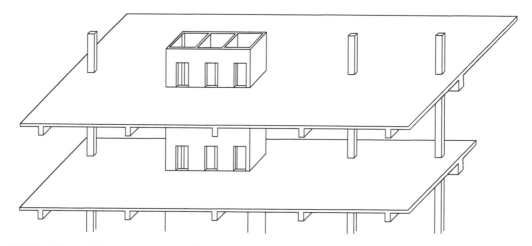

FIGURE 4.22 Reinforced concrete building elements.

4.14.1 Capacity of Columns under Pure Compression

Under pure compression (i.e., no moment) the axial capacity of columns reinforced with hoops and ties as transverse reinforcement is the sum of the axial capacity of the concrete and the steel:

$$\phi P_{n,max} = \phi\phi_{ecc}[0.85f'_c(A_g - A_{st}) + A_{st}f_y] \qquad (4.38)$$

The strength reduction factor ϕ for tied columns is 0.65. The additional reduction factor ϕ_{ecc} shown in the equation accounts for accidental eccentricity from loading or due to construction tolerances that will induce moment. For tied column $\phi_{ecc} = 0.80$. For spiral columns, $\phi = 0.75$ and $\phi_{ecc} = 0.85$. Columns reinforced with spiral reinforcement are more ductile and reliable in sustaining axial load after spalling of concrete cover. Hence, lower reduction factors are assigned by ACI.

4.14.2 Preliminary Sizing of Columns

For columns that are expected to carry no or low moment, the previous equation can be rearranged to estimate the required gross cross-sectional area to resist the axial force demand P_u:

$$A_g > \frac{(P_u/\phi\phi_{ecc}) - A_{st}f_y}{0.85f_c'} \tag{4.39}$$

The ACI Code limits the column reinforcement area A_{st} to 1 to 8% of A_g. Reinforcement percentages less than 4% are usually more practical in terms of avoiding congestion and to ease fabrication. If a column is expected to carry significant moment, the A_g estimated by the above expression would not be adequate. To obtain an initial trial size in that case, the above A_g estimate may be increased by an appropriate factor (e.g., doubling or more).

4.14.3 Capacity of Columns under Combined Axial Force and Moment

Under the combined actions of axial force and moment, the capacity envelope of a column is generally described by an interaction diagram (see Figure 4.23). Load demand points (M_u, P_u) from all load combinations must fall inside the $\phi P_n - \phi M_n$ capacity envelope; otherwise, the column is considered inadequate and should be redesigned. Computer software are typically used in design practice to generate column interaction diagrams.

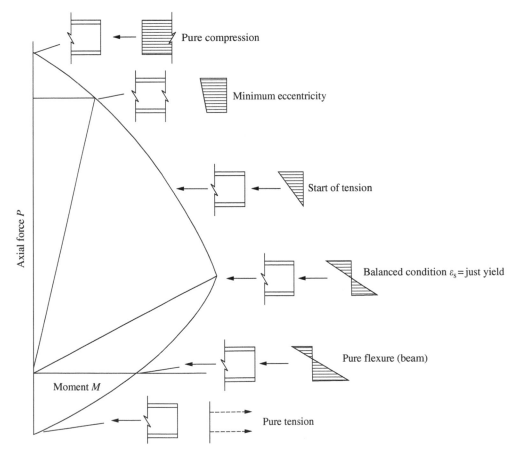

FIGURE 4.23 Column interaction diagram.

The upper point of an interaction curve is the case of pure axial compression. The lowest point is the case of pure axial tension, $\phi P_{n,tension} = \phi A_{st} f_y$ (it is assumed that the concrete section cracks and supplies no tensile strength). Where the interaction curve intersects with the moment axis, the column is under pure bending, in which case the column behaves like a beam. The point of maximum moment on the interaction diagram coincides with the balanced condition. The extreme concrete fiber strain reaches ultimate strain (0.003) simultaneously with yielding of the extreme layer of steel on the opposite side ($f_y/E_s = 0.002$).

Each point of the column interaction curve represents a unique strain distribution across the column section. The axial force and moment capacity at each point is determined by a strain compatibility analysis, similar to that presented for beams (see Section 4.12) but with an additional axial force component. The strain at each steel level i is obtained from similar triangles $\varepsilon_{si} = 0.003(c - d_i)/c$. Then, the steel stress at each level is $f_{st} = \varepsilon_{si} E_s$, but not greater in magnitude than the yield stress f_y. The steel force at each level is computed by $F_{si} = A_{si} f_{si}$. The depth of the equivalent concrete compressive stress block a is approximated by the relationship $a = \beta_1 c$. β_1 is the concrete stress block factor given in Figure 4.5. Hence, the resultant concrete compression force may be expressed as $C_c = 0.85 f_c' a b$. To satisfy equilibrium, summing forces of the concrete compression and the n levels of the steel, the axial capacity is obtained as

$$\phi P_n = \phi \left(C_c + \sum_{i=1}^{n} F_{si} \right) \tag{4.40}$$

The flexural capacity is obtained from summation of moments about the plastic centroid of the column

$$\phi M_n = \phi \left[C_c \left(\frac{h}{2} - \frac{a}{2} \right) + \sum_{i=1}^{n} F_{si} \left(\frac{h}{2} - d_i \right) \right] \tag{4.41}$$

The strength reduction ϕ factor is not a constant value over the column interaction curve. For points above the balanced point ϕ is 0.65 for tied columns and 0.70 for spiral columns. In this region the column section is compression controlled (extreme level steel strain is at or below yield) and has less ductility. Below the balanced point the column section becomes tension controlled (extreme steel strain greater than yield) and the behavior is more ductile, hence ϕ is allowed to increase linearly to 0.90. This transition occurs between the balanced point and where the extreme steel strain is at 0.005.

4.14.4 Detailing of Column Longitudinal Reinforcement

Longitudinal bars in a column are generally detailed to run continuous by through the story height without cutoffs. In nonseismic regions, column bars are generally spliced above the floor slab to ease construction. In seismic design, column splice should be located at midstory height, away from the section of maximum stress. See Section 4.17 on column splice lengths.

Where the column cross-section dimensions change, longitudinal bars need to be offset. The slope of the offset bar should not exceed 1 in 6. Horizontal ties are needed within the offset to resist 1.5 times the horizontal component of the offset bars. Offsets bents are not allowed if the column face is offset by 3 in. or more.

4.14.5 Shear Design of Columns

The general shear design procedure for selecting transverse reinforcement for columns is similar to that for beams (see Section 4.12). In columns, the axial compression load N_u enhances the concrete shear strength, hence, in lieu of the simplified $V_c = 2\sqrt{f_c'} b_w d$, alternative formulas may be used:

$$V_c = 2 \left(1 + \frac{N_u}{2000 A_g} \right) \sqrt{f_c'} b_w d \tag{4.42a}$$

The quantity N_u/A_g must be in units of pounds per square inch. A second alternative formula for concrete shear strength V_c is

$$V_c = \left(1.9\sqrt{f_c'} + 2500\rho_w \frac{V_u d}{M_m}\right) b_w d \leq 3.5\sqrt{f_c'}b_w d\sqrt{1 + \frac{N_u}{500A_g}} \qquad (4.42b)$$

where

$$M_m = M_u - N_u \frac{(4h - d)}{8} \qquad (4.43)$$

If M_m is negative, the upper bound expression for V_c is used.

Under seismic conditions, additional transverse reinforcement is required to confine the concrete to enhance ductile behavior. See Section IV of this book on earthquake design.

4.14.6 Detailing of Column Hoops and Ties

The main transverse reinforcement should consist of one or a series of perimeter hoops (see Figure 4.24), which not only serve as shear reinforcement, but also prevent the longitudinal bars from buckling out through the concrete cover. Every corner and alternate longitudinal bar should have a hook support (see Figure 4.24). The angle of the hook must be less than 135°. All bars should be hook supported if the clear spacing between longitudinal bars is more than 6 in. The transverse reinforcement must be at least a No. 3 size if the longitudinal bars are No. 10 or smaller, and at least a No. 4 size if the longitudinal bars are greater than No. 10.

To prevent buckling of longitudinal bars, the vertical spacing of transverse reinforcement in columns should not exceed 16 longitudinal bar diameters, 48 transverse bar diameters, or the least dimension of the column size.

4.14.7 Design of Spiral Columns

Columns reinforced with spirals provide superior confinement for the concrete core. Tests have shown that spiral columns are able to carry their axial load even after spalling of the concrete cover. Adequate confinement is achieved when the center-to-center spacing s of the spiral of diameter d_b and yield strength f_y satisfies the following:

$$s \leq \frac{\pi f_y d_b^2}{0.45 h_c f_c' \left[(A_g/A_c) - 1\right]} \qquad (4.44)$$

where h_c is the diameter of the concrete core measured out-to-out of the spiral.

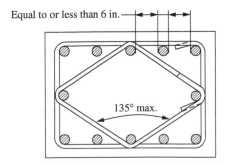

FIGURE 4.24 Column transverse reinforcement detailing.

4.14.8 Detailing of Columns Spirals

Spiral columns require a minimum of six longitudinal bars. Spacers should be used to maintain the design spiral spacing and to prevent distortions. The diameter of the spiral d_b should not be less than $\frac{3}{8}$ in. The clear spacing between spirals should not exceed 3 in. or be less than 1 in. Spirals should be anchored at each column end by providing an extra one and one-half turns of spiral bar. Spirals may be spliced by full mechanical or welded splices or by lap splices with lap lengths not less than 12 in. or $45d_b$ ($72d_b$ if plain bar). While spirals are not required to run through the column-to-floor connection zones, ties should be inserted in those zones to maintain proper confinement, especially if horizontal beams do not frame into these zones.

4.14.9 Detailing of Column to Beam Joints

Joints will perform well if they are well confined. By containing the joint concrete, its structural integrity is ensured under cyclic loading, which allows the internal force capacities, as well as the splices and anchorages detailed within the joint, to develop. Often, confinement around a joint will be provided by the beams or other structural elements that intersect at the joint, if they are of sufficient size. Otherwise, some closed ties, spirals, or stirrups should be provided within the joint to confine the concrete. For nonseismic design, the ACI has no specific requirements on joint confinement.

4.14.10 Columns Subject to Biaxial Bending

If a column is subject to significant moments biaxially, for example, a corner column at the perimeter of a building, the column capacity may be defined by an interaction surface. This surface is essentially an extension of the 2-D interaction diagram described in Figure 4.23 to three coordinate axes $\phi P_n - \phi M_{nx} - \phi M_{ny}$. For rectangular sections under biaxial bending the resultant moment axis may not coincide with the neutral axis. (This is never the case for a circular cross-section because of point symmetry.) An iterative procedure is necessary to determine this angle of deviation. Hence, an accurate generation of the biaxial interaction surface generally requires computer software. Other approximate methods have been proposed. The ACI Code Commentary (R10.3.7) presents the Reciprocal Load Method in which the biaxial capacity of a column ϕP_{ni} is related in a reciprocal manner to its uniaxial capacities, ϕP_{nx} and ϕP_{ny}, and pure axial capacity P_0:

$$\frac{1}{\phi P_{ni}} = \frac{1}{\phi P_{nx}} + \frac{1}{\phi P_{ny}} - \frac{1}{\phi P_0} \qquad (4.45)$$

4.14.11 Slender Columns

When columns are slender the internal forces determined by a first-order analysis may not be sufficiently accurate. The change in column geometry from its deflection causes secondary moments to be induced by the column axial force, also referred to as the P–Δ effect. In stocky columns these secondary moments are minor. For columns that are part of a nonsway frame, for which analysis shows limited side-sway deflection, the effects of column slenderness can be neglected if the column slenderness ratio

$$\frac{kl_u}{r} \le 34 - 12(M_1/M_2) \qquad (4.46)$$

The effective length factor k can be obtained from Figure 4.25 or be conservatively assumed to be 1.0 for nonsway frames. The radius of gyration r may be taken to be 0.30 times the overall dimension of a rectangular column (in the direction of stability) or 0.25 times the diameter for circular columns. The ratio of the column end moments (M_1/M_2) is taken as positive if the column is bent in single curvature, and negative in double curvature.

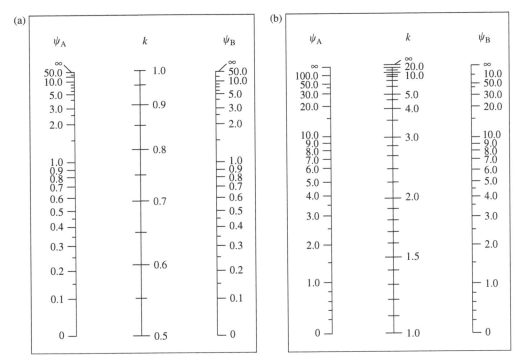

FIGURE 4.25 Effective length factor k: (a) nonsway frames and (b) sway frames.

Note: ψ is the ratio of the summation of column stiffness $[\sum(EI/L)]$ to beam stiffness at the beam–column joint.

For a building story, a frame is considered to be nonsway if its stability index

$$Q = \frac{\sum P_u \Delta_0}{V_u l_c} \le 0.05 \qquad (4.47)$$

where Δ_0 is the first-order relative deflection between the top and bottom of the story and $\sum P_u$ and V_u are the total vertical load and story shear, respectively.

For sway frames, slenderness may be neglected if the slenderness ratio $k l_u / r \le 22$. The k factor must be taken as greater than or equal to 1.0 (see Figure 4.25).

For structural design, it is preferable to design reinforced concrete structures as nonsway systems and with stocky columns. Structural systems should be configured with stiff lateral resistant elements such as shear walls to control sway. Column cross-sectional dimensions should be selected with the slenderness criteria in mind.

If slender columns do exist in a design, adopting a computerized second-order analysis should be considered so that the effects of slenderness will be resolved internally by the structural analysis (see Section 4.7). Then, the internal force demands from the computer output can be directly checked against the interaction diagram in like manner as a nonslender column design. Alternatively, the ACI code provides a manual method called the Moment Magnifier Method to adjust the structural analysis results of a first-order analysis.

4.14.12 Moment Magnifier Method

The Moment Magnifier Method estimates the column moment M_c in a slender column by magnifying the moment obtained from a first-order analysis M_2. For the nonsway case, the factor δ_{ns} magnifies the column moment:

$$M_c = \delta_{ns} M_2 \qquad (4.48)$$

where

$$\delta_{ns} = \frac{C_m}{1 - (P_u/0.75P_c)} \geq 1.0 \qquad (4.49)$$

and

$$P_c = \frac{\pi^2 EI}{(kl_u)^2} \qquad (4.50)$$

The column stiffness may be estimated as

$$EI = \frac{(0.20E_c I_g + E_s I_{se})}{1 + \beta_d} \qquad (4.51)$$

or a more simplified expression may be used:

$$EI = \frac{0.4E_c I_g}{1 + \beta_d} \qquad (4.52)$$

In the sway case, the nonsway moments M_{ns} (e.g., gravity loads) are separated from the sway moments M_s (e.g., due to wind, unbalanced live loads). Only the sway moment is magnified:

$$M_c = M_{ns} + \delta_s M_s \qquad (4.53)$$

$$\delta_s M_s = \frac{M_s}{1 - Q} \geq M_s \qquad (4.54)$$

where Q is the stability index given by Equation 4.47.

4.15 Walls

If tall walls (or shear walls) and combined walls (or core walls) subjected to axial load and bending behave like a column, the design procedures and formulas presented in the previous sections are generally applicable. The reinforcement detailing of wall differs from that of columns. Boundary elements, as shown in Figure 4.26, may be attached to the wall ends or corners to enhance moment capacity. The ratio ρ_n of vertical shear reinforcement to gross area of concrete of horizontal section should not be less than

$$\rho_n = 0.0025 + 0.5\left(2.5 - \frac{h_w}{l_w}\right)(\rho_h - 0.0025) \geq 0.0025 \qquad (4.55)$$

The spacing of vertical wall reinforcement should not exceed $l_w/3$, $3h$, or 18 in. To prevent buckling, the vertical bars opposite each other should be tied together with lateral ties if the vertical reinforcement is greater than 0.01 the gross concrete area.

4.15.1 Shear Design of Walls

The general shear design procedure given in Section 4.12 for determining shear reinforcement in columns applies to walls. For walls in compression, the shear strength provided by concrete V_c may be taken as $2\sqrt{f_c'}hd$. Alternatively, V_c may be taken from the lesser of

$$3.3\sqrt{f_c'}hd + \frac{N_u d}{4l_w} \qquad (4.56)$$

and

$$\left[0.6\sqrt{f_c'} + \frac{l_w(1.25\sqrt{f_c'} + 0.2(N_u/l_w h))}{(M_u/V_u) - (l_w/2)}\right]hd \qquad (4.57)$$

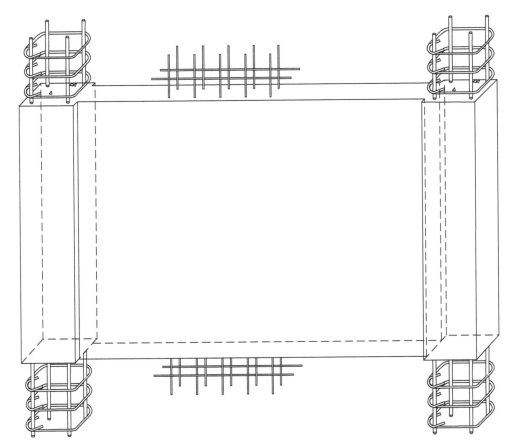

FIGURE 4.26 Reinforced concrete wall with boundary columns.

In lieu of a strain compatibility analysis, the depth of walls d may be assumed to be $0.8l_w$. Shear strength provided by the horizontal reinforcement in walls is also calculated by the equation $V_s = A_v f_y d/s$. The shear capacity of walls $\phi V_n = \phi(V_c + V_s)$ should not be greater than $\phi 10\sqrt{f_c'}hd$.

The spacing of horizontal wall reinforcement should not exceed $l_w/5$, $3h$, or 18 in. The minimum ratio of horizontal wall reinforcement should be more than 0.0025 (or 0.0020 for bars not larger than No. 5). The vertical and horizontal wall bars should be placed as close to the two faces of the wall as cover allows.

4.16 Torsion Design

Torsion will generally not be a serious design issue for reinforced concrete structures if the structural scheme is regular and symmetrical in layout and uses reasonable member sizes. In building floors, torsion may need to be considered for edge beams and members that sustain large unbalanced loading. Concrete members are relatively tolerant of torsion. The ACI permits torsion design to be neglected if the factored torsional moment demand T_u is less than

$$\phi\sqrt{f_c'}\left(\frac{A_{cp}^2}{P_{cp}}\right) \tag{4.58}$$

which corresponds to about one quarter of the torsional cracking capacity. For hollow sections the gross area of section A_g should be used in place of A_{cp}. If an axial compressive or tensile force N_u exists, the torsion design limit becomes

$$\phi\sqrt{f_c'}\left(\frac{A_{cp}^2}{p_{cp}}\right)\sqrt{1+\frac{N_u}{4A_g\sqrt{f_c'}}} \tag{4.59}$$

If the torsional moment demands are higher than the above limits, the redistribution of torque after cracking may be taken into account, which occurs if the member is part of an indeterminate structural system. Hence, in torsion design calculations, the torsional moment demand T_u need not be taken greater than

$$\phi 4\sqrt{f_c'}\left(\frac{A_{cp}^2}{p_{cp}}\right) \tag{4.60}$$

If axial force is present, the upper bound on the design torque T_u is

$$\phi 4\sqrt{f_c'}\left(\frac{A_{cp}^2}{p_{cp}}\right)\sqrt{1+\frac{N_u}{4A_g\sqrt{f_c'}}} \tag{4.61}$$

4.16.1 Design of Torsional Reinforcement

The torsional moment capacity may be based on the space truss analogy (see Figure 4.27). The space truss formed by the transverse and longitudinal reinforcement forms a mechanism that resists torsion. To be effective under torsion, the transverse reinforcement must be constructed of closed hoops (or closed ties) perpendicular to the axis of the member. Spiral reinforcement or welded wire fabric may be used.

To prevent failure of the space truss from concrete crushing and to control diagonal crack widths, the cross-section dimensions must be selected to satisfy the following criteria. For solid sections

$$\sqrt{\left(\frac{V_u}{b_w d}\right)^2+\left(\frac{T_u p_h}{1.7A_{oh}^2}\right)^2}\le\phi\left(\frac{V_c}{b_w d}+8\sqrt{f_c'}\right) \tag{4.62}$$

and for hollow sections

$$\left(\frac{V_u}{b_w d}\right)+\left(\frac{T_u p_h}{1.7A_{oh}^2}\right)\le\phi\left(\frac{V_c}{b_w d}+8\sqrt{f_c'}\right) \tag{4.63}$$

After satisfying these criteria, the torsional moment capacity is determined by

$$\phi T_n = \phi\frac{2A_o A_t f_{yv}}{s}\cot\theta \tag{4.64}$$

The shear flow area A_o may be taken as $0.85A_{oh}$, where A_{oh} is the area enclosed by the closed hoop (see Figure 4.28). The angle θ may be assumed to be 45°. More accurate values of A_o and θ may be used from analysis of the space truss analogy.

To determine the additional transverse torsional reinforcement required to satisfy ultimate strength, that is, $\phi T_n \ge T_u$, the transverse reinforcement area A_t and its spacing s must satisfy the following:

$$\frac{A_t}{s} > \frac{T_u}{\phi 2A_o f_{yv}\cot\theta} \tag{4.65}$$

The area A_t is for one leg of reinforcement. This torsional reinforcement area should then be combined with the transverse reinforcement required for shear demand A_v (see Section 4.11). The total transverse reinforcement required for the member is thus

$$\frac{A_v}{s}+2\frac{A_t}{s} \tag{4.66}$$

The above expression assumes that the shear reinforcement consists of two legs. If more than two legs are present, only the legs adjacent to the sides of the cross-section are considered effective for torsional resistance.

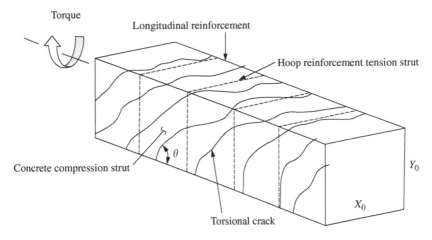

FIGURE 4.27 Truss analogy for torsion.

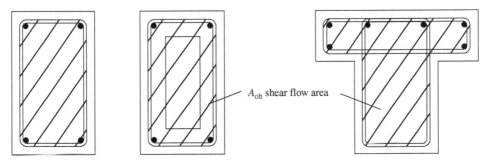

FIGURE 4.28 Torsional reinforcement and shear flow area.

The total transverse reinforcement must exceed the following minimum amounts:

$$0.75\sqrt{f_c'}\frac{b_w}{f_{yv}} \geq \frac{50b_w}{f_{yv}} \tag{4.67}$$

A minimum amount of longitudinal reinforcement is also required:

$$A_l = \frac{A_t}{s}p_h\left(\frac{f_{yv}}{f_{yt}}\right)\cot^2\theta \tag{4.68}$$

The reinforcement area A_l is additional to that required for resisting flexure and axial forces and should not be less than

$$\frac{5\sqrt{f_c'}A_{cp}}{f_{yl}} - \left(\frac{A_t}{s}\right)p_h\frac{f_{yv}}{f_{yl}} \tag{4.69}$$

where A_t/s should not be less than $25b_w/f_{yv}$. The torsional–longitudinal reinforcement should be distributed around the section in a uniform manner.

4.16.2 Detailing of Torsional Reinforcement

The spacing of closed transverse reinforcement under torsion must not exceed $p_h/8$ or 12 in. Torsion reinforcement should be provided for a distance of at least $(b_t + d)$ beyond the point theoretically required. Torsional stresses cause unrestrained corners of the concrete to spall off. Transverse torsion reinforcement needs to be anchored by 135° hooks. In hollow cross-sections, the closed hoops should be placed near the outer surface of the wall. The distance from the centerline of the hoop reinforcement to the inside wall face should not be less than $0.5A_{oh}/p_h$.

The longitudinal torsion reinforcement should be distributed so that its centroid is near the centroid of the cross-section. It should be distributed around the perimeter and be positioned inside the closed hoop with a maximum spacing of 12 in. There should be at least one longitudinal bar at each corner of the hoop. The longitudinal reinforcement must have a diameter of at least 0.042 times the hoop spacing. The ends of the longitudinal reinforcement must be fully developed for yielding. It is permitted to reduce the area of the longitudinal reinforcement by an amount equal to $M_u/(0.9df_{yl})$ since flexural compression offsets the longitudinal tension due to torsion.

4.17   Reinforcement Development Lengths, Hooks, and Splices

The various ultimate capacity formulas presented in the previous sections are premised on the assumption that the reinforcement will reach its yield strength f_y. This is not assured unless the reinforcement has (1) sufficient straight embedment length on each side of the point of yielding, (2) a hook of sufficient anchorage capacity, or (3) a qualified mechanical anchor device.

4.17.1   Tension Development Lengths

The ACI development length equation for bars in tension l_d is expressed in terms of a multiple of the bar diameter d_b (inch unit):

$$l_d = \left(\frac{3}{40} \frac{f_y}{\sqrt{f_c'}} \frac{\alpha\beta\gamma\lambda}{(c + K_{tr})/d_b} \right) d_b \geq 12 \text{ in.} \tag{4.70}$$

where the transverse reinforcement index $K_{tr} = A_{tr}f_{yt}/1500sn$, which may be assumed to be zero for simplicity. Table 4.13 gives the development length for the case of normal weight concrete ($\lambda = 1.0$) and uncoated reinforcement ($\beta = 1.0$). Development lengths need to be increased under these conditions: beam reinforcement positioned near the top surface, epoxy coating, lightweight concrete, and bundling of bars (see ACI Section 12.2.4).

TABLE 4.13   Development Lengths in Tension

| | Tension development length (in.) | |
| | Concrete strength (psi) | |
Bar size	4000	8000
3	12	12
4	12	12
5	15	12
6	21	15
7	36	26
8	47	34
9	60	43
10	77	54
11	94	67
14	136	96
18	242	171

Note: Normal-weight concrete, Grade 60 reinforcement. $\alpha = 1.0$, $\beta = 1.0$, $c = 1.5$ in., and $K_{tr} = 0$.

4.17.2 Compression Development Lengths

For bars under compression, such as in columns, yielding is assured if the development length meets the largest value of $(0.02 f_y / \sqrt{f_c'}) d_b$, $(0.0003 f_y) d_b$, and 8 in. Compression development lengths l_{dc} are given in Table 4.14. Compression development length may be reduced by the factor $(A_s$ required$)/(A_s$ provided$)$ if reinforcement is provided in excess of that required by the load demand. Reinforcement within closely spaced spirals or tie reinforcement may be reduced by the factor 0.75 (spiral not less than $\frac{1}{4}$ in. in diameter and not more than 4 in. in pitch; column ties not less than No. 4 in size and spaced not more than 4 in.).

4.17.3 Standard Hooks

The standard (nonseismic) hook geometry as defined by ACI is shown in Figure 4.9. The required hook length l_{dh} is given in Table 4.15 and is based on the empirical formula $(0.02 f_y / \sqrt{f_c'}) d_b$. Hook lengths may be reduced by 30% when the side and end covers over the hook exceed 2.5 and 2 in., respectively. A 20% reduction is permitted if the hook is within a confined concrete zone where the transverse

TABLE 4.14 Development Lengths in Compression

| | Compression development length (in.) | |
| | Concrete strength (psi) | |
Bar size	4000	8000
3	8	8
4	9	9
5	12	11
6	14	14
7	17	16
8	19	18
9	21	20
10	24	23
11	27	25
14	32	30
18	43	41

Note: Grade 60 reinforcement.

TABLE 4.15 Development Lengths of Hooks in Tension

| | Development length of standard hook (in.) | |
| | Concrete strength (psi) | |
Bar size	4000	8000
3	7	6
4	9	7
5	12	8
6	14	10
7	17	12
8	19	13
9	21	15
10	24	17
11	27	19
14	32	23
18	43	30

Note: Grade 60 steel. $\beta = 1.0$, $\lambda = 1.0$, l_{dh} not less than $8d_b$ nor 6 in.

reinforcement spacing is less than three times the diameter of the hooked bar. Note that whether the standard hook is detailed to engage over a longitudinal bar has no influence on the required hook length.

When insufficient hook length is available or in regions of heavy bar congestion, mechanical anchors may be used. There are a number of proprietary devices that have been tested and prequalified. These generally consist of an anchor plate attached to the bar end.

4.17.4 Splices

There are three choices for joining bars together: (1) mechanical device, (2) welding, and (3) lap splices. The mechanical and welded splices must be tested to show the development in tension or compression of at least 125% of the specified yield strength f_y of the bar. Welded splices must conform to ANSI/AWS D1.4, "Structural Welding Code — Reinforcing Steel." Since splices introduce weak leaks into the structure, they should be located as much as possible away from points of maximum force and critical locations.

4.17.4.1 Tension Lap Splices

Generally, bars in tension need to be lapped over a distance of $1.3l_d$ (Class B splice, see Section 4.17.1 for l_d), unless laps are staggered or more than twice the required steel is provided (Class A splice $= 1.0l_d$).

4.17.4.2 Compression Lap Splices and Column Splices

Compression lap splice lengths shall be $0.0005f_y d_b$, but not less than 12 in. If any of the load demand combinations is expected to introduce tension in the column reinforcement, column bars should be lapped as tension splices. Class A splices ($1.0l_d$) are allowed if half or fewer of the bars are spliced at any section and alternate lap splices are staggered by l_d. Column lap lengths may be multiplied by 0.83 if the ties provided through the lap splice length have an effective area not less than $0.0015hs$. Lap lengths within spiral reinforcement may be multiplied by 0.75.

4.18 Deflections

The estimation of deflections for reinforced concrete structures is complicated by the cracking of the concrete and the effects of creep and shrinkage. In lieu of carrying out a refined nonlinear analysis involving the moment curvature analysis of member sections, an elastic analysis may be used to incorporate a reduced or effective moment of inertia for the members. For beam elements an effective moment of inertia may be taken as

$$I_e = \left(\frac{M_{cr}}{M_a}\right)^3 I_g + \left[1 - \left(\frac{M_{cr}}{M_a}\right)^3\right] I_{cr} \le I_g \qquad (4.71)$$

where the cracking moment of the section

$$M_{cr} = \frac{f_r I_g}{y_t} \qquad (4.72)$$

The cracking stress or modulus of rupture of normal weight concrete is

$$f_r = 7.5\sqrt{f_c'} \qquad (4.73)$$

For all-lightweight concrete f_r should be multiplied by 0.75, for sand-lightweight concrete, by 0.85.

For estimating the deflection of prismatic beams, it is generally satisfactory to take I_e at the section at midspan to represent the average stiffness for the whole member. For cantilevers, the I_e at the support should be taken. For nonprismatic beams, an average I_e of the positive and negative moment sections should be used.

Long-term deflections may be estimated by multiplying the immediate deflections of sustained loads (e.g., self-weight, permanent loads) by

$$\lambda = \frac{\xi}{1 + 50\rho'} \tag{4.74}$$

The time-dependent factor ξ is plotted in Figure 4.29. More refined creep and shrinkage deflection models are provided by ACI Committee 209 and the CEP-FIP Model Code (1990).

Deflections of beams and one-way slab systems must not exceed the limits in Table 4.16. Deflection control of two-way floor systems is generally satisfactory by following the minimum slab thickness

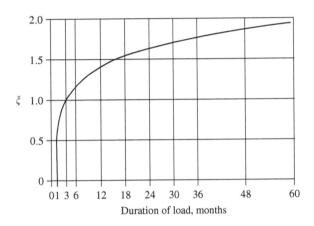

FIGURE 4.29   Time-dependent factor ξ.

TABLE 4.16   Deflection Limits of Beams and One-Way Slab Systems

Type of member	Deflection to be considered	Deflection limitation
Flat roots not supporting or attached to nonstructural elements likely to be damaged by large deflections	Immediate deflection due to live load L	$l/180$[a]
Floors not supporting or attached to nonstructural elements likely to be damaged by large deflections	Immediate deflection due to live load L	$l/360$
Roof or floor construction supporting or attached to nonstructural elements likely to be damaged by large deflections	That part of the total deflection occurring after attachment of nonstructural elements (sum of the long-term deflection due to all sustained loads and the immediate deflection due to any additional live load)[b]	$l/480$[c]
Roof or floor construction supporting or attached to nonstructural elements not likely to be damaged by large deflections		$l/240$[d]

[a] Limit not intended to safeguard against ponding. Ponding should be checked by suitable calculations of deflection, including added deflections due to ponded water, and consideration of long-term effects of all sustained loads, camber, construction tolerances, and reliability of provisions for drainage.

[b] Long-term deflection should be determined in accordance with Equation 4.74, but may be reduced by the amount of deflection calculated to occur before attachment of nonstructural elements. This amount should be determined on the basis of accepted engineering date relating to time deflection characteristics of members similar to those being considered.

[c] Limit may be exceeded if adequate measures are taken to prevent to supported or attached elements.

[d] Limit should be greater than the tolerance provided for nonstructural elements. Limit may be exceeded if camber is provided so that total deflection minus camber does not exceed limit.

requirements (see Table 4.8). Lateral deflections of columns may be a function of occupancy comfort under high wind or seismic drift criteria (e.g., $H/200$).

4.19 Drawings, Specifications, and Construction

Although this chapter has focused mainly on the structural mechanics of design, design procedures and formulas, and rules that apply to reinforced concrete construction, the importance of drawings and specifications as part of the end products for communicating the structural design must not be overlooked. Essential information that should be included in the drawings and specifications are: specified compressive strength of concrete at stated ages (e.g., 28 days) or stage of construction; specified strength or grade of reinforced (e.g., Grade 60); governing design codes (e.g., IBC, AASHTO); live load and other essential loads; size and location of structural elements and locations; development lengths, hook lengths, and their locations; type and location of mechanical and welded splices; provisions for the effects of temperature, creep, and shrinkage; and details of joints and bearings.

The quality of the final structure is highly dependent on material and construction quality measures that improve durability, construction formwork, quality procedures, and inspection of construction. Although many of these aspects may not fall under the direct purview of the structural designer, attention and knowledge are necessary to help ensure a successful execution of the structural design. Information and guidance on these topics can be found in the *ACI Manual of Concrete Practice*, which is a comprehensive five-volume compendium of current ACI standards and committee reports: (1) Materials and General Properties of Concrete, (2) Construction Practices and Inspection, Pavements, (3) Use of Concrete in Buildings — Design, Specifications, and Related Topics, (4) Bridges, Substructures, Sanitary, and Other Special Structures, Structural Properties, and (5) Masonry, Precast Concrete, Special Processes.

Notation

a = depth of concrete stress block

A_s' = area of compression reinforcement

A_b = area of an individual reinforcement

A_c = area of core of spirally reinforced column measured to outside diameter of spiral

A_c = area of critical section

A_{cp} = area enclosed by outside perimeter of concrete cross-section

A_g = gross area of section

A_l = area of longitudinal reinforcement to resist torsion

A_o = gross area enclosed by shear flow path

A_{oh} = area enclosed by centerline of the outermost closed transverse torsional reinforcement

A_s = area of tension reinforcement

$A_{s,min}$ = minimum area of tension reinforcement

A_{st} = total area of longitudinal reinforcement

A_t = area of one leg of a closed stirrup resisting torsion within a distance s

A_{tr} = total cross-sectional area of all transverse reinforcement that is within

the spacing s and that crosses the potential place of splitting through the reinforcement being developed

A_v = area of shear reinforcement

$A_{v,min}$ = minimum area of shear reinforcement

b = width of compression face

b_1 = width of critical section in l_1 direction

b_2 = width of critical section in l_2 direction

b_0 = perimeter length of critical section

b_t = width of that part of the cross-section containing the closed stirrups resisting torsion

b_w = web width

C = cross-sectional constant to define torsional properties $= \sum(1 - 0.63(x/y))/(x^3y/3)$ (total section is divided into separate rectangular parts, where x and y are the shorter and longer dimensions of each part, respectively).

c = distance from centroid of critical section to its perimeter (Section 4.13.2.1)

c = spacing or cover dimension

c_1 = dimension of column or capital support in l_1 direction

c_2 = dimension of column or capital support in l_2 direction

c_c = clear cover from the nearest surface in tension to the surface of the flexural reinforcement

C_c = resultant concrete compression force

C_m = factor relating actual moment diagram to an equivalent uniform moment

d = distance from extreme compression fiber to centroid of tension reinforcement

d' = distance from extreme compression fiber to centroid of compression reinforcement

d_b = nominal diameter of bar

d_i = distance from extreme compression fiber to centroid of reinforcement layer i

E_c = modulus of elasticity of concrete

E_{cb} = modulus of elasticity of beam concrete

E_{cs} = modulus of elasticity of slab concrete

EI = flexural stiffness of column

E_s = modulus of elasticity of steel reinforcement

f'_c = specified compressive strength of concrete

F_n = nominal structural strength

f_r = modulus of rupture of concrete

f_s = reinforcement stress

F_{si} = resultant steel force at bar layer i

f_y = specified yield stress of reinforcement

f_{yl} = specified yield strength of longitudinal torsional reinforcement

f_{yt} = specified yield strength of transverse reinforcement

f_{yv} = specified yield strength of closed transverse torsional reinforcement

h = overall thickness of column or wall

h_c = diameter of concrete core measured out-to-out of spiral

h_w = total height of wall

I_b = moment of inertia of gross section of beam

I_{cr} = moment of inertia of cracked section transformed to concrete

I_e = effective moment of inertia

I_s = moment of inertia of gross section of slab

I_{se} = moment of inertia of reinforcement about centroidal axis of cross-section

J_c = equivalent polar moment of inertia of critical section

k = effective length factor for columns

K_m = material constant

K_{tr} = transverse reinforcement index

L = member length

l_1 = center-to-center span length in the direction moments are being determined

l_2 = center-to-center span length transverse to l_1

l_c = center-to-center length of columns

l_d = development length of reinforcement in tension

l_{dc} = development length of reinforcement in compression

l_{dh} = development length of standard hook in tension, measured from critical section to outside end of hook

l_n = clear span length, measured from face-to-face of supports

l_u = unsupported length of columns

l_w = horizontal length of wall

M_1 = smaller factored end moment in a column, negative if bent in double curvature

M_2 = larger factored end moment in a column, negative if bent in double curvature

M_a = maximum moment applied for deflection computation

M_c = factored magnified moment in columns

M_{cr} = cracking moment

M_m = modified moment

M_n = nominal or theoretical moment strength

M_{ns} = factored end moment of column due to loads that do not cause appreciable side sway

M_0 = total factored static moment

M_s = factored end moment of column due to loads that cause appreciable side-ways

M_u = moment demand

M_{unb} = unbalanced moment at slab–column connections

n = modular ratio $= E_s/E_c$

N_C = resultant compressive force of concrete

N_T = resultant tensile force of reinforcement

N_u = factored axial load occurring simultaneously with V_u or T_u, positive sign for compression

P_c = critical load

p_{cp} = outside perimeter of concrete cross-section

p_h = perimeter of centerline of outermost concrete cross-section

P_n = nominal axial load strength of column

$P_{n,max}$ = maximum nominal axial load strength of column

P_{ni} = nominal biaxial load strength of column

P_{nx} = nominal axial load strength of column about x-axis

P_{ny} = nominal axial load strength of column about y-axis

P_0 = nominal axial load strength of column at zero eccentricity

P_u = axial load demand

Q = stability index

r = radius of gyration of cross-section

s = spacing of shear or torsional reinforcement along longitudinal axis of member

S_C = structural capacity

S_D = structural demand

T_n = nominal torsional moment strength

T_u = torsional moment demand

V_c = nominal shear strength provided by concrete

V_n = nominal shear strength

v_n = nominal shear stress strength of critical section

V_s = nominal shear strength provided by shear reinforcement

V_u = shear demand

v_u = shear stress at critical section

w_u = factored load on slab per unit area

y_t = distance from centroidal axis of gross section to extreme tension fiber

α = ratio of flexural stiffness of beam section to flexural stiffness of width of a slab bounded laterally by centerlines of adjacent panels on each side of beam = $E_{cb}I_b/E_{cs}I_s$

α = reinforcement location factor (Table 4.13)

α_i = angle between inclined shear reinforcement and longitudinal axis of member

α_m = average value of α for all beams on edges of a panel

α_s = shear strength factor

α_1 = α in direction of l_1

β = ratio of clear spans in long to short direction of two-way slabs

β = reinforcement coating factor (Section 4.17.1)

β_c = ratio of long side to short side dimension of column

β_d = ratio of maximum factored sustained axial load to maximum factored axial load

β_t = ratio of torsional stiffness of edge beam section to flexural stiffness of a width of slab equal to span length of beam, center-to-center of supports

β_1 = equivalent concrete stress block factor defined in Figure 4.5

δ_{ns} = nonsway column moment magnification factor

δ_s = sway column moment magnification factor

Δ_0 = first-order relative deflection between the top and bottom of a story

ε_c = concrete strain

ε_t = steel strain

γ = reinforcement size factor = 0.8 for No. 6 and smaller bars; = 1.0 for No. 7 and larger

γ_f = fraction of unbalanced moment transferred by flexure at slab–column connections

γ_v = fraction of unbalanced moment transferred by eccentricity of shear at slab–column connections

λ = lightweight aggregate concrete factor (Section 4.17); = 1.3 for light weight concrete

λ = multiplier for additional long-term deflection

ϕ_{ecc} = strength reduction factor for accidental eccentricity in columns = 1.3 for lightweight concrete

ϕ_u = curvature at ultimate

ϕ_y = curvature at yield

ρ = ratio of tension reinforcement = A_s/bd

ρ' = ratio of compression reinforcement = A'_s/bd

ρ_h = ratio of horizontal wall reinforcement area to gross section area of horizontal section

ρ_n = ratio of vertical wall reinforcement area to gross section area of horizontal section

ρ_w = ratio of reinforcement = $A_s/b_w d$

ξ = time-dependent factor for sustained load

ϕ = strength reduction factor, see Table 4.4

θ = angle of compression diagonals in truss analogy for torsion

Useful Web Sites

American Concrete Institute: www.aci-int.org
Concrete Reinforcing Steel Institute: www.crsi.org
Portland Cement Association: www.portcement.org
International Federation of Concrete Structures: http://fib.epfl.ch
Eurocode 2: www.eurocode2.info
Reinforced Concrete Council: www.rcc-info.org.uk
Japan Concrete Institute: www.jci-net.or.jp
Emerging Construction Technologies: www.new-technologies.org

5

Prestressed Concrete*

Edward G. Nawy
Department of Civil and Environmental Engineering
Rutgers University — The State University of New Jersey,
Piscataway, NJ

---

*This chapter is a condensation from several chapters of *Prestressed Concrete — A Fundamental Approach*, 4th edition, 2003, 944 pp., by E. G. Nawy, with permission of the publishers, Prentice Hall, Upper Saddle River, NJ.

5.1   Introduction

Concrete is strong in compression, but weak in tension: its tensile strength varies from 8 to 14% of its compressive strength. Due to such a low tensile capacity, flexural cracks develop at early stages of loading. In order to reduce or prevent such cracks from developing, a concentric or eccentric force is imposed in the longitudinal direction of the structural element. This force prevents the cracks from developing by eliminating or considerably reducing the tensile stresses at the critical midspan and support sections at service load, thereby raising the bending, shear, and torsional capacities of the sections. The sections are then able to behave elastically, and almost the full capacity of the concrete in compression can be efficiently utilized across the entire depth of the concrete sections when all loads act on the structure.

Such an imposed longitudinal force is termed a *prestressing force*, that is, a compressive force that prestresses the sections along the span of the structural element prior to the application of the transverse gravity dead and live loads or transient horizontal live loads. The type of prestressing force involved, together with its magnitude, are determined mainly on the basis of the type of system to be constructed and the span length. As a result, permanent stresses in the prestressed structural member are created before the full dead and live loads are applied, in order to eliminate or considerably reduce the net tensile stresses caused by these loads.

With reinforced concrete, it is assumed that the tensile strength of the concrete is negligible and disregarded. This is because the tensile forces resulting from the bending moments are resisted by the bond created in the reinforcement process. Cracking and deflection are therefore essentially irrecoverable in reinforced concrete once the member has reached its limit state at service load. In prestressed concrete elements, cracking can be controlled or totally eliminated at the service load level. The reinforcement required to produce the prestressing force in the prestressed member actively preloads the member, permitting a relatively high controlled recovery of cracking and deflection.

5.2   Concrete for Prestressed Elements

Concrete, particularly high-strength concrete, is a major constituent of all prestressed concrete elements. Hence, its strength and long-term endurance have to be achieved through proper quality control and quality assurance at the production stage. The mechanical properties of hardened concrete can be classified into two categories: short-term or instantaneous properties, and long-term properties. The short-term properties are strength in compression, tension, and shear; and stiffness, as measured by the modulus of elasticity. The long-term properties can be classified in terms of creep and shrinkage. The following subsections present some details on these properties.

5.2.1   Compressive Strength

Depending on the type of mix, the properties of aggregate, and the time and quality of the curing, compressive strengths of concrete can be obtained up to 20,000 psi or more. Commercial production of concrete with ordinary aggregate is usually in the range 4,000 to 12,000 psi, with the most common concrete strengths being in the 6,000 psi level.

The compressive strength f'_c is based on standard 6 in. by 12 in. cylinders cured under standard laboratory conditions and tested at a specified rate of loading at 28 days of age. The standard specifications used in the United States are usually taken from American Society for Testing and Materials (ASTM) C-39. The strength of concrete in the actual structure may not be the same as that of the cylinder because of the difference in compaction and curing conditions.

5.2.2   Tensile Strength

The tensile strength of concrete is relatively low. A good approximation for the tensile strength f_{ct} is $0.10f'_c < f_{ct} < 0.20f'_c$. It is more difficult to measure tensile strength than compressive strength because of the gripping problems with testing machines. A number of methods are available for tension testing, the most commonly used method being the cylinder splitting, or Brazilian, test.

For members subjected to bending, the value of the modulus of rupture f_r rather than the tensile splitting strength f_t' is used in design. The modulus of rupture is measured by testing to failure plain concrete beams 6 in.[2] in cross-section, having a span of 18 in., and loaded at their third points (ASTM C-78). The modulus of rupture has a higher value than the tensile splitting strength. The American Concrete Institute (ACI) specifies a value of 7.5 for the modulus of rupture of normal-weight concrete.

In most cases, lightweight concrete has a lower tensile strength than does normal-weight concrete. The following are the code stipulations for lightweight concrete:

1. If the splitting tensile strength f_{ct} is specified

$$f_r = 1.09 f_{ct} \le 7.5 \sqrt{f_c'} \tag{5.1}$$

2. If f_{ct} is not specified, use a factor of 0.75 for all-lightweight concrete and 0.85 for sand-lightweight concrete. Linear interpolation may be used for mixtures of natural sand and lightweight fine aggregate. For high-strength concrete, the modulus of rupture can be as high as 11–$12\sqrt{f_c'}$.

5.2.3 Shear Strength

Shear strength is more difficult to determine experimentally than the tests discussed previously because of the difficulty in isolating shear from other stresses. This is one of the reasons for the large variation in shear-strength test values reported in the literature, varying from 20% of the compressive strength in normal loading to a considerably higher percentage of up to 85% of the compressive strength in cases where direct shear exists in combination with compression. Control of a structural design by shear strength is significant only in rare cases, since shear stresses must ordinarily be limited to continually lower values in order to protect the concrete from the abrupt and brittle failure in diagonal tension

$$E_c = 57{,}000\sqrt{f_c'} \text{ psi } (4{,}700\sqrt{f_c'} \text{ MPa}) \tag{5.2a}$$

or

$$E_c = 0.043 w^{1.5}\sqrt{f_c'} \text{ MPa} \tag{5.2b}$$

5.2.4 High-Strength Concrete

High-strength concrete is termed as such by the ACI 318 Code when the cylinder compressive strength exceeds 6,000 psi (41.4 MPa). For concrete having compressive strengths 6,000 to 12,000 psi (42 to 84 MPa), the expressions for the modulus of concrete are [1–3]

$$E_c \text{ (psi)} = \left[40{,}000\sqrt{f_c'} + 10^6\right]\left(\frac{w_c}{145}\right)^{1.5} \tag{5.3a}$$

where $f_c' = $ psi and $w_c = $ lb/ft^3

$$E_c \text{ (MPa)} = \left[3.32\sqrt{f_c'} + 6895\right]\left(\frac{w_c}{2320}\right)^{1.5} \tag{5.3b}$$

where $f_c' = $ MPa and $w_c = $ kg/m^3.

Today, concrete strength up to 20,000 psi (138 MPa) is easily achieved using a maximum stone aggregate size of $\frac{3}{8}$ in. (9.5 mm) and pozzolamic cementitious partial replacements for the cement such as silica fume. Such strengths can be obtained in the field under strict quality control and quality assurance conditions. For strengths in the range of 20,000 to 30,000 (138 to 206 MPa), other constituents such as steel or carbon fibers have to be added to the mixture. In all these cases, mixture design has to be made by several field trial batches (five or more), modifying the mixture components for the workability needed in concrete placement. Steel cylinder molds size 4 in. (diameter) × 8 in. length have to be used, applying the appropriate dimensional correction.

PHOTO 5.1   A rendering of the new Maumee River Bridge, Toledo, Ohio. This cable-stayed bridge spans the Maumee River in downtown Toledo as a monument icon for the city. The design includes single pylon, single plane of stays, and a main span with a horizontal clearance of 612 ft in both directions. The main pylon is clad on four of its eight sides with a glass curtain wall system, symbolizing the glass industry and heritage of Toledo. This glass prismatic system and stainless steel clad cables create a sleek and industrial look during the day. At night, the glass becomes very dynamic with the use of LED arrays back-lighting the window wall. Owner: Ohio Department of Transportation (courtesy of the Designer, Figg Engineering Group, Linda Figg, President, Tallahassee, Florida).

5.2.5 Initial Compressive Strength and Modulus

Since prestressing is performed in most cases prior to concrete's achieving its 28-day strength, it is important to determine the concrete compressive strength f'_{ct} at the prestressing stage as well as the concrete modulus E_c at various stages in the loading history of the element. The general expression for the compressive strength as a function of time [4] is

$$f'_{ci} = \frac{t}{\alpha + \beta t} f'_c \tag{5.4a}$$

where

f'_c = 28-day compressive strength
t = time in days
α = factor depending on type of cement and curing conditions
    = 4.00 for moist-cured type-I cement and 2.30 for moist-cured type-III cement
    = 1.00 for steam-cured type-I cement and 0.70 for steam-cured type-III cement
β = factor depending on the same parameters for α giving corresponding values of 0.85, 0.92, 0.95, and
    0.98, respectively

Hence, for a typical moist-cured type-I cement concrete

$$f'_{ci} = \frac{t}{4.00 + 0.85t} f'_c \tag{5.4b}$$

5.2.6 Creep

Creep, or lateral material flow, is the increase in strain with time due to a sustained load. The initial deformation due to load is the *elastic strain*, while the additional strain due to the same sustained load is the *creep strain*. This practical assumption is quite acceptable, since the initial recorded deformation includes few time-dependent effects.

The ultimate creep coefficient, C_u, is given by

$$C_u = \rho_u E_c \tag{5.5}$$

or average $C_u \cong 2.35$.

Branson's model, verified by extensive tests, relates the creep coefficient C_t at any time to the ultimate creep coefficient (for standard conditions) as

$$C_t = \frac{t^{0.6}}{10 + t^{0.6}} C_u \tag{5.6}$$

or, alternatively,

$$\rho_t = \frac{t^{0.6}}{10 + t^{0.6}} \tag{5.7}$$

where t is the time in days and ρ_t is the time multiplier. Standard conditions as defined by Branson pertain to concretes of slump 4 in. (10 cm) or less and a relative humidity of 40%.

When conditions are not standard, creep correction factors have to be applied to Equation 5.6 or 5.7 as follows:

1. For moist-cured concrete loaded at an age of 7 days or more

$$k_a = 1.25t^{-0.118} \tag{5.8a}$$

2. For steam-cured concrete loaded at an age of 1 to 3 days or more

$$k_a = 1.13t^{-0.095} \tag{5.8b}$$

For greater than 40% relative humidity, a further multiplier correction factor of

$$k_{c_1} = 1.27 - 0.0067H \tag{5.9}$$

5.2.7 Shrinkage

Basically, there are two types of shrinkage: plastic shrinkage and drying shrinkage. *Plastic shrinkage* occurs during the first few hours after placing fresh concrete in the forms. Exposed surfaces such as floor slabs are more easily affected by exposure to dry air because of their large contact surface. In such cases, moisture evaporates faster from the concrete surface than it is replaced by the bleed water from the lower layers of the concrete elements. *Drying shrinkage*, on the other hand, occurs after the concrete has already attained its final set and a good portion of the chemical hydration process in the cement gel has been accomplished.

Drying shrinkage is the decrease in the volume of a concrete element when it loses moisture by evaporation. The opposite phenomenon, that is, volume increase through water absorption, is termed *swelling*. In other words, shrinkage and swelling represent water movement out of or into the gel structure of a concrete specimen due to the difference in humidity or saturation levels between the specimen and the surroundings irrespective of the external load.

Shrinkage is not a completely reversible process. If a concrete unit is saturated with water after having fully shrunk, it will not expand to its original volume. The rate of the increase in shrinkage strain decreases with time since older concretes are more resistant to stress and consequently undergo less shrinkage, such that the shrinkage strain becomes almost asymptotic with time.

Branson recommends the following relationships for the shrinkage strain as a function of time for standard conditions of humidity ($H \cong 40\%$):

1. For moist-cured concrete any time t after 7 days

$$\varepsilon_{SH,t} = \frac{t}{35+t}\left(\varepsilon_{SH,u}\right) \tag{5.10a}$$

where $\varepsilon_{SH,u} = 800 \times 10^{-6}$ in./in. if local data are not available.

2. For steam-cured concrete after the age of 1 to 3 days

$$\varepsilon_{SH,t} = \frac{t}{55+t}\left(\varepsilon_{SH,u}\right) \tag{5.10b}$$

For other than standard humidity, a correction factor has to be applied to Equations 5.10a and 5.10b as follows:

1. For $40 < H \leq 80\%$

$$k_{SH} = 1.40 - 0.010H \tag{5.11a}$$

2. For $80 < H \leq 100\%$

$$k_{SH} = 3.00 - 0.30H \tag{5.11b}$$

5.3 Steel Reinforcement Properties

5.3.1 Non-Prestressing Reinforcement

Steel reinforcement for concrete consists of bars, wires, and welded wire fabric, all of which are manufactured in accordance with ASTM standards (see Table 5.1).

TABLE 5.1 Weight, Area, and Perimeter of Individual Bars

Bar designation number	Weight per foot (lb)	Standard nominal dimensions		
		Diameter, d_b [in. (mm)]	Cross-sectional area, A_b (in.^{2})	Perimeter (in.)
3	0.376	0.375 (10)	0.11	1.178
4	0.668	0.500 (13)	0.2	1.571
5	1.043	0.625 (16)	0.31	1.963
6	1.502	0.750 (19)	0.44	2.356
7	2.044	0.875 (22)	0.6	2.749
8	2.670	1.000 (25)	0.79	3.142
9	3.400	1.128 (29)	1	3.544
10	4.303	1.270 (32)	1.27	3.99
11	5.313	1.410 (36)	1.56	4.43
14	7.65	1.693 (43)	2.25	5.32
18	13.6	2.257 (57)	4	7.09

5.3.2 Prestressing Reinforcement

Because of the high creep and shrinkage losses in concrete, effective prestressing can be achieved by using very high strength steels in the range of 270,000 psi or more (1,862 MPa or higher). Such highly stressed steels are able to counterbalance these losses in the surrounding concrete and have adequate leftover stress levels to sustain in the long term the required prestressing force (see Table 5.2).

A typical reinforcement stress–strain plot is shown in Figure 5.1.

TABLE 5.2 Seven-Wire Standard Strand for Prestressed Concrete

Nominal diameter of strand (in.)	Breaking strength of strand (min. lb)	Nominal steel area of strand (sq in.)	Nominal weight of strands (lb per 1000 ft)[a]	Minimum load at 1% extension (lb)
Grade 250				
0.25	9,000	0.036	122	7,650
0.313	14,500	0.058	197	12,300
0.375	20,000	0.08	272	17,000
0.438	27,000	0.108	367	23,000
0.500	36,000	0.144	490	30,600
0.600	54,000	0.216	737	45,900
Grade 270				
0.375	23,000	0.085	290	19,550
0.438	31,000	0.115	390	26,350
0.500	41,300	0.153	520	35,100
0.600	58,600	0.217	740	49,800

[a] 100,000 psi = 689.5 MPa.
0.1 in. = 2.54 mm; 1 in.^{2} = 645 mm^2.
Weight: multiply by 1.49 to obtain weight in kg per 1,000 m.
1,000 lb = 4,448 Newton.
  Source: Post-tensioning Institute.

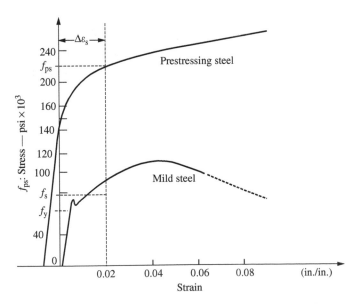

FIGURE 5.1 Stress–strain diagram for steel prestressing strands in comparison with mild steel bar reinforcement [5].

5.4 Maximum Permissible Stresses

The following are definitions of some important mathematical terms used in this section:

f_{py} is the specified yield strength of prestressing tendons, in psi.
f_y is the specified yield strength of nonprestressed reinforcement, in psi.
f_{pu} is the specified tensile strength of prestressing tendons, in psi.
f_c' is the specified compressive strength of concrete, in psi.
f_{ct}' is the compressive strength of concrete at time of initial prestress.

5.4.1 Concrete Stresses in Flexure

Stresses in concrete immediately after prestress transfer (before time-dependent prestress losses) shall not exceed the following:

1. Extreme fiber stress in compression, $0.60f_{ci}'$.
2. Extreme fiber stress in tension except as permitted in (3), $3\sqrt{f_{ci}'}$.
3. Extreme fiber stress in tension at ends of simply supported members $6\sqrt{f_{ci}'}$.

Where computed tensile stresses exceed these values, bonded auxiliary reinforcement (nonprestressed or prestressed) has to be provided in the tensile zone to resist the total tensile force in concrete computed under the assumption of an uncracked section.

Stresses in concrete at service loads (after allowance for all prestress losses) should not exceed the following:

1. Extreme fiber stress in compression due to prestress plus sustained load, where sustained dead load and live load are a large part of the total service load, $0.45f_c'$.
2. Extreme fiber stress in compression due to prestress plus total load, if the live load is transient, $0.60f_c'$.
3. Extreme fiber stress in tension in precompressed tensile zone, $6\sqrt{f_c'}$.
4. Extreme fiber stress in tension in precompressed tensile zone of members (except two-way slab systems), where analysis based on transformed cracked sections and on bilinear moment–deflection relationships shows that immediate and long-time deflections comply with the ACI definition requirements and minimum concrete cover requirements, $12\sqrt{f_c'}$.

5.4.2 Prestressing Steel Stresses

Tensile stress in prestressing tendons shall not exceed the following:

1. Due to tendon jacking force, $0.94f_{py}$; but not greater than the lesser of $0.80f_{pu}$ and the maximum value recommended by the manufacturer of prestressing tendons or anchorages.
2. Immediately after prestress transfer, $0.82f_{py}$; but not greater than $0.74f_{pu}$.
3. Posttensioning tendons, at anchorages and couplers, immediately after tendon anchorage, $0.70f_{pu}$.

5.5 Partial Loss of Prestress

Essentially, the reduction in the prestressing force can be grouped into two categories:

- Immediate elastic loss during the fabrication or construction process, including elastic shortening of the concrete, anchorage losses, and frictional losses.
- Time-dependent losses such as creep and shrinkage and those due to temperature effects and steel relaxation, all of which are determinable at the service-load limit state of stress in the prestressed concrete element.

An exact determination of the magnitude of these losses, particularly the time-dependent ones, is not feasible, since they depend on a multiplicity of interrelated factors. Empirical methods of estimating losses differ with the different codes of practice or recommendations, such as those of the Prestressed Concrete Institute, the ACI–ASCE joint committee approach, the AASHTO lump-sum approach, the Comité Eurointernationale du Béton (CEB), and the FIP (Federation Internationale de la Précontrainte). The degree of rigor of these methods depends on the approach chosen and the accepted practice of record (see Table 5.3 to Table 5.5).

In Table 5.5, $\Delta f_{pR} = \Delta f_{pR}(t_0, t_{tr}) + \Delta f_{pR}(t_{tr}, t_s)$, where t_0 is the time at jacking, t_{tr} is the time at transfer, and t_s is the time at stabilized loss. Hence, computations for steel relaxation loss have to be performed for the time interval t_1 through t_2 of the respective loading stages.

As an example, the transfer stage, say, at 18 h, would result in $t_{tr} = t_2 = 18\,\mathrm{h}$ and $t_0 = t_1 = 0$. If the next loading stage is between transfer and 5 years (17,520 h), when losses are considered stabilized, then $t_2 = t_s = 17,520\,\mathrm{h}$ and $t_1 = 18\,\mathrm{h}$. Then, if f_{pi} is the initial prestressing stress that the concrete element is subjected to and f_{pj} is the jacking stress in the tendon.

5.5.1 Steel Stress Relaxation (R)

The magnitude of the decrease in the prestress depends not only on the duration of the sustained prestressing force, but also on the ratio f_{pi}/f_{py} of the initial prestress to the yield strength of the

TABLE 5.3 AASHTO Lump-Sum Losses

	Total loss	
Type of prestressing steel	$f_c' = 4{,}000$ psi (27.6 N/mm^2)	$f_c' = 5{,}000$ psi (34.5 N/mm^2)
Pretensioning strand		45,000 psi (310 N/mm^2)
Posttensioning[a] wire or strand	32,000 psi (221 N/mm^2)	33,000 psi (228 N/mm^2)
Bars	22,000 psi (152 N/mm^2)	23,000 psi (159 N/mm^2)

[a] Losses due to friction are excluded. Such losses should be computed according to Section 6.5 of the AASHTO specifications.

TABLE 5.4 Approximate Prestress Loss Values for Posttensioning

	Prestress loss, psi	
Posttensioning tendon material	Slabs	Beams and joists
Stress-relieved 270-K strand and stress-relieved 240-K wire	30,000 (207 N/mm^2)	35,000 (241 N/mm^2)
Bar	20,000 (138 N/mm^2)	25,000 (172 N/mm^2)
Low-relaxation 270-K strand	15,000 (103 N/mm^2)	20,000 (138 N/mm^2)

TABLE 5.5 Types of Prestress Losses

	Stage of occurrence		Tendon stress loss	
Type of prestress loss	Pretensioned members	Posttensioned members	During time interval (t_i, t_j)	Total or during life
Elastic shortening of concrete (ES)	At transfer	At sequential jacking	...	Δf_{pES}
Relaxation of tendons (R)	Before and after transfer	After transfer	$\Delta f_{pR}\,(t_i, t_j)$	Δf_{pR}
Creep of concrete (CR)	After transfer	After transfer	$\Delta f_{pCR}\,(t_i, t_j)$	Δf_{pCR}
Shrinkage of concrete (SH)	After transfer	After transfer	$\Delta f_{pSH}\,(t_i, t_j)$	Δf_{pSH}
Friction (F)	...	At jacking	...	Δf_{pF}
Anchorage seating loss (A)	...	At transfer	...	Δf_{pA}
Total	Life	Life	$\Delta f_{pT}\,(t_i, t_j)$	Δf_{pT}

reinforcement. Such a loss in stress is termed *stress relaxation.* The ACI 318-02 Code limits the tensile stress in the prestressing tendons to the following:

1. For stresses due to the tendon jacking force, $f_{pj} = 0.94f_{py}$, but not greater than the lesser of $0.80f_{pu}$ and the maximum value recommended by the manufacturer of the tendons and anchorages.
2. Immediately after prestress transfer, $f_{pi} = 0.82f_{py}$, but not greater than $0.74f_{pu}$.
3. In posttensioned tendons, at the anchorages and couplers immediately after force transfer $= 0.70f_{pu}$.

The range of values of f_{py} is given by the following:

Prestressing bars: $f_{py} = 0.80f_{pu}$
Stress-relieved tendons: $f_{py} = 0.85f_{pu}$
Low-relaxation tendons: $f_{py} = 0.90f_{pu}$

If f_{pR} is the remaining prestressing stress in the steel after relaxation, the following expression defines f_{pR} for stress relieved steel:

$$\frac{f_{pR}}{f_{pi}} = 1 - \left(\frac{\log t_2 - \log t_1}{10}\right)\left(\frac{f_{pi}}{f_{py}} - 0.55\right) \tag{5.12}$$

In this expression, $\log t$ in hours is to the base 10, f_{pi}/f_{py} exceeds 0.55, and $t = t_2 - t_1$. Also, for low-relaxation steel, the denominator of the log term in the equation is divided by 45 instead of 10.

An approximation of the term $(\log t_2 - \log t_1)$ can be made in Equation 5.13 so that $\log t = \log(t_2 - t_1)$ without significant loss in accuracy. In that case, the stress-relaxation loss becomes

$$\Delta f_{pR} = f'_{pi}\frac{\log t}{10}\left(\frac{f'_{pi}}{f_{py}} - 0.55\right) \tag{5.13}$$

where f'_{pi} is the initial stress in steel to which the concrete element is subjected.

If a step-by-step loss analysis is necessary, the loss increment at any particular stage can be defined as

$$\Delta f_{pR} = f'_{pi}\left(\frac{\log t_2 - \log t_1}{10}\right)\left(\frac{f_{pi}}{f_{py}} - 0.55\right) \tag{5.14}$$

where t_1 is the time at the beginning of the interval and t_2 is the time at the end of the interval from jacking to the time when the loss is being considered.

For low relaxation steel, change the divider to 45 instead of 10 in Equation 5.14.

5.5.2 Creep Loss (CR)

The creep coefficient at any time t in days can be defined as

$$C_t = \frac{t^{0.60}}{10 + t^{0.60}}C_u \tag{5.15}$$

As discussed earlier, the value of C_u ranges between 2 and 4, with an average of 2.35 for ultimate creep. The loss in prestressed members due to creep can be defined for bonded members as

$$\Delta f_{CR} = C_t\frac{E_{ps}}{E_c}f_{cs} \tag{5.16}$$

where f_{cs} is the stress in the concrete at the level of the centroid of the prestressing tendon. In general, this loss is a function of the stress in the concrete at the section being analyzed. In posttensioned, nonbonded members, the loss can be considered essentially uniform along the whole span. Hence, an average value of the concrete stress f_{cs} between the anchorage points can be used for calculating the creep in posttensioned members.

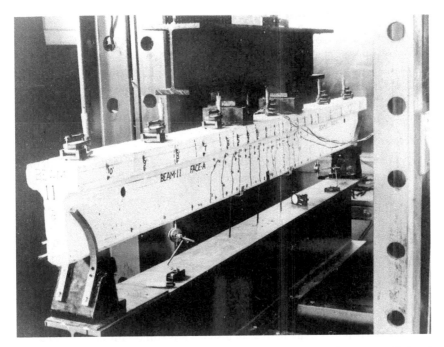

PHOTO 5.2 Prestressed concrete pretensioned T-beam tested to failure at the Rutgers University Concrete Research Laboratory (courtesy E.G. Nawy).

5.5.3 Shrinkage Loss (SH)

As with concrete creep, the magnitude of the shrinkage of concrete is affected by several factors. They include mixture proportions, type of aggregate, type of cement, curing time, time between the end of external curing and the application of prestressing, size of the member, and the environmental conditions. Size and shape of the member also affect shrinkage. Approximately 80% of shrinkage takes place in the first year of life of the structure. The average value of ultimate shrinkage strain in both moist-cured and steam-cured concrete is given as 780×10^{-6} in./in. in the ACI 209 R-92 Report. This average value is affected by the length of initial moist curing, ambient relative humidity, volume–surface ratio, temperature, and concrete composition. To take such effects into account, the average value of shrinkage strain should be multiplied by a correction factor γ_{SH} as follows:

$$\varepsilon_{SH} = 780 \times 10^{-6}\gamma_{SH} \tag{5.17}$$

Components of γ_{SH} are factors for various environmental conditions and tabulated in Ref. [11, Section 2].

The Prestressed Concrete Institute (PCI) stipulates for standard conditions an average value for nominal ultimate shrinkage strain $(\varepsilon_{SH})_u = 820 \times 10^{-6}$ in./in. (mm/mm) [7]. If ε_{SH} is the shrinkage strain after adjusting for relative humidity at volume-to-surface ratio V/S, the loss in prestressing in pretensioned member is

$$\Delta f_{pSH} = \varepsilon_{SH} \times E_{ps} \tag{5.18}$$

For posttensioned members, the loss in prestressing due to shrinkage is somewhat less since some shrinkage has already taken place before posttensioning. If the relative humidity is taken as a percent value and the V/S ratio effect is considered, the PCI general expression for loss in prestressing due to shrinkage becomes

$$\Delta f_{pSH} = 8.2 \times 10^{-6} K_{SH} E_{ps}\left(1 - 0.06\frac{V}{S}\right)(100 - RH) \tag{5.19}$$

where RH is the relative humidity (see Table 5.6).

5.5.4 Losses Due to Friction (F)

Loss of prestressing occurs in posttensioning members due to friction between the tendons and the surrounding concrete ducts. The magnitude of this loss is a function of the tendon form or alignment, called the *curvature effect*, and the local deviations in the alignment, called the *wobble effect*. The values of the loss coefficients are often refined while preparations are made for shop drawings by varying the types of tendons and the duct alignment. Whereas the curvature effect is predetermined, the wobble effect is the result of accidental or unavoidable misalignment, since ducts or sheaths cannot be perfectly placed.

5.5.4.1 Curvature Effect

As the tendon is pulled with a force F_1 at the jacking end, it will encounter friction with the surrounding duct or sheath such that the stress in the tendon will vary from the jacking plane to a distance L along the span.

The frictional loss of stress Δf_{pF} is then given by

$$\Delta f_{pF} = f_1 - f_2 = f_1 \left(1 - e^{-\mu\alpha - KL}\right) \tag{5.20}$$

Assuming that the prestress force between the start of the curved portion and its end is small ($\cong 15\%$), it is sufficiently accurate to linearize Equation 5.20 into the following form:

$$\Delta f_{pF} = -f_1 (\mu\alpha + KL) \tag{5.21}$$

where L is in feet.

Since the ratio of the depth of beam to its span is small, it is sufficiently accurate to use the projected length of the tendon for calculating α, giving

$$\alpha = 8y/x \text{ radian} \tag{5.22}$$

Table 5.7 gives the design values of the curvature friction coefficient μ and the wobble or length friction coefficient K adopted from the ACI 318 Code.

TABLE 5.6 Values of K_{SH} for Posttensioned Members

Time from end of moist curing to application of prestress, days	1	3	5	7	10	20	30	60
K_{SH}	0.92	0.85	0.8	0.77	0.73	0.64	0.58	0.45

Source: Prestressed Concrete Institute.

TABLE 5.7 Wobble and Curvature Friction Coefficients

Type of tendon	Wobble coefficient, K per foot	Curvature coefficient, μ
Tendons in flexible metal sheathing		
Wire tendons	0.0010–0.0015	0.15–0.25
Seven-wire strand	0.0005–0.0020	0.15–0.25
High-strength bars	0.0001–0.0006	0.08–0.30
Tendons in rigid metal duct		
Seven-wire strand	0.0002	0.15–0.25
Mastic-coated tendons		
Wire tendons and seven-wire strand	0.0010–0.0020	0.05–0.15
Pregreased tendons		
Wire tendons and seven-wire strand	0.0003–0.0020	0.05–0.15

Source: Prestressed Concrete Institute.

5.5.5 Example 1: Prestress Losses in Beams

A simply supported posttensioned 70-ft-span lightweight steam-cured double T-beam as shown in Figure 5.2 is prestressed by twelve $\frac{1}{2}$-in. diameter (twelve 12.7 mm diameter) 270-K grade stress-relieved strands. The tendons are harped, and the eccentricity at midspan is 18.73 in. (476 mm) and at the end 12.98 in. (330 mm). Compute the prestress loss at the critical section in the beam of 0.40 span due to dead load and superimposed dead load at

1. stage I at transfer
2. stage II after concrete topping is placed
3. two years after concrete topping is placed

Suppose the topping is 2 in. (51 mm) normal-weight concrete cast at 30 days. Suppose also that prestress transfer occurred 18 h after tensioning the strands. Given

$$f_c' = 5000 \text{ psi, lightweight (34.5 MPa)}$$
$$f_{ci}' = 3500 \text{ psi (24.1 MPa)}$$

and the following noncomposite section properties:

$$A_c = 615 \text{ in.}^2 \ (3{,}968 \text{ cm}^2)$$
$$I_c = 59{,}720 \text{ in.}^4 \ (2.49 \times 10^6 \text{ cm}^4)$$
$$c_b = 21.98 \text{ in. (55.8 cm)}$$
$$c^t = 10.02 \text{ in. (25.5 cm)}$$
$$S_b = 2{,}717 \text{ in.}^3 \ (44{,}520 \text{ cm}^3)$$
$$S^t = 5{,}960 \text{ in.}^3 \ (97{,}670 \text{ cm}^3)$$
$$W_D \text{ (no topping)} = 491 \text{ plf (7.2 kN/m)}$$
$$W_{SD} \text{ (2-in. topping)} = 250 \text{ plf (3.65 kN/m)}$$

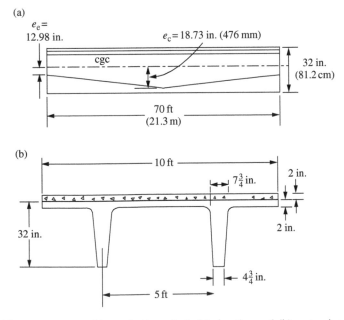

FIGURE 5.2 Double tee pretensioned beam in Example 1: (a) elevation and (b) pretensioned section [5].

$$W_L = 40 \text{ psf } (1{,}915 \text{ Pa}) \text{ — transient}$$

$$f_{pu} = 270{,}000 \text{ psi } (1{,}862 \text{ MPa})$$

$$f_{py} = 0.85 f_{pu} \approx 230{,}000 \text{ psi } (1{,}589 \text{ MPa})$$

$$f_{pi} = 0.70 f_{pu} = 0.82 f_{py} = 0.82 \times 0.85 f_{pu} \cong 0.70 f_{pu}$$

$$= 189{,}000 \text{ psi } (1{,}303 \text{ MPa})$$

$$E_{ps} = 28 \times 10^6 \text{ psi } (193.1 \times 10^6 \text{ MPa})$$

$$18\text{-day modular ratio} = 9.72$$

Solution

1. *Anchorage seating loss*

$$\Delta_A = \tfrac{1}{4} \text{ in.} = 0.25 \text{ in.}, \quad L = 70 \text{ ft}$$

The anchorage slip stress loss is

$$\Delta f_{PA} = \frac{\Delta_A}{L} E_{ps} = \frac{0.25}{70 \times 12} \times 28 \times 10^6 \cong 8333 \text{ psi } (40.2 \text{ MPa})$$

2. *Elastic shortening.* Since all jacks are simultaneously posttensioned, the elastic shortening will precipitate during jacking. As a result, no elastic shortening stress loss takes place in the tendons. Hence, $\Delta f_{pES} = 0$.

3. *Frictional loss.* Assume that the parabolic tendon approximates the shape of an arc of a circle. Then, from Equation 5.22

$$\alpha = \frac{8y}{x} = \frac{8(18.73 - 12.98)}{70 \times 12} = 0.0548 \text{ radian}$$

From Table 5.7, use $K = 0.001$ and $\mu = 0.25$.

From Equation 5.21, the stress loss in prestress due to friction is

$$\Delta f_{pF} = f_{pi}(\mu\alpha + KL)$$

$$= 189{,}000(0.25 \times 0.0548 + 0.001 \times 70)$$

$$= 15{,}819 \text{ psi } (109 \text{ MPa})$$

The stress remaining in the prestressing steel after all initial instantaneous losses is

$$f_{pi} = 189{,}000 - 8{,}333 - 0 - 15{,}819 = 164{,}848 \text{ psi } (1{,}136 \text{ MPa})$$

Hence, the net prestressing force is

$$P_i = 164{,}848 \times 12 \times 0.153 = 296{,}726 \text{ lb}$$

Stage I: Stress at Transfer

1. *Anchorage seating loss*

$$\text{Loss} = 8{,}333 \text{ psi}$$

$$\text{Net stress} = 164{,}848 \text{ psi}$$

2. *Relaxation loss*

$$\Delta f_{pR} = 164{,}848 \left(\frac{\log 18}{10}\right) \left(\frac{164{,}848}{230{,}000} - 0.55\right)$$

$$\cong 3{,}450 \text{ psi } (23.8 \text{ MPa})$$

3. *Creep loss*

$$\Delta f_{pCR} = 0$$

4. *Shrinkage loss*

$$\Delta f_{pSH} = 0$$

So the tendon stress f_{pi} at the end of stage I is

$$164,848 - 3,450 = 161,398 \text{ psi } (1,113 \text{ MPa})$$

Stage II: Transfer to Placement of Topping after 30 Days

1. *Creep loss*

$$P_i = 161,398 \times 12 \times 0.153 = 296,327 \text{ lb}$$

$$\bar{f}_{cs} = -\frac{P_i}{A_c}\left(1 + \frac{e^2}{r^2}\right) + \frac{M_D e}{I_c}$$

$$= -\frac{296,327}{615}\left(1 + \frac{(17.58)^2}{97.11}\right) + \frac{3,464,496 \times 17.58}{59,720}$$

$$= -2016.2 + 1020.0 = 996.2 \text{ psi } (6.94 \text{ MPa})$$

Hence, the creep loss for lightweight concrete, K_{CR} is reduced by 20%, hence $= 1.6 \times 0.80 = 1.28$

$$\Delta f_{pCR} = nK_{CR}\left(\bar{f}_{cs} - \bar{f}_{csd}\right)$$
$$= 9.72 \times 1.28(996.2 - 519.3) \cong 5933 \text{ psi } (41 \text{ MPa})$$

2. *Shrinkage loss.* $K_{SH} = 0.58$ at 30 days, Table 5.6

$$\Delta f_{pSH} = 6190 \times 0.58 = 3590 \text{ psi } (24.8 \text{ MPa})$$

3. *Steel relaxation loss at 30 days*

$$f_{ps} = 161,398 \text{ psi}$$

The relaxation loss in stress becomes

$$\Delta f_{pR} = 161,398\left(\frac{\log 720 - \log 18}{10}\right)\left(\frac{161,398}{230,000} - 0.55\right)$$
$$\cong 3,923 \text{ psi } (27.0 \text{ MPa})$$

Stage II: Total Losses

$$\Delta f_{pT} = \Delta f_{pCR} + \Delta f_{pSH} + \Delta f_{pR}$$
$$= 5,933 + 3,590 + 3,923 = 13,446 \text{ psi } (93 \text{ MPa})$$

The increase f_{SD} in stress in the strands due to the addition of topping is

$$f_{SD} = n\bar{f}_{csd} = n(M_{SD} \times e/I_c) = n \times 519.3 = 9.72 \times 519.3 = 5048 \text{ psi } (34.8 \text{ MPa}) \text{ in tension}$$

$f_{SD} = 5048$ psi (34.8 MPa); hence, the strand stress at the end of stage II is

$$f_{pe} = f_{ps} - \Delta f_{pT} + \Delta f_{SD} = 161,398 - 13,446 + 5,048 = 153,000 \text{ psi } (1,055 \text{ MPa})$$

Stage III: At the End of 2 Years

$$f_{pe} = 151,516 \text{ psi}$$
$$t_1 = 720 \text{ h}$$
$$t_2 = 17,520 \text{ h}$$

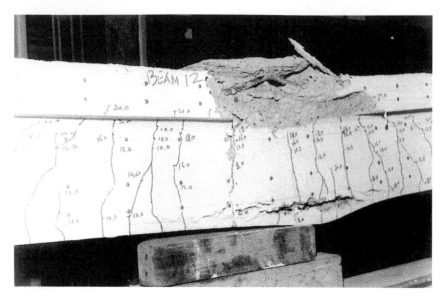

PHOTO 5.3 Crushing of the concrete at the top fibers and yield of reinforcement at the bottom fibers of the prestressed concrete T-beam in Photo 4.2 at Ultimate State of Failure. Tested at the Rutgers University Concrete Research Laboratory (courtesy E.G. Nawy).

The steel relaxation stress loss is

$$\Delta f_{pR} = 153,000 \left(\frac{\log 17,520 - \log 720}{10} \right) \left(\frac{153,000}{230,000} - 0.55 \right)$$
$$\cong 2,444 \text{ psi } (16.9 \text{ MPa})$$

Hence, the strand stress f_{pe} at the end of stage III is approximately
$$153,000 - 2,444 = 150,536 \text{ psi } (1,038 \text{ MPa})$$

5.5.6 Example 2: Prestressing Losses Evaluation Using SI Units

Solve Example 1 using SI units for losses in prestress, considering self-weight and superimposed dead load only.

    Data

$$f'_c = 34.5 \text{ MPa}$$
$$f'_{ci} = 24.1 \text{ MPa}$$

$$A_c = 3,968 \text{ cm}^2, \qquad S^t = 97,670 \text{ cm}^3$$
$$I_c = 2.49 \times 10^6 \text{ cm}^4, \quad S_b = 44,520 \text{ cm}^3$$
$$r^2 = I_c/A_c = 626 \text{ cm}^2$$
$$c_b = 55.8 \text{ cm}, \qquad c^t = 25.5 \text{ cm}$$
$$e_c = 47.6 \text{ cm}, \qquad e_e = 33.0 \text{ cm}$$
$$f_{pu} = 1,860 \text{ MPa}$$
$$f_{py} = 0.85 f_{pu} = 1,580 \text{ MPa}$$
$$f_{pi} = 0.82 f_{py} = (0.82 \times 0.85) f_{pu}$$
$$= 0.7 f_{pu} = 1,300 \text{ MPa}$$
$$E_{ps} = 193,000 \text{ MPa}$$

Span $l = 21.3$ m

$$A_{ps} = 12 \text{ strands, } 12.7\text{-mm diameter } (99 \text{ mm}^2)$$
$$= 12 \times 99 = 1{,}188 \text{ mm}^2$$
$$M_D = 391 \text{ kN m}, \quad M_{SD} = 199 \text{ kN m}$$
$$\Delta_A = 0.64 \text{ cm}$$
$$V/S = 1.69, \quad \text{RH} = 70\%$$
$$E_c = w^{1.5} 0.043 \sqrt{f_c'}, \quad w \text{ (lightweight)} \approx 1830 \text{ kg/m}^3$$
$$E_{ci} = w^{1.5} 0.043 \sqrt{f_{ci}'}$$
$$\text{MPa} = 10^6 \text{ N/m}^2 = \text{N/mm}^2$$
$$(\text{psi}) \ 0.006895 = \text{MPa}$$
$$(\text{lb/ft}) \ 14.593 = \text{N/m}$$
$$(\text{in. lb}) \ 0.113 = \text{N m}$$

Solution

1. *Anchorage seating loss*

$$\Delta_A = 0.64 \text{ cm}, \quad l = 21.3 \text{ m}$$

$$\Delta f_{PA} = \frac{\Delta_A}{L} E_{ps} = \frac{0.64}{21.3 \times 100} \times 193{,}000 = 58.0 \text{ MPa}$$

2. *Elastic shortening*
   Since all jacks are simultaneously tensioned, the elastic shortening will simultaneously precipitate during jacking. As a result, no elastic shortening loss takes place in the tendons.

$$E_c = w^{1.5} 0.043 \sqrt{34.5}$$
$$= 1{,}830^{1.5} \times 0.043 \sqrt{34.5} = 19{,}770 \text{ MPa}$$

$$n = \frac{E_{ps}}{E_c} = \frac{193{,}000}{19{,}770} = 9.76$$

$\bar{f}_{csd}$ = stress in concrete at cgs due to all superimposed dead loads after prestressing is accomplished

$$\bar{f}_{csd} = \frac{M_{SD} e}{I_c} = \frac{1.99 \times 10^7 \text{ N cm} \times 44.7}{2.49 \times 10^6} \times \frac{1}{100} \text{ N/mm}^2$$
$$= 3.57 \text{ MPa}$$
$$K_{CR} = 1.6 \text{ for posttensioned beam}$$
$$\Delta f_{pCR} = n K_{CR}(\bar{f}_{cs} - \bar{f}_{csd})$$
$$= 9.76 \times 1.6(6.90 - 3.57) = 52.0 \text{ MPa}$$

3. *Shrinkage loss at 30 days*
   From Equation 5.19

$$\Delta f_{pSH} = 8.2 \times 10^{-6} K_{SH} E_{ps} \left(1 - 0.06 \frac{V}{S}\right)(100 - \text{RH})$$

K_{SH} at 30 days $= 0.58$ (Table 5.6):

$$\Delta f_{pSH} = 8.2 \times 10^{-6} \times 0.58 \times 193,000(1 - 0.06 \times 1.69)(100 - 70)$$
$$= 24.7 \text{ MPa}$$

4. *Relaxation loss at 30 days (720 h)*

$$f_{ps} = 1108 \text{ MPa}$$

$$\Delta f_{pR} = 1108 \left(\frac{\log 720 - \log 18}{10}\right) \left(\frac{1108}{1580} - 0.55\right)$$
$$= 110.8(2.85 - 1.25)0.151 = 26.8 \text{ MPa}$$

Stage II: Total Losses

$$\Delta f_{pT} = \Delta f_{pCR} + \Delta f_{pSH} + \Delta f_{pR}$$
$$= 52.0 + 24.7 + 26.8 = 104 \text{ MPa}$$

Increase of tensile stress at bottom cgs fibers due to addition of topping is from before

$$\Delta f_{SD} = n f_{CSD} = 9.76 \times 3.57 = 34.8 \text{ MPa}$$
$$f_{pe} = f_{ps} - \Delta f_{pT} + \Delta f_{SD}$$
$$= 1108 - 103.5 + 34.5 = 1039 \text{ MPa}$$

Stage III: At End of 2 Years

$$f_{pe} = 1039 \text{ MPa}$$
$$t_1 = 720 \text{ h}, \qquad t_2 = 17,520 \text{ h}$$

$$\Delta f_{pR} = 1,039 \left(\frac{\log 17,520 - \log 720}{10}\right) \left(\frac{1,039}{1,580} - 0.55\right)$$
$$= 103.9(4.244 - 2.857)0.108 = 15.6 \text{ MPa}$$

On the assumption that Δf_{pCR} and Δf_{pSH} were stable in this case, the stress in the tendons at end of stage III can approximately be $f_{ps} = 1039 - 15.6 \cong 1020 \text{ MPa}$ (see Figure 5.3).

5.6 Flexural Design of Prestressed Concrete Elements

Unlike the case of reinforced concrete members, the external dead load and partial live load are applied to the prestressed concrete member at varying concrete strengths at various loading stages. These loading stages can be summarized as follows:

- Initial prestress force P_i is applied; then, at transfer, the force is transmitted from the prestressing strands to the concrete.
- The full self-weight W_D acts on the member together with the initial prestressing force, provided that the member is simply supported, that is, there is no intermediate support.
- The full superimposed dead load W_{SD}, including topping for composite action, is applied to the member.
- Most short-term losses in the prestressing force occur, leading to a reduced prestressing force P_{eo}.
- The member is subjected to the full service load, with long-term losses due to creep, shrinkage, and strand relaxation taking place and leading to a net prestressing force P_e.
- Overloading of the member occurs under certain conditions up to the limit state at failure.

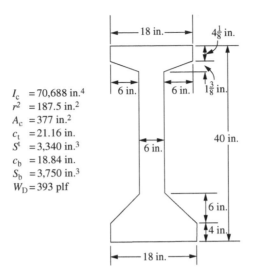

FIGURE 5.3 I-Beam section in Example 2 [5].

Stress at transfer

$$f^{t} = -\frac{P_i}{A_c}\left(1 - \frac{ec_t}{r^2}\right) - \frac{M_D}{S^t} \le f_{ti} \qquad (5.23a)$$

$$f_b = -\frac{P_i}{A_c}\left(1 + \frac{ec_b}{r^2}\right) + \frac{M_D}{S_b} \le f_{ci} \qquad (5.23b)$$

where P_i is the initial prestressing force. While a more accurate value to use would be the horizontal component of P_i, it is reasonable for all practical purposes to disregard such refinement.

Effective stresses after losses

$$f^{t} = -\frac{P_e}{A_c}\left(1 - \frac{ec_t}{r^2}\right) - \frac{M_D}{S^t} \le f_t \qquad (5.24a)$$

$$f_b = -\frac{P_e}{A_c}\left(1 + \frac{ec_b}{r^2}\right) + \frac{M_D}{S_b} \le f_c \qquad (5.24b)$$

Service-load final stresses

$$f^{t} = -\frac{P_e}{A_c}\left(1 - \frac{ec_t}{r^2}\right) - \frac{M_T}{S^t} \le f_c \qquad (5.25a)$$

$$f_b = -\frac{P_e}{A_c}\left(1 + \frac{ec_b}{r^2}\right) + \frac{M_T}{S_b} \le f_c \qquad (5.25b)$$

where $M_T = M_D + M_{SD} + M_L$; P_i is the initial prestress; P_e is the effective prestress after losses, where t denotes the top and b denotes the bottom fibers; e is the eccentricity of tendons from the concrete section center of gravity, cgc; r^2 is the square of radius of gyration; and S^t/S_b is the top/bottom section modulus value of concrete section.

The *decompression stage* denotes the increase in steel strain due to the increase in load from the stage when the effective prestress P_e acts *alone* to the stage when the additional load causes the compressive stress in the concrete at the cgs level to reduce to zero. At this stage, the *change* in concrete stress due to decompression is

$$f_{decomp} = \frac{P_e}{A_c}\left(1 + \frac{e^2}{r^2}\right) \qquad (5.25c)$$

This relationship is based on the assumption that the strain between the concrete and the prestressing steel bonded to the surrounding concrete is such that the gain in the steel stress is the same as the decrease in the concrete stress.

5.6.1 Minimum Section Modulus

To design or choose the section, a determination of the required minimum section modulus, S_b and S^t, has to be made first.

1. *For variable tendon eccentricity:*

$$S^t \geq \frac{(1 - \gamma)M_D + M_{SD} + M_L}{\gamma f_{ti} - f_c} \tag{5.26a}$$

and

$$S_b \geq \frac{(1 - \gamma)M_D + M_{SD} + M_L}{f_t - \gamma f_{ci}} \tag{5.26b}$$

2. *For constant tendon eccentricity:*

$$S^t \geq \frac{M_D + M_{SD} + M_L}{\gamma f_{ti} - f_c} \tag{5.27a}$$

and

$$S_b \geq \frac{M_D + M_{SD} + M_L}{f_t - \gamma f_{ci}} \tag{5.27b}$$

The required eccentricity value at the critical section, such as the support for an ideal beam section having properties close to those required by Equations 5.27a and 5.27b, is

$$e_e = (f_{ti} - \bar{f}_{ci}) \frac{S^t}{P_i} \tag{5.28}$$

Table 5.8 and Table 5.9 list the properties of standard sections.

TABLE 5.8 Geometrical Outer Dimensions and Section Moduli of Standard AASHTO Bridge Sections

Designation	AASHTO sections					
	Type 1	Type 2	Type 3	Type 4	Type 5	Type 6
Area A_c, in.^{2}	276	369	560	789	1,013	1,085
Moment of inertia						
$I_g(x - x)$, in.^{4}	22,750	50,979	125,390	260,741	521,180	733,320
$I_g(y - y)$, in.^{4}	3,352	5,333	12,217	24,347	61,235	61,619
Top-/bottom-section modulus, in.^{3}	1,476	2,527	5,070	8,908	16,790	20,587
	1,807	3,320	6,186	10,544	16,307	20,157
Top flange width, b_f (in.)	12	12	16	20	42	42
Top flange average thickness, t_f (in.)	6	8	9	11	7	7
Bottom flange width, b_2 (in.)	16	18	22	26	28	28
Bottom flange average thickness, t_2 (in.)	7	9	11	12	13	13
Total depth, h (in.)	28	36	45	54	63	72
Web width, b_w (in.)	6	6	7	8	8	8
c_t/c_b (in.)	15.41	20.17	24.73	29.27	31.04	35.62
	12.59	15.83	20.27	24.73	31.96	36.38
r^2, in.^{2}	82	132	224	330	514	676
Self-weight w_D, lb/ft	287	384	583	822	1,055	1,130

TABLE 5.9 Geometrical Outer Dimensions and Section Moduli of Standard PCI Double T-Sections

Designation	Top-/bottom-section modulus, in.³	Flange width b_f, in.	Flange depth t_f, in.	Total depth h, in.	Web width $2b_w$, in.
8DT12	1,001/315	96	2	12	9.5
8DT14	1,307/429	96	2	14	9.5
8DT16	1,630/556	96	2	16	9.5
8DT20	2,320/860	96	2	20	9.5
8DT24	3,063/1,224	96	2	24	9.5
8DT32	5,140/2,615	96	2	32	9.5
10DT32	5,960/2,717	120	2	32	12.5
12DT34[a]	10,458/3,340	144	4	34	12.5
15DT34[a]	13,128/4,274	180	4	34	12.5

[a] Pretopped.

5.6.2 Example 3: Flexural Design of Prestressed Beams at Service Load Level

Design an I-section for a beam having a 65-ft (19.8 m) span to satisfy the following section modulus values:

$$\text{Required } S^t = 3,570 \text{ in.}^3 \ (58,535 \text{ cm}^3)$$

$$\text{Required } S_b = 3,780 \text{ in.}^3 \ (61,940 \text{ cm}^3)$$

Given

$$W_{SD} = 100 \text{ plf}$$

$$W_L = 1,100 \text{ plf}$$

$$f_c' = 5,000 \text{ psi } (34.5 \text{ MPa})$$

$$f_{ci}' \text{ at transfer} = 75\% \text{ of cylinder strength}$$

$$f_{pu} = 270,000 \text{ psi } (1,862 \text{ MPa})$$

$$f_t = 12\sqrt{f_c'}$$

Solution

Since the section moduli at the top and bottom fibers are almost equal, a symmetrical section is adequate. Next, analyze the section in Figure 5.3 chosen by trial and adjustment.

Analysis of Stresses at Transfer

$$\bar{f}_{ci} = f_{ti} - \frac{c_t}{h}(f_{ti} - f_{ci})$$

$$= +184 - \frac{21.16}{40}(+184 + 2,250) \cong -1,104 \text{ psi } (C) \ (7.6 \text{ MPa})$$

$$P_i = A_c \bar{f}_{ci} = 377 \times 1,104 = 416,208 \text{ lb } (1,851 \text{ kN})$$

$$M_D = \frac{393(65)^2}{8} \times 12 = 2,490,638 \text{ in. lb } (281 \text{ kN m})$$

The eccentricity required at the section of maximum moment at midspan is

$$e_c = (f_{ti} - \bar{f}_{ci})\frac{S_t}{P_i} + \frac{M_D}{P_i}$$

$$= (184 + 1,104)\frac{3,572}{416,208} + \frac{2,490,638}{416,208}$$

$$= 11.05 + 5.98 = 17.04 \text{ in. } (433 \text{ mm})$$

Since $c_b = 18.84$ in., and assuming a cover of 3.75 in., try $e_c = 18.84 - 3.75 = 15.0$ in. (381 mm):

$$\text{Required area of strands } A_p = \frac{P_i}{f_{pi}} = \frac{416{,}208}{189{,}000} = 2.2 \text{ in.}^2 \ (14.2 \text{ cm}^2)$$

$$\text{Number of strands} = \frac{2.2}{0.153} = 14.38$$

Try thirteen $\frac{1}{2}$-in. strands, $A_p = 1.99$ in.^{2} (12.8 cm^2), and an actual $P_i = 189{,}000 \times 1.99 = 376{,}110$ lb (1,673 kN), and check the concrete extreme fiber stresses. From Equation 5.23a

$$f^t = -\frac{P_i}{A_c}\left(1 - \frac{ec_t}{r^2}\right) - \frac{M_D}{S^t}$$

$$= -\frac{376{,}110}{377}\left(1 - \frac{15.0 \times 21.16}{187.5}\right) - \frac{2{,}490{,}638}{3{,}340}$$

$$= +691.2 - 745.7 = -55 \text{ psi } (C), \text{ no tension at transfer, OK.}$$

From Equation 5.23b

$$f_b = -\frac{P_i}{A_c}\left(1 + \frac{ec_b}{r^2}\right) + \frac{M_D}{S_b}$$

$$= -\frac{376{,}110}{377}\left(1 + \frac{15.0 \times 18.84}{187.5}\right) + \frac{2{,}490{,}638}{3{,}750}$$

$$= -2501.3 + 664.2 = -1837 \text{ psi } (C) < f_{ci} = 2{,}250 \text{ psi, OK.}$$

Analysis of stresses at service load. From Equation 5.25a

$$f^t = -\frac{P_e}{A_c}\left(1 - \frac{ec_t}{r^2}\right) - \frac{M_T}{S^t}$$

$$P_e = 13 \times 0.153 \times 154{,}980 = 308{,}255 \text{ lb } (1{,}371 \text{ kN})$$

$$M_{SD} + M_L = (100 + 1{,}100)(65)^2 \times 12/8 = 7{,}605{,}000 \text{ in. lb}$$

$$\begin{aligned}\text{Total moment } M_T = M_D + M_{SD} + M_L &= 2{,}490{,}638 + 7{,}605{,}000 \\ &= 10{,}095{,}638 \text{ in. lb } (1{,}141 \text{ kN m})\end{aligned}$$

$$f^t = -\frac{308{,}255}{377}\left(1 - \frac{15.0 \times 21.16}{187.5}\right) - \frac{10{,}095{,}638}{3{,}340}$$

$$= +566.5 - 3022.6 = -2{,}456 \text{ psi } (C) > f_c = -2{,}250 \text{ psi}$$

Hence, either enlarge the depth of the section or use higher strength concrete. Using $f_c' = 6{,}000$ psi

$$f_b = -\frac{P_e}{A_c}\left(1 + \frac{ec_b}{r^2}\right) + \frac{M_T}{S_b} = -\frac{308{,}255}{377}\left(1 + \frac{15.0 \times 18.84}{187.5}\right) + \frac{10{,}095{,}638}{3{,}750}$$

$$= -2{,}050 + 2692.2 = 642 \text{ psi } (T), \text{ OK.}$$

Check support section stresses. Allowable

$$f_{ci}' = 0.75 \times 6000 = 4500 \text{ psi}$$

$$f_{ci} = 0.60 \times 4500 = 2700 \text{ psi}$$

$$f_{ti} = 3\sqrt{f_{ci}'} = 201 \text{ psi} \quad \text{for midspan}$$

$$f_{ti} = 6\sqrt{f_{ci}'} = 402 \text{ psi} \quad \text{for support}$$

$$f_c = 0.45 f_c' = 2700 \text{ psi}$$

$$f_{t1} = 6\sqrt{f_c'} = 465 \text{ psi}$$
$$f_{t2} = 12\sqrt{f_c'} = 930 \text{ psi}$$

1. *At transfer.* Support section compressive fiber stress

$$f_b = -\frac{P_i}{A_c}\left(1 + \frac{e c_b}{r^2}\right) + 0$$

For $P_i = 376{,}110$ lb

$$-2{,}700 = -\frac{376{,}110}{377}\left(1 + \frac{e \times 18.84}{187.5}\right)$$

so that

$$e_e = 16.98 \text{ in.}$$

Accordingly, try $e_e = 12.49$ in.

$$f^t = -\frac{376{,}110}{377}\left(1 - \frac{12.49 \times 21.16}{187.5}\right) - 0$$
$$= 409 \text{ psi } (T) > f_{ti} = 402 \text{ psi}$$
$$f_b = 2{,}250 \text{ psi}$$

Thus, use mild steel at the top fibers at the support section to take all tensile stresses in the concrete, or use a higher strength concrete for the section, or reduce the eccentricity.

2. *At service load*

$$f^t = -\frac{308{,}255}{377}\left(1 - \frac{12.49 \times 21.16}{187.5}\right) - 0 = 335 \text{ psi } (T) < 930 \text{ psi, OK.}$$

$$f_b = -\frac{308{,}255}{377}\left(1 + \frac{12.49 \times 18.84}{187.5}\right) + 0 = -1{,}844 \text{ psi } (C) < -2{,}700 \text{ psi, OK.}$$

Hence, adopt the 40-in. (102-cm)-deep I-section prestressed beam of f_c' equal to 6,000 psi (41.4 MPa) normal-weight concrete with thirteen $\frac{1}{2}$-in. strands tendon having midspan eccentricity $e_c = 15.0$ in. (381 mm) and end section eccentricity $e_e = 12.5$ in. (318 m).

An alternative to this solution is to continue using $f_c' = 5000$ psi, but change the number of strands and eccentricities.

5.6.3 Development and Transfer Length in Pretensioned Members and Design of their Anchorage Reinforcement

As the jacking force is released in pretensioned members, the prestressing force is dynamically transferred through the bond interface to the surrounding concrete. The interlock or adhesion between the prestressing tendon circumference and the concrete over a finite length of the tendon gradually transfers the concentrated prestressing force to the entire concrete section at planes away from the end block and toward the midspan. The length of embedment determines the magnitude of prestress that can be developed along the span: the larger the embedment length, the higher is the prestress developed.

As an example for $\frac{1}{2}$-in. seven-wire strand, an embedment of 40 in. (102 cm) develops a stress of 180,000 psi (1,241 MPa), whereas an embedment of 70 in. (178 cm) develops a stress of 206,000 psi (1,420 MPa). The embedment length l_d that gives the full development of stress is a combination of the transfer length l_t and the flexural bond length l_f. These are given, respectively, by

$$l_t = \frac{1}{1000}\left(\frac{f_{pe}}{3}\right)d_b \tag{5.29a}$$

or

$$l_t = \frac{f_{pe}}{3000} d_b \qquad (5.29b)$$

and

$$l_f = \frac{1}{1000} (f_{ps} - f_{pe}) d_b \qquad (5.29c)$$

where f_{ps} is the stress in prestressed reinforcement at nominal strength (psi), f_{pe} is the effective prestress after losses (psi), and d_b is the nominal diameter of prestressing tendon (in.).

Combining Equations 5.29b and 5.29c gives

$$\text{Min } l_d = \frac{1}{1000} \left(f_{ps} - \frac{2}{3} f_{pe} \right) d_b \qquad (5.29d)$$

Equation 5.29d gives the minimum required development length for prestressing strands. If part of the tendon is sheathed toward the beam end to reduce the concentration of bond stresses near the end, the stress transfer in that zone is eliminated and an increased adjusted development length l_d is needed.

5.6.3.1 Design of Transfer Zone Reinforcement in Pretensioned Beams

Based on laboratory tests, empirical expressions developed by Mattock et al. give the total stirrup force F as

$$F = 0.0106 \frac{P_i h}{l_t} \qquad (5.30)$$

where h is the pretensioned beam depth and l_t is the transfer length. If the average stress in a stirrup is taken as *half* the maximum permissible steel f_s, then $F = \frac{1}{2} A_t f_s$. Substituting this for F in Equation 5.30 gives

$$A_t = 0.021 \frac{P_i h}{f_s l_t} \qquad (5.31)$$

where A_t is the total area of the stirrups and $f_s \leq 20{,}000$ psi (138 MPa) for crack-control purposes.

5.6.4 Posttensioned Anchorage Zones: Strut-and-Tie Design Method

The anchorage zone can be defined as the volume of concrete through which the concentrated pre-stressing force at the anchorage device spreads transversely to a linear distribution across the entire cross-section depth along the span. The length of this zone follows St Venant's principle, namely, that the stress becomes uniform at an approximate distance ahead of the anchorage device equal to the depth, h, of the section. The entire prism that would have a transfer length, h, is the total anchorage zone.

This zone is thus composed of two parts:

1. *General zone:* The general extent of the zone is identical to the total anchorage zone. Its length extent along the span is therefore equal to the section depth, h, in standard cases.
2. *Local zone:* This zone is the insert prism of concrete surrounding and immediately ahead of the anchorage device and the confining reinforcement it contains.

After significant cracking is developed, compressive stress trajectories in the concrete tend to congregate into straight lines that can be idealized as straight compressive struts in uniaxial compression. These struts would become part of truss units where the principal tensile stresses are idealized as tension ties in the truss unit with the nodal locations determined by the direction of the idealized compression struts. Figure 5.4 sketches standard strut-and-tie idealized trusses for concentric and eccentric cases both for solid and flanged sections as given in ACI 318-02 Code.

Simplified equations can be used to compute the magnitude of the bursting force, T_{burst}, and its centroid distance, d_{burst}, from the major bearing surface of the anchorage [9]. The member has to have

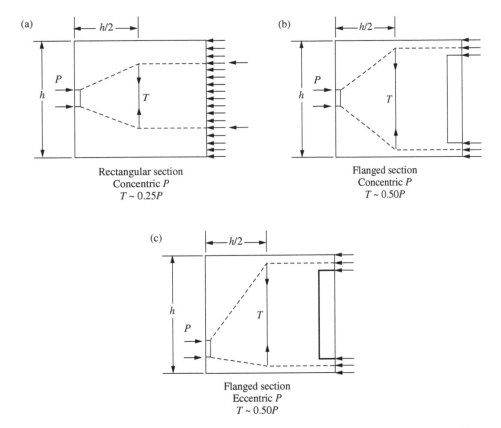

FIGURE 5.4   Strut-and-tie idealized trusses in standard concentric and eccentric cases, ACI 318-02 [8].

a rectangular cross-section with no discontinuities along the span. The bursting force, T_{burst} and its distance, d_{burst}, can be computed from the following expressions:

$$T_{burst} = 0.25 \sum P_{su}\left(1 - \frac{a}{h}\right) \tag{5.32a}$$

$$d_{burst} = 0.5(h - 2e) \tag{5.32b}$$

where $\sum P_{su}$ is the sum of the total factored prestress loads for the stressing arrangement considered (lb), a is the plate width of anchorage device or single group of closely spaced devices in the direction considered (in.), e is the eccentricity (always taken positive) of the anchorage device or group closely spaced devices with respect to the centroid of the cross-section (in.), and h is the depth of the cross-section in the direction considered (in.).

The ACI 318 Code requires that the design of confining reinforcement in the end anchorage block of posttensioned members be based on the *factored* prestressing force P_{su} for both the general and local zones. A load factor of 1.2 is to be applied to an end anchorage stress level $f_{pi} = 0.80f_{pu}$ for low-relaxation strands at the *short time interval* of jacking, which can reduce to an average value of $0.70f_{pu}$ for the total group of strands at the completion of the jacking process. For stress-relieved strands, a lower $f_{pi} = 0.70f_{pu}$ is advised. The maximum force P_{su} stipulated in the ACI 318 Code for designing the confining reinforcement at the end-block zone is as follows for the more widely used low-relaxation strands:

$$P_{su} = 1.2A_{ps}(0.8f_{pu}) \tag{5.32c}$$

The AASHTO Standard for the case where P_{su} acts at an inclined angle α in the direction of the beam span adds the term $[0.5\sum(P_{su}\sin\alpha)]$ to Equation 5.32a and $5e(\sin\alpha)$ to Equation 5.32b. For horizontal $P_{su}\sin\alpha = 0$.

5.6.4.1 Allowable Bearing Stresses

The maximum allowable bearing stress at the anchorage device seating should not exceed the smaller of the two values obtained from Equations 5.33a and 5.33b as follows:

$$f_b \leq 0.7f'_{ci}\sqrt{A/A_g} \tag{5.33a}$$

$$f_b \leq 2.25f'_{ci} \tag{5.33b}$$

where f_b is the maximum factored tendon load, P_u, divided by the effective bearing area A_b; f'_{ci} is the concrete compressive strength at initial stressing; A is the maximum area of the portion of the supporting surface that is geometrically similar to the loaded area and concentric with it, contained wholly within the section, with the upper base being the loaded surface area of the concrete and sloping sideway with a slope of 1 vertical to 2 horizontal; and A_g is the gross area of the bearing plate.

Equations 5.33a and 5.33b are valid only if general zone reinforcement is provided and if the extent of concrete along the tendon axis ahead of the anchorage device is at least twice the length of the local zone.

5.6.5 Example 4: End Anchorage Design by the Strut-and-Tie Method

Design an end anchorage reinforcement for the posttensioned beam in Example 3, giving the size, type, and distribution of reinforcement. Use $f'_{ci} = 5,000$ psi (34.5 MPa) normal-weight concrete. $f_{pu} = 270,000$ psi low-relaxation steel.

Assume that the beam ends are rectangular blocks extending 40 in. (104 cm.) into the span beyond the anchorage devices which then transitionally reduce to the 6-in. thick web.

1. *Establish the configuration of the tendons to give eccentricity $e_e = 12.49$ in. (317 mm).* From Example 3, $c_b = 18.84$ in.; hence, distance from the beam fibers $= c_b - e_e = 6.35$ in. (161 mm). For a centroidal distance of the $13\frac{1}{2}$-in. size strands $= 6.35$ in. from the beam bottom fibers, try the following row arrangement of tendons with the indicated distances from the bottom fibers:

   first row: five tendons at 2.5 in.
   second row: five tendons at 7.0 in.
   third row: three tendons at 11.5 in.

   $$\text{Distance of the centroid of tendons} = \frac{(5\times2.5 + 5\times7.0 + 3\times11.5)}{13} \cong 6.35 \text{ in., OK.}$$

2. *Factored forces in tendon rows and bearing capacity of the concrete.* From Equation 5.32c for low-relaxation strands, $f_{pi} = 0.80f_{pu}$. The factored jacking force at the short jacking time interval is

   $$P_{su} = 1.2A_{ps}(0.80f_{pu}) = 0.96f_{pu}\,A_{ps} = 0.96\times270,000 = 259,200 \text{ psi.}$$

   If stress-relieved strands are used, it would have been advisable to use $f_{pi} = 0.70f_{pu}$:

   first row force: $P_u1 = 5\times0.153\times259,200 = 198,286$ lb (882 kN)
   second row force: $P_u2 = 5\times0.153\times259,200 = 198,286$ lb (882 kN)
   third row force: $P_u3 = 3\times0.153\times259,200 = 118,973$ lb (529 kN)

   The total ultimate compressive force $= 198,286 + 198,286 + 118,973 = 515,545$ lb (2,290 kN). The total area of rigid bearing plates supporting the Supreme 13-chucks anchorage devices $= 14\times11 + 6\times4 = 178$ in.^{2} (113 cm^2). The actual bearing stress

   $$f_b = \frac{515,545}{178} = 2,896 \text{ psi (19.9 MPa)}$$

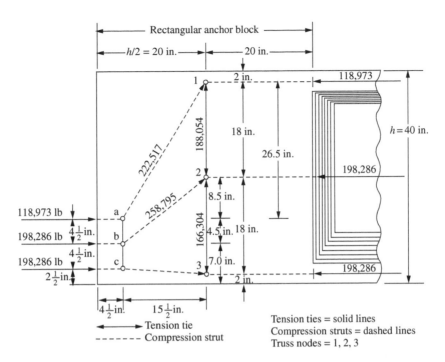

FIGURE 5.5 Struts-and-ties in Example 4 [5].

From Equations 5.33a and 5.33b, the maximum allowable bearing pressure on the concrete is the lesser of

$$f_b \leq 0.7 f'_{ci} \sqrt{A/A_g}$$
$$f_b \leq 2.25 f'_{ci}$$

Assume that the initial concrete strength at stressing is $f'_{ci} = 0.75 f'_c = 0.75 \times 5000 = 3750$ psi. The concentric area, A, of concrete with the bearing plates $\cong (14+4)(11+4) + (6+4)(4+4) = 350$ in.^{2} Allowable bearing stress, $f_b = 0.7 \times 0.90 \times 3750 \sqrt{322/178} = 3178$ psi > 3020 psi, OK. The bearing stress does not control. It should be noted that the area A_g of the rigid steel plate or plates, and the corresponding concrete pyramid base area A within the end block assumed to receive the bearing stress, are purely determined by engineering judgment. The areas are based on the geometry of the web and bottom flange of the section, the rectangular dimension of the beam end, and the arrangement and spacing of the strand anchorages in contact with the supporting steel end bearing plates.

3. *Draw the strut-and-tie model.* Total length of distance a between forces $P_{u1} - P_{u3} = 11.5 - 2.5 = 9.0$ in. Hence depth $a/2$ ahead of the anchorages $= 9.0/2 = 4.5$ in. Construct the strut-and-tie model assuming it to be as shown in Figure 5.5. The geometrical dimensions for finding the horizontal force components from the ties 1–2 and 3–2 have cotangent values of 26.5/15.5 and 13.0/15.5, respectively. From statics, truss analysis in Figure 5.5 gives the member forces as follows:

$$\text{tension tie } 1\text{--}2 = 118{,}973 \times \frac{24.5}{15.5} = 188{,}054 \text{ lb } (836 \text{ kN})$$

$$\text{tension tie } 3\text{--}2 = 198{,}286 \times \frac{13}{15.5} = 166{,}304 \text{ lb } (699 \text{ kN})$$

Use the larger of the two values for choice of the closed tension tie stirrups. Try No. 3 closed ties, giving a tensile strength per tie $= \phi f_y A_v = 0.90 \times 60{,}000 \times 2(0.11) = 11{,}880$ lb

$$\text{required number of stirrup ties} = \frac{188{,}054}{11{,}880} = 15.8$$

For the tension tie a–b–c in Figure 5.5, use the force $P_u = 166{,}304$ lb to concentrate additional No. 4 vertical ties ahead of the anchorage devices. Start the first tie at a distance of $1\frac{1}{2}$ in. from the end rigid steel plate transferring the load from the anchorage devices to the concrete

$$\text{number of ties} = \frac{166{,}304}{0.90 \times 60{,}000 \times 2 \times 0.20} = 7.7$$

Use eight No. 4 closed ties @ $1\frac{1}{4}$ in. (12.7 mm @ 32 mm) center to center with the first tie to start at $1\frac{1}{2}$ in. *ahead* of the anchorage devices.

Only 13 ties in lieu of the 15.0 calculated are needed since part of the zone is covered by the No. 4 ties. Use 13 No. 3 closed ties @ $2\frac{1}{2}$ in. (9.5 mm @ 57 mm) center to center beyond the last No. 4 tie so that a total distance of 40 in. (104 cm) width of the rectangular anchor block is confined by the reinforcement closed ties.

Adopt this design of the anchorage zone.

It should also be noted that the idealized paths of the compression struts for cases where there are several layers of prestressing strands should be such that at each layer level a stress path is assumed in the design.

5.6.6 Ultimate-Strength Flexural Design

5.6.6.1 Cracking-Load Moment

One of the fundamental differences between prestressed and reinforced concrete is the continuous shift in the prestressed beams of the compressive C-line away from the tensile cgs line as the load increases. In other words, the moment arm of the internal couple continues to increase with the load without any appreciable change in the stress f_{pe} in the prestressing steel. As the flexural moment continues to increase when the full superimposed dead load and live load act, a loading stage is reached where the concrete compressive stress at the bottom-fibers reinforcement level of a simply supported beam becomes zero.

This stage of stress is called the limit state of *decompression*. Any additional external load or overload results in cracking at the bottom face, where the modulus of rupture of concrete f_r is reached due to the cracking moment M_{cr} caused by the first cracking load. At this stage, a sudden increase in the steel stress takes place and the tension is dynamically transferred from the concrete to the steel. It is important to evaluate the first cracking load, since the section stiffness is reduced and hence an increase in deflection has to be considered. Also, the crack width has to be controlled in order to prevent reinforcement corrosion or leakage in liquid containers.

The concrete fiber stress at the tension face is

$$f_b = -\frac{P_e}{A_c}\left(1 + \frac{ec_b}{r^2}\right) + \frac{M_{cr}}{S_b} = f_r \tag{5.34}$$

where the modulus of rupture $f_r = 7.5\sqrt{f_c'}$ and the cracking moment M_{cr} is the moment due to all loads at that load level $(M_D + M_{SD} + M_L)$. From Equation 5.34

$$M_{cr} = f_r S_b + P_e\left(e + \frac{r^2}{c_b}\right) \tag{5.35}$$

5.6.6.2 ACI Load Factors Equations

The ACI 318 Building Code for concrete structures is an international code. As such, it has to conform to the International Building Codes, IBC 2000 and IBC 2003 [10] and be consistent with the ASCE-7 Standard on Minimum Design Loads for Buildings and Other Structures. The effect of one or more loads not acting simultaneously has to be investigated. Structures are seldom subjected to dead and live loads alone. The following equations present combinations of loads for situations in which wind, earthquake, or lateral pressures due to earthfill or fluids should be considered:

$$U = 1.4(D + F) \tag{5.36a}$$
$$U = 1.2(D + F + T) + 1.6(L + H) + 0.5(L_r \text{ or } S \text{ or } R) \tag{5.36b}$$
$$U = 1.2D + 1.6(L_r \text{ or } S \text{ or } R) + (1.0L \text{ or } 0.8W) \tag{5.36c}$$
$$U = 1.2D + 1.6W + 0.5L + 1.0(L_r \text{ or } S \text{ or } R) \tag{5.36d}$$
$$U = 1.2D + 1.0E + 1.0L + 0.2S \tag{5.36e}$$
$$U = 0.9D + 1.6W + 1.6H \tag{5.36f}$$
$$U = 0.9D + 1.0E + 1.6H \tag{5.36g}$$

where D is the dead load, E is the earthquake load, F is the lateral fluid pressure load and maximum height; H is the load due to the weight and lateral pressure of soil and water in soil; L is the live load, L_r is the roof load, R is the rain load, S is the snow load; T is the self-straining force such as creep, shrinkage, and temperature effects; and W is the wind load.

It should be noted that the philosophy used for combining the various load components for earthquake loading is essentially similar to that used for wing loading.

5.6.6.2.1 Exceptions to the Values in These Expressions
1. The load factor on L in Equations 5.36c to 5.36e is allowed to be reduced to 0.5 except for garages, areas occupied as places of public assembly, and all areas where the live load L is greater than 100 lb/ft^2.
2. Where wind load W has not been reduced by a directionality factor, the code permits to use $1.3W$ in place of $1.6W$ in Equations 5.36d and 5.36f.
3. Where earthquake load E is based on service-level seismic forces, $1.4E$ shall be used in place of $1.0E$ in Equations 5.36e and 5.36g.
4. The load factor on H is to be set equal to zero in Equations 5.36f and 5.36g if the structural action due to H counteracts that due to W or E. Where lateral earth pressure provides resistance to structural actions from other forces, it should not be included in H but shall be included in the design resistance.

Due regard has to be given to sign in determining U for combinations of loadings, as one type of loading may produce effects of opposite sense to that produced by another type. The load combinations with $0.9D$ are specifically included for the case where a higher dead load reduces the effects of other loads.

5.6.6.3 Design Strength versus Nominal Strength: Strength-Reduction Factor ϕ

The strength of a particular structural unit calculated using the current established procedures is termed *nominal strength*. For example, in the case of a beam, the resisting moment capacity of the section calculated using the equations of equilibrium and the properties of concrete and steel is called the *nominal strength moment M_n* of the section. This nominal strength is reduced using a strength reduction factor ϕ to account for inaccuracies in construction, such as in the dimensions or position of reinforcement or variations in properties. The reduced strength of the member is defined as the design strength of the member.

For a beam, the design moment strength ϕM_n should be at least equal to, or slightly greater than the external factored moment M_u for the worst condition of factored load U. The factor ϕ varies for the

PHOTO 5.4 Paramount Apartments, San Francisco, CA: completed in 2002, it is the first hybrid precast prestressed concrete moment-resistant 39 floor high-rise frame building in a high seismicity zone. Based on tests of large-scale prototype of the moment-resisting connections, it was determined that the performance of the system was superior to cast-in-place concrete, both in cracking behavior and ductility. Hence, as a system, this new development succeeds in enhancing the performance of high-rise buildings and bridges in high seismicity earthquake zones. Since completion of this structure, several other buildings in California have been completed using the same principles (courtesy Charles Pankow Ltd, Design/Build contractors. Structural Engineers: Robert Englekirk Inc.).

different types of behavior and for the different types of structural elements. For beams in flexure, for instance, ϕ is 0.9.

For tied columns that carry dominant compressive loads, the factor ϕ equals 0.65. The smaller strength-reduction factor used for columns is due to the structural importance of the columns in supporting the total structure compared to other members, and to guard against progressive collapse and brittle failure with no advance warning of collapse. Beams, on the other hand, are designed to undergo excessive deflections before failure. Hence, the inherent capability of the beam for advanced warning of failure permits the use of a higher strength reduction factor or resistance factor. Table 5.10 summarizes the resistance factors ϕ for various structural elements as given in the ACI code.

TABLE 5.10 Resistance or Strength Reduction Factor ϕ

Structural element	Factor ϕ
Beam or slab: bending or flexure[a]	0.9
Columns with ties	0.65
Columns with spirals	0.7
Columns carrying very small axial loads (refer to Chapter 5 for more details)	0.65 – 0.9 or 0.70 – 0.9
Beam: shear and torsion[b]	0.75

[a] Flexure: for factory-produced precast prestressed concrete members, $\phi = 1.0$. For posttensioned cast-in-place concrete members, $\phi = 0.95$ by AASHTO.
[b] Shear and Torsion: Reduction factor for prestressed members, $\phi = 0.90$ by AASHTO.

PHOTO 5.5 Walt Disney World Monorail, Orlando, Florida: a series of hollow prestressed concrete 100-ft box girders individually posttensioned to provide a six-span continuous structure. Design by ABAM Engineers and owned by Walt Disney World Company (courtesy E.G. Nawy).

5.6.7 Limit States in Bonded Members from Decompression to Ultimate Load

The effective prestress f_{pe} at service load due to all loads results in a strain ε_1 such that

$$\varepsilon_1 = \varepsilon_{pe} = \frac{f_{pe}}{E_{ps}} \qquad (5.37a)$$

At decompression, that is, when the compressive stress in the surrounding concrete at the level of the prestressing tendon is neutralized by the tensile stress due to overload, a decompression strain $\varepsilon_{decomp} = \varepsilon_2$ results such that

$$\varepsilon_2 = \varepsilon_{decomp} = \frac{P_e}{A_c E_c}\left(1 + \frac{e^2}{r^2}\right) \qquad (5.37b)$$

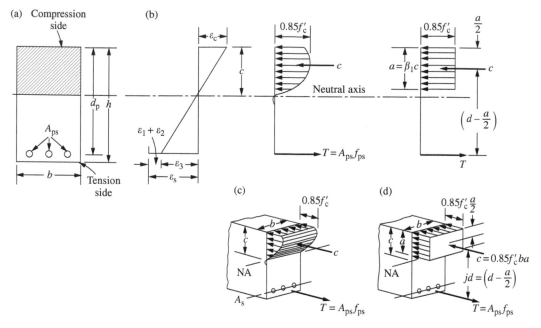

FIGURE 5.6 Stress and strain distribution across beam depth: (a) beam cross-section; (b) strains; (c) actual stress block; and (d) equivalent stress block [11].

Figure 5.6 illustrates the stress distribution in a prestressed concrete member at and after the decompression stage where the behavior of the prestressed beam starts to resemble that of a reinforced concrete beam.

As the load approaches the limit state at ultimate, the additional strain ε_3 in the steel reinforcement follows the linear triangular distribution shown in Figure 5.6b, where the maximum compressive strain at the extreme compression fibers is $\varepsilon_c = 0.003$ in./in. In such a case, the steel strain increment due to overload above the decompression load is

$$\varepsilon_3 = \varepsilon_c \left(\frac{d - c}{c} \right) \tag{5.37c}$$

where c is the depth of the neutral axis. Consequently, the total strain in the prestressing steel at this stage becomes

$$\varepsilon_s = \varepsilon_1 + \varepsilon_2 + \varepsilon_3 \tag{5.37d}$$

The corresponding stress f_{ps} at nominal strength can be easily obtained from the stress–strain diagram of the steel supplied by the producer. If $f_{pe} < 0.50 f_{pu}$, the ACI Code allows computing f_{ps} from the following expression:

$$f_{ps} = f_{pu} \left(1 - \frac{\gamma_p}{\beta_1} \left[\rho_p \frac{f_{pe}}{f_c'} + \frac{d}{d_p} (\omega - \omega') \right] \right) \tag{5.37e}$$

where γ_p can be used as 0.40 for $f_{py}/f_{pu} > 0.85$.

5.6.7.1 The Equivalent Rectangular Block and Nominal Moment Strength

1. The strain distribution is assumed to be linear. This assumption is based on Bernoulli's hypothesis that plane sections remain plane before bending and perpendicular to the neutral axis after bending.

2. The strain in the steel and the surrounding concrete is the same prior to cracking of the concrete or yielding of the steel as after such cracking or yielding.
3. Concrete is weak in tension. It cracks at an early stage of loading at about 10% of its compressive strength limit. Consequently, concrete in the tension zone of the section is neglected in the flexural analysis and design computations and the tension reinforcement is assumed to take the total tensile force.

5.6.7.2 Strain Limits Method for Analysis and Design

5.6.7.2.1 General Principles

In this approach, sometimes referred to as the "unified method," since it is equally applicable to flexural analysis of prestressed concrete elements, the nominal flexural strength of a concrete member is reached when the net compressive strain in the extreme compression fibers reaches the ACI code-assumed limit 0.003 in./in. It also stipulates that when the net tensile strain in the extreme tension steel, ε_t, is sufficiently large, as discussed in the previous section, at a value equal to or greater than 0.005 in./in., the behavior is fully ductile. The concrete beam section is characterized as *tension controlled*, with ample warning of failure as denoted by excessive cracking and deflection.

If the net tensile strain in the extreme tension fibers, ε_t, is small, such as in compression members, being equal to or less than a *compression-controlled* strain limit, a brittle mode of failure is expected, with little warning of such an impending failure. Flexural members are usually tension controlled. Compression members are usually compression controlled. However, some sections, such as those subjected to small axial loads, but large bending moments, the net tensile strain, ε_t, in the extreme tensile fibers, will have an intermediate or transitional value between the two strain limit states, namely, between the compression-controlled strain limit $\varepsilon_t = f_y/E_s = 60{,}000/29 \times 10^6 = 0.002$ in./in., and the tension-controlled strain limit $\varepsilon_t = 0.005$ in./in. Figure 5.7 delineates these three zones as well as the variation in the strength reduction factors applicable to the total range of behavior.

For the tension-controlled state, the strain limit $\varepsilon_t = 0.005$ corresponds to reinforcement ratio $\rho/\rho_b = 0.63$, where ρ_b is the balanced reinforcement ratio for the balanced strain $\varepsilon_t = 0.002$ in the extreme tensile reinforcement. The net tensile strain $\varepsilon_t = 0.005$ for a tension-controlled state is a single value that applies to all types of reinforcement regardless of whether mild steel or prestressing steel. High reinforcement ratios that produce a net tensile strain less than 0.005 result in a ϕ-factor value lower than 0.90, resulting in less economical sections. Therefore, it is more efficient to add compression

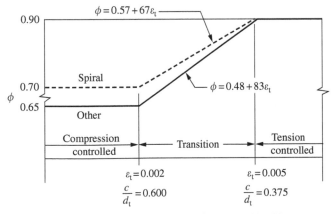

FIGURE 5.7 Strain limit zones and variation of strength reduction factor ϕ with the net tensile strain ε_t [1,5,8].

reinforcement if necessary or deepen the section in order to make the strain in the extreme tension reinforcement, $\varepsilon_t \geq 0.005$.

5.6.7.2.1.1 Variation of ϕ as a Function of Strain — Variation of the ϕ value for the range of strain between $\varepsilon_t = 0.002$ and $\varepsilon_t = 0.005$ can be linearly interpolated to give the following expressions:

Tied sections:

$$0.65 \leq [\phi = 0.48 + 83\varepsilon_t] \leq 0.90 \tag{5.38a}$$

Spirally reinforced sections:

$$0.70 \leq [\phi = 0.57 + 67\varepsilon_t] \leq 0.90 \tag{5.38b}$$

5.6.7.2.1.2 Variation of ϕ as a Function of Neutral Axis depth Ratio c/d_t — Equations 5.38a and 5.38b can be expressed in terms of the ratio of the neutral axis depth c to the effective depth d_t of the layer of reinforcement closest to the tensile face of the section as follows:

Tied sections:

$$0.65 \leq \left[\phi = 0.23 + \frac{0.25}{c/d_t}\right] \leq 0.90 \tag{5.39a}$$

Spirally reinforced sections:

$$0.70 \leq \left[\phi = 0.37 + \frac{0.20}{c/d_t}\right] \leq 0.90 \tag{5.39b}$$

For balanced strain, where the reinforcement at the tension side yields at the same time as the concrete crushes at the compression face ($f_s = f_y$), the neutral axis depth ratio for a limit strain $\varepsilon_t = 0.002$ in./in. can be defined as

$$\frac{c_b}{d_t} = \left(\frac{87,000}{87,000 + f_y}\right) \tag{5.40}$$

The Code permits decreasing the negative elastic moment at the supports for continuous members by not more than $[1000\varepsilon_t]\%$, with a maximum of 20%. The reason is that for ductile members plastic hinge regions develop at points of maximum moment and cause a shift in the elastic moment diagram. In many cases, the result is a reduction of the negative moment and a corresponding increase in the positive moment. The redistribution of the negative moment as permitted by the code can only be used when ε_t is equal or greater than about 0.0075 in./in. at the section at which the moment is reduced. Redistribution is inapplicable in the case of slab systems proportioned by the direct design method.

A minimum strain of 0.0075 at the tensile face is comparable to the case where the reinforcement ratio for the combined prestressed and mild steel reinforcement has a reinforcement index of ω not exceeding $0.24\beta_1$ as an upper limit for ductile design. A maximum strain 0.005 for the tension-controlled state is comparable to a reinforcement index $\omega_p = 0.32\beta_1$ or $\omega_T = 0.36\beta_1$ as described in the Code commentary.

The ACI 318 Code stipulates a maximum strength reduction factor $\phi = 0.90$ for tension-controlled bending, to be used in computing the design strength of flexural members. This corresponds to neutral axis depth ratio $c/d_t = 0.375$ for a strain $\varepsilon_t = 0.005$, with a lower c/d_t ratio recommended. For a useful redistribution of moment in continuous members, this neutral axis depth ratio should be considerably lower, so that the net tensile strain is within the range of $\varepsilon_t = 0.0075$, giving a 7.5% redistribution, and 0.020, giving 20% redistribution, as shown. The tensile strain at the extreme tensile reinforcement has the value

$$\varepsilon_t = 0.003\left(\frac{d_t}{c} - 1\right) \tag{5.41}$$

Although the code allows a maximum redistribution of 20% or $1000\varepsilon_t$, it is more reasonable to limit the redistribution percentage to about 10 to 15%. Summarizing, the ACI 318-02 code stipulates that a redistribution (reduction) of the moments at supports of continuous flexural members *not to exceed*

$1000\varepsilon_t\%$, with a maximum of 20% while increasing the positive midspan moment accordingly. But inelastic moment redistribution should only be made when ε_t is equal or greater than 0.0075 at the section for which the moment is reduced. An adjustment in one span should also be applied to all the other spans in flexure, shear, and bar cutoffs.

5.6.7.3 Nominal Moment Strength of Rectangular Sections

The compression force C can be written as $0.85 f'_c ba$ — that is, the *volume* of the compressive block at or near the ultimate limit state when the tension steel has yielded ($\varepsilon_s > \varepsilon_y$). The tensile force T can be written as $A_{ps}f_{ps}$ and equating C to T gives

$$a = \beta_1 c = \frac{A_{ps}f_{ps}}{0.85f'_c b} \tag{5.42}$$

where for f'_c range of 4000 to 8000 psi

$$\beta_1 = 0.85 - 0.05\left(\frac{f'_c - 4000}{1000}\right) \tag{5.43}$$

The maximum value of β_1 is 0.85 and its minimum value is 0.65.

The nominal moment strength is obtained by multiplying C or T by the moment arm $(d_p - a/2)$, yielding

$$M_n = A_{ps}f_{ps}\left(d_p - \frac{a}{2}\right) \tag{5.44}$$

where d_p is the distance from the compression fibers to the center of the prestressed reinforcement. The steel percentage $\rho_p = A_{ps}/bd_p$ gives the nominal strength of the prestressing steel only as follows if the mild tension and compression steel is accounted for:

$$a = \frac{A_{ps}f_{ps} + A_s f_y - A'_s f_y}{0.85f'_c b} \tag{5.45}$$

where b is the section width of the compression face of the beam.

Taking moments about the center of gravity of the compressive block in Figure 5.6, the nominal moment strength becomes

$$M_n = A_{ps}f_{ps}\left(d_p - \frac{a}{2}\right) + A_s f_y\left(d - \frac{a}{2}\right) + A'_s f_y\left(\frac{a}{2} - d'\right) \tag{5.46}$$

5.6.7.4 Nominal Moment Strength of Flanged Sections

When the compression flange thickness h_f is less than the neutral axis depth c and equivalent rectangular block depth a, the section can be treated as a flanged section

$$T_p + T_s = T_{pw} + T_{pf} \tag{5.47a}$$

where

T_p = total prestressing force = $A_{ps}f_{ps}$
T_s = ultimate force in the nonprestressed steel = $A_s f_y$
T_{pw} = part of the total force in the tension reinforcement required to develop the web = $A_{pw}f_{ps}$
A_{pw} = total reinforcement area corresponding to the force T_{pw}
T_{pf} = part of the total force in the tension reinforcement required to develop the flange

On the basis of these definitions

$$A_{pw}f_{ps} = A_{ps}f_{ps} + A_s f_y - 0.85f'_c(b - b_w)h_f \tag{5.47b}$$

Hence

$$a = \frac{A_{pw}f_{ps}}{0.85f_c'b_w} = \frac{A_{ps}f_{ps} + A_sf_y - 0.85f_c'(b - b_w)h_f}{0.85f_c'b_w} \tag{5.48}$$

$$M_n = A_{pw}f_{ps}\left(d_p - \frac{a}{2}\right) + A_sf_y(d - d_p) + 0.85f_c'(b - b_w)h_f\left(d_p - \frac{h_f}{2}\right) \tag{5.49}$$

The design moment in all cases would be

$$M_u = \phi M_n \tag{5.50}$$

where $\phi = 0.90$ for flexure for the tension zone in Figure 5.7, and reduced in the transition and compression zone.

The value of the stress f_{ps} of the prestressing steel at failure is not readily available. However, it can be determined by *strain compatibility* through the various loading stages up to the limit state at failure. Such a procedure is required if

$$f_{pe} = \frac{P_e}{A_{ps}} < 0.50f_{pu} \tag{5.51}$$

Approximate determination is allowed by the ACI 318 building code provided that

$$f_{pe} = \frac{P_e}{A_{ps}} \geq 0.50f_{pu} \tag{5.52}$$

with separate equations for f_{ps} given for bonded and nonbonded members.

In order to ensure ductility of behavior, the percentage of reinforcement such that the reinforcement index, ω_p, does not exceed $0.36\beta_1$, noting that $0.32\beta_1$ is comparable to 0.005 in the strain limits approach. The ACI Code requires instead to compute the zone in Figure 5.7 applicable to the analyzed section so as to determine the appropriate strength reduction factor ϕ to be used for getting the design moment $M_u = \phi M_n$ as an indirect measure for limiting the reinforcement percentage. It also requires for ensuring ductility behavior that the maximum moment M_u should be equal or greater than 1.2 times the cracking moment M_{cr}.

5.6.8 Example 5: Ultimate Limit State Design of Prestressed Concrete Beams

Design the bonded beam in Example 3 by the ultimate-load theory using nonprestressed reinforcement to *partially* carry part of the factored loads. Use strain compatibility to evaluate f_{ps}, given the modified section in Figure 5.8 with a composite 3 in. top slab and

$f_{pu} = 270,000$ psi (1,862 MPa)
$f_{py} = 0.85f_{pu}$ for stress-relieved strands
$f_y = 60,000$ psi (414 MPa)
$f_c' = 5,000$ psi for normal-weight concrete (34.5 MPa)

Use seven-wire $\frac{1}{2}$-in. dia. strands. The nonprestressed partial mild steel is to be placed with a $1\frac{1}{2}$-in. clear cover, and no compression steel is to be accounted for. No wind or earthquake is taken into consideration.

Solution

From Example 3

Service $W_L = 1100$ plf (16.1 kN/m)
Service $W_{SD} = 100$ plf (1.46 kN/m)
Assumed $W_D = 393$ plf (5.74 kN/m)
Beam span = 65 ft (19.8 m)

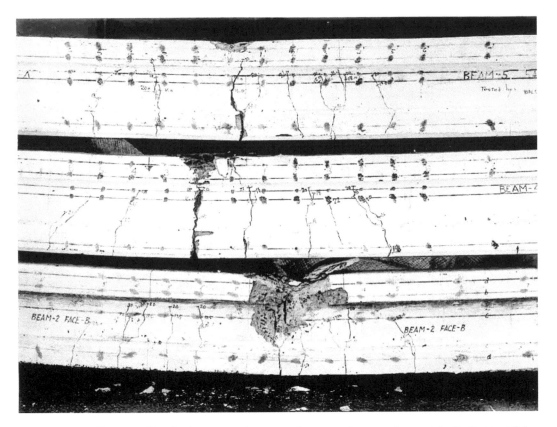

PHOTO 5.6 Different cracking development and patterns of prestressed concrete beams at the limit state at failure as a function of the reinforcement percentage. Tests made at the Rutgers University Concrete Research Laboratory (courtesy E.G. Nawy).

1. *Factored moment*

$$W_u = 1.2(W_D + W_{SD}) + 1.6W_L$$
$$= 1.2(100 + 393) + 1.6(1100) = 2352 \text{ plf } (34.4 \text{ kN/m})$$

The factored moment is given by

$$M_{\hat{u}} = \frac{w_u l^2}{8} = \frac{2{,}352(65)^2 12}{8} = 14{,}905{,}800 \text{ in. lb } (1{,}684 \text{ kN m})$$

and the required nominal moment strength is

$$M_u = \frac{M_u}{\phi} = \frac{14{,}905{,}800}{0.90} = 16{,}562{,}000 \text{ in. lb } (1{,}871 \text{ kN m})$$

2. *Choice of preliminary section.* Assuming a depth of 0.6 in./ft of span as a reasonable practical guideline, we can have a trial section depth $h = 0.6 \times 65 \cong 40$ in. (102 cm). Then assume a mild partial steel $4\#6 = 4 \times 0.44 = 1.76$ in.^{2} (11.4 cm^2). The empirical area of the concrete segment in compression up to the neutral axis can be empirically evaluated from

$$A'_c = \frac{M_n}{0.68 f'_c h} = \frac{16{,}562{,}000}{0.68 \times 5{,}000 \times 40} = 121.8 \text{ in.}^2 \ (786 \text{ cm}^2)$$

Assume on trial basis a total area of the beam section as $3A'_c = 3 \times 121.8 \approx 362.0$ in.^{2}

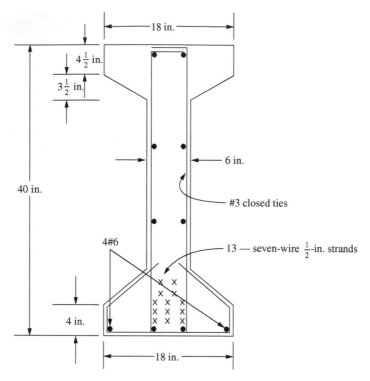

FIGURE 5.8 Midspan section of the beam in Example 5 [5].

Assume a flange width of 18 in. Then the average flange thickness = 121.8/18 ≅ 7.0 in. (178 mm). So suppose the web $b_w = 6$ in. (152 mm), to be subsequently verified for shear requirements. An empirical expression for the area of the prestressing steel can be expressed as

$$A_{ps} = \frac{M_n}{0.72f_{pu}h} = \frac{16{,}562{,}000}{0.72 \times 270{,}000 \times 40} = 2.13 \text{ in.}^2 \ (13.3 \text{ cm}^2)$$

Number of $\frac{1}{2}$-in. stress-relieved wire strands = 2.13/0.153 = 13.9. So try thirteen $\frac{1}{2}$-in. tendons

$$A_{ps} = 13 \times 0.153 = 1.99 \text{ in.}^2 \ (12.8 \text{ cm}^2)$$

Try the section in Figure 5.8 for analysis.
3. *Calculate the stress f_{ps} in the prestressing tendon at nominal strength using the strain-compatibility approach.* The geometrical properties of the trial section are very close to the assumed dimensions for the depth h and the top flange width b. Hence, use the following data for the purpose of the example:

$A_c = 377$ in.^{2}
$c_t = 21.16$ in.
$d_p = 15 + c_t = 15 + 21.16 = 36.16$ in.
$r^2 = 187.5$ in.^{2}
$e = 15$ in. at midspan
$e^2 = 225$ in.^{2}
$e^2/r^2 = 225/187.5 = 1.20$

$E_c = 57,000\sqrt{5,000} = 4.03 \times 10^6$ psi $(27.8 \times 10^3$ MPa$)$
$E_{ps} = 28 \times 10^6$ psi $(193 \times 10^3$ MPa$)$

The maximum allowable compressive strain ε_c at failure $= 0.003$ in./in. Assume that the effective prestress at service load is $f_{pe} = 155,000$ psi $(1,069$ MPa$)$.

A.

$$\varepsilon_1 = \varepsilon_{pe} = \frac{f_{pe}}{E_{ps}} = \frac{155,000}{28 \times 10^6} = 0.0055 \text{ in./in.}$$

$$P_e = 13 \times 0.153 \times 155,000 = 308,295 \text{ lb}$$

The increase in prestressing steel strain as the concrete is decompressed by the increased external load (see Figure 5.6 and Equation 5.37b) is given as

$$\varepsilon_2 = \varepsilon_{decomp} = \frac{P_e}{A_c E_c}\left(1 + \frac{e^2}{r^2}\right)$$

$$= \frac{308,295}{377 \times 4.03 \times 10^6}(1 + 1.20) = 0.0004 \text{ in./in.}$$

B. Assume that the stress $f_{ps} = 205,000$ psi as a first trial. Suppose the neutral axis inside the flange is verified on the basis of $h_f = 3 + 4\frac{1}{2} + 3\frac{1}{2}/2 = 9.25$ in. then, from Equation 5.37c

$$a = \frac{A_{ps}f_{ps} + A_s f_y}{0.85 f_c' b} = \frac{1.99 \times 205,000 + 1.76 \times 60,000}{0.85 \times 5,000 \times 18}$$

$$= 6.71 \text{ in. } (17 \text{ cm}) < h_f = 9.25 \text{ in.}$$

Hence, the equivalent compressive block is inside the flange and the section has to be treated as rectangular.

Accordingly, for 5000 psi concrete

$$\beta_1 = 0.85 - 0.05 = 0.80$$

$$c = \frac{a}{\beta_1} = \frac{6.71}{0.80} = 8.39 \text{ in. } (22.7 \text{ cm})$$

$$d = 40 - (1.5 + \tfrac{1}{2} \text{ in. for stirrups} + \tfrac{5}{16} \text{ in. for bar}) \cong 37.6 \text{ in.}$$

From Equation 1.37c, the increment of strain due to overload to the ultimate is

$$\varepsilon_3 = \varepsilon_c\left(\frac{d-c}{c}\right) = 0.003\left(\frac{37.6 - 8.39}{8.39}\right) = 0.0104 \text{ in./in.} \gg 0.005 \text{ in./in. OK.}$$

The total strain is

$$\varepsilon_{ps} = \varepsilon_1 + \varepsilon_2 + \varepsilon_3$$

$$= 0.0055 + 0.0004 + 0.0104 = 0.0163 \text{ in./in.}$$

From the stress–strain diagram in Figure 5.1, the f_{ps} corresponding to $\varepsilon_{ps} = 0.0163$ is 230,000 psi.

Second trial for f_{ps} value. Assume

$f_{ps} = 229,000$ psi

$$a = \frac{1.99 \times 229,000 + 1.76 \times 60,000}{0.85 \times 5,000 \times 18} = 7.34 \text{ in., consider section as a rectangular beam}$$

$$c = \frac{7.34}{0.80} = 9.17 \text{ in.}$$

$$\varepsilon_3 = 0.003\left(\frac{37.6 - 9.17}{9.17}\right) = 0.0093 \text{ in./in.}$$

Then the total strain is $\varepsilon_{ps} = 0.0055 + 0.0004 + 0.0093 = 0.0152$ in./in. From Figure 5.1, $f_{ps} = 229,000$ psi (1.579 MPa), OK. Use

$$A_{ps} = 4\#6 = 1.76 \text{ in.}^2$$

4. *Available moment strength.* From Equation 5.46, if the neutral axis were to fall within the flange

$$M_n = 1.99 \times 229,000\left(36.16 - \frac{7.34}{2}\right) + 1.76 \times 60,000\left(37.6 - \frac{7.34}{2}\right)$$

$$= 14,806,017 + 3,583,008 = 18,389,025 \text{ in. lb } (2,078 \text{ kN m})$$

$$> \text{required } M_n = 16,562,000 \text{ in. lb, OK.}$$

5. *Check of the reinforcement allowable limits.*
   A. Min $A_s = 0.004A$, where A is the area of the segment of the concrete section between the tension face and the centroid of the entire section

$$A = 377.0 - 18\left(4.125 + \frac{1.375}{2}\right) - 6(21.16 - 5.5) = 201.0 \text{ in.}^2$$

   Min $A_s = 0.004 \times 201.0 = 0.80$ in.$^2 < 1.76$ in.^{2} used, hence alright.
   B. Maximum reinforcement index $\omega_T = d/d_p(\omega_p - \omega) \le 0.36 \, \beta_1 < 0.29$ for $\beta_1 = 0.80$
   Actual

$$\omega_T = \frac{1.99 \times 229,000}{18 \times 36.16 \times 5,000} + \frac{37.6}{36.16}\left(\frac{1.76 \times 60,000}{18 \times 37.6 \times 5,000}\right)$$

$$= 0.14 + 0.03 = 0.17 < 0.29, \text{ hence alright.}$$

Alternatively, the ACI Code limit strain provisions as given in Figure 5.7 do not prescribe a maximum percentage of reinforcement. They require that a check be made of the strain ε_t at the level of the extreme tensile reinforcement to determine whether the beam is in the tensile, the transition, or the compression zone for verifying the appropriate ϕ value. In this case, for $c = 9.17$ and $d_t = 37.6$ in., and from similar triangles in the strain distribution across the beam depth

$$\varepsilon_t = 0.003 \times (37.6 - 9.17)/9.17 = 0.0093 > 0.005$$

Hence, the beam is in the tensile zone of Figure 5.7, with $\phi = 0.90$ as used in the solution, and the design is alright in terms of ductility and reinforcement limits.

5.6.9 Example 6: Ultimate Limit State Design of Prestressed Beams in SI Units

Solve Example 5 using SI units. Strands are bonded.

Data

$A_c = 5045$ cm^2, $b = 45.7$ cm, $b_w = 15.2$ cm

$I_c = 7.04 \times 10^6$ cm^4

$r^2 = 1394$ cm^2

$c_b = 89.4$ cm, $c^t = 32.5$ cm

$e_e = 84.2$ cm, $e_e = 60.4$ cm

$S_b = 78,707$ cm^3, $S^t = 216,210$ cm^3

$w_D = 11.9 \times 10^3$ kN/m, $w_{SD} = 1,459$ N/m, $w_L = 16.1$ kN/m

$l = 19.8$ m

$$f'_c = 34.5 \text{ MPa}, \qquad f_{pi} = 1300 \text{ MPa}$$

$$f_{pu} = 1860 \text{ MPa}, \qquad f_{py} = 1580 \text{ MPa}$$

Prestress loss $\gamma = 18\%, \qquad f_y = 414 \text{ MPa}$

$A_{ps} = 13$ strands, diameter 12.7 mm $(A_{ps} = 99 \text{ mm}^2)$

$$= 13 \times 99 = 1287 \text{ mm}^2$$

Required $M_n = 16.5 \times 10^6$ kN m

Solution

Assume $\phi = 0.90$ to be subsequently verified. Hence, $M_u = \phi M_n = 0.9 \times 10^6 = 14.9 \times 10^6$ kN m

1. *Section properties:*
   Flange width $b = 18$ in. $= 45.7$ cm
   Average thickness $h_f = 4.5 + \frac{1}{2}(3.5) \cong 6.25$ in. $= 15.7$ cm

   Try 4 No. 20 M mild steel bars for partial prestressing (diameter $= 19.5$ mm, $A_s = 300 \text{ mm}^2$)

   $$A_s = 4 \times 300 = 1200 \text{ mm}^2$$

2. *Stress f_{ps} in the prestressing steel at nominal strength and neutral axis position:*

$$f_{pe} = \gamma f_{pi} = 0.82 \times 1300 \cong 1066 \text{ MPa}$$

Verify neutral axis position. If outside flange, its depth has to be greater than $a = A_{pw}f_{ps}/0.85f'_c b_w$; $0.5f_{pu} = 0.50 \times 1860 = 930 \text{ MPa} < 1066$. Hence, one can use the ACI approximate procedure for determining f_{ps}. From Equation 5.37e.

$$f_{ps} = f_{pu}\left(1 - \frac{\gamma_p}{\beta_1}\left[\rho_p \frac{f_{pu}}{f'_c} + \frac{d}{d_p}(\omega - \omega')\right]\right)$$

$d_p = 36.16$ in. $= 91.8$ cm, $\quad d = 37.6$ in. $= 95.5$ cm

$$\frac{f_{py}}{f_{pu}} = \frac{1580}{1860} = 0.85, \quad \text{use } \gamma_p = 0.40$$

$$\rho_p = \frac{A_{ps}}{bd_p} = \frac{1287}{457 \times 918} = 0.00306$$

$$\rho = \frac{A_s}{bd} = \frac{1200}{457 \times 955} = 0.00275$$

$$\omega_p = \frac{A_{ps}}{bd_p} \times \frac{f_{ps}}{f'_c} = 0.00306 \times \frac{1674}{34.5} = 0.14$$

$$\omega = \frac{A_s}{bd} \times \frac{f_y}{f'_c} = 0.00275 \times \frac{414}{34.5} = 0.033$$

$$\omega' = 0$$

$f'_c = 34.5 \text{ MPa}, \quad \beta_1 = 0.80$

$$f_{ps} = 1860\left(1 - \frac{0.40}{0.80}\left[0.00306 \times \frac{1860}{34.5} + \frac{955}{918} \times 0.033\right]\right)$$

$$= 1860(1 - 0.1) \cong 1674 \text{ MPa}$$

From Equation 5.48

$$a = \frac{A_{pw}f_{ps}}{0.85f_c'b_w}$$

where $A_{pw}f_{ps} = A_{ps}f_{ps} + A_sf_y - 0.85f_c'(b - b_w)h_f$

$$A_{pw}f_{ps} = 1287 \times 1674 + 1200 \times 414 - 0.85 \times 34.5 \times (45.7 - 15.2)15.7 \times 10^2$$

$$= 10^6(2.15 + 0.5 - 1.14) \text{ N} = 1240 \text{ kN}$$

$$a = \frac{1240 \times 10}{0.85 \times 34.7 \times 15.2} = 24.7 \text{ cm} > h_f = 15.7 \text{ cm}$$

Hence, the neutral axis is outside the flange and analysis has to be based on a T-section.

3. *Available nominal moment strength.* Checking the limits of reinforcement:

$$\omega_T = \omega_p + \omega = 0.14 + 0.033 = 0.173 < 0.36\beta_1$$

hence, the maximum allowable reinforcement index is not exceeded. Alternatively by the ACI Code, $c/d_t = a/\beta_1 d_t = 15.7/(0.80 \times 91.8) = 0.22 < 0.375$. From Figure 5.7, the beam is in the tensile zone, giving $\phi = 0.90$ for determining the design moment M_u as assumed; hence, the section satisfies the maximum reinforcement limit, as an indirect Code to ensure ductility of the member.

The nominal moment strength is as follows for this flanged section:

$$M_n = A_{pw}f_{ps}\left(d - \frac{a}{2}\right) + A_sf_y(d - d_p) + 0.85f_c'(b - b_w)h_f\left(d_p - \frac{h_f}{2}\right)$$

$$\text{Available } M_n = 1.24 \times 10^6\left(91.8 + \frac{27.7}{2}\right) + 1200 \times 414(95.5 - 91.8)$$

$$+ 0.85 \times 34.5(45.7 - 15.2)15.7\left(91.8 - \frac{15.2}{2}\right) \times 10^2$$

$$= 10^6(96.6 + 1.83 + 118.2)\text{N cm} = 2166 \text{ kN m}$$

$$> \text{Required } M_n = 1871 \text{ kN m}$$

Hence, the section is alright for the design moment M_u on the basis of the $\phi = 0.90$ used for determining the required nominal strength M_n.

5.7 Shear and Torsional Strength Design

Two types of shear control the behavior of prestressed concrete beams: flexure shear (V_{ci}) and web shear (V_{cw}). To design for shear, it is necessary to determine whether flexure shear or web shear controls the choice of concrete shear strength V_c

$$V_{ci} = \frac{M_{cr}}{M/V - d_p/2} + 0.6b_wd_p\sqrt{f_c'} + V_d \qquad (5.53)$$

where V_d is the vertical shear due to self-weight. The vertical component V_p of the prestressing force is disregarded in Equation 5.53, since it is small along the span sections where the prestressing tendon is not too steep.

The value of V in Equation 5.53 is the factored shear force V_i at the section under consideration due to externally applied loads occurring simultaneously with the maximum moment M_{max} occurring at that section, that is

$$V_{ci} = 0.6\lambda\sqrt{f_c'}b_w d_p + V_d + \frac{V_i}{M_{max}}M_{cr} \geq 1.7\lambda\sqrt{f_c'}b_w d_p$$
$$\leq 5.0\lambda\sqrt{f_c'}b_w d_p \tag{5.54}$$

where

λ = 1.0 for normal-weight concrete
= 0.85 for sand-lightweight concrete
= 0.75 for all-lightweight concrete
V_d = shear force at section due to unfactored dead load
V_{ci} = nominal shear strength provided by the concrete when diagonal tension cracking results from combined vertical shear and moment
V_i = factored shear force at section due to externally applied load occurring simultaneously with M_{max}

For lightweight concrete, $\lambda = f_{ct}/6.7\sqrt{f_c'}$ if the value of the tensile splitting strength f_{ct} is known. Note that the value $\sqrt{f_c'}$ should not exceed 100.

The equation for M_{cr}, the moment causing flexural cracking due to external load, is given by

$$M_{cr} = \frac{I_c}{y_t}\left(6\sqrt{f_c'} + f_{ce} - f_d\right) \tag{5.55}$$

where f_{ce} is the concrete compressive stress due to effective prestress after losses at *extreme* fibers of section where tensile stress is caused by external load, psi. At the centroid, $f_{ce} = \bar{f}_c$; f_d is the stress due to unfactored dead load at extreme fiber of section resulting from self-weight only where tensile stress is caused by externally applied load, psi; y_t is the distance from centroidal axis to extreme fibers in tension; and M_{cr} is the portion of the applied *live load* moment that causes cracking. For simplicity the section modulus S_b may be substituted by I_c/y_t.

The web-shear crack in the prestressed beam is caused by an indeterminate stress that can best be evaluated by calculating the principal tensile stress at the critical plane. The shear stress v_c can be defined as the web-shear stress v_{cw} and is maximum near the centroid cgc of the section where the actual diagonal crack develops, as extensive tests to failure have indicated. If v_{cw} is substituted for v_c and $\bar{f}_c$, which denotes the concrete stress f_c due to effective prestress *at the cgc level*, is substituted for f_c in the equation, the expression equating the principal tensile stress in the concrete to the direct tensile strength becomes

$$f_t' = \sqrt{(\bar{f}_c/2) + v_{cw}^2} - \frac{\bar{f}_c}{2} \tag{5.56}$$

where $v_{cw} = V_{cw}/(b_w d_p)$ is the shear stress in the concrete due to all loads causing a nominal strength vertical shear force V_{cw} in the web. Solving for v_{cw} in Equation 5.56 gives

$$v_{cw} = f_t'\sqrt{1 + \bar{f}_c/f_t'} \tag{5.57a}$$

Using $f_t' = 3.5\sqrt{f_c'}$ as a reasonable value of the tensile stress on the basis of extensive tests, Equation 5.57a becomes

$$v_{cw} = 3.5\sqrt{f_c'}\left(\sqrt{1 + \bar{f}_c/3.5\sqrt{f_c'}}\right) \tag{5.57b}$$

which can be further simplified to

$$v_{cw} = 3.5\sqrt{f_c'} + 0.3\bar{f}_c \tag{5.57c}$$

In the ACI code, $\bar{f}_c$ is termed f_{pc}. The notation used herein is intended to emphasize that this is the stress in the concrete, and not the prestressing steel. The nominal shear strength V_{cw} provided by the concrete when diagonal cracking results from *excessive principal tensile stress* in the web becomes

$$V_{cw} = \left(3.5\lambda\sqrt{f_c'} + 0.3\bar{f}_c\right)b_w d_p + V_p \tag{5.58}$$

where V_p is the vertical component of the effective prestress at the particular section contributing to added nominal strength, λ is equal to 1.0 for normal-weight concrete, and less for lightweight concrete, and d_p is the distance from the extreme compression fiber to the centroid of prestressed steel, or $0.8h$, whichever is greater.

The ACI code stipulates the value of $\bar{f}_c$ to be the resultant concrete compressive stress at either the centroid of the section or the junction of the web and the flange when the centroid lies within the flange. In case of composite sections, $\bar{f}_c$ is calculated on the basis of stresses caused by prestress and moments resisted by the precast member acting *alone*.

The spacing of the web shear reinforcement is determined from

$$s = \frac{A_v f_y d}{(V_u/\phi) - V_c} = \frac{A_v \phi f_y d}{V_u - \phi V_c} \tag{5.59}$$

with the following limitations:

1. $s_{max} \le \frac{3}{4}h \le 24$ in., where h is the total depth of the section.
2. If $V_s > 4\sqrt{f_c'}b_w d_p$, the maximum spacing in (1) shall be reduced by half.
3. If $V_s > 8\lambda\sqrt{f_c'}b_w d_p$, enlarge the section.

5.7.1 Composite-Action Dowel Reinforcement

Ties for horizontal shear may consist of single bars or wires, multiple leg stirrups, or vertical legs of welded wire fabric. The spacing cannot exceed four times the least dimension of the support element or 24 in., whichever is less. If μ is the coefficient of friction, then the nominal horizontal shear force F_h can be defined as

$$F_h = \mu A_{vf} f_y \le V_{nh} \tag{5.60}$$

The ACI values of μ are based on a limit shear-friction strength of 800 psi, a quite conservative value as demonstrated by extensive testing. The Prestressed Concrete Institute recommends, for concrete placed against an intentionally roughened concrete surface, a maximum $\mu_e = 2.9$ instead of $\mu = 1.0\lambda$, and a maximum design shear force

$$V_u \le 0.25\lambda^2 f_c' A_c \le 1000\lambda^2 A_{cc} \tag{5.61a}$$

with a required area of shear-friction steel of

$$A_{vf} = \frac{V_{uh}}{\phi f_y \mu_e} \tag{5.61b}$$

The PCI less conservative approach stipulates

$$\mu_e = \frac{1000\lambda^2 b_v I_{vh}}{F_h} \le 2.9$$

The minimum required reinforcement area is

$$A_{vf} = \frac{50b_v s}{f_y} = \frac{50b_v I_{vh}}{f_y} \tag{5.62}$$

where $b_v I_{vh} = A_{cc}$, wherein A_{cc} is the concrete contact surface area.

5.7.2 Example 7: Design of Web Reinforcement for Shear

Design the bonded beam of Example 3 to be safe against shear failure, and proportion the required web reinforcement.

Solution

Data and nominal shear strength determination

$$f_{pu} = 270{,}000 \text{ psi } (1{,}862 \text{ MPa})$$
$$f_y = 60{,}000 \text{ psi } (1{,}862 \text{ MPa})$$
$$f_{pe} = 155{,}000 \text{ psi } (1{,}862 \text{ MPa})$$
$$f_c' = 5{,}000 \text{ psi normal-weight concrete}$$
$$A_{ps} = 13 \text{ seven-wire } \tfrac{1}{2}\text{-in. strands tendon}$$
$$= 1.99 \text{ in.}^2 \ (12.8 \text{ cm}^2)$$
$$A_s = 4\#6 \text{ bars} = 1.76 \text{ in.}^2 \ (11.4 \text{ cm}^2)$$
$$\text{Span} = 65 \text{ ft } (19.8 \text{ m})$$
$$\text{Service } W_L = 1{,}100 \text{ plf } (16.1 \text{ kN/m})$$
$$\text{Service } W_{SD} = 100 \text{ plf } (1.46 \text{ kN/m})$$
$$\text{Service } W_D = 393 \text{ plf } (5.7 \text{ kN/m})$$

Section properties

$$h = 40 \text{ in. } (101.6 \text{ cm})$$
$$d_p = 36.16 \text{ in. } (91.8 \text{ cm})$$
$$d = 37.6 \text{ in. } (95.5 \text{ cm})$$
$$b_w = 6 \text{ in. } (15 \text{ cm})$$
$$e_c = 15 \text{ in. } (38 \text{ cm})$$
$$e_e = 12.5 \text{ in. } (32 \text{ cm})$$
$$I_c = 70{,}700 \text{ in.}^4 \ (18.09 \times 10^6 \text{ cm}^4)$$
$$A_c = 377 \text{ in.}^2 \ (2{,}432 \text{ cm}^2)$$
$$r^2 = 187.5 \text{ in.}^2 \ (1{,}210 \text{ cm}^2)$$
$$c_b = 18.84 \text{ in. } (48 \text{ cm})$$
$$c_t = 21.16 \text{ in. } (54 \text{ cm})$$
$$p_e = 308{,}255 \text{ lb } (1{,}371 \text{ kN})$$
$$\text{Factored load } W_u = 1.2D + 1.6L$$
$$= 1.2\,(100 + 393) + 1.6 \times 1{,}100 = 2{,}352 \text{ plf}$$
$$\text{Factored shear force at face of support} = V_u = W_u L/2 = (2{,}352 \times 65)/2 = 76{,}440 \text{ lb}$$
$$\text{Req. } V_n = V_u/\phi = 76{,}440/0.75 = 101{,}920 \text{ lb at support}$$

Plane at $\tfrac{1}{2} d_p$ from face of support

Nominal shear strength V_c of web

$$\frac{1}{2} d_p = \frac{36.16}{2 \times 12} \cong 1.5 \text{ ft}$$
$$V_c = 101{,}920 \times \frac{[(65/2) - 1.5]}{65/2} = 97{,}216 \text{ lb}$$

Flexure-shear cracking, V_{ci}

From Equation 5.54

$$V_{ci} = 0.6\lambda\sqrt{f_c'}b_w d_p + V_d + \frac{V_i}{M_{max}}(M_{cr}) \geq 1.7\lambda\sqrt{f_c'}b_w d$$

From Equation 5.55, the cracking moment is

$$M_{cr} = \frac{I_c}{y_t}\left(6\sqrt{f_c'} + f_{ce} - f_d\right)$$

where $I_c/y_t = S_b$, since y_t is the distance from the centroid to the extreme tension fibers. Thus

$$I_c = 70,700 \text{ in.}^4$$
$$c_b = 18.84 \text{ in.}$$
$$P_e = 308,255 \text{ lb}$$
$$S_b = 3,753 \text{ in.}^3$$
$$r^2 = 187.5 \text{ in.}^2$$

The concrete stress at the extreme bottom fibers *due to prestress only* is

$$f_{ce} = -\frac{P_e}{A_c}\left(1 + \frac{ec_b}{r^2}\right)$$

and the tendon eccentricity at $d_p/2 \cong 1.5$ ft from the face of the support is

$$e \cong 12.5 + (15 - 12.5)\frac{1.5}{65/2} = 12.62 \text{ in.}$$

Thus

$$f_{ce} = -\frac{308,255}{377}\left(1 + \frac{12.62 \times 18.84}{187.5}\right) \cong -1,855 \text{ psi } (12.8 \text{ MPa})$$

From Example 3, the unfactored dead load due to self-weight $W_D = 393$ plf (5.7 kN/m) is

$$M_{d/2} = \frac{W_D x(l - x)}{2} = \frac{393 \times 1.5(65 - 1.5) \times 12}{2} = 224,600 \text{ in. lb } (25.4 \text{ kN m})$$

and the stress due to the unfactored dead load at the extreme concrete fibers where tension is created by the external load is

$$f_d = \frac{M_{d/2}c_b}{I_c} = \frac{224,600 \times 18.84}{70,700} = 60 \text{ psi}$$

Also

$$M_{cr} = 3,753(6 \times 1.0 \times \sqrt{6,000} + 1855 - 60)$$
$$= 8,480,872 \text{ in. lb } (958 \text{ kN m})$$

$$V_d = W_D\left(\frac{l}{2} - x\right) = 393\left(\frac{65}{2} - 1.5\right) = 12,183 \text{ lb } (54.2 \text{ kN})$$

$$W_{SD} = 100 \text{ plf}$$
$$W_L = 1,100 \text{ plf}$$
$$W_U = 1.2 \times 100 + 1.6 \times 1,100 = 1,880 \text{ plf}$$

The factored shear force at the section due to externally applied loads occurring simultaneously with M_{max} is

$$V_i = W_U\left(\frac{l}{2} - x\right) = 1,880\left(\frac{65}{2} - 1.5\right) = 58,280 \text{ lb } (259 \text{ kN})$$

and

$$M_{max} = \frac{W_U x(l-x)}{2} = \frac{1,880 \times 1.5(65-1.5)}{2} \times 12$$
$$= 1,074,420 \text{ in. lb } (122 \text{ kN m})$$

Hence

$$V_{ci} = 0.6 \times 1.0\sqrt{6,000} \times 6 \times 36.16 + 12,183 + \frac{58,280}{1,074,420}(8,480,872)$$
$$= 482,296 \text{ lb } (54.5 \text{ kN m})$$
$$1.7\lambda\sqrt{f_c'}b_w d_p = 1.7 \times 1.0\sqrt{6,000} \times 6 \times 36.16 = 28,569 \text{ lb } (127 \text{ kN}) < V_{ci} = 482,296 \text{ lb}$$

Hence, $V_{ci} = 482,296$ lb (214.5 kN).

Web-shear cracking, V_{cw}
From Equation 5.58

$$V_{cw} = (3.5\sqrt{f_c'} + 0.3\bar{f}_c)b_w d_p + V_p$$
$$\bar{f}_c = \text{compressive stress in concrete at the cgc}$$
$$= \frac{P_e}{A_c} = \frac{308,255}{377} \cong 818 \text{ psi } (5.6 \text{ MPa})$$
$$V_p = \text{vertical component of effective prestress at section}$$
$$= P_e \tan\theta \text{ (more correctly } P_e \sin\theta)$$

where θ is the angle between the inclined tendon and the horizontal. So

$$V_p = 308,255\frac{(15-12.5)}{(65/2) \times 12} = 1,976 \text{ lb } (8.8 \text{ kN})$$

Hence

$$V_{cw} = \left(3.5\sqrt{6,000} + 0.3 \times 818\right) \times 6 \times 36.16 + 1,976 = 114,038 \text{ lb } (507 \text{ kN})$$

In this case, web-shear cracking controls [i.e., $V_c = V_{cw} = 114,038$ lb (507 kN)] is used for the design of web reinforcement

$$V_s = \frac{V_u}{\phi = 0.75} - V_c = (97,216 - 114,038) \text{ lb, namely } V_c > V_n$$

So no web steel is needed unless $V_u/\phi > \frac{1}{2}V_c$. Accordingly, we evaluate the latter:

$$\frac{1}{2}V_c = \frac{114,038}{2} = 57,019 \text{ lb } (254 \text{ kN}) < 97,216 \text{ lb } (432 \text{ kN})$$

Since $V_u/\phi > \frac{1}{2}V_c$ but $< V_c$, use minimum web steel in this case.

Minimum web steel

$$\text{Req. } \frac{A_v}{s} = 0.0077 \text{ in.}^2/\text{in.}$$

So, trying #3 U stirrups, we get $A_v = 2 \times 0.11 = 0.22$ in.^{2}, and it follows that

$$s = \frac{A_v}{\text{Req. } A_v/s} = \frac{0.22}{0.0077} = 28.94 \text{ in. } (73 \text{ cm})$$

We then check for the minimum A_v as the lesser of the two values given by

$$A_v = 0.75\sqrt{f_c'}\left(\frac{b_w s}{f_y}\right), \quad A_v = 50b_w s/f_y, \quad \text{whichever is larger}$$

and

$$A_v = \frac{A_{ps}\, f_{pu}\, s}{80\, f_y\, d_p} \sqrt{\frac{d_p}{b_w}}$$

So the maximum allowable spacing $\leq 0.75\,h \leq 24$ in. Use #3 U stirrups at 22 in. center to center over a stretch length of 84 in. from the face of the support (see Figure 5.8).

5.7.3 SI Expressions for Shear in Prestressed Concrete Beams

$$V_{ci} = \left[\frac{\lambda\sqrt{f_c'}}{20} b_w d + V_d + V_i\left(\frac{M_{cr}}{M_{max}}\right)\right] \geq \left(\frac{\sqrt{f_c'}}{7}\right) b_w d \tag{5.63}$$

$$M_{cr} = S_b\left(0.5\lambda\sqrt{f_c'} + f_{ce} - f_d\right) \tag{5.64}$$

$$V_{cw} = 0.3\left(\lambda\sqrt{f_c'} + 0.3\bar{f_c}\right) b_w d + V_p \tag{5.65}$$

where M_{cr} for shear analysis is equal to the moment causing flexural cracking at section due to *externally* applied load

$$V_c = \left(\frac{\lambda\sqrt{f_c'}}{20} + f\frac{V_u d}{M_u}\right) b_w d; \quad \frac{V_u d}{M_u} \leq 1.0$$

$$\geq \left[\lambda\frac{\sqrt{f_c'}}{5}\right] b_w d$$

$$\leq \left[0.4\lambda\sqrt{f_c'}b_w d\right] \tag{5.66}$$

$$s = \frac{A_v f_y d}{(V_u/\phi) - V_c} = \frac{A_y f_y d}{V_s} \tag{5.67}$$

Min. A_v: the smaller of

$$A_v \geq \frac{0.35 b_w s}{f_y} \quad \text{or} \quad \frac{A_{ps}\, f_{pu}\, s}{80\, f_y\, d_p} \sqrt{\frac{d}{b_w}}$$

where b_w, s, and d are in millimeters and f_y is in MPa.

5.7.4 Design of Prestressed Concrete Beams Subjected to Combined Torsion, Shear, and Bending in Accordance with the ACI 318-02 Code

5.7.4.1 Compatibility Torsion

In statically indeterminate systems, stiffness assumptions, compatibility of strains at the joints, and redistribution of stresses may affect the stress resultants, leading to a reduction of the resulting torsional shearing stresses. A reduction is permitted in the value of the factored moment used in the design of the member if part of this moment can be redistributed to the intersecting members. The ACI Code permits a maximum factored torsional moment at the critical section $h/2$ from the face of the supports for prestressed concrete members as follows:

A_{cp} = area enclosed by outside perimeter of concrete cross-section = $x_0 y_0$
p_{cp} = outside perimeter of concrete cross-section A_{cp}, in. = $2(x_0 + y_0)$

$$T_u = \phi 4\sqrt{f_c'}\left(\frac{A_{cp}^2}{p_{pc}}\right)\sqrt{1 + \frac{\bar{f_c}}{4\sqrt{f_c'}}} \tag{5.68}$$

where $\bar{f_c}$ is the average compressive stress in the concrete at the centroidal axis due to effective prestress only after allowing for all losses.

Neglect of the full effect of the total value of external torsional moment in this case does not, in effect, lead to failure of the structure but may result in excessive cracking if $\phi 4\sqrt{f_c'}(A_{cp}^2/p_{pc})$ is considerably smaller in value than the actual factored torque.

If the actual factored torque is less than that given in Equation 5.68, the beam has to be designed for the lesser torsional value. Torsional moments are neglected however if for prestressed concrete

$$T_u < \phi\sqrt{f_c'}\left(\frac{A_{cp}^2}{p_{pc}}\right)\sqrt{1+\frac{\bar{f_c}}{4\sqrt{f_c'}}} \tag{5.69}$$

5.7.4.2 Torsional Moment Strength

The size of the cross-section is chosen on the basis of reducing unsightly cracking and preventing the crushing of the surface concrete caused by the inclined compressive stresses due to shear and torsion defined by the left-hand side of the expressions in Equation 5.70. The geometrical dimensions for torsional moment strength in both reinforced and prestressed members are limited by the following expressions:

1. *Solid sections*

$$\sqrt{\left(\frac{V_u}{b_w d}\right)^2+\left(\frac{T_u p_h}{1.7A_{oh}^2}\right)} \le \phi\left(\frac{V_c}{b_w d}+8\sqrt{f_c'}\right) \tag{5.70}$$

2. *Hollow sections*

$$\left(\frac{V_u}{b_w d}\right)+\left(\frac{T_u p_h}{1.7A_{oh}^2}\right) \le \phi\left(\frac{V_c}{b_w d}+8\sqrt{f_c'}\right) \tag{5.71}$$

where A_{oh} is the area enclosed by the centerline of the outermost closed transverse torsional reinforcement (sq. in.) and p_h is the perimeter of the centerline of the outermost closed transverse torsional reinforcement (in.).

The sum of the stresses at the left-hand side of Equation 5.71 should not exceed the stresses causing shear cracking plus $8\sqrt{f_c'}$. This is similar to the limiting strength for $V_s \le 8\sqrt{f_c'}$ for shear without torsion. The strength of the plain concrete in the web is taken as

$$V_c = \left(0.6\lambda\sqrt{f_c'}+700\frac{V_u d_p}{M_u}\right)b_w d_p; \quad \frac{V_u d_p}{M_u} \le 1.0$$
$$\ge 1.7\lambda\sqrt{f_c'}b_w d_p$$
$$\le 5.0\lambda\sqrt{f_c'}b_w d_p \tag{5.72}$$

where $f_{pe} > 0.4f_{pu}$.

5.7.4.3 Torsional Web Reinforcement

Meaningful additional torsional strength due to the addition of torsional reinforcement can be achieved only by using both stirrups and longitudinal bars. Ideally, *equal* volumes of steel in both the closed stirrups and the longitudinal bars should be used so that both participate equally in resisting the twisting moments. This principle is the basis of the ACI expressions for proportioning the torsional web steel. If s is the spacing of the stirrups, A_l is the total cross-sectional area of the longitudinal bars, and A_t is the cross-section of one stirrup leg, the transverse reinforcement for torsion has to be based on the full external torsional moment strength value T_n, namely, T_u/ϕ

$$T_n = \frac{2A_0 A_t f_{yv}}{s}\cot\theta \tag{5.73}$$

where A_0 is the gross area enclosed by the shear flow path (sq. in.), A_t is the cross-sectional area of one leg of the transverse closed stirrups (sq. in.), f_{yv} is the yield strength of closed transverse torsional reinforcement not to exceed 60,000 psi, and θ is the angle of the compression diagonals (struts) in the space truss analogy for torsion.

Transposing terms in Equation 5.73, the transverse reinforcement area becomes

$$\frac{A_t}{s} = \frac{T_n}{2A_0 f_{yv} \cot\theta} \tag{5.74}$$

The area A_0 is determined by analysis, except that the ACI 318 Code permits taking $A_0 = 0.85A_{0h}$ in lieu of the analysis.

The factored torsional resistance ϕT_n must equal or exceed the factored external torsional moment T_u. All the torsional moments are assumed in the ACI 318-02 Code to be resisted by the closed stirrups and the longitudinal steel with the torsional resistance, T_c, of the concrete disregarded, namely $T_c = 0$. The shear V_c resisted by the concrete is assumed to be unchanged by the presence of torsion. The angle θ subtended by the concrete compression diagonals (struts) should not be taken smaller than 30° nor larger than 60°. It can be obtained by analysis as detailed by Hsu [6]. The additional longitudinal reinforcement for torsion should not be less than

$$A_l = \frac{A_t}{s} p_h \left(\frac{f_{yv}}{f_{yl}}\right) \cot^2\theta \tag{5.75}$$

where f_{yl} is the yield strength of the longitudinal torsional reinforcement, not to exceed 60,000 psi.

The same angle θ should be used in both Equations 5.73 and 5.74. It should be noted that as θ gets smaller, the amount of required stirrups required by Equation 5.74 decreases. At the same time the amount of longitudinal steel required by Equation 5.75 increases.

In lieu of determining the angle θ by analysis, the ACI Code allows a value of θ equal to

1. 45° for nonprestressed members or members with less prestress than in 2,
2. 37.5° for prestressed members with an effective prestressing force larger than 40% of the tensile strength of the longitudinal reinforcement.

The PCI recommends computing the value of θ from the expression:

$$\cot\theta = \frac{T_u/\phi}{1.7A_{0h}(A_t/S)f_{yv}} \tag{5.76}$$

5.7.4.3.1 Minimum Torsional Reinforcement

It is necessary to provide a minimum area of torsional reinforcement in all regions where the factored torsional moment T_u exceeds the value given by Equation 5.69. In such a case, the minimum area of the required transverse closed stirrups is

$$A_v + 2A_t \geq \frac{50b_w s}{f_{yv}} \tag{5.77}$$

The maximum spacing should not exceed the smaller of $p_n/8$ or 12 in.

The minimum total area of the additional longitudinal torsional reinforcement should be determined by

$$A_{l,\min} = \frac{5\sqrt{f_c'}A_{cp}}{f_{yl}} - \left(\frac{A_t}{s}\right) p_h \frac{f_{yv}}{f_{yl}} \tag{5.78}$$

where A_t/s should not be taken less than $25b_w/f_{yv}$. The additional longitudinal reinforcement required for torsion should be distributed around the perimeter of the closed stirrups with a maximum spacing of 12 in. The longitudinal bars or tendons should be placed inside the closed stirrups and at least one longitudinal bar or tendon in each corner of the stirrup. The bar diameter should be at least

$\frac{1}{16}$ of the stirrup spacing but not less than a No. 3 bar. Also, the torsional reinforcement should extend for a minimum distance of $(b_t + d)$ beyond the point theoretically required for torsion because torsional diagonal cracks develop in a helical form extending beyond the cracks caused by shear and flexure. b_t is the width of that part of cross-section containing the stirrups resisting torsion. The critical section in beams is at a distance d from the face of the support for reinforced concrete elements and at $h/2$ for prestressed concrete elements, d being the effective depth and h the total depth of the section.

5.7.4.4 SI-Metric Expressions for Torsion Equations

In order to design for combined torsion and shear using the SI (System International) method, the following equations replace the corresponding expressions in the PI (Pound–Inch) method:

$$T_u \leq \frac{\phi\sqrt{f_c'}}{3}\left(\frac{A_{cp}^2}{p_{cp}}\right)\sqrt{1 + \frac{3\bar{f_c}}{\sqrt{f_c'}}} \tag{5.79}$$

$$T_u \leq \frac{\phi\sqrt{f_c'}}{12}\left(\frac{A_{cp}^2}{p_{cp}}\right)\sqrt{1 + \frac{3\bar{f_c}}{\sqrt{f_c'}}} \tag{5.80}$$

$$\sqrt{\left(\frac{V_u}{b_w d}\right)^2 + \left(\frac{T_u p_h}{1.7 A_{oh}^2}\right)^2} \leq \phi\left(\frac{V_c}{b_w d} + \frac{8\sqrt{f_c'}}{12}\right) \tag{5.81}$$

$$\left(\frac{V_u}{b_w d}\right) + \left(\frac{T_u p_h}{1.7 A_{oh}^2}\right) \leq \phi\left(\frac{V_c}{b_w d} + \frac{8\sqrt{f_c'}}{12}\right) \tag{5.82}$$

$$V_c = \left(\frac{\lambda\sqrt{f_c'}}{20} + \frac{5 V_u d_p}{M_u}\right) b_w d_p$$

$$\geq \left(0.17\lambda\sqrt{f_c'}\right) b_w d_p$$

$$\leq \left(0.4\lambda\sqrt{f_c'}\right) b_w d_p \tag{5.83}$$

$$\frac{V_u d_p}{M_u} \leq 1.0$$

$$T_n = \frac{2 A_0 A_t f_{yv}}{s}\cot\theta \tag{5.84}$$

where f_{yv} is in MPa, s is in millimeter, A_0, A_t are in mm^2, and T_n is in kN m

$$\frac{A_t}{s} = \frac{T_n}{2 A_0 f_{yv}\cot\theta} \tag{5.85}$$

$$A_l = \frac{A_t}{s} p_h \left(\frac{f_{yv}}{f_{yl}}\right)\cot^2\theta \tag{5.86}$$

where f_{yv} and f_{yl} are in MPa, p_h and s are in millimeters, and A_l and A_t are in mm^2

$$A_v = \frac{0.35 b_w s}{f_y} \tag{5.87}$$

$$A_{l,\min} = \frac{5\sqrt{f_c'} A_{cp}}{12 f_{yl}} - \left(\frac{A_t}{s}\right) p_h \left(\frac{f_{yv}}{f_{yl}}\right) \tag{5.88}$$

PHOTO 5.7 Sunshine Skyway Bridge, Tampa, Florida. Designed by Figg and Muller Engineers, Inc. The bridge has a 1200 ft cable-stayed main span with a single pylon, 175 ft vertical clearance, a total length of 21,878 ft and twin 40-ft roadways (courtesy Portland Cement Association).

where A_t/s should not be taken less than $0.175b_w/f_{yv}$. Maximum allowable spacing of transverse stirrups is the smaller of $\frac{1}{8}p_h$ or 300 mm, and bars should have a diameter of at least $\frac{1}{16}$ of the stirrups spacing but not less than No. 10 M bar size. Max. f_{yv} or f_{yl} should not exceed 400 MPa. Min. A_{vt} the smaller of

$$\frac{A_{vt}}{s} \geq \frac{0.35b_w}{f_y} \quad \text{or} \quad \frac{A_{vt}}{s} = \frac{1}{16}\sqrt{f_c'}\left(\frac{b_w}{f_y}\right)$$

whichever is larger, where b_w, d_p, and s are in millimeters

$$\geq \frac{A_{ps}f_{pu}}{80f_y d_p}\sqrt{\frac{d_p}{b_w}}$$

Use the lesser of the two sets.

5.8   Camber, Deflection, and Crack Control

5.8.1   Serviceability Considerations

Prestressed concrete members are continuously subjected to sustained eccentric compression due to the prestressing force, which seriously affects their long-term creep deformation performance. Failure to predict and control such deformations can lead to high reverse deflection, that is, camber, which can produce convex surfaces detrimental to proper drainage of roofs of buildings, to uncomfortable ride characteristics in bridges and aqueducts, and to cracking of partitions in apartment buildings, including misalignment of windows and doors.

The difficulty of predicting very accurately the total long-term prestress losses makes it more difficult to give a precise estimate of the magnitude of expected camber. Accuracy is even more difficult in partially prestressed concrete systems, where limited cracking is allowed through the use of additional nonprestressed reinforcement. Creep strain in the concrete increases camber, as it causes a negative increase in curvature that is usually more dominant than the decrease produced by the decrease in prestress losses due to creep, shrinkage, and stress relaxation. A best estimate of camber increase should be based on accumulated experience, span-to-death ratio code limitations, and a correct choice of the modulus E_c of the concrete. Calculation of the moment–curvature relationships at the major incremental stages of loading up to the limit state at failure would also assist in giving a more accurate evaluation of the stress-related load deflection of the structural element.

The cracking aspect of serviceability behavior in prestressed concrete is also critical. Allowance for limited cracking in "partial prestressing" through the additional use of nonprestressed steel is prevalent. Because of the high stress levels in the prestressing steel, corrosion due to cracking can become detrimental to the service life of the structure. Therefore, limitations on the magnitudes of crack widths and their spacing have to be placed, and proper crack width evaluation procedures used. The presented discussion of the state of the art emphasizes the extensive work of the author on cracking in pretensioned and posttensioned prestressed beams.

Prestressed concrete flexural members are classified into three classes in the new ACI 318 Code:

1. *Class U:*

$$f_t \leq 7.5\sqrt{f_c'} \qquad (5.89a)$$

In this class, the gross section is used for section properties when both stress computations at service loads and deflection computations are made. No skin reinforcement needs to be used in the vertical faces.

2. *Class T:*

$$7.5\sqrt{f_c'} \leq f_t \leq 12\sqrt{f_c'} \qquad (5.89b)$$

This class is a transition between uncracked and cracked sections. For stress computations at service T loads, the gross section is used. The cracked bilinear section is used in the deflection computations. No skin reinforcement needs to be used in the vertical faces.

3. *Class C:*

$$f_t > 12\sqrt{f_c'} \qquad (5.89c)$$

This class denotes cracked sections. Hence, a cracked section analysis has to be made for evaluation of the stress level at service and for deflection. Computation of Δf_{ps} or f_s for crack control is necessary, where Δf_{ps} is the stress increase beyond the decompression state and f_s is the stress in the mild reinforcement when mild steel reinforcement is also used. Prestressed two-way slab systems are to be designed as Class U.

Ideally, the load–deflection relationship is trilinear, as shown in Figure 5.9. The three regions prior to rupture are:

Region I — Precracking stage, where a structural member is crack free.
Region II — Postcracking stage, where the structural member develops acceptable controlled cracking in both distribution and width.
Region III — Postserviceability cracking stage, where the stress in the tensile reinforcement reaches the limit state of yielding.

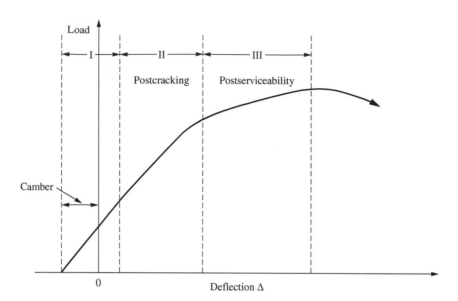

FIGURE 5.9 Beam load–deflection relationship: Region I, precracking stage; Region II, postcracking stage; Region III, postserviceability stage [5,11].

The precracking segment of the load–deflection curve is essentially a straight line defining full elastic behavior, as in Figure 5.9. The maximum tensile stress in the beam in this region is less than its tensile strength in flexure, that is, it is less than the modulus of rupture f_r of concrete. The flexural stiffness EI of the beam can be estimated using Young's modulus E_c of concrete and the moment of inertia of the uncracked concrete cross-section.

The precracking region ends at the initiation of the first crack and moves into region II of the load–deflection diagram in Figure 5.9. Most beams lie in this region at service loads. A beam undergoes varying degrees of cracking along the span corresponding to the stress and deflection levels at each section. Hence, cracks are wider and deeper at midspan, whereas only narrow, minor cracks develop near the supports in a simple beam.

The load–deflection diagram in Figure 5.9 is considerably flatter in region III than in the preceding regions. This is due to substantial loss in stiffness of the section because of extensive cracking and considerable widening of the stabilized cracks throughout the span. As the load continues to increase, the strain ε_s in the steel at the tension side continues to increase beyond the yield strain ε_y with no additional stress. The beam is considered at this stage to have structurally failed by initial yielding of the tension steel. It continues to deflect without additional loading, the cracks continue to open, and the neutral axis continues to rise toward the outer compression fibers. Finally, a secondary compression failure develops, leading to total crushing of the concrete in the maximum moment region followed by rupture. Figure 5.10 gives the deflection expressions for the most common loading cases in terms of both load and curvature.

5.8.1.1 Strain and Curvature Evaluation

The distribution of strain across the depth of the section at the controlling stages of loading is linear, as is shown in Figure 5.11, with the angle of curvature dependent on the top and bottom concrete extreme fiber strains ε_{ct} and ε_{cb}. From the strain distributions, the curvature at the various stages of loading can be expressed as follows:

1. *Initial prestress:*

$$\phi_i = \frac{\varepsilon_{cbi} - \varepsilon_{cti}}{h} \tag{5.90a}$$

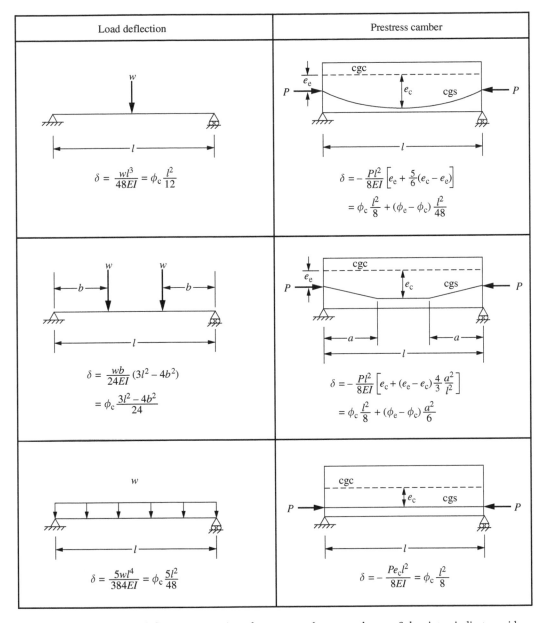

FIGURE 5.10 Short-term deflection expressions for prestressed concrete beams. Subscript c indicates midspan and subscript e the support [5].

2. *Effective prestress after losses:*

$$\phi_e = \frac{\varepsilon_{cbe} - \varepsilon_{cte}}{h} \tag{5.90b}$$

3. *Service load:*

$$\phi = \frac{\varepsilon_{cb} - \varepsilon_{ct}}{h} \tag{5.90c}$$

4. *Failure:*

$$\phi_u = \frac{\varepsilon_u}{c} \tag{5.90d}$$

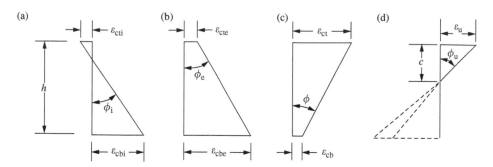

FIGURE 5.11 Strain distribution and curvature at controlling stages: (a) initial prestress, (b) effective prestress after losses, (c) service load, and (d) limit state at failure [5].

Use a *plus* sign for tensile strain and a *minus* sign for compressive strain. Figure 5.11c denotes the stress distribution for an uncracked section. It has to be modified to show tensile stress at the bottom fibers if the section is cracked.

The cracking moment due to that portion of the applied *live load* that causes cracking is

$$M_{cr} = S_b \left[6.0\lambda \sqrt{f'_c} + f_{ce} - f_d \right] \tag{5.91a}$$

where f_{ce} is the compressive stress at the center of gravity of the concrete section due to effective prestress only after losses when the tensile stress is caused by applied external load and f_d is the concrete stress at extreme tensile fibers due to unfactored dead load when tensile stresses and cracking are caused by the external load.

A factor of 7.5 can also be used instead of 6.0 for deflection purposes for beams

$$\frac{M_{cr}}{M_a} = 1 - \left(\frac{f_{tl} - f_r}{f_L} \right) \tag{5.91b}$$

where M_a is the maximum service unfactored live load moment, f_{tl} is the final calculated total service load concrete stress in the member, and f_r is the modulus of rupture.

5.8.1.2 Effective-Moment-of-Inertia for Cracked Sections

As the prestressed element is overloaded, or in the case of partial prestressing where limited controlled cracking is allowed, the use of the gross moment of inertia I_g underestimates the camber or deflection of the prestressed beam. Theoretically, the cracked moment of inertia I_{cr} should be used for the section across which the cracks develop while the gross moment of inertia I_g should be used for the beam sections between the cracks. However, such refinement in the numerical summation of the deflection increases along the beam span is sometimes unwarranted because of the accuracy difficulty of deflection evaluation. Consequently, an effective moment of inertia I_e can be used as an average value along the span of a simply supported bonded tendon beam, a method developed by Branson. According to this method

$$I_e = I_{cr} + \left(\frac{M_{cr}}{M_a} \right)^3 (I_g - I_{cr}) \le I_g \tag{5.92a}$$

Equation 5.92a can also be written in the form

$$I_e = \left(\frac{M_{cr}}{M_a} \right)^3 I_g + \left[1 - \left(\frac{M_{cr}}{M_a} \right)^3 \right] I_{cr} \le I_g \tag{5.92b}$$

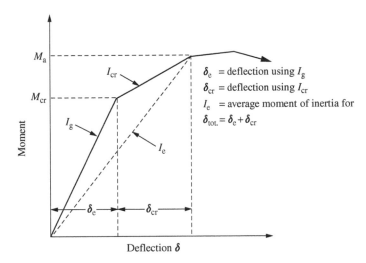

FIGURE 5.12 Moment–deflection relationship [1,5].

The ratio (M_{cr}/M_a) from Equation 5.91b can be substituted into Equations 5.92a and 5.92b to get the effective moment of inertia. Figure 5.12 gives the bilinear moment–deflection relationship for I_g, I_{cr}, and I_e, where

$$I_{cr} = n_p A_{ps} d_p^2 (1 - 1.6\sqrt{n_p \rho_p}) \tag{5.93a}$$

where $n_p = E_{ps}/E_c$. If nonprestressed reinforcement is used to carry tensile stresses, namely, in "partial prestressing," Equation 5.93a can be modified to give

$$I_{cr} = (n_p A_{ps} d_p^2 + n_s A_s d^2)(1 - 1.6\sqrt{n_p \rho_p + n_s \rho_s}) \tag{5.93b}$$

5.8.2 Long-Term Effects on Deflection and Camber

5.8.2.1 PCI Multipliers Method

The PCI multipliers method provides a multiplier C_1 that takes account of long-term effects in prestressed concrete members as presented in Table 5.11. This table, based on Ref. [7], can provide reasonable multipliers of immediate deflection and camber provided that the upward and downward components of the initial calculated camber are separated in order to take into account the effects of loss of prestress, *which only apply to the upward component.* Substantial reduction can be achieved in long-term camber by the addition of nonprestressed steel. In that case, a reduced multiplier C_2 can be used, given by

$$C_2 = \frac{C_1 + A_s/A_{ps}}{1 + A_s/A_{ps}} \tag{5.94}$$

where C_1 is the multiplier from Table 5.11, A_s is the area of nonprestressed reinforcement, and A_{ps} is the area of prestressed strands.

5.8.3 Permissible Limits of Calculated Deflection

The ACI Code requires that the calculated deflection has to satisfy the serviceability requirement of maximum permissible deflection for the various structural conditions listed in Table 5.12. Note that long-term effects cause measurable increases in deflection and camber with time and result in excessive overstress in the concrete and the reinforcement, *requiring* computation of deflection and camber.

TABLE 5.11 C_1 Multipliers for Long-Term Camber and Deflection

	Without composite topping	With composite topping
At erection		
Deflection (downward) component — apply to the elastic deflection due to the member weight at release of prestress	1.85	1.85
Camber (upward) component — apply to the elastic camber due to prestress at the time of release of prestress	1.80	1.80
Final		
Deflection (downward) component — apply to the elastic deflection due to the member weight at release of prestress	2.70	2.40
Camber (upward) component — apply to the elastic camber due to prestress at the time of release of prestress	2.45	2.20
Deflection (downward) — apply to the elastic deflection due to the superimposed dead load only	3.00	3.00
Deflection (downward) — apply to the elastic deflection caused by the composite topping	—	2.30

TABLE 5.12 ACI Minimum Permissible Ratios of Span (l) to Deflection (δ) (l = Longer Span)

Type of member	Deflection δ to be considered	$(l/\delta)_{min}$
Flat roofs not supporting and not attached to nonstructural elements likely to be damaged by large deflections	Immediate deflection due to live load L	180[a]
Floors not supporting and not attached to nonstructural elements likely to be damaged by large deflections	Immediate deflection due to live load L	360
Roof or floor construction supporting or attached to nonstructural elements likely to be damaged by large deflections	That part of total deflection occurring after attachment of nonstructural elements; sum of long-term deflection due to all sustained loads (dead load plus any sustained portion of live load) and immediate deflection due to any additional live load[b]	240[c]
Roof or floor construction supporting or attached to nonstructural elements not likely to be damaged by large deflections		480

[a] Limit not intended to safeguard against ponding. Ponding should be checked by suitably calculating deflection, including added deflections due to ponded water, and considering long-term effects of all sustained loads, camber, construction tolerances, and reliability of provisions for drainage.
[b] Long-term deflection has to be determined, but may be reduced by the amount of deflection calculated to occur before attachment of nonstructural elements. This reduction is made on the basis of accepted engineering data relating to time-deflection characteristics of members similar to those being considered.
[c] Ratio limit may be lower if adequate measures are taken to prevent damage to supported or attached elements, but should not be lower than tolerance of nonstructural elements.

AASHTO permissible deflection requirements, shown in Table 5.13, are more rigorous because of the dynamic impact of moving loads on bridge spans.

5.8.3.1 Approximate Time-Steps Method

The approximate time-steps method is based on a simplified form of summation of constituent deflections due to the various time-dependent factors. If C_u is the long-term creep coefficient, the

TABLE 5.13 AASHTO Maximum Permissible Deflection (l = Longer Span)

		Maximum permissible deflection	
Type of member	Deflection considered	Vehicular traffic only	Vehicular and pedestrian traffic
Simple or continuous spans	Instantaneous due to service live load plus impact	$l/800$	$l/1000$
Cantilever arms		$l/300$	$l/375$

curvature at effective prestress P_e can be defined as

$$\phi_e = \frac{P_i e_x}{E_c I_c} + (P_i - P_e)\frac{e_x}{E_c I_c} - \left(\frac{P_i + P_e}{2}\right)\frac{e_x}{E_c I_c}C_u \tag{5.95}$$

The following expression can predict the time-dependent increase in deflection $\Delta\delta$:

$$\Delta\delta = -\left[\eta + \frac{(1+\eta)}{2}k_r C_t\right]\delta_{i(P_i)} + k_r C_t \delta_{i(D)} + K_a k_r C_t \delta_{i(SD)} \tag{5.96}$$

where
$\eta = P_e/P_i$
C_t = creep coefficient at time t
K_a = factor corresponding to age of concrete at superimposed load application
  = $1.25t^{-0.118}$ for moist-cured concrete
  = $1.13t^{-0.095}$ for steam-cured concrete
t = age, in days, at loading
$k_r = 1/(1 + A_s/A_{ps})$ when $A_s/A_{ps} \ll 1.0$
  $\cong 1$ for all practical purposes

For the final deflection increment, C_u is used in place of C_t in Equation 5.96.
For noncomposite beams, the total deflection $\delta_{T,t}$ becomes [4]

$$\delta_{T,t} = -\delta_{pi}\left[1 - \frac{\Delta P}{P_0} + \lambda(k_r C_t)\right] + \delta_D[1 + k_r C_t] + \delta_{SD}[1 + K_a k_r C_t] + \delta_L \tag{5.97}$$

where δ_p is the deflection due to prestressing, ΔP is the total loss of prestress excluding initial elastic loss, and $\lambda = 1 - (\Delta P/2P_0)$, in which P_0 is the prestress force at transfer after elastic loss and P_i less than the elastic loss.
For composite beams, the total deflection is

$$\delta_T = -\delta_{pi}\left[1 - \frac{\Delta P}{P_0} + K_a k_r C_u \lambda\right] + \delta_D[1 + K_a k_r C_u]$$

$$+ \delta_{pi}\frac{I_e}{I_{comp.}}\left[1 - \frac{\Delta P - \Delta P_c}{P_0} + k_r C_u(\lambda - \alpha\lambda')\right]$$

$$+ (1-\alpha)k_r C_u \delta_D \frac{I_c}{I_{comp.}} + \delta_D\left[1 + \alpha k_r C_u \frac{I_c}{I_{comp.}}\right] + \delta_{df} + \delta_L \tag{5.98}$$

where
$\lambda' = 1 - (\Delta P_c/2P_0)$
ΔP_c = loss of prestress at time composite topping slab is cast, excluding initial elastic loss
$I_{comp.}$ = moment of inertia of composite section
δ_{df} = deflection due to differential shrinkage and creep between precast section and composite topping slab
  = $F y_{cs}l^2/8E_{cc}I_{comp.}$ for simply supported beams (for continuous beams, use the appropriate factor in the denominator)

y_{cs} = distance from centroid of composite section to centroid of slab topping
F = force resulting from differential shrinkage and creep
E_{cc} = modulus of composite section
α = creep strain at time t divided by ultimate creep strain
    = $t^{0.60}/(10 + t^{0.60})$

5.8.4 Long-Term Deflection of Composite Double-Tee Cracked Beam [2,5]

5.8.4.1 Example 8: Deflection Computation of a Double-Tee Beam

A 72-ft (21.9 m) span simply supported roof normal weight concrete double-T-beam (Figure 5.13) is subjected to a superimposed topping load $W_{SD} = 250$ plf (3.65 kN/m) and a service live load $W_L = 280$ plf (4.08 kN/m). Calculate the short-term (immediate) camber and deflection of this beam by (a) the I_e method and (b) the bilinear method as well as the time-dependent deflections after 2-in. topping is cast (30 days) and the final deflection (5 years), using the PCI multipliers method. Given prestress losses 18%.

	Noncomposite	Composite
A_c, in.^{2}	615 (3,968 cm^2)	855 (5,516 cm^2)
I_c, in.^{4}	59,720 (24.9 × 10^5 cm^4)	77,118 (32.1 × 10^5 cm^4)
r^2, in.^{2}	97 (625 cm^2)	90 (580 cm^2)
c_b, in.	21.98 (558 mm)	24.54 (623 mm)
c_t, in.	10.02 (255 mm)	9.46 (240 mm)
S_b, in.^{3}	2,717 (4.5 × 10^4 cm^3)	3,142 (5.1 × 10^4 cm^3)
S^t, in.^{3}	5,960 (9.8 × 10^4 cm^3)	8,152 (13.4 × 10^4 cm^3)
W_d, plf	641 (9.34 kN/m)	891 (13.0 kN/m)

$$V/S = 615/364 = 1.69 \text{ in. (43 mm)}$$

$$RH = 75\%$$

$$e_c = 18.73 \text{ in. (476 mm)}$$

$$e_{\hat{e}} = 12.81 \text{ in. (325 mm)}$$

$$f_c' = 5000 \text{ psi (34.5 MPa)}$$

$$f_{ci}' = 3750 \text{ psi (25.9 MPa)}$$

Topping $f_c' = 3000$ psi (20.7 MPa)

f_t at bottom fibers $= 12\sqrt{f_c'} = 849$ psi (5.9 MPa)

A_{ps} = twelve $\frac{1}{2}$-in. diameter low-relaxation prestressing steel depressed at midspan only

$$f_{pu} = 270,000 \text{ psi (1,862 MPa), low relaxation}$$

$$f_{pi} = 189,000 \text{ psi (1,303 MPa)}$$

$$f_{pj} = 200,000 \text{ psi (1,380 MPa)}$$

$$f_{py} = 260,000 \text{ psi (1,793 MPa)}$$

$$E_{ps} = 28.5 \times 10^6 \text{ psi (19.65 × 10}^4 \text{ MPa)}$$

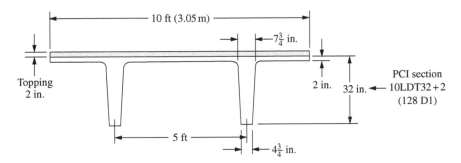

FIGURE 5.13 Double-tee composite beam in Example 8 [5].

Solution by the I_e Method

1. *Midspan section stresses*

$$f_{pj} = 200,000 \text{ psi at jacking}$$

$$f_{pi} \text{ assumed} = 0.945 f_{pj} = 189,000 \text{ psi at transfer}$$

$$e_c = 18.73 \text{ in. } (475 \text{ mm})$$

$$P_i = 12 \times 0.153 \times 189,000 = 347,004 \text{ lbs } (1,540 \text{ kN})$$

Self-weight moment

$$M_D = \frac{641(72)^2}{8} \times 12 = 4,984,416 \text{ in.-lb}$$

A. *At transfer*

$$
\begin{aligned}
f^t &= -\frac{P_i}{A_c}\left(1 - \frac{e_c c_t}{r^2}\right) - \frac{M_D}{S^t} \\
&= -\frac{347,004}{615}\left(1 - \frac{18.73 \times 10.02}{97}\right) - \frac{4,984,416}{5,960} \\
&= +527.44 - 836.31 \\
&= -308.87 \text{ psi } (C), \text{ say } 310 \text{ psi } (C)(2.1 \text{ MPa}) < 0.60 f'_{ci} = 0.60(3,750) \\
&= 2,250 \text{ psi, OK}
\end{aligned}
$$

$$
\begin{aligned}
f_b &= -\frac{P_i}{A_c}\left(1 - \frac{e_c c_b}{r^2}\right) - \frac{M_D}{S_b} \\
&= -\frac{347,004}{615}\left(1 + \frac{18.73 \times 21.98}{97}\right) + \frac{4,984,416}{2,717} \\
&= -2958.95 + 1834.53 \\
&= -1124.42 \text{ psi } (C), \text{ say } 1125 \text{ psi } (C) < -2,250 \text{ psi, OK}
\end{aligned}
$$

B. *After slab is cast.* At this load level assume 18% prestress loss

$$f_{pe} = 0.82 f_{pi} = 0.82 \times 189,000 = 154,980 \text{ psi}$$

$$P_e = 12 \times 0.153 \times 154,980 = 284,543 \text{ lb}$$

For the 2-in. slab

$$W_{SD} = \frac{2}{12} \times 10 \text{ ft} \times 150 = 250 \text{ plf } (3.6 \text{ kN/m})$$

$$M_{SD} = \frac{250(72)^2}{8} \times 12 = 1,944,000 \text{ in.-lb}$$

$$M_D + M_{SD} = 4,984,416 + 1,944,000 = 6,928,416 \text{ in. lb } (783 \text{ kN m})$$

$$f^t = -\frac{P_e}{A_c}\left(1 - \frac{e_c c_t}{r^2}\right) - \frac{M_D + M_{SD}}{S^t}$$

$$= -\frac{284,543}{615}\left(1 + \frac{18.73 \times 10.02}{97}\right) + \frac{6,928,416}{5.960}$$

$$= +432.5 - 1,162.5 = -730 \text{ psi } (5.0 \text{ MPa}) < 0.45f_c' = -2.250 \text{ psi, OK}$$

$$f_b = -\frac{P_e}{A_c}\left(1 + \frac{e_c c_b}{r^2}\right) - \frac{M_D + M_{SD}}{S^t}$$

$$= -\frac{284,543}{615}\left(1 + \frac{18.73 \times 21.98}{97}\right) + \frac{6,928,416}{2,717}$$

$$= -2426.33 + 2550.02 = +123.7 \ (0.85 \text{ MPa}), \text{ say } 124 \text{ psi } (T), \text{ OK}$$

This is a very low tensile stress when the unshored slab is cast and before the service load is applied, $\ll 12\sqrt{f_c'} = 849$ psi.

C. *At service load for the precast section.* Section modulus for composite section at the top of the precast section is

$$S_c^t = \frac{77,118}{9.46 - 2} = 10,337 \text{ in.}^3$$

$$M_L = \frac{280(72)^2}{8} \times 12 = 2,177,288 \text{ in.-lb } (246 \text{ kN m})$$

$$f^t = -\frac{P_e}{A_c}\left(1 - \frac{e_c c_t}{r^2}\right) - \frac{M_D + M_{SD}}{S^t} - \frac{M_{CSD} + M_L}{S_c^t}$$

$$M_{CSD} = \text{superimposed dead load} = 0 \text{ in this case}$$

$$f^t = -730 - \frac{2,177,288}{10,337}$$

$$= -730 - 210 = -940 \text{ psi } (6.5 \text{ MPa}) (C), \text{ OK}$$

$$f_b = +123.7 + \frac{2,177,288}{3,142} = +123.7 + 693.0$$

$$= +816.7, \text{ say } 817 \text{ psi } (T) \ (5.4 \text{ MPa}) < f_t = 849 \text{ psi, OK}$$

D. *Composite slab stresses.* Precast double-T concrete modulus is

$$E_c = 57,000\sqrt{f_c'} = 57,000\sqrt{5,000} = 4.03 \times 10^6 \text{ psi } (3.8 \times 10^4 \text{ MPa})$$

Situ-cast slab concrete modulus is

$$E_c = 57,000\sqrt{3,000} = 3.12 \times 10^6 \text{ psi } (2.2 \times 10^4 \text{ MPa})$$

Modular ratio

$$n_p = \frac{3.12 \times 10^6}{4.03 \times 10^6} = 0.77$$

S_c^t for 2-in. slab top fibers $= 8,152$ in.^{3} from data.

S_{cb} for 2-in. slab bottom fibers $= 10,337$ in.^{3} from before for top of precast section.

Stress f_{cs}^t at top slab fibers $= n\dfrac{M_L}{S_c^t}$

$$= -0.77 \times \frac{2{,}177{,}288}{8{,}152} = -207 \text{ psi } (1.4 \text{ MPa}) \ (C)$$

Stress f_{csb} at bottom slab fibers

$$= -0.77 \times \frac{2{,}177{,}288}{10{,}377} = -162 \text{ psi } (1.1 \text{ MPa}) \ (C)$$

2. *Support section stresses*

Check is made at the support face (a slightly less conservative check can be made at $50d_b$ from end)

$$e_c = 12.81 \text{ in.}$$

A. *At transfer*

$$f^t = -\frac{347{,}004}{615}\left(1 - \frac{12.81 \times 10.02}{97}\right) - 0$$

$$= +182 \text{ psi } (T) \ (1.26 \text{ MPa}) \ll -2{,}250 \text{ psi, OK}$$

$$f_b = -\frac{347{,}004}{615}\left(1 + \frac{12.81 \times 21.98}{97}\right) + 0$$

$$= -2{,}202 \text{ psi } (C) \ (15.2 \text{ MPa}) < 0.60 f'_{ci} = -2{,}250 \text{ psi, OK}$$

B. After the slab is cast and at service load, the support section stresses both at top and bottom extreme fibers were found to be below the allowable; hence, OK.

Summary of midspan stresses (psi)

	f^t	f^b
Transfer P_e only	+433	−2426
W_D at transfer	−1163	+2550
Net at transfer	−730	+124
External load (W_L)	−210	+693
Net total at service	−940	+817

3. *Camber and deflection calculation*

At transfer

Initial

$$E_{ci} = 57{,}000\sqrt{3{,}570} = 3.49 \times 10^6 \text{ psi } (2.2 \times 10^4 \text{ MPa})$$

From before, 28 days

$$E_c = 4.03 \times 10^6 \text{ psi } (2.8 \times 10^4 \text{ MPa})$$

Due to initial prestress only, from Figure 5.10

$$\delta_i = \frac{P_i e_c l^2}{8 E_{ci} I_g} + \frac{P_i(e_e - e_c)l^2}{24 E_{ci} I_g}$$

$$= \frac{(-347{,}004)(18.73)(72 \times 12)^2}{8(3.49 \times 10^6)59{,}720}$$

$$+ \frac{(-347{,}004)(12.81 - 18.73)(72 \times 12)^2}{24(3.49 \times 10^6)59{,}720}$$

$$= -2.90 + 0.30 = -2.6 \text{ in. } (66 \text{ mm}) \ \uparrow$$

Self-weight intensity $w = 641/12 = 53.42$ lb/in.

$$\text{Self-weight } \delta_D = \frac{5wl^4}{384 E_{ci} I_g} \text{ for uncracked section}$$

$$= \frac{5 \times 53.42(72 \times 12)^4}{384(3.49 \times 10^6)59{,}720} = 1.86 \text{ in. (47 mm) } \downarrow$$

Thus, the net camber at transfer is

$$-2.6 + 1.86 = -0.74 \text{ in. (19 mm) } \uparrow$$

4. *Immediate service load deflection*
   A. *Effective I_e method*
   Modulus of rupture

$$f_r = 7.5\sqrt{f_c'} = 7.5\sqrt{5000} = 530 \text{ psi}$$

f_b at service load $= 817$ psi (5.4 MPa) in tension (from before). Hence, the section is cracked and the effective I_e from Equation 5.92a or 5.92b should be used

$$d_p = 18.73 + 10.02 + 2(\text{topping}) = 30.75 \text{ in. (780 mm)}$$

$$\rho_p = \frac{A_{ps}}{bd_p} = \frac{12(0.153)}{120 \times 30.75} = 4.98 \times 10^{-4}$$

From Equation 5.93

$$I_{cr} = n_p A_{ps} d_p^2 (1 - 1.6\sqrt{n_p \rho_p})$$

$$n_p = 28.5 \times 10^6 / 4.03 \times 10^6 = 7 \text{ to be used in Equation 5.93}$$

Equation 5.93a gives $I_{cr} = 11{,}110$ in.^{4} (4.63×10^5 cm^4), use. From Equation 5.93a and the stress f_{pe} and f_d values already calculated for the bottom fibers at midspan with $f_r = 7.5\sqrt{f_c'} = 530$ psi. Moment M_{cr} due to that portion of live load that causes cracking is

$$M_{cr} = S_b \left(7.5\sqrt{f_c'} + f_{ce} - f_d\right)$$

$$= 3{,}142(530 + 2{,}426 - 2{,}550)$$

$$= 1{,}275{,}652 \text{ in.-lb}$$

M_a, unfactored maximum live load moment $= 2{,}177{,}288$ in.-lb

$$\frac{M_{cr}}{M_a} = \frac{1{,}275{,}652}{2{,}177{,}288} = 0.586$$

where M_{cr} is the moment due to that portion of the *live load* that causes cracking and M_a is the maximum service *unfactored live load*.

Using the preferable PCI expression of (M_{cr}/M_a) from Equation 5.91b, and the stress values previously tabulated

$$\frac{M_{cr}}{M_a} = 1 - \frac{f_{tl} - f_r}{f_L} = 1 - \left(\frac{817 - 530}{693}\right) = 0.586$$

$$\left(\frac{M_{cr}}{M_a}\right)^3 = (0.586)^3 = 0.20$$

Hence, from Equation 5.92b

$$I_e = \left(\frac{M_{cr}}{M_a}\right)^3 I_g + \left[1 - \left(\frac{M_{cr}}{M_a}\right)^3\right] I_{cr} \leq I_g$$

$$I_e = 0.2(77,118) + (1 - 0.2)11,110$$

$$= 15,424 + 8,888 = 24,312 \text{ in.}^4$$

$$w_{SD} = \tfrac{1}{12}(891 - 641) = 20.83 \text{ lb/in.}$$

$$w_L = \tfrac{1}{12} \times 280 = 23.33 \text{ lb/in.}$$

$$\delta_L = \frac{5wl^4}{384E_c I_e} = \frac{5 \times 23.33(72 \times 12)^4}{384(4.03 \times 10^6)24,312}$$

$$= +1.73 \text{ in. } (45 \text{ mm}) \downarrow \text{ (as an average value)}$$

When the concrete 2-in. topping is placed on the precast section, the resulting topping deflection with $I_g = 59,720$ in.^{4} becomes

$$\delta_{SD} = \frac{5 \times 20.83(72 \times 12)^4}{384(4.03 \times 10^6)59,720} = +0.63 \text{ in. } \downarrow$$

5. *Long-term deflection (camber) by PCI multipliers*
   Using PCI multipliers at slab topping completion stage (30 days) and at the final service load (5 years), the following are the tabulated deflection values:

Load	Transfer δ_p, in. (1)	PCI multipliers	δ_{30}, in. (2)	PCI multiplier (composite)	δ_{Final}, in. (3)
Prestress	−2.60	1.80	− 4.68	2.20	−5.71 ↑
w_D	+1.86	1.85	+3.44	2.40	+4.46 ↓
	−0.74 ↑		− 1.24 ↑		−1.25 ↑
w_{SD}			+0.63 ↓	2.30	+1.45 ↓
w_L			+1.89 ↓		+1.89 ↓
Final δ	−0.74 ↑		+1.28 ↓		+2.09 ↓

Hence, final deflection ≈ 2.1 in. (53 mm) $\downarrow$

$$\text{Allowable deflection} = \text{span}/180 = \frac{72 \times 12}{180} = 4.8 \text{ in.} > 2.1 \text{ in., OK}$$

5.8.5 Cracking Behavior and Crack Control in Prestressed Beams

If Δf_s is the net stress in the prestressed tendon or the magnitude of the tensile stress in the normal steel at any crack width load level in which the decompression load (decompression here means $f_c = 0$ at the level of the reinforcing steel) is taken as the reference point, then for the prestressed tendon

$$\Delta f_s = f_{nt} - f_d \text{ ksi } (=1000 \text{ psi})$$
(5.99)

where f_{nt} is the stress in the prestressing steel at any load level beyond the decompression load and f_d is the stress in the prestressing steel corresponding to the decompression load.

The unit strain $\varepsilon_s = \Delta f_s / E_s$. Because it is logical to disregard as insignificant the unit strains in the concrete due to the effects of temperature, shrinkage, and elastic shortening, the maximum crack width can be defined as

$$w_{max} = k a_{cs} \varepsilon_s^\alpha \tag{5.100}$$

where k and α are constants to be established by tests, a_{cs} is the spacing of cracks, and ε_s is the strain in the reinforcement.

For *pretensioned beams*, the maximum crack width can be evaluated from the maximum crack width at the reinforcing steel level

$$w_{max} = 5.85 \times 10^{-5} \frac{A_t}{\sum o} (\Delta f_s) \tag{5.101a}$$

and the maximum crack width (in.) at the concrete tension face

$$w'_{max} = 5.85 \times 10^{-5} R_i \frac{A_t}{\sum o} (\Delta f_s) \tag{5.101b}$$

where R_i is the ratio of distance from neutral axis to tension face to the distance from neutral axis to centroid of reinforcement.

For *posttensioned bonded beams*, the expression for the maximum crack width at the reinforcement level is

$$w_{max} = 6.51 \times 10^{-5} \frac{A_t}{\sum o} (\Delta f_s) \tag{5.102a}$$

At the tensile face, the crack width is

$$w'_{max} = 6.51 \times 10^{-5} R_i \frac{A_t}{\sum o} (\Delta f_s) \tag{5.102b}$$

For nonbonded beams, the factor 6.51 in Equations 5.102a and 5.102b becomes 6.83. The crack spacing stabilizes itself beyond an incremental stress Δf_s of 30,000 to 35,000 psi, depending on the *total* reinforcement percent ρ_T of both prestressed and nonprestressed steels.

Recent work by Nawy et al. [13] on the cracking performance of high-strength prestressed concrete beams, both pretensioned and posttensioned, has shown that the factor 5.85 in Equation 5.101a is considerably reduced. For concrete strengths in the range of 9,000 to 14,000 psi (60 to 100 MPa), this factor reduces to 2.75, so that the expression for the maximum crack width at the reinforcement level (inch) becomes

$$w_{max} = 2.75 \times 10^{-5} \frac{A_t}{\sum o} (\Delta f_s) \tag{5.103a}$$

In SI units, the expression is

$$w_{max} = 4.0 \times 10^{-5} \frac{A_t}{\sum o} (\Delta f_s) \tag{5.103b}$$

where A_t, cm^2; $\sum o$, cm; Δf_s, MPa.

For more refined values in cases where the concrete cylinder compressive strength ranges between 6,000 and 12,000 psi or higher, a modifying factor for particular f'_c values can be obtained from the following expressions:

$$\lambda_r = \frac{2}{\left(0.75 + 0.06 \sqrt{f'_c}\right) \sqrt{f'_c}} \tag{5.104a}$$

For posttensioned beams, the reduction multiplier λ_0 is

$$\lambda_0 = \frac{1}{0.75 + 0.06\sqrt{f_c'}} \tag{5.104b}$$

where f_c' and the reinforcement stress are in ksi.

5.8.6 ACI Expression for Cracking Mitigation

The ACI expression used for crack control in reinforced concrete structural elements through bar spacing is extended to prestressed concrete bonded beams, on the assumption of the desirability of a "seamless transition" between serviceability requirements for nonprestressed members and fully prestressed members. However, the mechanism of crack generation differs in the prestressed beam from that in reinforced concrete due to initially imposed precompression. Also, effects of environmental conditions are considerably more serious in the case of prestressed concrete elements due to the corrosion risks to the tendons. These provisions stipulate that the spacing of the bonded tendons should not exceed $\frac{2}{3}$ of the maximum spacing permitted for nonprestressed reinforcement. The ACI expression for prestressed members becomes

$$s = \frac{2}{3}\left(\frac{540}{\Delta f_s} - 2.5 c_s\right) \tag{5.105}$$

but not to exceed $8(36/\Delta f_{s0})$.

In SI units, the expression becomes

$$s = \frac{2}{3}\left(\frac{95,000}{\Delta f_s} - 2.5 c_s\right) \tag{5.106}$$

but not to exceed $200(252/\Delta f_s$, where Δf_s is in MPa and c_s is in mm)

Δf_s = difference between the stress computed in the prestressing tendon at service load based on cracked section analysis, and the decompression stress f_{dc} in the prestressing tendon. The code permits using the effective prestress f_{pe} in lieu of f_{dc}, ksi. A limit $\Delta f_s = 36$ ksi, and no check needed if Δf_s is less than 20 ksi.

c_c = clear cover from the nearest surface in tension to the flexural tension reinforcement, in.

While the code follows the author's definition of Δf_s given in Section 5.8.5, Equations 5.105 and 5.106 still lack the practicability of use as a crack control measure and the $\frac{2}{3}$ factor used in the expressions is arbitrary and not substantiated by test results. It should be emphasized that beams have finite web widths. Such spacing provisions as presented in the Code are essentially unworkable, since actual spacing of the tendons in almost all practical cases is *less* than the code equation limits, hence almost all beams satisfy the code, though cracking levels may be detrimental in bridge decks, liquid containment vessels, and other prestressed concrete structures in severe environment or subject to overload. They require additional mild steel reinforcement to control the crack width. Therefore, the expressions presented in Section 5.8.5 in conjunction with Table 5.14 from the ACI 224 Report [14] should be used for safe mitigation of cracking in prestressed concrete members.

TABLE 5.14   Maximum Tolerable Flexural Crack Widths

Exposure condition	Crack width	
	in.	mm
Dry air or protective membrane	0.016	0.41
Humidity, moist air, soil	0.012	0.30
De-icing chemicals	0.007	0.18
Seawater and seawater spray; wetting and drying	0.006	0.15
Water-retaining structures (excluding nonpressure pipes)	0.004	0.10

5.8.7 Long-Term Effects on Crack-Width Development

Limited studies on crack-width development and increase with time show that both sustained and cyclic loadings increase the amount of microcracking in the concrete. Also, microcracks formed at service-load levels in partially prestressed beams do not seem to have a recognizable effect on the strength or serviceability of the concrete element. Macroscopic cracks, however, do have a detrimental effect, particularly in terms of corrosion of the reinforcement and appearance. Hence, an increase of crack width due to sustained loading significantly affects the durability of the prestressed member regardless of whether prestressing is circular, such as in tanks, or linear, such as in beams. Information obtained from sustained load tests of up to 2 years and fatigue tests of up to one million cycles indicates that a *doubling* of crack width with time can be expected. Therefore, engineering judgment has to be exercised as regards the extent of tolerable crack width under long-term loading conditions.

5.8.8 Tolerable Crack Widths

The maximum crack width that a structural element should tolerate depends on the particular function of the element and the environmental conditions to which the structure is liable to be subjected. Table 5.14, from the ACI Committee 224 report on cracking, serves as a reasonable guide on the acceptable crack widths in concrete structures under the various environmental conditions encountered.

5.8.9 Example 9: Crack Control Check

A pretensioned prestressed concrete beam has a T-section as shown in Figure 5.14. It is prestressed with fifteen $\frac{7}{16}$-in. diameter seven-wire strand 270-K grade. The locations of the neutral axis and center of

PHOTO 5.8 West Kowloon Expressway Viaduct, Hong Kong, during construction, comprising 4.2 km dual three-lane causeway connecting Western Harbor Crossing to new airport (courtesy Institution of Civil Engineers, London, and [5]).

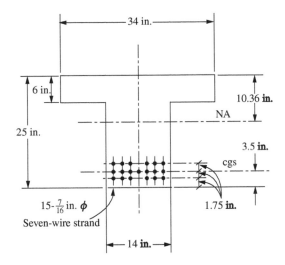

FIGURE 5.14   Beam cross-section in Example 9 [5].

gravity of steel are shown in the figure. $f'_c = 5,000$ psi, $E_c = 57,000\sqrt{f'_c}$, and $E_s = 28 \times 10^6$ psi. Find the mean stabilized crack spacing and the crack widths at the steel level as well as at the tensile face of the beam at $\Delta f_s = 30 \times 10^3$ psi. Assume that no failure in shear or bond takes place.

Solution

$$\Delta f_s = 30,000 \text{ psi} = 30 \text{ ksi}$$

Mean stabilized crack spacing

$$A_t = 7 \times 14 = 98 \text{ sq in.}$$

$$\sum o = 15\pi D = 15\pi \left(\frac{7}{16}\right) = 20.62 \text{ in.}$$

$$a_{cs} = 1.2 \left(\frac{A_t}{\sum o}\right) = 1.2 \left(\frac{98}{20.62}\right) = 5.7 \text{ in. (145 mm)}$$

Maximum crack width at steel level

$$w_{max} = 5.85 \times 10^{-5} \frac{A_t}{\sum o} (\Delta f_s) = 5.85 \times 10^{-5} \left(\frac{98}{20.62}\right) 30$$

$$= 834.1 \times 10^{-5} \text{ in.} \cong 0.0083 \text{ in. (0.21 mm)}$$

Maximum crack width at tensile face of beam

$$R_i = \frac{25 - 10.36}{25 - 10.36 - 3.5} = 1.31$$

$$w'_{max} = w_{max} R_i = 0.0083 \times 1.31 = 0.011 \text{ in. (0.28 mm)}$$

By the ACI method

$$\Delta f_s = 30 \text{ ksi}$$
$$c_c = 1.5 \text{ in.}$$

From Equation 5.105

$$s = \frac{2}{3}\left(\frac{540}{30} - 2 \times 1.5\right) = 10 \text{ in.} < 12 \text{ in., OK}$$

From this solution, it is evident that every prestressed concrete beam would satisfy the ACI Code requirements for crack control regardless of the loading conditions and/or overloading, or environmental conditions. It is rare that prestressed or mild steel reinforcement would ever be spaced within a flange that can violate the code spacing requirements. Hence, the code provisions are not effective, and probably rarely would they be effective for crack control even in two-way prestressed concrete plates.

5.8.10 SI Deflection and Cracking Expressions

$$E_c = w_c^{1.5} 0.043 \sqrt{f_c'} \text{ MPa} \tag{5.107}$$

where f_c' is in MPa units and w_c is in kg/m³ ranging between 1500 and 2500 kg/m³. For $f_c' > 35$ MPa, < 80 MPa

$$E_c = 3.32\sqrt{f_c'} + 6895\left(\frac{w_c}{2320}\right)^{1.5} \text{ MPa}$$

For normal-weight concrete, $E_c = 3.32\sqrt{f_c'} + 6895$ MPa

$$f_r = 0.62\sqrt{f_c'} \tag{5.108}$$

$$I_e = \left(\frac{M_{cr}}{M_a}\right)^3 I_g + \left[1 - \left(\frac{M_{cr}}{M_a}\right)^3\right] I_{cr} \tag{5.109}$$

$$\left(\frac{M_{cr}}{M_a}\right) = \left[1 + \left(\frac{f_{tl} - f_r}{f_L}\right)\right] \tag{5.110}$$

$$I_{cr} = n_p A_{ps} d_p^2 \left(1 - 1.6\sqrt{n_p \rho_p}\right) \tag{5.111}$$

$$= \left(n_p A_{ps} d_p^2 + n_s A_s d^2\right)\left(1 - 1.6\sqrt{n_p \rho_p + n_s \rho}\right) \tag{5.112}$$

$$w_{max} = \alpha_w \times 10^{-5} \frac{A_t}{\sum o}(\Delta f_s), \text{ mm} \tag{5.113}$$

where A_t, cm²; $\sum o$, cm; Δf_s, MPa

$$\alpha_w = 8.48 \times 10^{-5} \text{ for pretensioned}$$
$$= 9.44 \times 10^{-5} \text{ for posttensioned}$$
$$= 4.0 \times 10^{-5} \text{ for concretes with } f_c' > 70 \text{ MPa}$$
$$\text{MPa} = \text{N/mm}^2$$

(psi) 0.006895 = MPa

(lb/ft) 14.593 = N/m

(in. lb) 0.113 = N m

Acknowledgments

The author wishes to acknowledge the permission granted by Prentice Hall, Upper Saddle River, New Jersey, and Ms Marcia Horton, its vice president and publications director, to use several chapters of his book, *Prestressed Concrete — A Fundamental Approach*, 4th Edition, 2003, 944 pp. [5] as the basis for extracting the voluminous material that has formed this chapter. Also, grateful thanks are due to

Ms Mayrai Gindy, PhD candidate at Rutgers University, who helped in putting together and reviewing the final manuscript, including processing the large number of equations and computations.

Glossary

ACI Committee 116 Report "Cement and Concrete Terminology" and other definitions pertinent to prestressed concrete are presented:

AASHTO — American Association of State Highway Officials.

ACI — American Concrete Institute.

Allowable stress — Maximum permissible stress used in design of members of a structure and based on a factor of safety against yielding or failure of any type.

Balanced strain — Deformation caused by combination of axial force and bending moment that causes simultaneous crushing of concrete at the compression side and yielding of tension steel at the tension side of concrete members.

Beam — A structural member subjected primarily to flexure due to transverse load.

Beam-column — A structural member that is subjected simultaneously to bending and substantial axial forces.

Bond — Adhesion and grip of concrete or mortar to reinforcement or to other surfaces against which it is placed; to enhance bond strength, ribs or other deformations are added to reinforcing bars.

Camber — Reverse deflection (convex upwards) that is intentionally built into a structural element or form to improve appearance or to offset the deflection of the element under the effects of loads, shrinkage, and creep. It is also reverse deflection (convex upward) in prestressed concrete beams.

Cast-in-place concrete — Concrete placed in its final or permanent location, also called *in situ* concrete, in contrast to precast concrete.

Column — A member that supports primarily axial compressive loads with a height of at least three times its least lateral dimension; the capacity of short columns is controlled by strength; the capacity of long columns is limited by buckling.

Column strip — The portion of a flat slab over a row of columns consisting of a width equal to quarter of the panel dimension on each side of the column centerline.

Composite construction — A type of construction using members made of different materials (e.g., concrete and structural steel), or combining members made of cast-in-place and precast concrete such that the combined components act together as a single member.

Compression member — A member subjected primarily to longitudinal compression; often synonymous with "column."

Compressive strength — Strength typically measured on a standard 6×12 in. cylinder of concrete in an axial compression test, 28 days after casting. For high-strength concrete, 4×8 in. cylinders are used.

Concrete — A composite material that consists essentially of a binding medium within which are embedded coarse and fine aggregates; in portland cement concrete, the binder is a mixture of portland cement and water.

Confined concrete — Concrete enclosed by closely spaced transverse reinforcement to restrain concrete expansion in directions perpendicular to the applied stresses.

Construction joint — The interaction surface between two successive placements of concrete across which it may be desirable to achieve bond, and through which reinforcement may be continuous.

Continuous beam or slab — A beam or slab that extends as a unit over three or more supports in a given direction and is provided with the necessary reinforcement to develop the negative moments over the interior supports; a redundant structure that requires a statically indeterminant analysis (opposite of simple supported beam or slab).

Cover — In reinforced and prestressed concrete, the shortest distance between the surface of the reinforcement and the outer surface of the concrete; minimum values are specified to protect the reinforcement against corrosion and to assure sufficient bond strength with the reinforcement.

Cracks — Fracture in concrete elements when tensile stresses exceed the concrete tensile strength; a design goal is to keep their widths small (hairline cracks well-distributed along a member).

Cracked section — A section designed or analyzed on the assumption that concrete has no resistance to tensile stress.

Cracking load — The load that causes tensile stress in a member equal to or exceeding the modulus of rupture of concrete.

Deformed bar — Reinforcing bar with a manufactured pattern of surface deformations intended to prevent slip when the embedded bar is subjected to tensile stress.

Design strength — Ultimate load and moment capacity of a member multiplied by a strength reduction factor.

Development length — The length of embedded reinforcement to develop the design strength of the reinforcement; a function of bond strength.

Diagonal crack — An inclined crack caused by diagonal tension, usually at about 45° to the neutral axis of a concrete member.

Diagonal tension — The principal tensile stress resulting from the combination of normal and shear stresses acting upon a structural element.

Drop panel — The portion of a flat slab in the area surrounding a column such as column capital that is thicker than the slab in order to reduce the intensity of shear stresses.

Ductility — Capability of a material or structural member to undergo large inelastic deformations without distress; opposite of brittleness; very important material property, especially for earthquake-resistant design; steel is naturally ductile, concrete is brittle but it can be made ductile if well confined.

Durability — The ability of concrete to maintain its design qualities long term while exposed to weather, freeze–thaw cycles, chemical attack, abrasion, and other service load environmental conditions.

Effective depth — Depth of a beam or slab section measured from the compression face to the centroid of the tensile reinforcement.

Effective flange width — Width of slab adjoining a beam stem or web assumed to function as the flange of a T-section or L-section.

Effective prestress — The stress remaining in the prestressing reinforcement after all losses have occurred.

Effective span — The lesser of the distance between centers of supports and the clear distance between supports plus the effective depth of the beam or slab.

End block — End segment of a prestressed concrete beam.

Equivalent lateral force method — Static method for evaluating the horizontal base shear due to seismic forces.

Flat slab — A concrete slab reinforced in two, generally without beams or girders to transfer the loads to supporting members, sometimes with drop panels or column capitals or both.

High-early strength cement — Cement producing strength in mortar or concrete earlier than regular cement.

Hoop — A one-piece closed reinforcing tie or continuously wound tie that encloses the longitudinal reinforcement.

Interaction diagram — Load–moment curve for a member subjected to both axial force and the bending moment, indicating the moment capacity for a given axial load and vice versa; used to develop design charts for reinforced and prestressed concrete compression members.

Lightweight concrete — Concrete of substantially lower unit weight than that made using normal-weight gravel or crushed stone aggregate.

Limit analysis — See **Plastic analysis**.

Limit design — A method of proportioning structural members based on satisfying certain strength and serviceability limit states.

Load and resistance factor design (LRFD) — See **Ultimate strength design**.

Load factor — A factor by which a service load is multiplied to determine the factored load used in ultimate strength design.

LRFD — Load Resistance Factor Design.

Modulus of elasticity — The ratio of normal stress to corresponding strain for tensile or compressive stresses below the proportional limit of the material; for steel typically, $E_s = 29,000$ ksi; for concrete it is a function of the cylinder compressive strength f'_c; for normal-weight concrete, a common approximation is $E_c = 57,000\sqrt{f'_c}$ for concrete having strength not exceeding 6,000 psi.

Modulus of rupture — The tensile strength of concrete at the first fracture load.

Mortar — A mixture of cement paste and fine aggregate; in fresh concrete, the material filling the voids between the coarse aggregate particles.

Nominal strength — The strength of a structural member based on its assumed material properties and sectional dimensions, before application of any strength reduction factor.

Partial loss of prestress — Loss in the reinforcement prestressing due to elastic shortening, creep, shrinkage, relaxation, and friction.

PCI — Precast/Prestressed Concrete Institute.

Plastic analysis — A method of structural analysis to determine the intensity of a specified load distribution at which the structure forms a collapse mechanism.

Plastic hinge — Region in a flexural member where the ultimate moment capacity can be developed and maintained with corresponding significant inelastic rotation, as main tensile steel is stressed beyond the yield point.

Posttensioning — A method of prestressing concrete elements where the tendons are tensioned after the concrete has hardened (opposite of pretensioning).

Precast concrete — Concrete cast at a location different than its final placement, usually in plants or sites close to the final site (contrary to cast-in-place concrete).

Prestressed concrete — Concrete in which longitudinal compressive stresses are induced in the member prior to the placement of external loads through the tensioning of prestressing strands that are placed over the entire length of the member.

Prestressing steel — High-strength steel used to apply the compressive prestressing force in the member, commonly seven-wire strands, single wires, bars, rods, or groups of wires or strands.

Pretensioning — A method of prestressing the concrete element whereby the tendons are tensioned before the concrete has been placed (opposite to posttensioning).

PTI — Posttensioning Institute.

Reinforced concrete — Concrete containing adequate reinforcement and designed on the assumption that the two materials act together in resisting the applied forces.

Reinforcement — Bars, wires, strands, and other slender elements that are embedded in concrete in such a manner that the reinforcement and the concrete act together in resisting forces.

Relaxation — Time loss in the prestressing steel due to creep effect in the reinforcement.

Safety factor — The ratio of a load producing an undesirable state (such as collapse) to an expected service load.

Service loads — Loads on a structure with high probability of occurrence, such as dead weight supported by a member or the live loads specified in building codes and specifications.

Shear span — The distance from a support face of a simply supported beam to the nearest concentrated load for concentrated loads and the clear span for distributed loads.

Shear wall — See **Structural wall**.

Shotcrete — Mortar or concrete pneumatically projected at high velocity onto a surface.

Silica fume — Very fine noncrystalline silica produced in electric arc furnaces as a by-product in the production of metallic silicon and various silicon alloys (also know as condensed silica fume); used as a mineral admixture in concrete.

Slab — A flat, horizontal cast layer of plain or reinforced concrete, usually of uniform thickness, either on the ground or supported by beams, columns, walls, or other frame work. See also **Flat slab**.

Slump — A measure of consistency of freshly mixed concrete equal to the subsidence of the molded specimen immediately after removal of the slump cone, expressed in inches.

Splice — Connection of one reinforcing bar to another by lapping, welding, mechanical couplers, or other means.

Tensile split cylinder test — Test for tensile strength of concrete in which a standard cylinder is loaded to failure in diametral compression applied along the entire length of the cylinder (also called Brazilian test).

Standard cylinder — Cylindrical specimen of 12-in. height and 6-in. diameter, used to determine standard compressive strength and splitting tensile strength of concrete. For high-strength concrete the size is usually 4×8 in.

Stiffness coefficient — The coefficient k_{ij} of stiffness matrix **K** for a multi-degree of freedom structure is the force needed to hold the ith degree of freedom in place, if the jth degree of freedom undergoes a unit of displacement, while all others are locked in place.

Stirrup — A type of reinforcement used to resist shear and diagonal tension stresses in a structural concrete member; typically a steel bar bent into a U or rectangular shape and installed perpendicular to the longitudinal reinforcement and properly anchored; the term "stirrup" is usually applied to lateral reinforcement in flexural members and the term "tie" to lateral reinforcement in compression members. See **Tie**.

Strength design — See **Ultimate strength design**.

Strength reduction factor — Capacity reduction factor (typically designated as ϕ) by which the nominal strength of a member is to be multiplied to obtain the design strength; specified by the ACI Code for different types of members and stresses.

Structural concrete — Concrete used to carry load or to form an integral part of a structure (opposite of, e.g., insulating concrete).

Structural wall — Reinforced or prestressed wall carrying loads and subjected to stress, particularly horizontally due to seismic loading.

Strut-and-tie procedure — Procedure for the design of confining reinforcement in prestressed concrete beam end blocks.

T-beam — A beam composed of a stem and a flange in the form of a "T," with the flange usually provided by the slab part of a floor system.

Tie — Reinforcing bar bent into a loop to enclose the longitudinal steel in columns; tensile bar to hold a form in place while resisting the lateral pressure of unhardened concrete.

Ultimate strength design (USD) — Design principle such that the actual (ultimate) strength of a member or structure, multiplied by a strength factor, is no less than the effects of all service load combinations, multiplied by the respective overload factors.

Unbonded tendon — A tendon that is not bonded to the concrete.

Under-reinforced beam — A beam in which the strain in the extreme tension reinforcement is less than the balanced strain.

Water–cement ratio — Ratio by weight of water to cement in a mixture, inversely proportional to concrete strength.

Water-reducing admixture — An admixture capable of lowering the mix viscosity, thereby allowing a reduction of water (and increase in strength) without lowering the workability (also called superplasticizer).

Whitney stress block — A rectangular area of uniform stress intensity $0.85f_c'$, whose area and centroid are similar to those of the actual stress distribution in a flexural member at failure.

Workability — General property of freshly mixed concrete that defines the ease with which the concrete can be placed into the forms without honeycombing; closely related to slump.

Yield-line theory — Method of structural analysis of concrete plate structures at the collapse load level.

References

[1] Nawy, E. G., *Reinforced Concrete — A Fundamental Approach*, 5th ed. Prentice Hall, Upper Saddle River, NJ, 2003, 864 pp.

[2] ACI Committee 435, *Control of Deflection in Concrete Structures*, ACI Committee Report R435-95, E. G. Nawy, Chairman, American Concrete Institute, Farmington Hills, MI, 1995, 77 pp.

[3] Nawy, E. G., *Fundamentals of High Performance Concrete*, 2nd ed. John Wiley & Sons, New York, 2001, 460 pp.

[4] Branson, D. E., *Deformation of Concrete Structures*, McGraw-Hill, New York, 1977.

[5] Nawy, E. G., *Prestressed Concrete — A Fundamental Approach*, 4th ed. Prentice Hall, Upper Saddle River, NJ, 2003, 944 pp.

[6] Hsu, T. T. C., *Unified Theory of Reinforced Concrete*, CRC Press, Boca Raton, FL, 1993, 313 pp.

[7] Prestressed Concrete Institute. *PCI Design Handbook*, 5th ed. PCI, Chicago, 1999.

[8] ACI Committee 318, *Building Code Requirements for Structural Concrete (ACI 318–02) and Commentary (ACI 318 R-02)*. American Concrete Institute, Farmington Hills, MI, 2002, 446 pp.

[9] Breen, J. E., Burdet, O., Roberts, C., Sanders, D., and Wollman, G. "Anchorage Zone Reinforcement for Posttensioned Concrete Girders." NCHRP Report, 356 pp.

[10] International Code Council. *International Building Codes 2000–2003 (IBC)*, Joint UBC, BOCA, SBCCI, Whittier, CA, 2003.

[11] Nawy, E. G., Editor-in-Chief, *Concrete Construction Engineering Handbook*. CRC Press, Boca Raton, FL, 1998, 1250 pp.

[12] Englekirk, R. E., Design-Construction of the Paramount — A 39-Story Precast Prestressed Concrete Apartment Building. *PCI J.*, Precast/Prestressed Concrete Institute, Chicago, IL, July–August 2002, pp. 56–69.

[13] Nawy, E. G., *Design For Crack Control in Reinforced and Prestressed Concrete Beams, Two-Way Slabs and Circular Tanks — A State of the Art*. ACI SP-204, Winner of the 2003 ACI Design Practice Award, American Concrete Institute, Farmington Hills, MI, 2002, pp. 1–42.

[14] ACI Committee 224, *Control of Cracking in Concrete Structures*. ACI Committee Report R224-01, American Concrete Institute, Farmington Hills, MI, 2001, 64 pp.

[15] Posttensioning Institute. *Posttensioning Manual*, 5th ed. PTI, Phoenix, AZ, 2003.

[16] Nawy, E. G. and Chiang, J. Y., "Serviceability Behavior of Posttensioned Beams." *J. Prestressed Concrete Inst.* 25 (1980): 74–85.

[17] Nawy, E. G. and Huang, P. T., "Crack and Deflection Control of Pretensioned Prestressed Beams." *J. Prestressed Concrete Inst.* 22 (1977): 30–47.

[18] Branson, D. E., *The Deformation of Non-Composite and Composite Prestressed Concrete Members*. ACI Special Publication SP-43, *Deflection of Concrete Structures*, American Concrete Institute, Farmington Hills, MI, 1974, pp. 83–127.

[19] PCI, *Prestressed Concrete Bridge Design Handbook*. Precast/Prestressed Concrete Institute, Chicago, 1998.

[20] AASHTO, *Standard Specifications for Highway Bridges*, 17th ed. and 2002 Supplements. American Association of State Highway and Transportation Officials, Washington, DC, 2002.

[21] Nawy, E. G., "Discussion — The Paramount Building." *PCI J.*, Precast/Prestressed Concrete Institute, Chicago, IL, November–December 2002, p. 116.

[22] Freyssinet, E., *The Birth of Prestressing*. Public Translation, Cement and Concrete Association, London, 1954.

[23] Guyon, Y. *Limit State Design of Prestressed Concrete*, vol. 1. Halsted-Wiley, New York, 1972.

[24] Gerwick, B. C., Jr. *Construction of Prestressed Concrete Structures*. Wiley-Interscience, New York, 1997, 591 pp.

[25] Lin, T. Y. and Burns, N. H., *Design of Prestressed Concrete Structures*, 3rd ed. John Wiley & Sons, New York, 1981.

[26] Abeles, P. W. and Bardhan-Roy, B. K., *Prestressed Concrete Designer's Handbook*, 3rd ed. Viewpoint Publications, London, 1981.

[27] American Concrete Institute, *ACI Manual of Concrete Practice*, 2003. *Materials*. American Concrete Institute, Farmington Hills, MI, 2003.

[28] Portland Cement Association. *Design and Control of Concrete Mixtures*, 13th ed. PCA, Skokie, IL, 1994.

[29] Nawy, E. G., Ukadike, M. M., and Sauer, J. A. "High Strength Field Modified concretes." *J. Struct. Div., ASCE* 103 (No. ST12) (1977): 2307–2322.

[30] American Society for Testing and Materials. "Standard Specification for Cold-Drawn Steel Wire for Concrete Reinforcement, A8 2-79." ASTM, Philadelphia, 1980.

[31] Chen, B. and Nawy, E. G., "Structural Behavior Evaluation of High Strength Concrete Reinforced with Prestressed Prisms Using Fiber Optic Sensors." *Proc., ACI Struct. J.*, American Concrete Institute, Farmington Hills, MI, (1994): pp. 708–718.

[32] Posttensioning Institute. *Posttensioning Manual*, 5th ed. PTI, Phoenix, AZ, 1991.

[33] Cohn, M. Z., *Partial Prestressing, From Theory to Practice*. NATO-ASI Applied Science Series, vols. 1 and 2. Martinus Nijhoff, Dordrecht, The Netherlands, 1986.

[34] Yong, Y. K., Gadebeku, C. and Nawy, E. G., "Anchorage Zone Stresses of Posttensioned Prestressed Beams Subjected to Shear Forces." *ASCE Struct. Div. J.* 113 (8) (1987): 1789–1805.

[35] Nawy, E. G., "Flexural Cracking Behavior of Pretensioned and Posttensioned Beams — The State of the Art." *J. Am. Concrete Inst.* December 1985: 890–900.

[36] Federal Highway Administration. "Optimized Sections for High Strength Concrete Bridge Girders." FHWA Publication No. RD-95-180, Washington, DC, August 1997, 156 pp.

[37] Collins, M. P. and Mitchell, D. "Shear and Torsion Design of Prestressed and Non-Prestressed Concrete Beams." *J. Prestressed Concrete Inst.* 25 (1980): 32–100.

[38] Zia, P. and Hsu, T. T. C., *Design for Torsion and Shear in Prestressed Concrete*. ASCE Annual Convention, Reprint No. 3423, 1979.

[39] International Conference of Building Officials, *Uniform Building Code (UBC)*, vol. 2, ICBO, Whittier, CA, 1997.

[40] Naja, W. M., and Barth, F. G., "Seismic Resisting Construction," Chapter 26, in E. G., Nawy, editor-in-chief, *Concrete Construction Engineering Handbook*. CRC Press, Boca Raton, FL, 1998, pp. 26-1–26-69.

6

Masonry Structures

Richard E. Klingner
*Department of Civil Engineering,
University of Texas,
Austin, TX*

6.1 Introduction

Masonry is traditionally defined as hand-placed units of natural or manufactured material, laid with mortar. In this chapter, the earthquake behavior and design of masonry structures is discussed, extending the traditional definition somewhat to include thin stone cladding.

Masonry makes up approximately 70% of the existing building inventory in the United States (TMS, 1989). U.S. masonry comprises Indian cliff dwellings, constructed of sandstone at Mesa Verde (Colorado); the adobe missions constructed by Spanish settlers in Florida, California, and the southwestern United States; bearing-wall buildings such as the 16-story Monadnock Building, completed in 1891 in Chicago; modern reinforced bearing-wall buildings; and many veneer applications. Clearly, the behavior and design of each type of masonry is distinct. In this chapter, fundamental applications and nomenclature of U.S. masonry are discussed, major construction categories are reviewed, historical seismic performance of masonry is presented, and principal design and retrofitting approaches are noted. Its purpose is to give designers, constructors, and building officials a basic foundation for further study of the behavior and design of masonry.

6.2 Masonry in the United States

6.2.1 Fundamentals of Masonry in the United States

Masonry can be classified according to architectural or structural function. Each is discussed later in this chapter. Regardless of how it is classified, U.S. masonry uses basically the same materials: units, mortar, grout, and accessory materials. In this section, those materials are discussed, with reference to the national consensus specifications of the American Society for Testing and Materials (ASTM). Additional information is available at the web sites of associations such as the National Concrete Masonry Association (NCMA), the Brick Industry Association (BIA), and The Masonry Society (TMS).

6.2.1.1 Masonry Units

Of the more than 20 different classifications of masonry units commercially available in the United States, only the most widely used are discussed here.

6.2.1.1.1 Clay or Shale Masonry Units

The most common structural clay or shale masonry units are Building Brick and Facing Brick. The former are specified using ASTM C62 Building Brick (Solid Masonry Units Made from Clay or Shale). The latter, specifically intended for use when appearance is important, are specified using ASTM C216 Facing Brick (Solid Masonry Units Made from Clay or Shale). Units are usually cored rather than being completely solid. The net cross-sectional area of the unit must be at least 75% of the gross area, that is, the cores occupy less than 25% of the area of the unit.

Many different sizes and shapes of clay or shale masonry units are available, varying widely from region to region of the United States. One common size is probably the "modular" unit, which measures $7\frac{5}{8}$ in. (194 mm) long by $2\frac{1}{4}$ in. (57 mm) high by $3\frac{5}{8}$ in. (92 mm) deep. Using mortar joints $\frac{3}{8}$ in. (9 mm) thick, this unit produces modules 8 in. (203 mm) wide by $2\frac{2}{3}$ in. (68 mm) high. That is, three courses of such units produce modules 8 in. (203 mm) wide by 8 in. high.

Clay or shale masonry units are sampled and tested using ASTM C67 (Methods of Sampling and Testing Brick and Structural Clay Tile). Specified properties include compressive strength and durability. Facing brick can have more restrictive dimensional tolerances and appearance requirements.

6.2.1.1.2 Concrete Masonry Units

The most common concrete masonry units are hollow load-bearing concrete masonry units, specified in ASTM C90 (Loadbearing Concrete Masonry Units). The units are typically made from low- or zero-slump concrete. In the eastern United States, these units are used for unreinforced inner wythes of cavity walls. In the western United States, these units are used for reinforced, fully grouted shear and bearing walls. The net area of the units is usually about 55 to 60% of their gross cross-sectional area. These units are commonly $15\frac{5}{8}$ in. (397 mm) long by $7\frac{5}{8}$ in. (194 mm) high by $7\frac{5}{8}$ in. (194 mm) thick. Using mortar joints $\frac{3}{8}$ in. (9 mm) thick, this unit produces modules 8 in. (203 mm) wide by 8 in. high. These modules are compatible with those of the modular clay brick discussed above. Concrete masonry units are sampled and tested using ASTM C140 (Methods of Sampling and Testing Concrete Masonry Units). Specified properties include shrinkage, compressive strength, and absorption.

6.2.1.2 Mortar

Mortar holds units together, and also compensates for their dimensional tolerances. In the United States, mortar for unit masonry is specified using ASTM C270 (Specification for Mortar for Unit Masonry), which addresses three cementitious systems: portland cement–lime, masonry cement, and mortar cement. These cementitious systems are combined with sand and water to produce mortar.

Portland cement–lime mortar consists of portland cement and other hydraulic cements, hydrated mason's lime, sand, and water. Masonry cement mortar consists of masonry cement, sand, and water. The contents of masonry cement and mortar cement, specified under ASTM C91 and ASTM C1329, respectively, vary from manufacturer to manufacturer, and are not disclosed. They typically include portland cement and other hydraulic cements, finely ground limestone, and air-entraining and water-retention admixtures. Mortar cement has a minimum specified tensile bond strength and a lower maximum air content than masonry cement. Model codes prohibit the use of masonry cement in seismic design categories C and higher. Portland cement–lime mortars and mortar cement mortars are not restricted in this respect.

Within each cementitious system, masonry mortar is also classified according to type. Types are designated as M, S, N, O, and K (derived from every other letter of the phrase "MaSoN wOrK"). These designations refer to the proportion of portland cement in the mixture. Type M has the most, S less, and so on. Higher proportions of portland cement result in faster strength gain, higher compressive strength, and higher tensile bond strength; they also result in lower long-term deformability. Mortar types S and N are typically specified.

Within each cementitious system, mortar can be specified by proportion or by property, with the former being the default. For example, Type S portland cement–lime mortar, specified by proportion, consists of one volume of portland cement, $\frac{1}{2}$ volume of hydrated mason's lime, about $4\frac{1}{2}$ volumes of masons' sand, and sufficient water for good workability. Type S masonry cement mortar or mortar cement mortar is made with one volume of masonry cement or mortar cement, respectively, three volumes of mason's sand, and sufficient water for good workability.

6.2.1.3 Grout

Masonry grout is essentially fluid concrete, used to fill spaces in masonry and to surround reinforcement and connectors. It is specified using ASTM C476 (Grout for Masonry). Grout for masonry is composed of portland cement and other hydraulic cements, sand, and (in the case of coarse grout) pea gravel. It is permitted to contain a small amount of hydrated mason's lime, but usually does not. It is permitted to be specified by proportion or by property, with the former being the default. A coarse grout specified by proportion would typically contain one volume of portland cement or other hydraulic cements, about three volumes of mason's sand, and about two volumes of pea gravel.

Masonry grout is placed with a slump of at least 8 in. (203 mm), so that it will flow freely into the cells of the masonry. Because of its high water–cement (w/c) ratio at the time of grouting, masonry grout undergoes considerable plastic shrinkage as the excess water is absorbed by the surrounding units. To prevent the formation of voids due to this process, the grout is consolidated during placement and reconsolidated after initial plastic shrinkage. Grouting admixtures, which contain plasticizers and water-retention agents, are also useful in the grouting process.

If grout is specified by property (compressive strength), the compressive strength must be verified using permeable molds, duplicating the loss of water and decreased w/c ratio that the grout would experience in actual use.

6.2.1.4 Accessory Materials

Accessory materials for masonry consist of reinforcement, connectors, sealants, flashing, coatings, and vapor barriers. In this section, each is briefly reviewed.

Reinforcement consists of deformed reinforcing bars or joint reinforcement. Deformed reinforcing bars are placed vertically in the cells of hollow units, horizontally in courses of bond-beam units, or vertically and horizontally between wythes of solid units. Model codes require that it be surrounded by grout. Joint reinforcement is placed in the bed (horizontal) joints of masonry and is surrounded by mortar.

Connectors are used to connect the wythes of a masonry wall (ties), to connect a masonry wall to a frame (anchors), or to connect something else to a masonry wall (fasteners).

Sealants are used to prevent the passage of water at places where gaps are intentionally left in masonry walls. Three basic kinds of gaps (joints) are used: expansion joints are used in brick masonry to accommodate expansion, control joints are used in concrete masonry to conceal cracking due to shrinkage, and construction joints are placed between different sections of a structure.

Flashing is a flexible waterproof barrier, intended to permit water that has penetrated the outer wythe to re-exit the wall. It is placed at the bottom of each story level (on shelf angles or foundations), over window and door lintels, and under window and door sills. Flashing should be lapped, and ends of flashing should be defined by end dams (flashing turned up at ends). Directly above the level of the flashing, weepholes should be provided at 24-in. spacing. Flashing is made of metal, polyvinyl chloride (PVC), or rubberized plastic (EPDM). Metallic flashing lasts much longer than plastic flashing. Nonmetallic flashings are subject to tearing. Modern EPDM self-adhering flashing is a good compromise between durability and ease of installation.

6.2.1.5 Masonry Nomenclature by Architectural Function

The architectural functions of masonry include acting as a building envelope to resist liquid water. Masonry walls are classified in terms of this function into barrier walls and drainage walls. Barrier walls act by a combination of thickness, coatings, and integral water-repellent admixtures. Drainage walls act by the above, plus drainage details. Examples of each are shown in Figure 6.1 and Figure 6.2. In drainage walls, an outer wythe (thickness of masonry) is separated from an inner wythe of masonry or from a backup system by a cavity with drainage details.

6.2.1.6 Masonry Nomenclature by Structural Function

From the viewpoint of structural function, U.S. masonry can be broadly classified as nonload-bearing and load-bearing. The former resists gravity loads from self-weight alone, and possibly out-of-plane wind loads or seismic forces from its own mass only. The latter may resist gravity and lateral loads from overlying floors or roof. Both classifications of masonry use the same materials.

Nonload-bearing masonry includes panel walls (an outer wythe of masonry connected to an inner wythe of masonry or a backup system), curtain walls (masonry spanning horizontally between columns), and interior partitions.

Load-bearing masonry walls resist out-of-plane loads by spanning as horizontal or vertical strips, in-plane gravity loads by acting as a shallow beam–column loaded perpendicular to the plane of the wall, and in-plane shear forces by acting as a deep beam–column loaded in the plane of the wall.

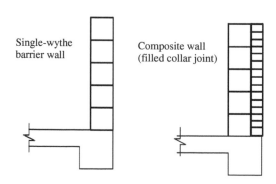

Single-wythe barrier wall

Composite wall (filled collar joint)

FIGURE 6.1   Examples of barrier walls.

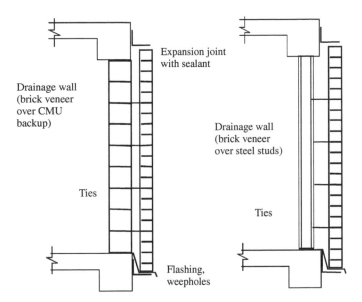

FIGURE 6.2   Examples of drainage walls.

6.2.1.7 Masonry Nomenclature by Design Intent

From the viewpoint of design intent, U.S. masonry can be broadly classified as unreinforced and reinforced. Unreinforced masonry is designed assuming that flexural tension is resisted by masonry alone, and neglecting stresses in reinforcement. Reinforced masonry is designed assuming that flexural tension is resisted by reinforcement alone, and neglecting the flexural tensile resistance of masonry. Both types of masonry are designed assuming that masonry has some diagonal tensile resistance, because both types permit some shear to be resisted without shear reinforcement.

To decipher design intent may be impossible by examination of the masonry alone, with no knowledge of its design process. Masonry elements, no matter how designed, are required to have minimum prescriptive reinforcement whose location and percentage depend on the seismic design category of the structure in which they are located.

Differences in historical tradition have led to potentially confusing differences in nomenclature. For example, in parts of the United States where the *Uniform Building Code* (UBC) has been dominant (roughly speaking, to the west of Denver), "partially reinforced" masonry referred to reinforced masonry whose reinforcement did not comply with UBC requirements for prescriptive reinforcement in zones of highest seismic risk. East of Denver, however, partially reinforced masonry referred to masonry reinforced with wire-type bed-joint reinforcement only, rather than deformed reinforcement placed in grouted cells or bond beams.

6.2.2 Modern Masonry Construction in the United States

A decade ago, it might have been possible to distinguish between modern masonry in the eastern versus the western United States, with the latter being characterized by more emphasis on seismic design. As model codes increasingly adopt the philosophy that almost all regions of the United States have some level of seismic risk, such regional distinctions are disappearing.

6.2.2.1 Modern Masonry Veneer

Modern masonry veneer resists vertical loads due to self-weight only, and transfers out-of-plane loads from wind or earthquake to supporting elements such as wooden stud walls, light-gage steel framing, or a backup wythe of masonry. Veneer is most commonly clay masonry units, but concrete masonry units, glass block, and glazed tile are also used. Stone cladding can be laid like manufactured masonry units, using masonry mortar. Thin stone can also be attached without mortar to a backup frame, using stainless steel connectors.

6.2.2.2 Modern Masonry Partition Walls

Modern masonry partition walls are interior elements designed to resist vertical loads due to self-weight only and out-of-plane loads due to inertial forces from their own mass only. They are of clay or concrete masonry units, glass block, or glazed tile.

6.2.2.3 Modern Masonry Panel Walls

Modern masonry panel walls are combinations of a veneer wythe and a backup system. They resist vertical loads due to self-weight only. The veneer wythe transfers out-of-plane loads from wind or earthquake to the backup system. The backup system is not intended to resist in-plane shear loads or vertical loads from overlying roofs or floors. If the space between the masonry veneer and the backup system is separated by a cavity at least 2 in. (50 mm) wide and is provided with drainage details, the result is a drainage wall.

6.2.2.4 Modern Masonry Curtain Walls

Curtain walls are multistory masonry walls that resist gravity loads from self-weight only and out-of-plane loads from wind or earthquake. Their most common application is for walls of industrial buildings, warehouses, gymnasiums, or theaters. They are most commonly single-wythe walls. Because they occupy multiple stories, curtain walls are generally designed to span horizontally between columns or pilasters. If a single wythe of masonry is used, horizontal reinforcement is often required for resistance to out-of-plane loads. This reinforcement is usually provided in the form of welded wire reinforcement, placed in the horizontal joints of the masonry.

6.2.2.5 Modern Masonry Bearing and Shear Walls

Bearing walls resist gravity loads from self-weight and overlying floor and roof elements, out-of-plane loads from wind or earthquake, and in-plane shears. If bearing walls are composed of hollow units, vertical reinforcement consists of deformed bars placed in vertical cells, and horizontal reinforcement consists of either deformed bars placed in grouted courses (bond beams) or bed-joint reinforcement. Bearing walls, whether designed as unreinforced or reinforced, must have reinforcement satisfying seismic requirements. If bearing walls are composed of solid units, vertical and horizontal reinforcement generally consist of deformed bars placed in a grouted space between two wythes of masonry.

Although model codes sometimes distinguish between bearing walls and shear walls, in practical terms every bearing wall is also a shear wall, because it is impossible in practical terms for a wall to resist gravity loads from overlying floor or roof elements and yet be isolated from in-plane shears transmitted from those same elements.

Reinforced masonry shear walls differ from reinforced concrete shear walls primarily in that their inelastic deformation capacity is lower. They are usually not provided with confined boundary elements because these are difficult or impossible to place. Their vertical reinforcement is generally distributed uniformly over the plane length of the wall. Sections of masonry shear wall that separate window or door openings are commonly referred to as "piers."

Using this type of construction, 30-story masonry bearing-wall buildings have been built in Las Vegas, NV, a region of seismic risk in the United States (Suprenant 1989).

Masonry infills are structural panels placed in a bounding frame of steel or reinforced concrete. This mode of structural action, though common in panel walls, is not addressed directly by design codes. Provisions are under development. Historical masonry infills are addressed later in this chapter.

6.2.2.6 Modern Masonry Beams and Columns

Other modern masonry elements are beams and columns. Masonry beams are most commonly used as lintels over window or door openings, but can also be used as isolated elements. They are reinforced horizontally for flexure. Although shear reinforcement is theoretically possible, it is difficult to install and is rarely used. Instead, masonry beams are designed deep enough so that shear can be resisted by masonry alone.

Isolated masonry columns are rare. The most common form of masonry beam–column is a masonry bearing wall subjected to a combination of axial load from gravity and out-of-plane moment from eccentric axial load or out-of-plane wind or seismic loads.

6.2.2.7 Role of Horizontal Diaphragms in Structural Behavior of Modern Masonry

Horizontal floor and roof diaphragms play a critical role in the structural behavior of modern masonry. In addition to resisting gravity loads, they transfer horizontal forces from wind or earthquake to the lateral force-resisting elements of a masonry building, which are usually shear walls. Modern horizontal diaphragms are usually composed of cast-in-place concrete or of concrete topping overlying hollow-core, prestressed concrete planks or corrugated metal deck supported on open-web joists. Distinctions between rigid and flexible diaphragms, and appropriate analytical approaches for each, are addressed later in this chapter. Performance of horizontal diaphragms in modern masonry is addressed by structural requirements for in-plane flexural and shear resistance and by detailing requirements for continuous chords and other embedded elements.

6.2.3 Historical Structural Masonry in the United States

6.2.3.1 Historical Unreinforced Masonry Bearing Walls in the United States

Unreinforced masonry bearing walls were constructed before 1933 in the western United States and as late as the 1950s elsewhere in the United States. They commonly consisted of two wythes of masonry, bonded by masonry headers, and sometimes also had an interior wythe of rubble masonry (pieces of masonry units surrounded by mortar).

6.2.3.2 Historical Masonry Infills in the United States

Masonry infills are structural panels placed in a bounding frame of steel or reinforced concrete. Before the advent of drywall construction, masonry infills of clay tile were often used to fill interior or exterior bays of steel or reinforced concrete frames. Although sometimes considered nonstructural, they have high elastic stiffness and are usually built tight against the bounding frame. As a result, they can significantly alter the seismic response of the frame in which they are placed.

6.2.3.3 Role of Horizontal Diaphragms in Structural Behavior of Historical Masonry in the United States

Horizontal diaphragms play a crucial role in the seismic resistance of historical as well as modern masonry construction. In contrast to their role in modern construction, however, the behavior of horizontal diaphragms in historical masonry is usually deficient. Historical diaphragms are usually composed of lumber, supported on wooden joists inserted in pockets in the inner wythe of unreinforced masonry walls. Such diaphragms are not strong enough, and not sufficiently well connected, to transfer horizontal seismic forces to the building's shear walls. Out-of-plane deformations of the bearing walls can cause the joists to slip out of their pockets, often resulting in collapse of the entire building. For this reason, horizontal diaphragms are among the elements addressed in the seismic rehabilitation of historical masonry.

6.3 Fundamental Basis for Design of Masonry in the United States

Design of masonry in the United States is based on the premise that reinforced masonry structures can perform well under combinations of gravity and lateral loads, including earthquake loads, provided that they meet the following conditions:

1. They must have engineered lateral-force-resisting systems, generally consisting of reinforced masonry shear walls distributed throughout their plan area and acting in both principal plan directions.
2. Their load–displacement characteristics under cyclic reversed loading must be consistent with the assumptions used to develop their design loadings:
   - If they are intended to respond primarily elastically, they must be provided with sufficient strength to resist elastic forces. Such masonry buildings are typically low-rise, shear-wall structures.
   - If they are intended to respond inelastically, their lateral-force-resisting elements must be proportioned and detailed to be capable of resisting the effects of the reversed cyclic deformations consistent with that inelastic response. They must be proportioned, and must have sufficient shear reinforcement, so that their behavior is dominated by flexure ("capacity design"). The most desirable structural system for such a response is composed of multiple masonry shear walls, designed to act in flexure and loosely coupled by floor slabs.

U.S. masonry has shown good performance under such conditions (TMS Northridge 1994).

Good load–displacement behavior has also been observed under laboratory conditions. This research has been described extensively in U.S. technical literature over the last two decades. A representative sample is given in the proceedings of North American Masonry conferences (NAMC 1985, 1987, 1990, 1993, 1996, 1999).

Of particular relevance is the U.S. Coordinated Program for Masonry Building Research, also known as the TCCMAR Program (Noland 1990). With the support of the National Science Foundation and the masonry industry, the Technical Coordinating Committee for Masonry Research (TCCMAR) was formed in February 1984 for the purpose of defining and performing both analytical and experimental research and development necessary to improve masonry structural technology, and specifically to lay the technical basis for modern, strength-based design provisions for masonry. Under the coordination of TCCMAR, research was carried out in the following areas:

1. Material properties and tests.
2. *Reinforced masonry walls.* In-plane shear and combined in-plane shear and vertical compression.
3. *Reinforced masonry walls.* Out-of-plane forces combined with vertical compression.
4. Floor diaphragms.
5. Bond and splicing of reinforcement in masonry.
6. Limit state design concepts for reinforced masonry.
7. Modeling of masonry components and building systems.
8. Large-scale testing of masonry building systems.
9. Determination of earthquake-induced forces on masonry buildings.

Work began on the initially scheduled research tasks in September 1985, and the program lasted for more than 10 years. Numerous published results include the work of Hamid et al. (1989) and Blondet and Mayes (1991), who studied masonry walls loaded out-of-plane, and He and Priestley (1992), Leiva and Klingner (1994), and Seible et al. (1994a,b), who studied masonry walls loaded in-plane. In all cases, flexural ductility was achieved without the use of confining reinforcement.

Using pseudodynamic testing procedures, Seible et al. (1994a,b) subjected a full-scale, five-story masonry structure to simulated earthquake input. The successful inelastic performance of this structure under global drift ratios exceeding 1% provided additional verification for field observations and previous TCCMAR laboratory testing. With proper proportioning and detailing, reinforced masonry assemblies can exhibit significant ductility.

Limited shaking-table testing has been conducted on reinforced masonry structures built using typical modern U.S. practice:

1. *Gulkan et al. (1990a,b)*. A series of single-story, one-third-scale masonry houses were constructed and tested on a shaking table. The principal objective of the testing was to verify prescriptive reinforcing details for masonry in zones of moderate seismic risk.
2. *Abrams and Paulson (1991)*. Two 3-story, quarter-scale reinforced masonry buildings were tested to evaluate the validity of small-scale testing.
3. *Cohen (2001)*. Two low-rise, half-scale, reinforced masonry buildings with flexible roof diaphragms were subjected to shaking-table testing. Results were compared with the results of static testing and analytical predictions.

Results of these tests have generally supported field observations of satisfactory behavior of modern reinforced masonry structures in earthquakes.

6.3.1 Design Approaches for Modern U.S. Masonry

Three design approaches are used for modern U.S. masonry: allowable-stress design, strength design, and empirical design. In this section, each approach is summarized.

6.3.1.1 Allowable-Stress Design

Allowable-stress design is the traditional approach of building codes for calculated masonry design. Stresses from unfactored loads are compared with allowable stresses, which are failure stresses reduced by a factor of safety that is usually between 2.5 and 4.

6.3.1.2 Strength Design

Within the past decade, strength-design provisions for masonry have been developed within the 1997 UBC, the 1997 and 2000 NEHRP documents, and the 2000 *International Building Code* (IBC). The 2002 edition of the Masonry Standards Joint Committee (MSJC) code includes strength-design provisions, and those provisions will be referenced by the 2003 IBC.

Strength-design provisions for masonry are generally similar to those for concrete. Factored design actions are compared with nominal capacities reduced by capacity reduction factors. Strength-design provisions for masonry differ from those for reinforced concrete, however, in three principal areas, unreinforced masonry, confining reinforcement, and maximum flexural reinforcement:

1. Some masonry can be designed as unreinforced (flexural tension resisted by masonry alone). For this purpose, nominal flexural tensile capacity is computed as the product of the masonry's tensile bond strength (modulus of rupture) and the section modulus of the section under consideration. This nominal strength is then reduced by a capacity reduction factor.
2. Because it is impractical to confine the compressive zones of masonry elements, the inelastic strain capacity of such elements is less than that of confined reinforced concrete elements. The available displacement ductility ratio of masonry shear walls is therefore lower than that of reinforced concrete shear walls with confined boundary elements, and corresponding *R* factors (response modification factors) are lower.
3. Maximum flexural reinforcement for masonry elements is prescribed in terms of the amount of steel required to equilibrate the compressive stress block of the element under a critical strain

FIGURE 6.3   Monadnock Building, Chicago (1891).

gradient, in which the maximum strain in masonry is the value used in design (0.0025 for concrete masonry and 0.0035 for clay masonry), and the maximum strain in the extreme tensile reinforcement is a multiple of the yield strain. That multiple depends on the ductility expected of the element. under reversed cyclic inelastic deformations. In practical terms, if inelastic response is possible, an element cannot be designed to work above its balanced axial load. The intent of these provisions is to ensure, for inelastic elements, that the flexural reinforcement can yield and begin to strain-harden before the compression toe crushes.

6.3.1.3 Empirical Design

At the end of the 19th century, masonry bearing-wall buildings were designed using empirical rules of thumb, such as using walls 12 in. (305 mm) thick at the top of a building and increasing the wall thickness by 4 in. (102 mm) for every story. The Monadnock Building, built in Chicago in 1891 (Figure 6.3), is 16 stories high, is of unreinforced masonry, and has bearing walls 6 ft (1.83 m) thick at the base. It is still in use today.

Today's empirical design is the descendant of those rules, adapted for the characteristics of modern structures. They involve primarily limitations on the length to thickness ratios of elements, with some rudimentary axial stress checks and limits on the arrangement of lateral-force-resisting elements and the plan aspect ratio of floor diaphragms.

6.4   Masonry Design Codes Used in the United States

6.4.1 Introduction to Masonry Design Codes in the United States

The United States has no national design code, primarily because the U.S. Constitution has been interpreted as delegating building code authority to the states, which in turn delegate it to municipalities

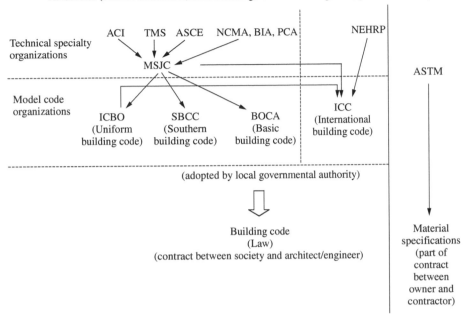

FIGURE 6.4 Schematic of code-development process for masonry in the United States.

and other local governmental agencies. Design codes used in the United States are developed by a complex process involving technical experts, industry representatives, code users, and building officials. As this process applies to the development of design provisions for masonry, it is shown in Figure 6.4 and described herein:

1. *Consensus design provisions* and *specifications for materials or methods of testing* are first drafted in mandatory language by *technical specialty organizations*, operating under consensus rules approved by the American National Standards Institute (ANSI). These consensus rules can vary from organization to organization, but must include the following requirements:
   - Balance of interests (producer, user, and general interest).
   - Written balloting of proposed provisions, with prescribed requirements for a successful ballot.
   - Resolution of negative votes. Negative votes must be discussed and found nonpersuasive before a ballot item can pass. A single negative vote, if found persuasive, can prevent an item from being passed.
   - Public comment. After being approved within the technical specialty organization, the mandatory-language provisions must be published for public comment. If significant public comments are received (usually more than 50 comments on a single item), the organization must respond to the comments.
2. These consensus design provisions and specifications are adopted, sometimes in modified form, by *model-code organizations*, and take the form of *model codes*.
3. These model codes are adopted, sometimes in modified form, by *local governmental agencies* (such as cities or counties). Upon adoption, but not before, they acquire legal standing as *building codes*.

6.4.1.1 Technical Specialty Organizations

Technical specialty organizations are open to designers, contractors, product suppliers, code developers, and end users. Their income (except for Federal Emergency Management Agency [FEMA], a U.S. government agency) is derived from member dues and the sale of publications. Technical specialty organizations active in the general area of masonry include the following:

1. *American Society for Testing and Materials (ASTM).* Through its many technical committees, ASTM develops consensus specifications for materials and methods of test. Although some model-code organizations use their own such specifications, most refer to ASTM specifications. Many ASTM specifications are also listed by the ISO (International Standards Organization).
2. *American Concrete Institute (ACI).* Through its many technical committees, this group publishes a variety of design recommendations dealing with different aspects of concrete design. ACI Committee 318 develops design provisions for concrete structures. ACI is also involved with masonry, as one of the three sponsors of the MSJC. This committee was formed in 1982 to combine the masonry design provisions then being developed by ACI, American Society of Civil Engineers (ASCE), TMS, and industry organizations. It currently develops and updates the MSJC design provisions and related specification (MSJC 2002a,b).
3. *American Society of Civil Engineers (ASCE).* ASCE is a joint sponsor of many ACI technical committees dealing with concrete or masonry. ASCE is the second of the three sponsoring societies of the MSJC (see above). ASCE publishes *ASCE 7-02* (2002), which prescribes design loadings and load factors for all structures, independent of material type.
4. *The Masonry Society (TMS).* Through its technical committees, this group influences different aspects of masonry design. TMS is the third of the three sponsoring societies of the MSJC (see above). TMS publishes a *Masonry Designers' Guide* to accompany the MSJC design provisions.

6.4.1.2 Industry Organizations

1. *Portland Cement Association (PCA).* This marketing and technical support organization is composed of cement producers. Its technical staff participates in technical committee work.
2. *National Concrete Masonry Association (NCMA).* This marketing and technical support organization is composed of producers of concrete masonry units. Its technical staff participates in technical committee work and also produces technical bulletins that can influence consensus design provisions.
3. *Brick Industry Association (BIA).* This marketing, distributing, and technical support organization consists of clay brick and tile producers. Its technical staff participates in technical committee work and also produces technical bulletins that can influence consensus design provisions.
4. *National Lime Association (NLA).* This marketing and technical support organization consists of hydrated lime producers. Its technical staff participates in technical committee work.
5. *Expanded Clay, Shale and Slate Institute (ECSSI).* This marketing and technical support organization consists of producers. Its technical staff participates in technical committee meetings.
6. *International Masonry Institute (IMI).* This is a labor-management collaborative supported by dues from union masons. Its technical staff participates in technical committee meetings.
7. *Mason Contractors' Association of America (MCAA).* This organization consists of nonunion mason contractors. Its technical staff participates in technical committee meetings.

6.4.1.3 Governmental Organizations

1. *Federal Emergency Management Agency (FEMA).* FEMA has jurisdiction over the National Earthquake Hazard Reduction Program (NEHRP) and develops and periodically updates the NEHRP provisions (NEHRP 2000), a set of recommendations for earthquake-resistant design. This document includes provisions for masonry design. The document is published by the Building Seismic Safety Council (BSSC), which operates under a contract with FEMA. BSSC is not an ANSI consensus organization. Its recommended design provisions are intended for consideration and possible adoption by consensus organizations. The 2000 NEHRP *Recommended Provisions* (NEHRP 2000) is the latest of a series of such documents, now issued at 3-year intervals

and pioneered by *ATC 3-06*, which was issued by the ATC in 1978 under contract to the National Bureau of Standards. The 2000 NEHRP *Recommended Provisions* addresses the broad issue of seismic regulations for buildings. It contains chapters dealing with the determination of seismic loadings on structures and with the design of masonry structures for those loadings.

6.4.1.4 Model-Code Organizations

Model-code organizations consist primarily of building officials, although designers, contractors, product suppliers, code developers, and end users can also be members. Their income is derived from dues and the sale of publications. The United States has three model-code organizations:

1. *International Conference of Building Officials (ICBO)*. In the past, this group developed and published the UBC. The latest and final edition is the 1997 UBC.
2. *Southern Building Code Congress (SBCC)*. In the past, this group developed and published the *Standard Building Code* (SBC). The latest and final edition is the 1999 SBC.
3. *Building Officials/Code Administrators (BOCA)*. In the past, this group developed and published the *Basic Building Code* (BBC). The latest and final edition is the 1999 BBC.

In the past, certain model codes were used more in certain areas of the country. The UBC has been used throughout the western United States and in the state of Indiana. The SBC has been used in the southern part of the United States. The BBC has been used in the eastern and northeastern United States.

Since 1996, intensive efforts have been under way in the United States to harmonize the three model building codes. The primary harmonized model building code is called the IBC. It has been developed by the International Code Council (ICC), consisting primarily of building code officials of the three model-code organizations. The first edition of the IBC (2000) was published in May 2000. In most cases, it references consensus design provisions and specifications. It is intended to take effect when adopted by local jurisdictions. It is intended to replace the three current model building codes. Although not all details have been worked out, it is generally understood that the IBC will continue to be administered by the three model-code agencies. Another harmonized model code is being developed by the National Fire Protection Association, consisting primarily of fire-protection officials.

6.4.2 Masonry Design Provisions of Modern Model Codes in the United States

Over the next 3 to 5 years, as the IBC and other harmonized model codes such as that of the National Fire Protection Association are adopted by local jurisdictions, their provisions will become minimum legal design requirements for masonry structures throughout the United States. The 2003 IBC will reference the masonry design provisions of the 2002 MSJC code in essentially direct form.

6.4.2.1 Strength-Design Provisions of the 2002 MSJC Code

Strength-design provisions of the 2002 MSJC code deal with unreinforced as well as reinforced masonry.

Strength-design provisions for unreinforced masonry address failure in flexural tension, in combined flexural and axial compression, and in shear. Nominal capacity in flexural tension is computed as the product of the masonry's flexural tensile strength and its section modulus. Nominal capacity in combined flexural and axial compression is computed using a triangular stress block, with an assumed failure stress of 0.80 f'_m (the specified compressive strength of the masonry). Nominal capacity in shear is computed as the least of several values, corresponding to different possible failure modes (diagonal tension, crushing of compression diagonal, sliding on bed joints). Each nominal capacity is multiplied by a capacity reduction factor.

Strength-design provisions for reinforced masonry address failure under combinations of flexure and axial loads, and in shear. Nominal capacity under combinations of flexure and axial loads is computed using a moment–axial force interaction diagram, determined assuming elasto-plastic behavior of tensile reinforcement and using an equivalent rectangular stress block for the masonry. The diagram also has an upper limit on pure compressive capacity. Shear capacity is computed as the summation of capacity from masonry plus capacity from shear reinforcement, similar to reinforced concrete. Nominal capacity of masonry in shear is computed as the least of several values, corresponding to different possible failure modes (diagonal tension, crushing of compression diagonal, sliding on bed joints). Each nominal capacity is multiplied by a capacity reduction factor.

Strength-design provisions of the 2002 MSJC impose strict upper limits on reinforcement, which are equivalent to requiring that the element remain below the balanced axial load. Nominal flexural and axial strengths are computed neglecting the tensile strength of masonry, using a linear strain variation over the depth of the cross section, a maximum usable strain of 0.0035 for clay masonry and 0.0025 for concrete masonry, and an equivalent rectangular compressive stress block in masonry with a stress of $0.80 \, f'_m$. These provisions are regarded as too conservative by many elements of the masonry technical community and are currently undergoing extensive review.

6.4.2.2 Allowable-Stress Design Provisions of the 2002 MSJC Code

The allowable-stress provisions of the 2002 MSJC Code are based on linear elastic theory.

Allowable-stress design provisions for unreinforced masonry address failure in flexural tension, in combined flexural and axial compression, and in shear. Flexural tensile stresses are computed elastically, using an uncracked section, and are compared with allowable flexural tensile stresses, which are observed strengths divided by a factor of safety of about 2.5. Allowable flexural tensile stresses for in-plane bending are zero. Flexural and axial compressive stresses are also computed elastically and are compared with allowable values using a so-called "unity equation." Axial stresses divided by allowable axial stresses, plus flexural compressive stresses divided by allowable flexural compressive stresses, must not exceed unity. Allowable axial stresses are one quarter of the specified compressive strength of the masonry, reduced for slenderness effects. Allowable flexural compressive stresses are one third of the specified compressive strength of the masonry. Shear stresses are computed elastically, assuming a parabolic distribution of shear stress based on beam theory. Allowable stresses in shear are computed corresponding to different possible failure modes (diagonal tension, crushing of compression diagonal, sliding on bed joints). The factor of safety for each failure mode is at least 3.

Allowable-stress design provisions for reinforced masonry address failure in combined flexural and axial compression, and in shear. Stresses in masonry and reinforcement are computed using a cracked transformed section. Allowable tensile stresses in deformed reinforcement are the specified yield strength divided by a safety factor of 2.5. Allowable flexural compressive stresses are one third of the specified compressive strength of the masonry. Allowable capacities of sections under combinations of flexure and axial force can be expressed using an allowable-stress moment–axial force interaction diagram, which also has a maximum allowable axial capacity as governed by compressive axial stress. Shear stresses are computed elastically, assuming a uniform distribution of shear stress. Allowable shear stresses in masonry are computed corresponding to different possible failure modes (diagonal tension, crushing of compression diagonal, sliding on bed joints). If those allowable stresses are exceeded, all shear must be resisted by shear reinforcement, and shear stresses in masonry must not exceed a second, higher set of allowable values. The factor of safety for shear is at least 3.

6.4.3 Seismic Design Provisions for Masonry in the 2000 International Building Code

In contrast to wind loads, which are applied forces, earthquake loading derives fundamentally from imposed ground displacements. Although inertial forces are applied to a building as a result, these inertial forces depend on the mass, stiffness, and strength of the building as well as the characteristics of the ground motion itself. This is also true for masonry buildings. In this section, the seismic design provisions of the 2000 IBC are summarized as they apply to masonry elements.

Modern model codes in the United States address the design of masonry for earthquake loads by first prescribing seismic design loads in terms of the building's geographic location, its function, and its underlying soil characteristics. These three characteristics together determine the building's "seismic design category." Seismic Design Category A corresponds to a low level of ground shaking, typical use, and typical underlying soil. Increasing levels of ground shaking, an essential facility, and unknown or undesirable soil types correspond to higher seismic design categories, with Seismic Design Category F being the highest. In addition to being designed for the seismic forces corresponding to their seismic design category, masonry buildings must comply with four types of prescriptive requirements, whose severity increases as the building's seismic design category increases from A to F:

1. Seismic-related restrictions on materials
2. Seismic-related restrictions on design methods
3. Seismic-related requirements for connectors
4. Seismic-related requirements for locations and minimum percentages of reinforcement

The prescriptive requirements are incremental; for example, a building in Seismic Design Category C must also comply with prescriptive requirements for buildings in Seismic Design Categories A and B.

6.4.3.1 Determination of Seismic Design Forces

For structural systems of masonry, as for other materials, seismic design forces are determined based on the structure's location, underlying soil type, and degree of structural redundancy, and the system's expected inelastic deformation capacity. The last characteristic is described indirectly in terms of shear wall types: ordinary plain, ordinary reinforced, detailed plain, intermediate reinforced, and special reinforced. As listed, these types are considered to have increasing inelastic deformation capacity, and a correspondingly increasing response modification coefficient, R, is applied to the structural systems comprising them. Higher values of R correspond, in turn, to lower seismic design forces.

6.4.3.2 Seismic-Related Restrictions on Materials

In Seismic Design Categories A through C, no additional seismic-related restrictions apply beyond those related to design in general. In Seismic Design Categories D and E, Type N mortar and masonry cement are prohibited, due to their relatively low tensile bond strength.

6.4.3.3 Seismic-Related Restrictions on Design Methods

In Seismic Design Category A, masonry structural systems can be designed by strength design, allowable-stress design, or empirical design. They are permitted to be designed including the flexural tensile strength of masonry.

In Seismic Design Category B, elements that are part of the lateral-force-resisting system can be designed by strength design or allowable-stress design only, but not by empirical design. Elements not part of the lateral-force-resisting system, however, can still be designed by empirical design.

No additional seismic-related restrictions on design methods apply in Seismic Design Category C. In Seismic Design Category D, elements that are part of the lateral-force-resisting system must be designed as reinforced, by either strength design or allowable-stress design.

No additional seismic-related restrictions on design methods apply in Seismic Design Categories E and F.

6.4.3.4 Seismic-Related Requirements for Connectors

In Seismic Design Category A, masonry walls are required to be anchored to the roof and floors that support them laterally. This provision is not intended to require a mechanical connection.

No additional seismic-related restrictions on connection forces apply in Seismic Design Category B.

In Seismic Design Category C, connectors for masonry partition walls must be designed to accommodate story drift. Horizontal elements and masonry shear walls must be connected by connectors capable of resisting the forces between those elements, and minimum connector capacity and maximum spacing are also specified.

No additional seismic-related requirements for connectors apply in Seismic Design Categories D and higher.

6.4.3.5 Seismic-Related Requirements for Locations and Minimum Percentages of Reinforcement

In Seismic Design Categories A and B, there are no seismic-related requirements for locations and minimum percentages of reinforcement.

In Seismic Design Category C, masonry partition walls must have reinforcement meeting requirements for minimum percentage and maximum spacing. The reinforcement is not required to be placed parallel to the direction in which the element spans. Masonry walls must have reinforcement with an area of at least 0.2 in.^{2} (129 mm^2) at corners, close to each side of openings, movement joints, and ends of walls, and spaced no farther than 10 ft (3 m) apart.

In Seismic Design Category D, masonry walls that are part of the lateral-force-resisting system must have uniformly distributed reinforcement in the horizontal and vertical directions, with a minimum percentage in each direction of 0.0007 and a minimum summation (in both directions) of 0.002. Maximum spacing in either direction is 48 in. (1.22 m). Closer maximum spacing requirements apply for stack bond masonry. Masonry shear walls have additional requirements for minimum vertical reinforcement and hooking of horizontal shear reinforcement.

In Seismic Design Categories E and F, stack-bonded masonry partition walls have minimum horizontal reinforcement requirements. Stack-bonded masonry walls that are part of the lateral-force-resisting system have additional requirements for spacing and percentage of horizontal reinforcement. Masonry shear walls must be "special reinforced."

6.4.4 Future of Design Codes for Masonry in the United States

The next decade is likely to witness increased harmonization of U.S. model codes, and increasing direct reference to the MSJC Code and Specification. In the MSJC design provisions, the Specification is likely to be augmented by a Code chapter dealing with construction requirements.

6.5 Seismic Retrofitting of Historical Masonry in the United States

The preceding sections of this chapter have dealt primarily with the seismic performance of masonry built under the reinforced masonry provisions that were introduced in the western United States as part of the reaction to the 1933 Long Beach Earthquake. Such masonry generally behaves well. Unreinforced masonry, in contrast, often collapses or experiences heavy damage. This is true whether the masonry was built in the western United States prior to 1933 or in other places either before or after 1933. As a result of this observed poor behavior, recent decades have witnessed significant interest in the seismic retrofitting of historical masonry in the United States. In this section, those retrofitting efforts are briefly reviewed.

6.5.1 Observed Seismic Performance of Historical U.S. Masonry

As noted earlier in this chapter, unreinforced masonry buildings have performed poorly in many U.S. earthquakes. The 1933 Long Beach, California, earthquake severely damaged many unreinforced masonry buildings, particularly schools. One consequence of this damage was the passage of California's Field Act, which prohibited the use of masonry (as it was then used) in all public buildings in the state. When masonry construction was revived in California during the mid-1940s, it was required to comply with newly developed UBC provisions. Those provisions, based on the reinforced concrete design practice of the time, required that minimum seismic lateral forces be considered in the design of masonry buildings, that tensile stresses in masonry be resisted by reinforcement, and that all masonry have at least a minimum percentage of horizontal and vertical reinforcement.

6.5.2 Laboratory Performance of Historical U.S. Masonry

Although laboratory testing of historical U.S. masonry is difficult, some testing has been carried out on masonry specimens cut from existing structures. Much more extensive testing has been carried out on reduced-scale replicas of unreinforced masonry construction. Some of the testing is typical of testing in the United States; some is typical of that found in other countries. That testing includes the following:

1. *Benedetti et al. (1998).* Twenty-four, simple, 2-story, half-scale, unreinforced masonry buildings were tested to varying degrees of damage, repaired and strengthened, and tested again.
2. *Tomazevic and Weiss (1994).* Two, 3-story, reduced-scale, plain and reinforced masonry buildings were tested. The efficacy of reinforcement was confirmed.
3. *Costley and Abrams (1996).* Two reduced-scale brick masonry buildings were tested on a shaking table. Observed and predicted behavior were compared and used to make recommendations for the use of FEMA retrofitting guidelines.

6.5.3 Basic Principles of Masonry Retrofitting

Over the past 15 years, efforts have focused on the seismic response and retrofitting of existing URM buildings. The goals of seismic retrofitting are

1. To correct deficiencies in the overall structural concept.
2. To correct deficiencies in behavior of structural elements.
3. To correct deficiencies in behavior of nonstructural elements.

The most basic elements of seismic retrofitting involve bracing parapets to roofs and connecting floor diaphragms to walls using through anchors (mechanical, grouted, or adhesive).

6.5.4 History of URM Retrofitting in the Los Angeles Area

6.5.4.1 Division 88

In 1949 the city of Los Angeles passed the Parapet Correction Ordinance, which required that unreinforced masonry or concrete parapets above exits, and parapets above public access, be retrofitted to minimize hazards. As a result, such parapets were either laterally braced or removed. Consequently, many unreinforced masonry buildings withstood the 1971 San Fernando earthquake better than previous earthquakes (Lew et al. 1971).

Following the February 1971 San Fernando Earthquake, the city of Los Angeles, the Federal Government, and the Structural Engineers Association of Southern California joined forces in a 10-year investigation. As a result of this investigation, Los Angeles adopted an ordinance known as Division 68 on February 13, 1981. Division 68 required seismic retrofitting of all unreinforced masonry bearing-wall buildings that were built, that were under construction, or for which a permit had been issued prior to October 6, 1933. The ordinance did not include one- or two-family dwellings or detached apartment houses comprising fewer than five dwelling units and used solely for residential purposes.

The 1985 edition of the Los Angeles Building Code revised Division 68 to Division 88 and included provisions for the testing and strengthening of mortar joints to meet minimum values for shear strength.

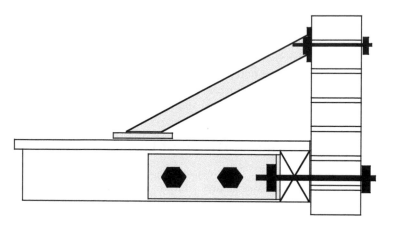

FIGURE 6.5 Schematic of Division 88 URM retrofitting techniques for URM.

Furthermore, Division 88 required that unreinforced masonry be positively anchored to floor and roof diaphragms with anchors spaced not more than 6 ft apart. There were also parapet height limitations, based on wall thickness. Continuous inspection was also required on the retrofitting work. These retrofitting measures are shown schematically in Figure 6.5.

Alternatives to these specific provisions were also possible. Division 88 was renamed Chapter 88 in the 1988 City of Los Angeles Code. In addition to masonry bearing walls, veneer walls constructed before October 6, 1933, were included. This edition also added Section 8811 ("Design Check — Compatibility of Roof Diaphragm Stiffness to Unreinforced Masonry Wall Out-of-Plane Stability").

At the time of the Northridge Earthquake, it is believed that essentially all URM buildings in the city of Los Angeles had their parapets either removed or laterally braced. Unconfirmed reports indicate that in the city of Los Angeles, about 80% of URM buildings had been retrofitted to comply with Division 88; however, the percentage was reported to be considerably lower in other cities in the Los Angeles area.

6.5.4.2 Other Retrofitting Guidelines

The NEHRP, in conjunction with the FEMA, has produced a series of documents dealing with the seismic evaluation and retrofitting of structures, including masonry structures:

1. FEMA 172 (1992), *Handbook of Techniques for the Seismic Rehabilitation of Existing Buildings*, provides a general list of retrofitting techniques.
2. FEMA 178 (1992) presents an overall method for engineers to identify buildings or building components that present unacceptable risks in case of an earthquake.
3. FEMA 273 (1997) and 274 (1997), *NEHRP Guidelines and Commentary* for the seismic rehabilitation of buildings, provide code-type procedures for the assessment, evaluation, analysis, and rehabilitation of existing building structures.
4. FEMA 306 (1998), *Evaluation of Earthquake-Damaged Concrete and Masonry Wall Buildings, Basic Procedures Manual*, provides guidance on evaluation of damage and on performance analysis and includes newly formulated Component Damage Classification Guides and Test and Investigation Guides. The procedures characterize the observed damage caused by an earthquake in terms of the loss in building performance capability.
5. FEMA 307 (1998), *Evaluation of Earthquake-Damaged Concrete and Masonry Wall Buildings, Technical Resources*, contains supplemental information, including results from a theoretical

analysis of the effects of prior damage on single-degree-of-freedom mathematical models, additional background information on the Component Damage Classification Guides, and an example of the application of the basic procedures.

6. FEMA 308 (1998), *The Repair of Earthquake-Damaged Concrete and Masonry Wall Buildings*, discusses the technical and policy issues pertaining to the repair of earthquake-damaged buildings and includes guidance on the specification of individual repair techniques and newly formulated Repair Guides.

7. FEMA 356 (2000), *Prestandard and Commentary for the Seismic Rehabilitation of Buildings*, is an attempt to encourage the use of FEMA 273 and to put the guidelines of that document into mandatory language.

6.6 Future Challenges

This chapter has presented an overview of current issues in the design of masonry for earthquake loads and of the historical process by which those issues have developed. It would not be complete without at least a brief mention of the challenges facing the masonry technical community in this area.

6.6.1 Performance-Based Seismic Design of Masonry Structures

Across the entire spectrum of construction materials, increased attention has been focused on performance-based seismic design, which can be defined as design whose objective is a structure that can satisfy different performance objectives under increasing levels of probable seismic excitation. For example, a structure might be designed to remain operational under a design earthquake with a relatively short recurrence interval, to be capable of immediate occupancy under a design earthquake with a longer recurrence interval, to ensure life safety under a design earthquake with a still longer recurrence interval, and to not collapse under a design earthquake with a long recurrence interval.

This design approach, accepted qualitatively since the 1970s, has been adopted quantitatively in recent documents related to seismic rehabilitation (FEMA 356 2000), and will probably be incorporated into future seismic design provisions for new structures as well. Because masonry structures are inherently composed of walls rather than frames, they tend to be laterally stiff, which is usually a useful characteristic in meeting performance objectives for seismic response.

6.6.2 Increased Consistency of Masonry Design Provisions

The 2003 IBC will reference the 2002 MSJC Code and Commentary essentially in its entirety. Other model codes will probably do the same. As a result, future development of seismic design provisions for masonry structures will take place almost exclusively within the MSJC, rather than in a number of different technical forums. This is expected to lead to increased rationality of design provisions, increased consistency among designs produced by different design methods (e.g., strength versus allowable-stress design), and possibly also simplification of design provisions. As an additional benefit, the emergence of a single set of ANSI-consensus design provisions for masonry is expected to encourage the production of computer-based tools for the analysis and design of masonry structures using those provisions.

References

Abrams, D. and Paulson, T. (1991). "Modeling Earthquake Response of Concrete Masonry Building Structures," *ACI Struct. J.*, 88(4), 475–485.

ACI SP-127 (1991). *Earthquake-Resistant Concrete Structures: Inelastic Response and Design (Special Publication SP-127)*, S.K. Ghosh, editor, 479–503.

ACI 318-02 (2002). ACI Committee 318, *Building Code Requirements for Reinforced Concrete (ACI 318-99)*, American Concrete Institute, Farmington Hills, MI.

ASCE 7-02 (2002). *Minimum Design Loads for Buildings and Other Structures (ASCE 7-02)*, American Society of Civil Engineers.

ATC 3-06 (1978). *Tentative Provisions for the Development of Seismic Regulations for Buildings (ATC 3-06)*, Applied Technology Council, National Bureau of Standards.

BBC (1999). *Basic Building Code*, Building Officials/Code Administrators International.

Benedetti, D., Carydis, P., and Pezzoli, P. (1998). "Shaking-Table Tests on 24 Simple Masonry Buildings," *Earthquake Eng. Struct. Dyn.*, 27(1), 67–90.

Binder, R.W. (1952). "Engineering Aspects of the 1933 Long Beach Earthquake," *Proceedings of the Symposium on Earthquake Blast Effects on Structures*, Berkeley, CA, 186–211.

Blondet, M. and Mayes, R.L. (1991). *The Transverse Response of Clay Masonry Walls Subjected to Strong Motion Earthquakes — Volume 1: General Information*, U.S.–Japan Coordinated Program for Masonry Building Research, TCCMAR Report No. 3.2(b)-2.

Cohen, G.L. (2001). "Seismic Response of Low-Rise Masonry Buildings with Flexible Roof Diaphragms," M.S. Thesis, The University of Texas at Austin, TX.

Costley, A.C. and Abrams, D.P. (1996). "Response of Building Systems with Rocking Piers and Flexible Diaphragms," *Proceedings, Structures Congress*, American Society of Civil Engineers, Chicago, IL, April 15–18, 1996, 135–140.

FEMA 172 (1992). *Handbook of Techniques for the Seismic Rehabilitation of Existing Buildings*, Building Seismic Safety Council.

FEMA 178 (1992). *NEHRP Handbook for the Seismic Evaluation of Existing Buildings*, Building Seismic Safety Council.

FEMA 273 (1997). *NEHRP Guidelines for the Seismic Rehabilitation of Buildings*, Building Seismic Safety Council.

FEMA 274 (1997). *NEHRP Commentary on the Guidelines for the Seismic Rehabilitation of Buildings*, Building Seismic Safety Council.

FEMA 306 (1998). *Evaluation of Earthquake-Damaged Concrete and Masonry Wall Buildings: Basic Procedures Manual*, Federal Emergency Management Agency.

FEMA 307 (1998). *Evaluation of Earthquake-Damaged Concrete and Masonry Wall Buildings: Technical Resources*, Federal Emergency Management Agency.

FEMA 308 (1998). *Repair of Earthquake-Damaged Concrete and Masonry Wall Buildings*, Federal Emergency Management Agency.

FEMA 356 (2000). *Prestandard and Commentary for the Seismic Rehabilitation of Buildings*, Federal Emergency Management Agency.

Gulkan, P., Clough, R.W., Mayes, R.L., and Manos, G. (1990a). "Seismic Testing of Single-Story Masonry Houses, Part I," *J. Struct. Eng.*, American Society of Civil Engineers, 116(1), 235–256.

Gulkan, P., Clough, R.W., Mayes, R.L., and Manos, G. (1990b). "Seismic Testing of Single-Story Masonry Houses, Part II," *J. Struct. Eng.*, American Society of Civil Engineers, 116(1), 257–274.

Hamid, A., Abboud, B., Farah, M., Hatem, K., and Harris, H. (1989). *Response of Reinforced Block Masonry Walls to Out-of-plane State Loads*, U.S.–Japan Coordinated Program for Masonry Building Research, TCCMAR Report No. 3.2(a)-1.

He, L. and Priestley, M.J.N. (1992). *Seismic Behavior of Flanged Masonry Shear Walls* U.S.–Japan Coordinated Program for Masonry Building Research, TCCMAR Report No. 4.1-2.

IBC (2000). *International Building Code*, International Code Council.

Leiva, G. and Klingner, R.E. (1994). "Behavior and Design of Multi-Story Masonry Walls under In-Plane Seismic Loading," *Masonry Soc. J.*, 13(1), 15–24.

Lew, H.S., Leyendecker, E.V., and Dikkers, R.D. (1971). *Engineering Aspects of the 1971 San Fernando Earthquake*, Building Science Series 40, United States Department of Commerce, National Bureau of Standards, 412 pp.

MSJC (2002a). *Building Code Requirements for Masonry Structures (ACI 530-02/ASCE 5-02/ TMS402-02)*, American Concrete Institute, American Society of Civil Engineers, and The Masonry Society.

MSJC (2002b). *Specifications for Masonry Structures (ACI 530.1-02/ASCE 6-02/TMS 602-02)*, American Concrete Institute, American Society of Civil Engineers, and The Masonry Society.

NAMC (1985). *Proceedings of the Third North American Masonry Conference*, University of Texas, Arlington, TX, June 3–5, 1985, The Masonry Society, Boulder, CO.

NAMC (1987). *Proceedings of the Fourth North American Masonry Conference*, University of California, Los Angeles, CA, August 16–19, 1987, The Masonry Society, Boulder, CO.

NAMC (1990). *Proceedings of the Fifth North American Masonry Conference*, University of Illinois, Urbana-Champaign, IL, June 3–6, 1990, The Masonry Society, Boulder, CO.

NAMC (1993). *Proceedings of the Sixth North American Masonry Conference*, Drexel University, Philadelphia, PA, June 7–9, 1993, The Masonry Society, Boulder, CO.

NAMC (1996). *Proceedings of the Seventh North American Masonry Conference*, University of Notre Dame, Notre Dame, IN, June 2–5, 1996, The Masonry Society, Boulder, CO.

NAMC (1999). *Proceedings of the Eighth North American Masonry Conference*, University of Texas at Austin, Austin, TX, June 6–9, 1999, The Masonry Society, Boulder, CO.

NEHRP (1997). *NEHRP (National Earthquake Hazards Reduction Program) Recommended Provisions for the Development of Seismic Regulations for New Buildings (FEMA 222)*, Building Seismic Safety Council.

NEHRP (2000). *NEHRP (National Earthquake Hazards Reduction Program) Recommended Provisions for the Development of Seismic Regulations for New Buildings (FEMA 368)*, Building Seismic Safety Council.

Noland, J.L. (1990). "1990 Status Report: US Coordinated Program for Masonry Building Research," *Proceedings, Fifth North American Masonry Conference*, University of Illinois at Urbana-Champaign, IL, June 3–6, 1990.

SBC (1999). *Standard Building Code*, Southern Building Code Congress International.

Seible, F., Hegemier, A., Igarashi, A., and Kingsley, G. (1994a). "Simulated Seismic-Load Tests on Full-Scale Five-Story Masonry Building," *J. Struct. Eng.*, American Society of Civil Engineers, 120(3), 903–924.

Seible, F., Priestley, N., Kingsley, G., and Kurkchubashe, A. (1994b). "Seismic Response of Full-Scale Five-Story Reinforced-Masonry Building," *J. Struct. Eng.*, American Society of Civil Engineers, 120(3), 925–947.

Suprenant, B.A. (1989). "A Floor a Week per Tower," *Masonry Constr.*, 2(11), 478–482.

TMS (1989). "The Masonry Society," *Proceedings of an International Seminar on Evaluating, Strengthening, and Retrofitting Masonry Buildings*, Construction Research Center, The University of Texas at Arlington, TX, October 1989.

TMS Northridge (1994). Klingner, R.E., editor, *Performance of Masonry Structures in the Northridge, California Earthquake of January 17, 1994*, Technical Report 301-94, The Masonry Society, Boulder, CO, 100 pp.

Tomazevic, M. and Weiss, P. (1994). "Seismic Behavior of Plain- and Reinforced-Masonry Buildings," *J. Struct. Eng.*, American Society of Civil Engineers, 120(2), 323–338.

7

Timber Structures

J. Daniel Dolan
Department of Civil and Environmental Engineering, Washington State University, Pullman, WA

7.1 Introduction

Stone and wood were the first materials used by man to build shelter, and in the United States wood continues to be the primary construction material for residential and commercial buildings today. In California, for example, wood accounts for 99% of residential buildings (Schierle 2000). Design and construction methods for wood currently used by the residential construction industry in North America have developed through a process of evolution and tradition. Historically, these construction methods have been sufficient to provide acceptable performance under seismic loading mainly due to the relatively light weight of wood and the historical high redundancy in single-family housing. However, in recent years architectural trends and society's demands for larger rooms, larger windows, and a more open, airy feel to the structure have resulted in a reduction in the structural redundancy of the typical house, as well as a reduction in symmetry of stiffness and strength that was inherent in traditional structures.

If one were to review the type of structure that was built in the 1930s, 1940s, and even into the 1950s, one would realize that the average house had a pedestrian door, small double-hung windows, and relatively small rooms with lots of walls. If one compares this typical construction to buildings that are being built in the early 2000s, newer buildings have large windows, if not four walls of primarily glass, large great rooms, and often a single room that pierces the first-story ceiling, becoming two stories high and causing a torsional irregularity in the second-floor structure. If one then includes multifamily construction, which includes apartments, condominiums, and townhouses, the structures become fairly stiff and strong in one direction while in the orthogonal direction (the side with the windows and the doors to the hallways or patio) they become very weak and flexible, due to the lack of structural wall space. In general, modern structures typically have more torsional irregularities, vertical irregularities such as soft stories, and uneven stiffness and strength in orthogonal directions when compared to traditional buildings. As a result, it was observed in the 1994 Northridge earthquake that "most demolished single-family dwellings and multifamily dwellings (as a percentage of existing buildings) were built 1977–1993" (Schierle 2000). The CUREE-Caltech woodframe project also illustrated the inherent soft-story response of light-frame construction.

An additional need for improved understanding of the material and structure used in modern timber buildings is the continued movement toward performance-based design methods and an increased concern over damage. If one considers that the house is the single largest investment that the average person makes in his or her lifetime, it should be no surprise that concern over accumulated damage due to moderate seismic events has begun to be discussed in the context of model building codes. To support the seismic design of timber structures, the wood industry has sponsored the development of the *Standard for Load and Resistance Factor Design (LRFD) for Engineered Wood Construction* (American Forest and Paper Association 1996; American Society of Civil Engineers 1996). However, most designers continue to use the *National Design Specification (NDS®) for Wood Construction* and its supplements (American Forest and Paper Association 2001) for designing wood structures in North America. The NDS is an allowable stress design methodology, while the LRFD is a strength-based design methodology that is intended to provide a better design for seismic concerns. The NDS has been repackaged into the *ASD* (Allowable Stress Design) *Manual for Engineered Wood Construction* (American Forest and Paper Association 2001).

This section reviews the types of wood products available for use in timber construction, the types of structures that are typically designed and built in North America, the design standards that are available for directing the design process, the industry resources that are available, the performance of wood buildings in recent earthquakes, and some of the restrictions that are placed on wood structures due to issues other than seismic concerns. While the rest of the chapter will focus primarily on strength-based design (i.e., LRFD), the NDS will be referred to from time to time where differences are significant. Since the average design firm continues to use ASD for wood structures, it is important that these differences be highlighted.

7.1.1 Types of Wood-Based Products

There are a wide variety and an increasing number of wood-based products available for use in building construction. While the largest volume of wood-based products includes dimensional lumber, plywood, and oriented strand board (OSB), new composites include structural composite lumber (SCL), I-joists, laminated veneer lumber (LVL), and plastic wood (Figure 7.1). In addition, when large sizes are required, glued–laminated lumber (*glulam*) and SCL, such as Paralam, can be used. The current trend is to move toward increased use of wood-based composites in building construction. This is due to the increased difficulty in obtaining large sizes of timber because of restrictions on logging and changes in the economic structure of manufacturing. Therefore, designers should become familiar with SCL, LVL, and glulam when long spans or heavy loads are anticipated. Many of the new composites can be custom manufactured to the size and strength required for a particular application. Designers must obtain the proprietary technical information required to design structures with most of the new composites from the suppliers of the products.

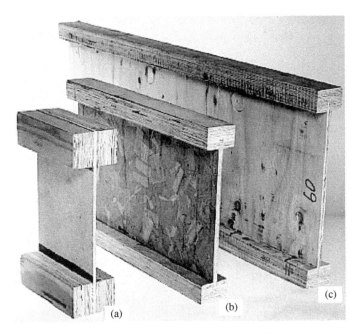

FIGURE 7.1 Prefabricated I-joists with laminated veneer lumber flanges and structural panel webs. One experimental product has (a) a hardboard web; the other two commercial products have (b) oriented strandboard, and (c) plywood webs (courtesy of U.S. Department of Agriculture. 1999. *Wood Handbook: Wood as an Engineering Material*, Agriculture Handbook 72, Forest Products Laboratory, U.S. Department of Agriculture (USDA), Madison, WI).

7.1.2 Types of Structures

Timber structures can be classified into two general categories:

1. *Heavy timber* construction includes buildings such as sports arenas, gymnasiums, concert halls, museums, office buildings, and parking garages. These heavier structures are usually designed to resist higher levels of loading and are therefore designed with the intent of using large section timbers that typically require the use of glulam timber, LVL, structural composite lumber, or similar products. These types of structures require a fairly high level of engineering to ensure the safe performance of the structure due to the lower redundancy of the structure.

2. *Light-frame* construction is by far the largest volume of timber construction in North America. These types of buildings include one- and two-family dwellings (Figure 7.2), apartments, town-houses, hotels, and other light-commercial buildings. These types of structures are highly redundant and indeterminate, and there are currently no computer analysis tools that provide a detailed analysis of these structures.

Light-frame construction consists of 2-in. nominal dimension lumber that ranges in size from 2×4 to 2×12. While the design specifications have geometric parameters that include 2×14, the availability of such large sizes is questionable. The lattice of framing comprising $2\times$ dimensional lumber is then typically sheathed with panel products that include plywood, OSB, fiberboard, gypsum, stucco, or other insulation-type products (Figure 7.2). This system of light-framing sheathed with load-distributing elements then acts to transmit the load horizontally through the roof and floors and vertically through the walls, constituting the lateral-force-resisting system (Figure 7.3). This results in a highly redundant and indeterminate structure that has a good history of performance in seismic loading. With modern architectural trends, the reduction in redundancy results in an increased need to involve structural engineering to ensure good performance.

Roof/floor span systems
1. Wood joist and rafter
2. Diagonal sheathing
3. Straight sheathing

Wall systems
4. Stud wall (platform or balloon frame)
5. Horizontal siding

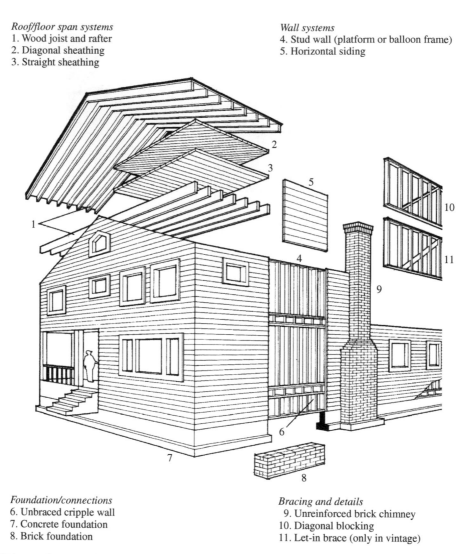

Foundation/connections
6. Unbraced cripple wall
7. Concrete foundation
8. Brick foundation

Bracing and details
 9. Unreinforced brick chimney
10. Diagonal blocking
11. Let-in brace (only in vintage)

FIGURE 7.2 Schematic of wood light-frame construction (courtesy of Federal Emergency Management Agency. 1988. *Rapid Visual Screening of Buildings for Potential Seismic Hazards: A Handbook*, FEMA 154 Federal Emergency Management Agency (FEMA), Washington, DC).

Light-frame construction can be broken into two principal categories: fully designed and prescriptive. Currently, the design requirements for using mechanics-based design methods are in the *International Building Code* (IBC) (International Code Council 2003a) or the National Fire Prevention Association *NFPA 5000 Building Code* (National Fire Prevention Association 2002). Requirements for prescriptive construction are contained in the *International Residential Code* (IRC) (International Code Council 2003b), which is a replacement for the Council of American Building Officials' (CABO) *One- and Two-Family Dwelling Code* (Council of American Building Officials 1995). The IRC allows for a mix between a fully designed and rationalized system and the prescriptive systems. In other words, a building that is primarily designed and constructed according to the prescriptive rules of the IRC can have elements that are rationally designed to eliminate such things as irregularities in the structure due to form. The IRC and IBC have consistent seismic provisions as far as the load determination is concerned.

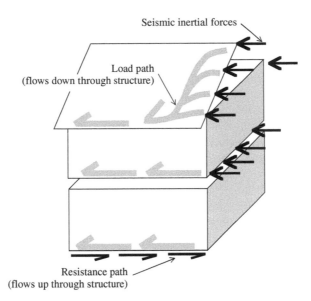

Seismic inertial forces

Load path
(flows down through structure)

Resistance path
(flows up through structure)

FIGURE 7.3 Lateral-force-resisting system: the load is transmitted horizontally through the roof and floors and vertically through the walls.

7.1.3 Design Standards

Design standards can be divided into two categories: the performance requirements, which are typically covered by the building codes, and the required design methodology to provide that performance, which is included in the design standards. In North America, buildings are typically governed by the IBC or NFPA 5000 for engineered systems. The IRC provides the performance requirements and the methodology to provide the required resistance for one- and two-family residential structures. All of the model building codes available in the United States, plus the *ASCE-7: Minimum Design Load for Buildings and Other Structures* (American Society of Civil Engineers 2002) base their seismic design requirements on the National Earthquake Hazard Reduction Program (*NEHRP*) *Recommended Provisions for Seismic Regulations for New Buildings and Other Structures* (Building Seismic Safety Council 2003a).

All model building codes in effect in the United States recognize either the LRFD standard (American Forest and Paper Association 1996; American Society of Civil Engineers 1996) for strength-based design or the NDS (American Forest and Paper Association 2001) for allowable stress design for engineered wood construction. To parallel the format of the LRFD design manual, the NDS has been incorporated into the *ASD Manual for Wood Construction* (American Forest and Paper Association 2001). In this format, all of the wood-based structural products available for use in the design and construction of timber structures are available to the designer. Most of the provisions of the LRFD standard are similar to those used in the NDS or ASD manual. The difference is that the ASD design methodology bases its requirements in terms of working stresses or allowable stresses, while the LRFD manual bases its design values on the nominal strength values. The two design methodologies also use different load combinations. Since the NEHRP provisions require the use of strength-based design methodology, this chapter will focus on LRFD.

In addition to the ASD and LRFD manuals, the designer may utilize industry documents that are available for most of the trade associations associated with the timber industry. Some of these references include the *Timber Construction Manual* (American Institute of Timber Construction 2004), the *Plywood Design Specification* (APA 1997), the *Engineered Wood Construction Guide* (APA 2003), and the *Wood Frame Construction Manual (WFCM) for One- and Two-Family Dwellings* (American Forest and Paper Association 2001). The *Timber Construction Manual* (American Institute of Timber Construction 2004) provides the designer with guidance on the use of glulam construction and associated issues such as heavy timber connectors, notching and drilling of beams, and the design of arches and other curved members. The Plywood

Design Specification and its supplements provide guidance on the use of plywood and other structural panel products, especially for nontypical applications such as plywood box beams, folded plate roofs, and spanning of the product in the weak direction across supports. The *Engineered Wood Construction Guide* is a general document that provides information on how structural panel products are graded and marked, and the design of floor and roof systems, and also provides design values for allowable stress design of shear walls and diaphragms. Additional documents and guidance can be obtained from various industry associations such as the American Forest and Paper Association, APA, the Engineered Wood Association, Canadian Wood Council, Wood Truss Council of America, Truss Plate Institute, Western Wood Products Association, Southern Forest Products Association, and others.

While the design and construction of relatively tall structures are technically achievable, a typical building code restricts the use of light-frame construction as well as heavy timber construction to low-rise buildings. Both light-frame and heavy-timber construction are classified as combustible materials. However, heavy-timber construction can be classified as fire resistant and can be built, according to most building codes, as high as five stories. Light-frame construction is typically limited to four stories. There is, however, a move to allow a higher number of stories for light-frame, provided fire suppression systems, such as sprinklers, are included in the building.

7.2 Wood as a Material

Wood in general can be considered a relatively brittle material from a structural standpoint. Wood is a natural material that is *viscoelastic* and *anisotropic*, but is generally considered and analyzed as an *orthotropic* material. Wood has a cellular form that enhances the *hygroscopic* or affinity to water response. From a structural engineering point of view, wood can be considered as similar to over-reinforced concrete, and one might consider designing a wood building as building with overreinforced concrete members that are connected with ductile connections.

One needs to differentiate between wood and *timber*. In this chapter, wood is considered as small clear specimens that one might idealize as perfect material. On the other hand, timber comes in the sizes that are typically used in construction and has growth characteristics such as knots, slope of grain, splits and checks, and other characteristics due to the conditions under which the tree grew or the *lumber* was manufactured. These growth characteristics contribute to significant differences in performance between wood and timber. They also provide the inherent weakness in the material that the structural engineer or designer needs to be aware of.

Since wood is a natural material and is hygroscopic and viscoelastic, certain environmental end-use conditions affect the long-term performance of the material. Some of these variables include moisture content, dimensional stability, bending strength, stiffness, and load duration. Each of these variables will be dealt with individually.

Moisture content is one of the most important variables that a designer needs to consider when designing timber structures. Any inspection of wood buildings should include testing of the moisture content with standard moisture meters that are available on the commercial market. Moisture content is determined as the weight of water in a given piece of timber divided by the weight of the woody material within that member in an oven-dried state. Moisture content affects virtually all mechanical properties of timber. One can consider wood as being similar to a sponge; as the water is absorbed from the dry state, it enters the walls of the individual cells of the material. At some point these walls become saturated and any additional water is then stored in the lumen of the cell. The point at which the walls become saturated is known as the fiber saturation point. The fiber saturation point varies for different species, and even within a single species, but ranges somewhere between 23 and 35%. The average fiber saturation point is usually assumed to be 30% for most general applications.

Dimensional stability of timber is affected directly by the moisture content. One can assume a linear variation of shrinkage or swelling with changes of moisture content between the oven-dry state and the fiber saturation point. Once the fiber saturation point is reached, any additional increase in moisture content only adds water to the air space of the member and does not affect the dimensional portions of the member. One can estimate the dimensional stability or dimensional changes of an individual wood member using shrinkage coefficients that are available in resources such as the *Wood Handbook* (U.S. Department of Agriculture 1999). However, one can usually assume that shrinkage in the perpendicular-to-grain directions (either radial or tangential) may be significant, while shrinkage parallel-to-grain can be considered as negligible. This is why connections that are restrained by steel plates through significant depths (even as little as 12-in. members) cause the wood to split, because as the wood dries out it will shrink while the steel and the connectors holding the wood to the steel prevent it from moving.

Moisture content changes have effects in addition to simply changing the perpendicular-to-grain dimension. Restrained wood subjected to the cycling of moisture content, can cause distortion of the members. General issues of warping can include *checking* or *splitting* of the lumber due to restraint of the connections, or wood members with grain oriented perpendicular to each other can cause the member to *bow*, *warp*, or *cup*. Cup is often seen in lumber deckboards that are laid with the growth rings on the cross-section oriented such that the center of the tree is facing up. All distortional responses can have adverse effects from a structural standpoint, and many cause serviceability concerns. Due to the effects of dimensional change on the straightness and mechanical properties of timber, it is strongly recommended that designers specify, and require without substitution, the use of dry material for timber structures.

Bending strength of timber is affected by moisture content as well. Bending strength may increase as much as 4% for each 1% decrease in moisture content below the fiber saturation point. While the cross-section of the member will also decrease due to shrinkage, the strengthening effect of drying out is significantly larger than the effects of the reduction in cross-section. Stiffness also increases with a decrease in moisture content, and again, the effect of stiffening of the material is greater than the effect of the reduction in cross-section due to shrinkage. This is why it is recommended that timber structures be constructed with dry material that will remain dry during its service life. The less moisture content change within the material at the time it is constructed and used, the less the shrinkage that occurs, and the maximum bending and stiffness values can be used by the designer.

The final two variables to be considered in this section have to do with viscoelastic properties. Because wood, and therefore timber, is viscoelastic, it tends to creep over time and also has a duration of load effect on strength. *Creep* is the continued increase in deflection that occurs in the viscoelastic material that sustains a constant load. In wood that is dry when installed and remains dry during service, the creep effect may cause a 50% increase in elastic deflection over a period of 1 year. However, if the material is unseasoned at the time of installation and allowed to dry out in service, the deflections can increase by 100% over the elastic response. Finally, if the moisture content is cycled between various moisture contents during its lifetime, the deflections can be increased to as much as 200% over the initial elastic response. A clear difference between wood and timber is that wood is affected by moisture content at all times, while timber is only affected by moisture content in the stronger grade levels. This is why Select Structural grade lumber is affected more than Number 3 grade lumber, and the design manuals have a check for minimum strength before the moisture effect or wet service factor is applied.

Duration load is a variable that is directly accounted for in both the LRFD and ASD manuals. Load duration factor, C_D, is based on experimental results from the 1950s at the U.S. Forest Products Laboratory. This curve is often referred to as the Madison curve. It forms the basis of the provisions in the LRFD and ASD manuals. There is one significant difference in how load duration is dealt with in the two design methodologies. In LRFD, the reference time is set at a duration of between 5 and 10 min. This duration is associated with a C_D of 1.0, with longer durations having factors less than 1. The ASD manual uses a reference duration of 10 years, with an associated value of C_D of 1.0. All values of C_D for load durations shorter than 10 years are given values greater than 1.0 for ASD. The designer is responsible for changing the reference time, depending on which design methodology is used.

Finally, ductility and energy dissipation, which are important concepts for seismic design, are beneficial for most wood structures. Provided the structure's connections are detailed properly, timber structures have relatively high values of ductility and energy dissipation. Provided the connections yield, such as would occur for nail and small-diameter bolted connections, ductilities in the range of 4 to 8 are not uncommon. Energy dissipation, on the other hand, from a material standpoint is very low for wood.

Typically, the material damping that occurs in timber structures is less than 1%. However, the hysteretic energy dissipation of structural assemblies such as shear walls is very high. Equivalent viscous damping values for timber assemblies can range from a low of 15% to a high of more than 45% critical damping.

7.3   Seismic Performance of Wood Buildings

7.3.1   General

As noted above, wood frame structures tend to be mostly low rise (one to three stories, occasionally four stories). The following discussion of the seismic performance of wood buildings is drawn from several sources. In the next few paragraphs, we provide an overview of wood construction and performance drawn from FEMA 154 (1988), followed by a limited discussion of performance in specific earthquakes drawn from various sources.

Vertical framing may be of several types: stud wall, braced post and beam, or timber pole. Stud wall structures ("stick-built") are by far the most common type of wood structure in the United States and are typically constructed of 2-in. by 4-in. nominal wood members vertically set about 16 in. apart. These walls are braced by plywood or by diagonals made of wood or steel. Most detached single and low-rise multiple family residences in the United States are of stud wall wood frame construction. Post-and-beam construction is not very common and is found mostly in older buildings. These buildings usually are not residential, but are larger buildings such as warehouses, churches, and theaters. This type of construction consists of larger rectangular (6 in. by 6 in. and larger) or sometimes round wood columns framed together with large wood beams or trusses.

Stud wall buildings have performed well in past earthquakes due to inherent qualities of the structural system and because they are lightweight and low rise. Cracks in the plaster and stucco (if any) may appear, but these seldom degrade the strength of the building and are therefore classified as nonstructural damage. In fact, this type of damage dissipates a lot of the earthquake-induced energy. The most common type of structural damage in older buildings results from a lack of connection between the superstructure and the foundation. Houses can slide off their foundations if they are not properly bolted to the foundation, resulting in major damage to the building as well as to plumbing and electrical connections. Overturning of the entire structure is usually not a problem because of the low-rise geometry. In many municipalities, modern codes require wood structures to be bolted to their foundations. However, the year that this practice was adopted will differ from community to community and should be checked.

Another problem in older buildings is the stability of cripple walls. Cripple walls are short stud walls between the foundation and the first floor level (Figure 7.4). Often these have no bracing and thus may collapse when subjected to lateral earthquake loading (Figure 7.5). If the cripple walls collapse, the house will sustain considerable damage and may also collapse. This type of construction is generally found in older homes. Plywood sheathing nailed to the cripple studs may have been used to strengthen the cripple walls.

Garages often have a very large door opening in one wall with little or no bracing. This wall has almost no resistance to lateral forces, which is a problem if a heavy load such as a second story sits on top of the garage. Homes built over garages have sustained significant amounts of damage in past earthquakes, with many collapses. Therefore the house-over-garage configuration, which is found commonly in low-rise apartment complexes and some newer suburban detached dwellings, should be examined more carefully and perhaps strengthened.

7.3.2   1971 San Fernando Earthquake, California

The San Fernando Earthquake occurred on February 9, 1971 and measured 6.6 on the Richter scale. The following commentary is excerpted from Yancey et al. (1998):

> There were approximately 300,000 wood-frame dwellings in the San Fernando Valley of which about 5% were located in the region of heaviest shaking (Steinbrugge et al. 1971). A survey of

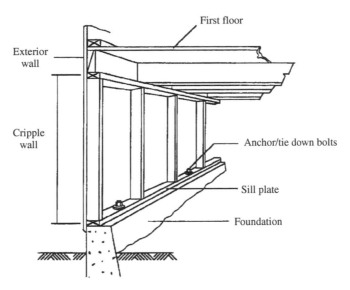

FIGURE 7.4 Cripple wall (courtesy of Benuska, L., Ed. 1990. *Earthquake Spectra*, 6 [Suppl.]).

FIGURE 7.5 Houses damaged due to cripple wall failure, 1983 Coalinga Earthquake (courtesy of EQE International).

12,000 single-family wood-frame houses was conducted by the Pacific Fire Rating Bureau (Steinbrugge et al. 1971). Most of the dwellings were constructed within the two decades prior to the earthquake. Typical types of foundations were either slab on-grade or continuous concrete foundation around the perimeter with concrete piers in the interior, with the former being more common. The majority of the houses were single-story. The survey showed that within the region of most intense shaking, 25% of the wood-frame dwellings sustained losses greater than 5% of the dwelling's value, with the remainder sustaining smaller losses. The number of houses with damage above the 5% threshold is equivalent to 1% of all the wood-frame dwellings in the San Fernando Valley.

7.3.3 1989 Loma Prieta Earthquake, California

The Loma Prieta Earthquake occurred on October 17, 1989, in the San Francisco Bay region and measured 7.1 on the Richter scale (Lew 1990). The following general observation and commentary are excerpted from Yancey et al. (1998):

> Property damage was estimated at over $6 billion and over 12,000 people were displaced from their homes. A survey of the damage to wood-framed structures was conducted by a group of three engineers from the American Plywood Association (APA) (Tissell 1990). Their main findings were
>
> 1. Damage was caused by failure of cripple walls. The failures of cripple walls were the result of inadequate nailing of plywood sheathing. When adequate nailing was provided, no failure was observed.
> 2. Lack of connection between the major framing members and the foundation was the cause of failure of two severely damaged houses.
> 3. Damage caused by soft stories was observed in the Marina District. The phenomenon of soft stories, first observed in this earthquake, results from garage door or large openings on the ground floor of apartment buildings and houses that reduce the lateral resistance of that story. The reduced lateral resistance causes severe racking to occur or increases lateral instability.
> 4. Chimney damage was common. Chimneys were typically unreinforced and not sufficiently tied to the structure.
> 5. Upward ground movements caused doors to be jammed and damage to basement floors.
> 6. Post-supported buildings were damaged because of inadequate connections of the floor to the post foundation and unequal stiffnesses of the posts due to unequal heights. Houses where the poles were diagonally braced were not damaged.

A particularly noteworthy concentration of damage was in the Marina section of San Francisco, where seven 1920s-era three- to five-story apartment buildings collapsed, and many were severely damaged (Figure 7.6). The excessive damage was due in large part to the man-made fill in the Marina, which liquefied and greatly increased ground motion accelerations and displacements during the shaking. However, the primary cause of the collapse was the soft-story nature of the buildings, due to required off-street parking. The buildings lacked adequate lateral-force-resisting systems and literally were a "house of cards."

7.3.4 1994 Northridge Earthquake, California

An earthquake with a magnitude of 6.8 struck the Northridge community in the San Fernando Valley on January 17, 1994. The effects of this earthquake were felt over the entire Los Angeles region. Approximately 65,000 residential buildings were damaged with 50,000 of those being single-family houses (U.S. Department of Housing and Urban Development 1995). The estimated damage based on insurance payouts was over $10 billion (Holmes and Somers 1996) for single- and multifamily residences.

City and county building inspectors estimated that 82% of all structures rendered uninhabitable by the earthquake were residential. Of these, 77% were apartments and condominiums, and the remaining 23% were single-family dwellings. A week after the earthquake, approximately 14,600 dwelling units were deemed uninhabitable (red or yellow tagged). Severe structural damage to residences was found as far away as the Santa Clarita Valley to the north, south-central Los Angeles to the south, Azusa to the east, and eastern Ventura County to the west.

7.3.4.1 Multifamily Dwellings

Particularly vulnerable were low-rise, multistory, wood-frame apartment structures with a soft (very flexible) first story and an absence of plywood shear walls. The soft first-story condition was most apparent in buildings with parking garages at the first-floor level (Figure 7.7). Such buildings, with large, often continuous, openings for parking, did not have enough wall area and strength to withstand the

(a)

(b)

FIGURE 7.6 Collapsed apartment buildings in Marina district of San Francisco, 1989 Loma Prieta Earthquake (courtesy of EQE International).

FIGURE 7.7 Typical soft story "tuck-under" parking, with apartments above (courtesy of EQE International).

earthquake forces. The lack of first-floor stiffness and strength led to collapse of the first floor of many structures throughout the valley. The main reason for failure was the lack of adequate bracing, such as plywood shear walls. Most older wood-frame structures had poor if any seismic designs and resisted lateral forces with stucco, plaster, and gypsum board wall paneling and diagonal let-in bracing.

7.3.4.2 Single-Family Dwellings

Widespread damage to unbolted houses and to older houses with cripple-stud foundations occurred. Newer houses on slab-on-grade foundations were severely damaged because they were inadequately anchored. Two-story houses without any plywood sheathing typically had extensive cracking of interior sheetrock, particularly on the second floor. Nine hillside houses built on stilts in Sherman Oaks collapsed. All but one of the homes were constructed in the 1960s — predating the major building code revisions made after the 1971 San Fernando Earthquake.

7.4 Design Considerations

Design considerations can be divided into essentially three categories:

1. Material choice
2. Performance requirements
3. Resistance determination

   These are governed by the location of the project, the adopted building code regulations, and the design process of choice. Once the choice of designing a wood structure is made, the designer must also consider what types and what grades of wood products are readily available in the region of the project. Both LRFD and ASD include a wide variety of products, product sizes, and grades of products. Local building supply companies rarely stock all of the available products. The economics of building with wood is often the reason for choosing the material for a project, and the choice of products within the broad spectrum of available products that are locally available and stocked greatly improves the economics of a project. If products are specified that must be specially ordered or shipped long distances relative to what is already available in the market, the cost of the project will increase.

   Performance requirements for a given building are usually determined by the building code that is enforced for the jurisdiction. However, there are also the local amendments and local conditions that must be considered when determining the performance requirements of a given project. Resistance determination and sizing of lumber for a given project are governed by the choice of design methodology used. If LRFD is used, one set of load combinations from the building code or ASCE 7 is required. If ASD is used, another set of load combinations should be used.

7.4.1 Building Code Loads and Load Combinations

Building codes that are in effect in the United States reference the ASCE load standard (American Society of Civil Engineers 2002) to provide guidance to the engineer on which load combination is to be used. The 2003 IBC and the NFPA 5000 Building Code both reference the ASCE 7–02 load standard. The following load combinations are to be considered when designing wood structures using the LRFD design methodology:

$$
\begin{aligned}
&1.\ 1.4(D+F)\\
&2.\ 1.2(D+F+T)\ +1.6(L+H)+0.5(L_r\ \text{or}\ S\ \text{or}\ R)\\
&3.\ 1.2D+1.6(L_r\ \text{or}\ S\ \text{or}\ R)+(L\ \text{or}\ 0.8W)\\
&4.\ 1.2D+1.6W+L+0.5(L_r\ \text{or}\ S\ \text{or}\ R)\\
&5.\ 1.2D+1.0E+L+0.2S\\
&6.\ 0.9D+1.6W+1.6H\\
&7.\ 0.9D+1.0E+1.6H
\end{aligned}
\tag{7.1}
$$

where D is the dead load, E is the earthquake load, F is the load due to fluids with well-defined pressures and maximum heights, H is the load due to lateral earth pressure, groundwater pressure, or pressure of bulk materials, L is the live load, L_r is the roof live load, R is the rain load, S is the snow load, T is the self-straining force, and W is the wind load.

If the designer is using ASD, the ASCE 7–02 standard provides different load combinations for use. The following load combinations should be used when designing timber structures following ASD:

1. $D + F$
2. $D + L + F + H + T$
3. $D + H + F + (L_r \text{ or } S \text{ or } R)$
4. $D + H + F + 0.75(L + T) + 0.75(L_r \text{ or } S \text{ or } R)$
5. $D + H + F + (W \text{ or } 0.7E)$
6. $D + H + F + 0.75(W \text{ or } 0.7E) + 0.75L + 0.75(L_r \text{ or } S \text{ or } R)$
7. $0.6D + W + H$
8. $0.6D + 0.7E + H$

$$(7.2)$$

Since components within a timber structure will be stressed to their capacity during a design seismic event, it is strongly recommended that designers use the LRFD format when considering seismic performance. This is the reason why the NEHRP provisions require that LRFD be used.

7.5 Resistance Determination

The *Load and Resistance Factor Design (LRFD) Manual for Engineered Wood Construction* (American Forest and Paper Association 1996) distinguishes between nominal design values for visually and mechanically graded lumber connections and has supplements to provide guidance for all other wood-based products. This document provides both the reference design mechanical property and the applicable adjustment factors to account for end-use and environmental conditions. The reference properties that are included in the document are F_b, bending strength, F_t, tension parallel-to-grain strength, F_s, shear strength, F_c, the compression parallel-to-grain strength, $F_{c\perp}$, the compression strength perpendicular-to-grain, and E, the modulus of elasticity. Two values for modulus of elasticity are typically provided in the supplements: the mean and fifth percentile values. It should be stressed that the resistance values provided in the LRFD manual should not be mixed or combined with values obtained from the ASD manuals. The two design methodologies use different reference conditions on which to base their design, and mixing the values may result in nonconservative designs. Similar resistance values for the mechanical properties of wood-based materials for ASD are provided in the *ASD Manual for Engineered Wood Construction* (American Forest and Paper Association 2001). The significant difference between ASD and LRFD is that ASD bases the design values for a normal duration of load of 10 years, while LRFD resistance values are based on a 5- to 20-min load duration. Since wood is a viscoelastic material, the duration of load has a significant effect on the strength and deflections of timber structures. Two exceptions for adjustment of mechanical properties for load duration are modulus elasticity and compression perpendicular-to-grain. Both values are based on mean mechanical properties and are not adjusted for load duration since compression perpendicular-to-grain is not associated with fracture-type failure and deflection is considered a serviceability rather than a safety criterion.

Since wood is a natural material that is affected by environmental conditions and has characteristics inherent to the material that cause size or volume effects, several adjustments are made to the design values to account for these variables. Reference strength is adjusted with factors that include the time effect factor (load duration factor), wet service factor, temperature factor, instability factor, size factor, volume factor, flat use factor, incising factor, repetitive member factor, curvature factor, form factor, calm stability factor, shear stress factor, buckling stiffness factor, and bearing area factor. While many of these variables affect virtually all members of design, most factors only affect special situations. Therefore, the more commonly used factors will be covered here, and the reader is directed to either the LRFD or ASD manuals for a full description of the less used factors. In addition to these, the LRFD specification includes adjustment factors for preservative treatments and fire retardant treatments. The values that should be used for these last two variables should be obtained from the supplier of the products, since each product used is proprietary in nature and each treatment affects the performance of the timber differently.

Values of the adjustments are all equal to 1.0 unless the application does not meet the reference conditions. Both LRFD and NDS use the following reference conditions as the basis:

1. Materials are installed having a maximum equilibrium moisture content (EMC) not exceeding 19% for solid wood and 16% for glued products.

2. Materials are new (not recycled or reused).
3. Members are assumed to be single members (not in a structural system, such as a wall or a floor).
4. Materials are untreated (except for poles or piles).
5. The continuous ambient temperature is not higher than 100°F, with occasional temperatures as high as 150°F. (If sustained temperatures are between 100°F and 150°F, adjustments are made. Timber should not be used at sustained temperatures higher than 150°F.)

The effect of load duration will be discussed later in this section.

The reference design values are adjusted for conditions other than the reference conditions using adjustment factors. The adjustment factors are applied in a cumulative fashion by multiplying the published reference design value by the appropriate values of the adjustments. The equation for making the adjustments is

$$R' = R \cdot C_1 \cdot C_2 \cdot C_3 \cdots C_n \tag{7.3}$$

where R' is the adjusted design value for all conditions, R is the tabulated reference design resistance, and C_1, C_2, C_3,...,C_n are the applicable adjustment factors.

Most factors for adjusting for end use are common between the LRFD and ASD methodologies, and many are a function of the species of lumber, strength grade, or width of lumber. Most adjustment factors are provided either by the LRFD or by the ASD manuals. However, some adjustments are associated with proprietary products (chemical treatments) and must be obtained from the product supplier.

Since wood is a viscoelastic material and therefore affected by time, the load duration variable is of particular interest to designers. The LRFD and ASD design methodologies handle the time effects or load duration differently. Regardless of which design methodology is used, the shortest duration load in the load combination will determine the value of the time effect factor. This is because the failure of timber is governed by a creep-rupture mechanism. This implies that timber can sustain higher magnitude loads for short periods of time. In the LRFD design methodology, the time effect factor λ is based on the load combinations considered and the design methodology specifically defines a time effect factor for each combination. These combinations, along with associated time effect factors, are shown in Table 7.1. The reference load duration for LRFD is 5 to 10 min, and this duration is given a value of 1.0. All others follow the Madison curve for load duration, and therefore longer duration loads have time effect factors that are less than 1.0.

The ASD design methodology assumes a load duration of 10 years accumulative, at the design level. Therefore, the load duration factor used in the ASD has a value of 1.0 for normal duration loads, which is 10 years. The shortest-duration load included in a given load combination determines which load duration factor is applicable. All load combinations with individual loads that have a cumulative duration of less than 10 years have load duration factor values greater than 1.0. The loads associated with various load duration factors for ASD are provided in Table 7.2.

Therefore, the adjusted values for either design methodology are determined essentially following the same process, only using the appropriate values. First, the reference design values are obtained from the appropriate table and then adjusted for the end-use conditions, size effects, etc. by multiplying the reference value by a string of adjustment factors. The final adjustment is for the time effect, and

TABLE 7.1   LRFD Load Combinations and Time Effect Factors

LRFD load combination	Time effect factor
$1.4D$	0.6
$1.2D + 1.6L + 0.5(L_r$ or S or $R)$	0.7 for L representing storage
	0.8 for L representing occupancy
	1.25 for L representing impact
$1.2D + 1.6(L_r$ or S or $R) + (0.5L$ or $0.8W)$	0.8
$1.2D + 1.3W + 0.5L + 0.5(L_r$ or S or $R)$	1.0
$1.2D + 1.0E + 0.5L + 0.2S$	1.0
$0.9D - (1.3W$ or $1.0E)$	1.0

TABLE 7.2 ASD Design Loads and Associated Load
Duration Factors

Design load type	Load duration factor value
Dead load	0.9
Occupancy live load	1.0
Snow load	1.15
Construction load	1.25
Wind or seismic load	1.6
Impact load	2.0

the corresponding time effect factor is determined depending on which load combination is being considered.

7.5.1 Bending Members

The most common application for sawn lumber, LVL, glulam timber structural composite lumber, and I-joists is to resist bending forces. In fact, since it is the most common application for dimensional timber members, the visual and machine stress-rated grading rules for lumber are developed around the concept that the members will be placed in bending about their strong axis. In this application, the bending members must account for size effects, duration load, and end-use conditions for the structure. Many times, load distribution elements such as sheathing on floors and walls can provide load sharing between bending members, which can be accounted for in design by use of the repetitive member factor (ASD) or load-sharing factor (LRFD).

Beams must be designed to resist moment, bending shear, bearing, and deflection criteria. The moment, shear, and bearing criteria are all designed using similar formats. The design format for LRFD is

$$L_u \leq \lambda \theta R' \tag{7.4}$$

where L_u is the resultant of the actions caused by the factored load combination, λ is the time effect factor associated with the load combination being considered, θ is the resistance factor (flexure $= 0.85$, stability $= 0.85$, shear $= 0.75$), and R' is the adjusted resistance.

The moment, shear, or bearing force being considered is determined using engineering mechanics and structural analysis. The assumption of linear elastic behavior is typically made for this analysis and is appropriate considering that timber behaves like a brittle material. Nonlinear analysis is acceptable; however, experimental data will probably be required to support such an analysis.

The resistance value involves consideration of the lateral support conditions and load-sharing conditions, and whether the member is or is not part of a larger assembly. The adjusted resistance is obtained by multiplying the published reference strength values for bending, shear, or compression perpendicular-to-grain by the appropriate adjustment factor. The adjustment factor includes the end use and adjustments based on product. The end-use adjustment factors include the wet service factor (C_M), temperature (C_t), preservative treatment factor (C_{pt}), and fire retardant treatment (C_{rt}). Adjustments for member configuration include composite action (C_E), load-sharing factor (C_r), size factor (C_F), beam stability factor (C_L), bearing area (C_b), and form factor (C_f). Additional adjustments for structural lumber and glulam timber include shear stress (C_H), stress interaction (C_I), buckling stiffness (C_T), volume effect (C_V), curvature (C_c), and flat use (C_{fu}). The form of the equation for determining the adjusted resistance is

$$R' = (G) \times F \times C_1 \cdot C_2 \cdot C_3 \cdots C_n \tag{7.5}$$

where R' is the adjusted resistance (e.g., adjusted strong-axis moment, adjusted shear), G is a geometry variable consistent with the resistance calculated (e.g., S_X), F is the reference design strength (e.g., F_b for

TABLE 7.3 Load-Sharing Factor, C_r, Values

Bending member product type	C_r
Dimensional lumber	1.15
Structural composite lumber	1.04
Prefabricated I-joists with visually graded lumber	1.07
Prefabricated I-joists with structural composite lumber flanges	1.04

bending strength, F_v for shear strength, F_c for perpendicular-to-grain compression strength), and C_i is the appropriate adjustment factor(s).

The reader is referred to the LRFD or ASD manuals for a full description of the adjustment factors and the values associated with the various conditions. Adjustment factors account for conditions that do not meet the reference conditions. For example, when lumber is being used for bending about its weak axis, the flat-use factor will be used, or if the beam is not fully supported against lateral movement along the compression edge, the beam stability factor C_L will be required.

One of the most commonly used variables for adjusting a bending strength is the load-sharing factor C_r, which is a variable that provides an increase in design bending resistance that accounts for the load sharing that occurs when beams are used in parallel. The condition under which the load-sharing factor is applicable and causes an effective increase in the bending resistance occurs when the spacing between beams is no more than 610 mm (24 in.) at the center and the beams are connected together using a load-distributing element, such as structural wood panel sheathing or lumber decking. This factor essentially accounts for the system effects that an effective connection between parallel beams provides for floor, roof, and wall systems. Because of the strong correlation between the variability of a given product and the magnitude of the system effect, the load-sharing factor has different values depending on the products used for the beams. Table 7.3 provides the load-sharing factors associated with the more common products used as beams in timber construction. The load-sharing factor C_r applies only to the moment resistance and is not applicable to any of the other design resistances.

When shear strength is being checked, the shear stress adjustment factor C_H may be used to increase the shear resistance. However, this is not recommended for general design because at the design stage, the designer is unable to guarantee that any given piece of dimensional lumber used as a beam will not have checks or splits on the wide face of the lumber. In addition, the designer is not able to guarantee that the member will not split during the lifetime of the structure. It is recommended that the shear stress factor be used only when analyzing existing structures, when inspection has been done to determine the length of the splits that are present.

7.5.2 Axial Force Members

Axial force members are considered to be either tension or compression members and are handled with single equations in both cases. Compression members, that is, columns, have been historically classified as short, intermediate, and long columns depending on whether material crushing or Euler elastic buckling was the controlling mechanism of failure. With the widespread use of computers, the ability to program complex equations has eliminated the need for the three classifications of columns.

7.5.2.1 Compression Members

Compression members are usually called columns, although the term may include drag struts and truss members. There are essentially three basic types of columns. The most common one is the solid or traditional column, which consists of a single member, usually dimensional lumber, post and timbers, poles and piles, or glulam timber. The second type of column is the spaced column, which is made of two or more parallel single-member columns that are separated by spacers, located at specific locations along the column and rigidly tied together at the ends of the column. The third type of column is the built-up column, which consists of two or more members that are mechanically fastened together, such as multiple nailed studs within a wall supporting a girder.

The slenderness ratio defines the primary mechanism of failure. Shorter columns obviously will be controlled by the material strength of the wood parallel-to-grain, while longer columns will be controlled by Euler buckling. The slenderness ratio is defined as the ratio of the effective length of the column, l_e, to the radius of gyration, which is

$$r = \frac{I}{A} \tag{7.6}$$

where I is the moment of inertia about the weak axis and A is the cross-sectional area. The effective length, l_e, is determined by multiplying the unbraced length by the buckling length coefficient:

$$l_e = K_e * l \tag{7.7}$$

where K_e is the buckling length coefficient, l is the unbraced length of the column. K_e is dependent on the end support conditions of the column and whether side sway of the top or bottom of the column is restrained or not. Theoretical values for ideal columns and empirical values for the buckling length coefficient, K_e, that are recommended by NDS and LRFD manuals are provided in Table 7.4. The LRFD and NDS specifications require that the maximum permitted slenderness ratio be 175.

The LRFD design equation that must be satisfied for solid columns has a similar form to the other equations:

$$P_u \leq \lambda \phi_c P_c' \tag{7.8}$$

where P_u is the compressive force due to the factored load combination being considered, λ is the applicable time effect factor for the load combination being considered, ϕ_c is the resistance factor for compression parallel-to-grain (0.9), and P_c' is the adjusted compressive resistance parallel-to-grain.

The adjusted compressive resistance P_c' is determined as the gross area times the adjusted compressive strength parallel-to-grain, F_c'. The adjusted compressive strength is determined in the same manner as all mechanical properties of wood, which is to multiply the reference compressive strength parallel-to-grain by all the applicable adjustment factors, such as duration load, temperature, etc. The one additional factor for columns is that the column stability factor C_p must be calculated. C_p accounts for the partial lateral support provided to a column and is determined by

$$C_p = \frac{1 + \alpha_c}{2c} \sqrt{\left(\frac{1 + \alpha_c}{2c}\right)^2 - \frac{\alpha_c}{c}} \tag{7.9}$$

where

$$\alpha_c = \frac{\phi_s P_e}{\lambda \phi_c P_0'} \tag{7.10}$$

$$P_e = \frac{\pi^2 E_{05}' I}{(K_e l)^2} = \frac{\pi^2 E_{05}' A}{[K_e (l/r)]^2} \tag{7.11}$$

and c is a coefficient based on the variability of the material being used for the column ($c = 0.8$ for dimensional lumber, 0.85 for round poles and piles, and 9.0 for glulam members and structural

TABLE 7.4 Buckling Length Coefficient for Compression in Wood Column Design

End support conditions	Side-sway restraint	Theoretical coefficients	Empirical recommended coefficients
Fixed–fixed	Restrained	0.5	0.65
Pin–fixed	Restrained	0.7	0.8
Fixed–fixed	Free	1.0	1.2
Pin–pin	Restrained	1.0	1.1
Pinned–fixed	Free	2.0	2.1
Fixed–pinned	Free	2.0	2.4

composite lumber). ϕ_s is the resistance factor for stability, which has a value of 0.85, E'_{05} is the adjusted modulus of elasticity at the fifth percentile level, A is the cross-sectional area, K_e is the effective length factor, and r is the radius of gyration.

If a prismatic column is notched at a critical location, then the factored compressive resistance is determined using the net section rather than the gross cross-section. In addition, C_p should be computed using the properties of the net area if the notches or holes are located in the middle half of the length between inflection points of the column and the net moment inertia is less than 80% of the gross moment of inertia or if the longitudinal dimension of the knot or hole is greater than the larger cross-sectional dimension of the column. If the notch is located in noncritical locations, then the adjusted compressive resistance can be computed using the gross cross-sectional area and C_p, or the net cross-sectional area times the factored compressive strength of the material.

Spaced columns are another common form of compression members that consist of two or more dimensional members connected together at a specific spacing using blocking to form a set of parallel columns that are restrained in a common manner. Figure 7.8 illustrates the concept of a spaced column. The general form of a spaced column has the ability to buckle in more than one manner. First, the overall column could buckle. Second, the individual members making up the column could buckle between the spacing blocks. Therefore, the design specifications restrict the geometry of spaced columns to prevent unexpected element buckling. The following maximum length to width ratios are imposed by the LRFD manual:

1. In a spaced column direction L_1/d_1 should not exceed 80.
2. In a spaced column direction l_3/d_1 should not exceed 40.
3. In solid column direction l_2/d_2 should not exceed 50.

Spaced columns that do not conform to these restrictions must be designed considering that each element within the column is acting as an independent column, unless a rational analysis can be used to account for the restraint conditions used.

Built-up columns are the final type of column. Built-up columns consist of two or more dimensional members that are mechanically fastened together to act as a single unit. Built-up columns can have the form of multiple studs nailed together to form a column to support a girder bearing on top of the wall or

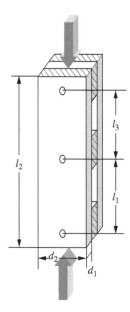

FIGURE 7.8   Illustration of variables associated with the design of spaced columns.

they may have the shape of a hollow column made of members that are nailed together to form the outer circumference of the shape. The capacity of this type of column can conservatively be estimated by considering each of the elements that make up the column as independent columns and adding their collective compressive strengths together. In all cases, the fasteners that connect the members together must be designed and spaced appropriately to transfer the shear and tension forces that occur within and between the members of the built-up column.

7.5.2.2 Tension Members

Compared to compression members, tension members are relatively easy to design and the adjustment factors for the factored resistance are all tabulated values and do not require independent calculations. The basic design equation for designing tension members in the LRFD manual has the form

$$T_u \le \lambda \phi_t T' \tag{7.12}$$

where T_u is the tension force due to the load combination being considered, λ is the time effect factor corresponding to the load combination being considered, ϕ_t is the resistance factor for tension ($\phi_t = 0.8$), and T' is the adjusted tension resistance parallel-to-grain, which is calculated based on the net section and the adjusted tension strength. The adjusted tension strength parallel-to-grain, F'_t, is determined by multiplying the reference tension strength parallel-to-grain by the appropriate adjustment factors, which include temperature, size effects, and load duration, among others. None of the adjustment factors require independent calculations.

If the tension forces are imparted into the member such that the eccentricity between the connection centroid and the centroid of the member is greater than 5% of the member dimension, then the member must be designed considering combined loading of tension and bending.

7.5.3 Combined Loading

Combined loading for member design in wood structures consists of three different categories:

1. Biaxial bending
2. Bending plus tension
3. Bending plus compression

Biaxial bending is a subset of either bending plus tension or bending plus compression. Essentially, biaxial bending uses the same equations, only setting the axial force being considered in those equations equal to 0. The simplified equation takes the form

$$\frac{M_{ux}}{\lambda \phi_b M'_s} + \frac{M_{uy}}{\lambda \phi_b M'_y} \le 1.0 \tag{7.13}$$

where M_u is the factored load moment for the load combination being considered, λ is the time effect factor for the load combination being considered, ϕ_b is the resistance factor for bending ($\phi_b = 0.85$), M'_s is the computed bending resistance moment about the x-axis with the beam stability factor $C_L = 1.0$, and any volume factor, C_v (used when designing glulam members) is included. M'_y is the adjusted moment resistance about the weak axis considering the lateral bracing conditions.

Combined bending and axial tension is the second easiest combination to calculate for combined loading. However, there are two conditions that must be considered: first, the condition along the tension face, for which lateral stability is not a concern and second, the condition along the compression face, for which lateral stability is of concern. Therefore, the following two equations must be satisfied for designing members with combined bending and axial tension:

Tension face:

$$\frac{T_u}{\lambda \phi_t T'} + \frac{M_{ux}}{\lambda \phi_b M'_s} + \frac{M_{uy}}{\lambda \phi_b M'_y} \le 1.0 \tag{7.14}$$

Compression face:

$$\frac{(M_{ux} - (d/6)T_u)}{\lambda\phi_b M'_x} + \frac{M_{uy}}{\lambda\phi_b M'_y(1 - (M_{ux}/\phi_b M_e))^2} \leq 1.0 \qquad (7.15)$$

where T_u is the tension force due to the factored load combination being considered, M_e is the elastic lateral buckling moment of the member, and d is the member depth.

If Equation 7.15 is used for nonrectangular cross-sections, the variable $d/6$ should be replaced by the ratio of the strong axis section modulus to the gross cross-sectional area, S_x/A. When the member is being designed for combined bending and compression, the following equation must be satisfied for the design conditions being considered:

$$\left(\frac{P_u}{\lambda\phi_c P'}\right) + \frac{M_{mx}}{\lambda\phi_b M'_x} + \frac{M_{my}}{\lambda\phi_b M'_y} \leq 1.0 \qquad (7.16)$$

where P_u is the axial compressive force due to the factored load combination being considered; P' is the adjusted resistance for axial compression parallel-to-grain acting alone for the axis of buckling providing the lower buckling strength; M_{mx} and M_{my} are the factored moment resistances, including any magnification for second-order effects for strong and weak axis bending, respectively; and M'_x and M'_y are the adjusted moment resistances for the strong and weak axes, respectively, from multiplying the section properties by the nominal bending resistance and appropriate adjustment factors.

The moments M_{mx} and M_{my} may be determined using second-order analysis for a simplified magnification method that is outlined in the design manual. The extent of the simplified method compels us to refer the reader to the design specification rather than including it here.

7.6 Diaphragms

When resisting lateral loads such as seismic forces, most light-frame wood buildings can be conceptualized as a box system. Forces are transmitted horizontally through diaphragms (i.e., roofs and floors) to reactions that are provided by the shear walls at the ends of the diaphragms. These forces are in turn transmitted to the lower stories and finally to the foundations. Some designers consider shear walls as vertical diaphragms, since the reaction to loading is similar to half of a diaphragm. However, this chapter will consider shear walls as separate elements.

A *diaphragm* is a structural unit that acts as a deep beam or girder that may or may not be able to transfer torsional loads, depending on the relative stiffness of the diaphragm and supporting shear walls. The analogy to a girder is somewhat more appropriate because girders and diaphragms can be made as assemblies. The sheathing acts as a web that is assumed to resist all of the shear loads applied to the diaphragm, and the framing members at the boundaries of the diaphragm are considered to act as flanges to resist tension and compression forces due to moment. The sheathing is stiffened by intermediate framing members to provide support for gravity loading and to transfer the shear load from one sheathing element to the adjacent sheathing element. Chords act as flanges and often consist of the top plates of the walls, ledgers that attach the diaphragm to concrete or masonry walls, bond beams that are part of the masonry walls, or any other continuous element at the parameter of the diaphragm. The third element of a diaphragm is the struts. Struts provide the load transfer mechanism from the diaphragm to the shear walls at the ends of the diaphragm and act parallel to the loading of the diaphragm. Chords act to resist internal forces acting perpendicular to the general loading in the direction of the diaphragm.

The chord can serve several functions at the same time, providing resistance to loads and forces from different sources and functioning as the tension or compression flange of the diaphragm. It is important that the connection to the sheathing be designed to accomplish the shear transfer since most diaphragm chords consist of many pieces. It is important that splices be designed to transmit the tension or compression occurring at the location of the splice. It is also important to recognize that the direction of application of the seismic forces will reverse. Therefore, it should be recognized that chords need to be designed for equal magnitude torsion and compression forces. When the seismic forces are acting at 90° to the original direction analyzed, the chords act as the struts for the diaphragm to transfer the reaction loads to the shear walls below. If the shear walls are not continuous along the length of the diaphragm, then the strut may act as the drag strut between the segments of the shear wall as well. Diaphragms often have openings to facilitate stairwells, great rooms that are more than one story high, or access to roof systems and ventilation. The transfer of forces around openings can be treated in a manner similar to the transfer around openings in the webs of steel girders. Members at the edges of the openings have forces due to flexure and the high web shear induced in them, and the resulting forces must be transferred into the body of the diaphragm beyond the opening.

In the past, wood sheathed diaphragms have been considered to be flexible by many registered design professionals and most code enforcement agencies. Recent editions of the model building codes recognize that diaphragms have a stiffness relative to the walls that are supporting them, and this relative rigidity determines how the forces will be distributed to the vertical resistance elements.

7.6.1 Stiffness versus Strength

Often in large diaphragms, as an economic measure, the designer will change the nailing or blocking requirements for the diaphragm to remove the high nail schedule or blocking requirements in the central section of the diaphragm where the shear loading is lowest. It is therefore imperative that the designer distinguish between the stiffness of the diaphragm and the strength of the diaphragm. However, at locations where blocking requirements change or the nail schedule is increased, the stiffness of the diaphragm also goes through a significant change. These locations result in potential stress concentrations and when a nail schedule or blocking schedule is changed, the designer should consider ensuring that the change occurs at locations where the loads are not simply equal to the resistance, but sufficiently below the changing resistance at that location. Timber-framed diaphragms and structures with light-frame shear walls are capable of being relatively rigid compared to the vertical resistance system. Added stiffness due to concrete toppings, blocking, and adhesives also makes the relative stiffness greater. Therefore, determination of the stiffness of the diaphragm relative to the vertical resistance system must be considered in the design to determine whether a flexible or rigid analysis is required.

The equation used to estimate the deflection of diaphragms was developed by APA and can be found in several references (Applied Technology Council 1981; APA 1997; National Fire Prevention Association 2002). The midspan deflection of a simple-span, blocked, structural panel sheathed diaphragm, uniformly nailed throughout, can be calculated using

$$\Delta = \frac{5vl^3}{8wEA} + \frac{vl}{4Gt} + 0.188\ le_n + \frac{\sum(\Delta_c X)}{2w} \tag{7.17}$$

where Δ is the calculated deflection, v is the maximum shear due to factored design loads in the direction under consideration, l is the diaphragm length, w is the diaphragm width, E is the elastic modulus of chords, A is the area of chord cross-section, Gt is the panel rigidity through the thickness, e_n is the nail deformation, and $\sum(\Delta_c X)$ is the sum of individual chord splice slip values on both sides of the diaphragm, each multiplied by its distance to the nearest support.

If the diaphragm is not uniformly nailed, the constant 0.188 in the third term must be modified accordingly. Guidance for using this equation can be found in ATC 7 (Applied Technology Council 1981). This formula was developed based on engineering principles and modified by testing. Therefore, it provides an estimate of diaphragm deflection due to loads applied in the factored resistance shear range. The effects of cyclic loading and energy dissipation may alter the values for nail deformation in the third term, as well as the chord splice effects in the fourth term, if mechanically spliced wood chords are used. The formula is not applicable to partially blocked or unblocked diaphragms.

Recent research, part of the wood frame project of the Consortium of Universities for Research in Earthquake Engineering (CUREE), was conducted as part of the shake table tests by Fischer et al. (2001) and Dolan et al. (2002). These new studies provide deflection equations that are broken into two components: one for bending deflection and one for shear deflection. The tests used nailed chord splices. If other types of splices are utilized in a design, additional terms to account for the deformation of the splice effects on the diaphragm need to be added. In addition, the CUREE equations are useful for working stress level loads and have not been validated for deflections approaching those associated with the capacity of the diaphragm. However, the equations are derived based on cyclic tests and the average cyclic stiffness within the reasonable deflection range.

The provisions are based on assemblies having energy dissipation capacities that were recognized in setting the R-factors included in the model building codes. For diaphragms utilizing timber framing, the energy dissipation is almost entirely due to nail bending. Fasteners other than nails and staples have not been extensively tested under cyclic load applications. When screws or adhesives have been tested in assemblies subjected to cyclic loading, they have had brittle failures in adhesives and provided minimal energy dissipation. For this reason, adhesives have been prohibited in light-frame shear wall assemblies in high seismic regions. However, the deformation range typically experienced by diaphragms during seismic events has not justified the restriction of using adhesives in the horizontal diaphragms. In fact, the addition of adhesives in most timber diaphragms provides significantly more benefits in the form of higher strength and stiffness to distribute loads to the horizontal members more efficiently. While in the Dolan et al. (2002) diaphragm study, the adhesives by themselves did not have as large an effect on stiffness as blocking, the use of adhesives provided a more uniform deflection pattern and enhanced the behavior of the diaphragm over a nailed-only diaphragm.

7.6.2 Flexible versus Rigid Diaphragms

The purpose of determining whether a diaphragm is flexible or rigid is to determine whether a diaphragm should have the loads proportioned according to the tributary area or the relative stiffness of the supports. For flexible diaphragms, the loads should be distributed according to the tributary area, whereas for rigid diaphragms, the load should be distributed according to the stiffness. The distribution of seismic forces to the vertical elements of the lateral force resistance system is dependent, first, on the relative stiffness of the vertical elements versus the horizontal elements and, second, on the relative stiffness of the vertical elements when they have varying deflection characteristics. The first issue defines when a diaphragm can be considered flexible or rigid. In other words, it sets limits on whether the diaphragms can act to transmit torsional resistance or cantilever. When the relative deflections of the diaphragm and shear walls are determined at the factored load resistance level, and the midspan deflection of the diaphragm is determined to be more than two times the average deflection of the vertical resistant elements, the diaphragms may be considered as being flexible. Conversely, a diaphragm should be considered rigid when the diaphragm deflection is equal to or less than two times the shear wall drift. Obviously, the performance of most diaphragms falls in a broad spectrum between perfectly rigid and flexible. However, at the current time, there are no design tools available to provide for analyzing diaphragms in the intermediate realm. Therefore, model building codes simply differentiate between the two extreme conditions.

The flexible diaphragm seismic forces should be distributed to the vertical resisting elements according to the tributary area and simple beam analysis. Although rotation of the diaphragm may occur because lines of vertical elements have different degrees of stiffness, the diaphragm is not considered sufficiently stiff to redistribute the seismic forces through rotation. The diaphragm may be visualized as a single-span beam supported on rigid supports in this instance.

For diaphragms defined as rigid, rotational or torsional behavior is expected and the action results in a redistribution of shear to the vertical force-resisting elements. Requirements for horizontal shear distribution involve a significantly more detailed analysis of the system than the assumption of flexibility. Torsional response of a structure due to an irregular stiffness at any level within the structure can be the potential cause of failure in the building. As a result, dimensional and diaphragm aspect ratio limitations are imposed for different categories of construction. Also, additional requirements are imposed on the diaphragm when the structure is deemed to have a general torsional irregularity such as when re-entering corners or when diaphragm discontinuities are present.

In an effort to form a frame of reference in which to judge when stiffness of the diaphragm may be critical to the performance of the building, one can consider two different categories of diaphragms. The first category includes rigid diaphragms that must rely on torsional response to distribute the loads to the building. A common example would be an open front structure with shear walls on three sides, such as a strip mall. This structurally critical category has the following limitations:

1. The diaphragm may not be used to resist the forces contributed by masonry or concrete in structures over one story in height.
2. The length of the diaphragm normal to the opening may not exceed 25 ft, and the aspect ratios are limited to being less than 1:1 for one-story structures or 1:1.5 for structures over one story in height. Where calculations show that the diaphragm deflections can be tolerated, the length will be permitted to increase so as to allow aspect ratios not greater than 1.5:1 when the diaphragm is sheathed with structural use panels.

The aspect ratio for diaphragms should not exceed 1:1 when sheathed with diagonal sheathing.

The second category of rigid diaphragms that may be considered is those that are supported by two or more shear walls in each of the two perpendicular directions, but have a center of mass that is not coincident with the center of rigidity, thereby causing a rotation in the diaphragm. This category of diaphragm may be divided into two different categories where category 2a would consist of diaphragms with a minimal eccentricity that may be considered on the order of incidental eccentricities, in which case the following restrictions would apply:

1. The diaphragm may not be used to resist forces contributed by masonry or concrete in structures over one story.
2. The aspect ratio of the diaphragm may not exceed 1:1 for one-story structures or 1:1.5 for structures greater than one story in height.

On the other hand, flexible diaphragms or nonrigid diaphragms have minimal capacity for distributing torsional forces to the shear walls. Therefore, limitations of aspect ratios are used to limit diaphragm deformation such that reasonable behavior will occur. The resulting deformation demand on the structure is also limited such that higher aspect ratios are allowed provided calculations demonstrate that the higher diaphragm deflections can be tolerated by the supporting structure. In this case, it becomes important to determine whether the diaphragm rigidity adversely affects the horizontal distribution and the ability of the other structural elements to withstand the resulting deformations.

Several proposals to prohibit wood diaphragms from acting in rotation have been advanced following the 1994 Northridge earthquake. However, the committees that have reviewed the reports to date have concluded that the collapses that occurred in that event were due in part to a lack of deformation compatibility between the various vertical resisting elements, rather than solely due to the inability of the diaphragm to act in rotation.

Often diaphragms are used to cantilever past the diaphragm's supporting structure. Limitations concerning diaphragms cantilevered horizontally past the outermost shear walls or other vertical support element are related to, but slightly different from, those imposed due to diaphragm rotation. Such diaphragms can be flexible or rigid, and the rigid diaphragms may be categorized in one of the three categories previously discussed. However, both the limitations based on the diaphragm rotation, if they are applicable, and a diaphragm limitation of not exceeding the lesser of 25 ft or two thirds of the diaphragm width must be considered in the design. This is due to the additional demand placed on a structure due to the irregularity resulting from the cantilever configuration of the diaphragm.

Further guidance on the design and detailing of diaphragms can be obtained from Breyer et al. (1998) and Faherty and Williamson (1999), both providing significant details on dealing with torsional irregularities of wood diaphragms.

7.6.3 Connections to Walls

When postevent reports are reviewed for many historical earthquakes, one finds that the principal location of failures in diaphragms occurs at the connection between the diaphragm and the supporting walls. Two of the most prevalent failures are due to cross-grain bending of ledger boards that are used to attach diaphragms to concrete and masonry walls, or to the use of the diaphragm sheathing to make the connection between the wall and the diaphragm. When one uses the ledger board to attach a wood diaphragm to a masonry or concrete wall, the diaphragm will tend to pull away from the wall, which causes cross-grain bending to occur in the ledger board. Cross-grain bending is not allowed in the design specifications for wood construction and is one of the weakest directions in which wood can be loaded. The same mechanism causes splitting of sill plates on shear walls due to the uplift forces of the sheathing.

When diaphragms are attached to walls, regardless of the construction of the wall, the connections must be made directly between the framing or reinforcement elements of the wall and the framing of the diaphragm. Straps should extend into the diaphragm a significant distance to provide adequate length in which to develop the forces experienced. Sheathing rarely provides sufficient capacity to transfer diaphragm forces and the sheathing nailing will inevitably fail. These straps also should extend down along the timber frame wall and be attached to the studs of the wall, or for masonry and concrete walls, be attached to reinforcement in the wall near the location.

7.6.4 Detailing around Openings

Openings in diaphragms may be designed similar to the way openings in steel girders are designed. After the bending and shear forces are determined, connections to transfer those loads can be designed. Construction of such connections often requires the use of blocking between the framing members and straps that extend past the opening, often for multiple joist spacings. It is imperative that connections to transfer such loads be made to the framing members and not simply to the sheathing. This is also an area which a designer should consider as requiring special inspection, to ensure that the nailing used to make the connection is actually located in the framing and not simply through the sheathing into air.

7.6.5 Typical Failure Locations

As stated earlier, the typical locations for failure of diaphragms are the connections between the diaphragm and the vertical force-resisting elements. However, additional areas need to be considered. These include openings in the diaphragm, where shear transfer is being designed, and locations where nail schedules or blocking requirements change. Often openings are placed in diaphragms without the design considering the force transfer required due to the absence of sheathing. This location also results in a difference or an anomaly in the stiffness distribution along the length of the diaphragm. These two actions combine to cause the sheathing nailing, and possibly the framing nailing, to separate, thus causing local failures.

7.7 Shear Walls

Shear walls are typically used as the vertical elements in the lateral-force-resisting system. The forces are distributed to the shear walls of the building by the diaphragms and the shear walls transmit the loads down to the next lower story or foundation. A shear wall can be defined as a vertical structural element that acts as a cantilever beam where the sheathing resists the shear forces and the end studs, posts, or chords act as the flanges of the beam in resisting the induced moment forces associated with the shear applied to the top of the wall. The sheathing is stiffened by the intermediate studs and is therefore braced against buckling. An exploded view of a typical light-frame shear wall is shown in Figure 7.9.

Light-frame shear walls can be divided into two categories: designed and prescriptive. Designed walls are sized and configured using a rational analysis of the loads and design resistance associated with the

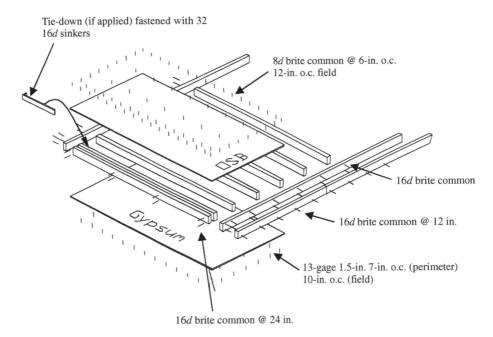

Tie-down (if applied) fastened with 32 16*d* sinkers

8*d* brite common @ 6-in. o.c. 12-in. o.c. field

16*d* brite common

16*d* brite common @ 12 in.

13-gage 1.5-in. 7-in. o.c. (perimeter) 10-in. o.c. (field)

16*d* brite common @ 24 in.

FIGURE 7.9 Typical shear wall assembly (shown lying on its side, rather than vertical) (courtesy of Heine, C.P. 1997. "Effect of Overturning Restraint on the Performance of Fully Sheathed and Perforated Timber Framed Shear Walls," M.S. thesis, Virginia Polytechnic and State University, Blacksburg, VA).

sheathing thickness, and fastener size and schedule. Prescriptive walls are often called shear panels rather than shear walls to differentiate them from the walls designed using rational analysis. Prescriptive walls are constructed according to a set of rules provided in the building code. In all cases, the current building code requires that all shear walls using wood structural panels as the sheathing material be fully blocked.

Adhesives are not allowed for attaching sheathing to the framing in shear walls. Whereas adhesives are advocated for attaching sheathing to roof and floor diaphragms as a method to improve their ability to distribute the loads to the shear walls, the use of adhesives in shear walls changes the ductile system usually associated with light-frame construction to a stiff, brittle system that would have to be designed with a significantly lower *R*-factor. Adhesives are allowed to be used in regions with low seismic hazard since they improve the performance of the building when wind loading is considered. An *R*-factor of 1.5 is recommended for seismic checks even in the low seismic hazard regions (Building Seismic Safety Council 2003a) when adhesives are used.

7.7.1 Rationally Designed Walls

Rationally designed light-frame shear walls can be further divided into two categories. The first is the traditional or segmented shear wall design, which involves the use of tables or mechanics. The second method of design is the perforated shear wall method, which is an empirically based method that accounts for the openings in a wall line. The principal difference between the two methods is the assumptions associated with the free-body diagrams used in each method. Each of these design methods will be discussed independently.

7.7.1.1 Segmented Wall Design

The segmented shear wall design method is the traditional method for designing light-frame shear walls. This method assumes a rigid free-body diagram and a uniform distribution of shear along the top of the wall line. Both this method and the perforated shear wall method assume that the tops of all wall segments in a wall line will displace the same amount. In other words, the assumption is that the tops of the wall segments are tied together with the platform above, the top plate of the wall, or collectors between wall segments, providing sufficient connectivity to ensure the wall segments will displace together as a unit. The segmented wall design method assumes that collectors will be detailed to transmit the shear forces distributed to the wall over the openings by the diaphragm to the adjoining wall segments. Each segment is assumed to resist the portion of the load according to its relative length in the wall line. In other words, the shear force per length of wall that is applied to the wall segment is determined by

$$v = \frac{V}{\sum L_i} \tag{7.18}$$

where v is the unit shear (force/length), V is the total shear load applied to the wall line, and $\sum L_i$ is the summation of all of the fully sheathed wall segments in the wall line. It is assumed that this unit shear is distributed uniformly to all of the fully sheathed wall segments. With this assumption, the individual wall segments are then designed assuming a rigid-body free-body diagram, similar to that shown in Figure 7.10.

The simple assumption of using rigid-body mechanics makes the determination of induced overturning forces in the chords an easy task of summing moments about one of the bottom corners of the wall segment. The required mechanical anchor to resist this overturning load can then be sized according to the uplift force determined. However, the assumption of a rigid body also opens an opportunity for error in the calculations. If one assumes that there is an imposed dead load due to the structure above (say, the floor of the story above), one might assume that the vertical forces due to this dead load may act to resist the imposed overturning action of the lateral load on the individual wall segment. In this case, the size of the mechanical anchor would be determined for the difference in the uplift force due to the overturning moment and the resisting force due to the dead load of the structure above. If the assumption of rigid-body action were valid, this would be an acceptable mechanism of resistance. However, if one investigates the construction of light-frame shear walls, the vertical load is applied to the wall across the top plate as a distributed load. However, the top plate of the wall is usually a double 2 × 4 nominal framing member, which has questionable ability to transmit vertical loads along the length of the wall through bending action. In addition, this top plate is supported by repetitive framing members called studs that would transmit the vertical load to the base of the wall rather than allow the top plate to distribute the load to the end stud or chord for the wall. Currently, there are therefore two schools of thought on how the mechanical anchors to resist overturning forces should be determined. One is to use the full dead load acting on the top of the wall to reduce the uplift forces at the chord. The other is to assume that little or none of the dead load acts to resist the uplift forces. The latter is obviously the more conservative assumption, but it also considers the top plate of the wall as a beam on an elastic foundation, for how vertical loads are transmitted along the length of the wall.

The final step in the design of the segmented shear wall is to determine the thickness of the sheathing and associated nail schedule to be used to attach the sheathing to the framing. This information is usually obtained using design tables available in the building code or design specification. However, it is permitted to use the properties of the individual nail connection and engineering mechanics to determine the resistance of a given sheathing thickness and nailing configuration.

7.7.1.2 Perforated Shear Wall Design

The perforated shear wall method is included in the *NEHRP Provisions* (Building Seismic Safety Council 2003a,b) and the 2003 IBC (International Code Council 2000a) for use in seismic design. It had been adopted earlier for wind design by the Building Officials and Code Administrators (BOCA) and Southern Building Code Congress International (SBCCI) building codes. The method is an empirical design method that accounts for the added resistance provided by the wall segments above and below window and door openings in the wall, if they are sheathed with equivalent sheathing to that used in the fully sheathed segments of the wall. The method was originally developed by Sugiyama and Matsumoto (1994) using reduced-scale light-frame wall specimens, and the method was

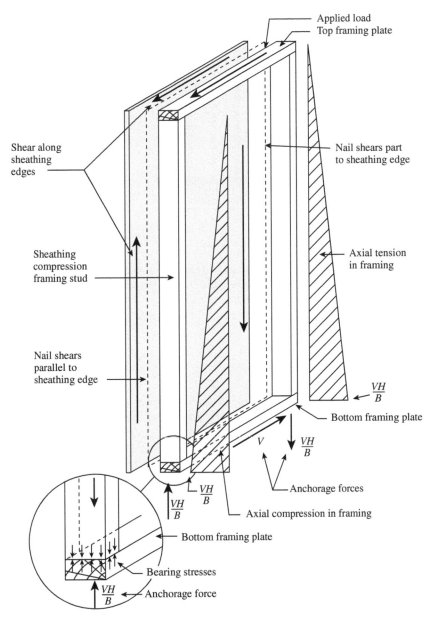

FIGURE 7.10 Rigid free-body diagram assumed for segmented shear wall design (courtesy of Stewart, W.G. 1987. "The Seismic Design of Plywood Sheathed Shearwalls," Ph.D. dissertation, University of Canterbury, New Zealand).

validated for full-scale wall construction under cyclic loads by Dolan and Heine (1997a,b) and Dolan and Johnson (1997a,b).

The perforated shear wall method of design assumes that the segments of wall above and below openings are not specifically designed for force transfer around the opening. Rather, the only assumptions are that the top of the wall line will displace uniformly (i.e., tied together with the top plate of the wall line) and the end full-height sheathed wall segments have mechanical overturning restraint at the extreme ends of the wall line. Other assumptions are that the bottom plate of the wall is attached to the floor platform or foundation sufficiently to resist the distributed shear force applied to the wall and a distributed uplift force equal to the distributed shear force is resisted along the wall length. This

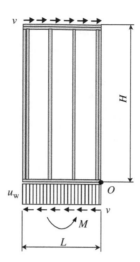

FIGURE 7.11 Free-body diagram for interior wall segment for perforated shear wall or prescriptive wall segment (courtesy of Salenikovich, A.J. 2000. "The Racking Performance of Light-Frame Shear Walls," Ph.D. dissertation, Virginia Polytechnic Institute and State University, Blacksburg, VA).

anchorage can be accomplished with nails, screws, lag screws, or other type of fastener capable of resisting the shear and uplift forces.

The basic difference in the assumed free-body diagram for the perforated shear wall method is that the shear force is not assumed to be uniform along the length of the wall line, and the individual wall segments are not assumed to act as rigid bodies. The end wall segment that has to resist an uplift force in the end post is assumed to reach the full design capacity as if it were a segmented wall since the uplift force will be resisted by a mechanical, overturning anchor. The rest of the wall segments are assumed to perform similar to a prescriptively constructed wall, with the overturning forces being resisted by the sheathing nails at the bottom of the wall. A free-body diagram for an interior wall segment of a perforated shear wall is illustrated in Figure 7.11. Due to the difference in overturning restraint between the different segments of the shear wall, the shear force cannot possibly be resisted as a uniformly distributed load, and the end wall segment must resist significantly more of the load than the interior wall segments. As an illustration, two examples from the commentary for the 2000 *NEHRP Provisions* (Building Seismic Safety Council 2000b) for applying the perforated shear wall method are included in this section. In addition, the American Iron and Steel Institute has introduced the perforated shear wall method for cold-formed steel framing as a change proposal for the 2003 edition of the IBC.

EXAMPLE 7.1: Perforated Shear Wall

Problem description

The perforated shear wall illustrated in Figure 7.12 is sheathed with $\frac{15}{32}$-in. wood structural panel with $10d$ common nails with 4-in. perimeter spacing. All full-height sheathed sections are 4 ft wide. The window opening is 4 ft high by 8 ft wide. The door opening is 6.67 ft high by 4 ft wide. Sheathing is provided above and below the window and above the door. The wall length and height are 24 and 8 ft, respectively. Hold-downs provide overturning restraint at the ends of the perforated shear wall and anchor bolts are used to restrain the wall against shear and uplift between perforated shear wall ends. Determine the shear resistance adjustment factor for this wall.

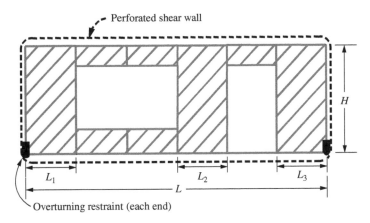

FIGURE 7.12 Perforated shear wall configuration (Example 7.1).

Solution

The wall defined in the problem description meets the application criteria outlined for the perforated shear wall design method. Hold-downs provide overturning restraint at perforated shear wall ends and anchor bolts provide shear and uplift resistance between perforated shear wall ends. Perforated shear wall height, factored shear resistances for the wood structural panel shear wall, and aspect ratio of full-height sheathing at perforated shear wall ends meet requirements of the perforated shear wall method.

The process of determining the shear resistance adjustment factor involves determining percent full-height sheathing and maximum opening height ratio. Once these are known, a shear resistance adjustment factor can be determined from Table 2305.3.7.2 of the 2003 IBC (International Code Council 2003a). From the problem description and Figure 7.12

$$\text{Percent full-height sheathing} = \frac{\text{sum of perforated shear wall segment widths, } \sum L}{\text{Length of perforated shear wall}} = \frac{4 \text{ ft} + 4 \text{ ft} + 4 \text{ ft}}{24 \text{ ft}} \times 100 = 50\%$$

$$\text{Maximum opening height ratio} = \frac{\text{maximum opening height}}{\text{Wall height, } h} = \frac{6.67 \text{ ft}}{8 \text{ ft}} = \frac{5}{6}$$

For a maximum opening height ratio of $\frac{5}{6}$ (or maximum opening height of 6.67 ft when wall height $h = 8$ ft) and percent full-height sheathing equal to 50%, a shear resistance adjustment factor $C_0 = 0.57$ is obtained from Table 2305.3.7.2 of the 2003 IBC (International Code Council 2003a). Note that if wood structural panel sheathing were not provided above and below the window or above the door, the maximum opening height would equal the wall height h.

EXAMPLE 7.2: Perforated Shear Wall

Problem Description

Figure 7.13 illustrates one face of a two-story building with the first- and second-floor walls designed as perforated shear walls. Window heights are 4 ft and door height is 6.67 ft. A trial design is performed in this example based on applied loads V. For simplification, dead load contribution to overturning and uplift restraint is ignored and the effective width for shear in each perforated shear wall segment is assumed to be the sheathed width. Framing is Douglas fir. After basic perforated shear wall resistance and force requirements are calculated, detailing options to provide for adequate shear, v, and uplift, t, transfer between perforated shear wall ends are covered. Method A considers the condition

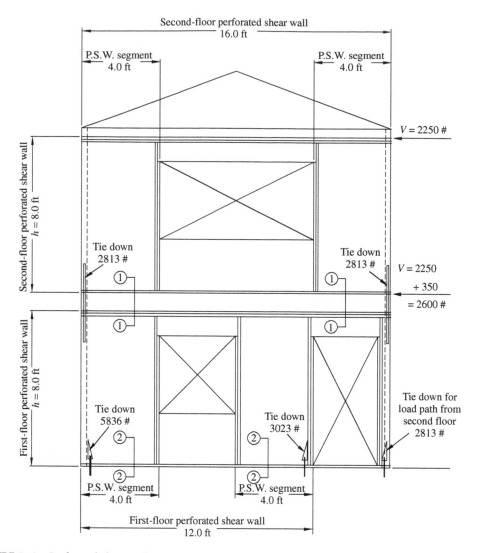

FIGURE 7.13   Perforated shear wall (Example 7.2), two-story building.

where a continuous rim joist is present at the second floor. Method B considers the case where a continuous rim joist is not provided, as when floor framing runs perpendicular to the perforated shear wall with blocking between floor framing joists.

Solution, second-floor wall

Determine the wood structural panel sheathing thickness and fastener schedule needed to resist applied load $V = 2.250$ kip, from the roof diaphragm, such that the shear resistance of the perforated shear wall is greater than the applied force. Also, determine anchorage and load path requirements for uplift force at ends, in plane shear, uplift between wall ends, and compression.

$$\text{Maximum opening height ratio} = \frac{4\,\text{ft}}{8\,\text{ft}} = \frac{1}{2}$$

$$\text{Percent full-height sheathing} = \frac{4\,\text{ft} + 4\,\text{ft}}{16\,\text{ft}} \times 100 = 50\%$$

Shear resistance adjustment factor $C_0 = 0.80$.

Try 15/32 rated sheathing with 8d common nails (0.131 by 2.5 in.) at 6-in. perimeter spacing.
Unadjusted shear resistance (Table 5.4 LRFD Structural-Use Panels Supplement) = 0.36 klf
(American Forest and Paper Association 1996).

Adjusted shear resistance = (unadjusted shear resistance)(C_0) = (0.36 klf)(0.80) = 0.288 klf.

Perforated shear wall resistance = (adjusted shear resistance)$(\sum L_i)$ = (0.288 klf)(4 ft + 4 ft) = 2.304 kip
2.304 kip > 2.250 kip.

Required resistance due to story shear forces V.

Overturning at shear wall ends

$$T = \frac{Vh}{C_0 \sum L_i} = \frac{2.25 \text{ kip}(8 \text{ ft})}{0.08(4 \text{ ft} + 4 \text{ ft})} = 2.813 \text{ kip}$$

In-plane shear

$$v = \frac{V}{C_0 \sum L_i} = \frac{2.25 \text{ kip}}{0.80(4 \text{ ft} + 4 \text{ ft})} = 0.352 \text{ klf}$$

Uplift t, between wall ends = $v = 0.352$ klf.

Compression chord force C, at each end of each perforated shear wall segment = $T = 2.813$ kip.

Solution, first-floor wall

Determine the wood structural panel sheathing thickness and fastener schedule needed to resist applied
load $V = 2.600$ kip, at the second-floor diaphragm, such that the shear resistance of the perforated shear
wall is greater than the applied force. Also, determine anchorage and load path requirements for uplift
force at ends, in plane shear, uplift between wall ends, and compression.

Percent full-height sheathing = [(4 ft + 4 ft)/12 ft] × 100 = 67%.

Shear resistance adjustment factor, $C_0 = 0.67$.

Unadjusted shear resistance (Table 5.4 LRFD Structural-Use Panels Supplement) = 0.49 klf (American
Forest and Paper Association 1996).

Adjusted shear resistance = (unadjusted shear resistance)(C_0) = (0.49 klf)(0.67) = 0.328 klf.

Perforated shear wall resistance = (adjusted shear resistance) $(\sum L_i)$ = (0.328 klf)(4 ft + 4 ft) =
2.626 kip, 2.626 kip > 2.600 kip.

Required resistance due to story shear forces = V.

Overturning at shear wall ends

$$T = \frac{Vh}{C_0 \sum L_i} = \frac{2.600 \text{ kip}(8 \text{ ft})}{0.67(4 \text{ ft} + 4 \text{ ft})} = 3.880 \text{ kip}$$

When maintaining load path from story above, $T = T$ from second floor + T from first floor =
2.813 kip + 3.880 kip = 6.693 kip.

In-plane shear

$$v = \frac{V}{C_0 \sum L_i} = \frac{2.600 \text{ kip}}{0.67(4 \text{ ft} + 4 \text{ ft})} = 0.485 \text{ klf}$$

Uplift t, between wall ends, = $v = 0.485$ klf.

Uplift t, can be cumulative with 0.352 klf from story above to maintain load path. Whether this occurs
depends on detailing for transfer of uplift forces between end walls.

Compression chord force C at each end of each perforated shear wall segment $= T = 3.880$ kip.

When maintaining load path from story above, $C = 3.880$ kip $+ 2.813$ kip $= 6.693$ kip.

Hold-downs and posts and the ends of perforated shear wall are sized using calculated force T. The compressive force, C, is used to size compression chords as columns and ensure adequate bearing.

Method A: continuous rim joist
See Figure 7.14.

Second floor: Determine fastener schedule for shear and uplift attachment between perforated shear wall ends. Recall that $v = t = 0.352$ klf.

Wall bottom plate (1.5-in. thickness) to rim joist: Use 20d box nail (0.148 by 4 in.). Lateral resistance $\phi\lambda Z' = 0.254$ kip per nail and withdrawal resistance $\phi\lambda W' = 0.155$ kip per nail.

Nails for shear transfer $=$ (shear force, v)$/\phi\lambda Z' = 0.352$ klf/0.254 kip per nail $= 1.39$ nails per foot.

Nails for uplift transfer $=$ (uplift force, t)$/\phi\lambda W' = 0.352$ klf/0.155 kip per nail $= 2.27$ nails per foot.

Net spacing for shear and uplift $= 3.3$ in. on center.

Rim joist to wall top plate: Use 8d box nails (0.113 by 2.5 in.) toe-nailed to provide shear transfer. Lateral resistance $\phi\lambda Z' = 0.129$ kip per nail.

Nails for shear transfer $=$ (shear force, v)$/\phi\lambda Z' = 0.352$ klf/0.129 kip per nail $= 2.73$ nails per foot.

Net spacing for shear $= 4.4$ in. on center.

See detail in Figure 7.14 for alternate means for shear transfer (e.g., metal angle or plate connector).

Transfer of uplift, t, from the second floor in this example is accomplished through attachment of second-floor wall to the continuous rim joist, which has been designed to provide sufficient strength to resist the induced moments and shears. Continuity of load path is provided by hold-downs at the ends of the perforated shear wall.

First floor: Determine the anchorage for shear and uplift attachment between perforated shear wall ends. Recall that $v = t = 0.485$ klf. Wall bottom plate (1.5-in. thickness) to concrete. Use 0.5-in. anchor bolt with lateral resistance $\phi\lambda Z' = 1.34$ kip.

Bolts for shear transfer $=$ (shear force, v)$/\phi\lambda Z' = 0.485$ klf/1.34 kip per bolt $= 0.36$ bolts per foot.

Net spacing for shear $= 33$ in. on center.

Bolts for uplift transfer: Check axial capacity of bolts for $t = v = 0.485$ klf and size plate washers accordingly. No interaction between axial and lateral load on anchor bolt is assumed (e.g., presence of axial tension does not affect lateral strength).

Method B: blocking between joists
See Figure 7.14.

Second floor: Determine fasteners schedule for shear and uplift attachment between perforated shear wall ends. Recall that $v = t = 0.352$ klf.

Wall bottom plate (1.5-in. thickness) to rim joist: Use 20d box nail (0.148 by 4 in.). Lateral resistance $\phi\lambda Z' = 0.254$ kip per nail.

Nails for shear transfer $=$ (shear force, v)$/\phi\lambda Z' = 0.352$ klf/0.254 kip per nail $= 1.39$ nails per foot.

Net spacing for shear $= 8.63$ in. on center.

Rim joist to wall top plate: Use 8d box nails (0.113 by 2.5 in.) toe-nailed to provide shear transfer. Lateral resistance $\phi\lambda Z' = 0.129$ kip per nail.

Nails for shear transfer $=$ (shear force, v)$/\phi\lambda Z' = 0.352$ klf/0.129 kip per nail $= 2.73$ nails per foot.

Net spacing for shear $= 4.4$ in. on center.

See detail in Figure 7.14 for alternative means for shear transfer (e.g., metal angle or plate connector).

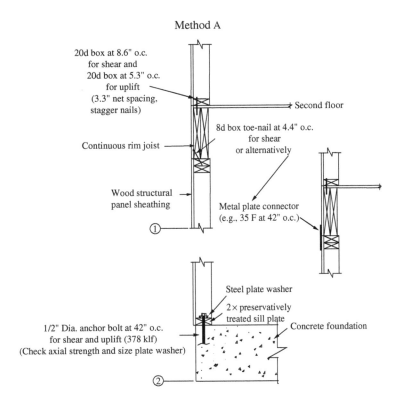

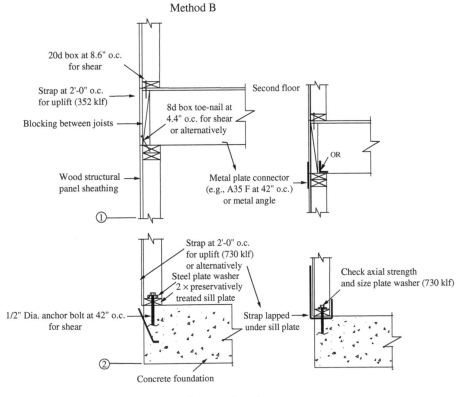

FIGURE 7.14 Details for perforated shear wall (Example 7.2).

Stud to stud: Provide a metal strap for transfer of uplift, t, from second-story wall studs to first-story wall studs. Size strap for 0.352-klf uplift and place at 2 ft on center to coincide with stud spacing. This load path will be maintained by transfer of forces through first-floor wall framing to the foundation.

First floor: Determine anchorage for shear and uplift attachment between perforated shear wall ends. Recall that $v = t = 0.485$ klf.

Wall bottom plate (1.5-in. thickness) to concrete: Use 0.5-in. anchor bolt with lateral resistance $\phi \lambda Z' = 1.34$ kip.

Bolts for shear transfer $= $ (shear force, v)$/\phi \lambda Z' = 0.485$ klf/1.34 kip per bolt $= 0.36$ bolts per foot.

Net spacing for shear $= 33$ in. on center.

Uplift transfer: A metal strap embedded in concrete at 2 ft on center and attached to first-story studs maintaining load path with second story is used. In this case all uplift forces, t, between perforated shear wall ends are resisted by the metal strap. Size metal strap and provide sufficient embedment for uplift force $t = 0.485$ klf $+ 0.352$ klf $= 0.837$ klf.

An alternative detail for uplift transfer uses a metal strap lapped under the bottom plate. Size metal strap, anchor bolt, and plate washers for uplift force $t = 0.485$ klf $+ 0.352$ klf $= 0.837$ klf are used to maintain load path from the second story. No interaction between axial and lateral load on anchor bolt is assumed (e.g., presence of axial tension does not affect lateral strength).

7.7.2 Prescriptive Construction

The rules for prescriptive wall construction are provided in the appropriate building code (IRC [International Code Council 2003b] or CABO [Council of American Building Officials 1995]). This construction method is often referred to as "conventional construction" and the individual wall segments are often called shear panels as a method of differentiating them from rationally designed shear walls. These rules set the required size of nail and spacing, along with the minimum percentage of wall area that must be sheathed, determined based on the location of the building and at which level of the building the wall panel is located. These provisions are not based on any rational analysis, but rather on the tradition of constructing wall systems under these rules. Using the analysis models and wall test results currently available to the engineering community, it is not possible to calculate sufficient resistance to resist the design loads expected for the seismic design categories designated in the building code. However, the code drafting and technical update committees have decided that the overall historic performance of buildings constructed following these rules has been sufficient to justify their use.

The basis of performance for these walls is that the overturning forces, associated with the lateral loading, are resisted by the sheathing nails at the bottom of the wall. Nailing for these walls is set at 6 in. around the perimeter of each sheet of sheathing and 12 in. along the intermediate supporting framing members. The free-body diagram for this type of wall is shown in Figure 7.11. The inherent weakness of this type of construction is the low overturning resistance supplied by the sheathing nails. Usually the assumed design value of this type of construction ranges from 140 to 300 plf, which is substantially below the capacities associated with the minimum nail schedule for segmented wall construction. This low resistance to lateral loading is why the masonry veneers are limited to one story in height when applied to prescriptive construction wall systems.

7.8 Connections

As stated earlier, timber structures rely on their connections to provide ductility and energy dissipation. Therefore, it is imperative that connections be given significant consideration when determining how to detail a structure for good seismic performance. In general, the concept for good performance of timber structures is to design connections where the steel yielding in the connection is the governing behavior, rather than the wood crushing. While both steel yielding and wood crushing occur in all connections that are loaded beyond their elastic range, the connection can be designed to favor the steel yielding as the dominant mechanism.

There are many types of connections used in wood. They range from nails, screws, and bolts (referred to as *dowel* connections) to metal plate connectors, shear plates, split rings, and other proprietary connectors. This chapter focuses on nails and bolts, since these are the most widely used connectors in timber structures. Connectors such as metal plate connectors and expanded tubes are proprietary connectors and the designer must contact the suppliers of these products to obtain the necessary information to safely design with them.

Dowel connections can be divided into two principal groups: small and large dowel connections. The designation refers to the relative length of the fastener in the wood member to the diameter, similar to the slenderness ratio for columns. This differentiation of dowel connections can be made because small dowel connections tend to be governed by the yield strength of the dowel and are usually considered to share the load equally. This is because as the individual fasteners yield, the load is redistributed to the other fasteners in the connection and all of the fasteners will yield and bend before the wood fails. On the other hand, large dowel connections tend to be increasingly governed by the crushing strength of the wood. Imperfections in the connection due to construction tolerances and variability of the wood material cause the load to be carried unequally between the individual fasteners in the connection (group action effects). The differentiation diameter for dimensional lumber connections is approximately $\frac{1}{4}$ in.

7.8.1 Design Methodology

In recent years, the design standards for timber construction in the United States changed the lateral design methodology for dowel connections from one with an empirical, restrictive basis to one based on mechanics. The new basis of lateral design is the yield strength of the fastener and the bearing strength of the materials being joined. This results in the ability to configure the connection assembly to provide a connection that is governed by the dowel yielding and bending, the dowel remaining straight and the wood crushing, or a combination of the two. This change also provides flexibility to the designer in choosing connection configurations that are more applicable to special situations, where the previous design methods were restricted to a few typical configurations unless special testing was done.

The new design method is called the yield theory and is best illustrated in Figure 7.15. As can be seen in the figure, there are basically four classes of yielding that are considered by the theory. Mode I is governed by crushing of the wood material by the dowel, and the bolt is held firmly by one member and does not even rotate as the connection displaces. Mode II is also governed by the crushing of the wood, but the dowel rotates as the connection displaces. Mode III is governed by a combination of the dowel yielding and the wood crushing. A single plastic hinge is formed in the dowel as the connection displaces. Finally, mode IV is governed by the dowel yielding, and two plastic hinges are formed in the dowel as the connection displaces.

The designer is referred to either the LRDF or ASD design standards for the complete set of applicable design equations for the particular type of dowel fastener being used. Each type of fastener (i.e., nails, spikes, wood screws, lag screws, bolts, and dowels) has a slightly different set of design equations that predict the yield mode and associated design value, since each type of fastener has different geometries for both the fastener itself and the connection as a whole.

7.8.2 Small-Diameter Dowel Connections

Small dowels, such as nails and screws, are by far the most prevalent fasteners used in wood construction. The framing in light-frame construction is typically nailed together and the sheathing applied to most timber structures is nailed to the framing. The driven fastener has several advantages over other options,

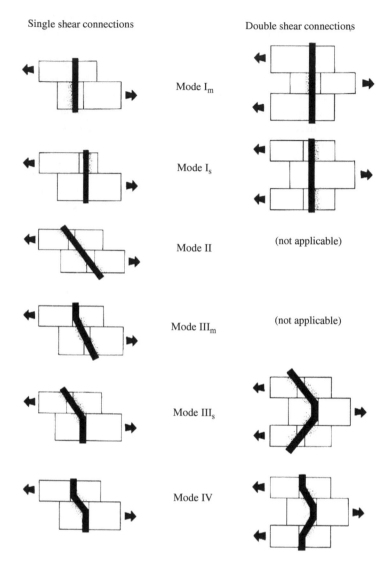

FIGURE 7.15 Yield modes considered for dowel connections in wood construction (courtesy of American Forest and Paper Association. 2001. *ASD Allowable Stress Design Manual for Engineered Wood Construction and Supplements*, Washington, DC).

in that it is easily installed (typically completed with pneumatic nailing tools and does not require predrilling), provides reasonable resistance to lateral and withdrawal loads, reduces splitting of the timber (unless the spacing is very close), and distributes the resistance over a larger area of the structure. In addition, these types of fasteners easily yield and provide significant levels of damping to the structure.

7.8.2.1 Lateral Resistance of Small-Diameter Fasteners

While these connections can be configured to yield in all of the modes, nails and screws are governed by a reduced set of equations due to the fact that certain subsets of the yield modes are not possible since the fastener does not pass through the entire connection for nails and screws. In dimensional lumber and larger sizes, small-diameter fasteners typically yield in Modes III and IV. These two yield modes provide the highest ductility and energy dissipation since the metal of the dowel yields in plastic yielding and the friction associated with the connection displacement is significant.

Since small dowel connections typically yield in the highly ductile yield modes, the displacement of the connection is such that by the time the assembly load reaches near capacity, all of the fasteners will have yielded and the load is assumed to be shared equally by all of the fasteners. In other words, there is no group action factor and the capacity of a multiple-fastener connection is equal to the sum of all of the connectors. In equation form, the resistance of a multiple nail or screw connection is

$$R' = \lambda \phi_z \sum Z' = \lambda \phi_z n Z' \tag{7.19}$$

where R' is the total factored resistance, λ is the applicable time effect factor, ϕ_z is the resistance factor for connections (0.65), Z' is the factored resistance for a single fastener, and n is the number of fasteners in the connection.

Small dowels are also not typically affected by grain direction. Nail and screw connections are typically governed almost completely by the yielding of the fastener. This implies that even the perpendicular-to-grain embedment strength of the wood is sufficiently high so that the yielding still is governed by the bending of the dowel.

7.8.2.2 Withdrawal Resistance of Small-Diameter Fasteners

Withdrawal of smooth shank nails can be problematic to the performance of timber structures. In some cases, such as when the nail is driven into the end-grain of a member (i.e., typically how the framing of light-frame timber walls are connected), the design resistance of the fastener is zero. On the other hand, driven into the side-grain of the member, the smooth shank nail may provide sufficient resistance to withdrawal.

Smooth shank fasteners are also significantly affected by changes in the moisture content of the wood during the life of the connection. If the connection is above 19% at any time during its life, and then dries out to below 19%, or vice versa, then 75% of the design withdrawal resistance is lost. If the connection is fabricated at one moisture condition and remains in this condition throughout its life, there is no reduction. Under these same changing moisture content conditions, the lateral design values for the connection are reduced by 30%. This implies that if a designer specifies, or allows, green lumber to be used to construct shear walls and roof systems, the design values used for the shear walls and roof systems should be reduced by a significant amount.

Hardened threaded and ring-shank nails overcome this weakness in performance by the way the deformed nail shank interlocks with the wood fibers. The result is that the design values of the threaded nails are not affected by changes in moisture content. This implies that if a designer allows green lumber to be used in a project, and does not wish to impose the design value reductions associated with moisture content changes, they should specify threaded or ring-shank nails. Helically threaded nails provide the best performance of all of the deformed shank nails available. This is because the deformation pattern allows the nail to withdraw a significant amount without a drop in resistance. This is due to the nail being able to withdraw without tearing the wood fibers around the nail itself. Other types of deformed nails such as ring-shank nails provide higher withdrawal resistance than smooth shank nails, but the resistance drops quickly as the nail withdraws due to the localized damage to the wood fibers.

7.8.3 Large-Diameter Dowel Connections

Large-diameter dowel connections can be designed to yield in any of the four modes shown in Figure 7.15. The diameter relative to the thickness of the timber member will determine the mode of yield that occurs. For a given thickness of timber member, the larger the diameter of the dowel, the lower the yield mode will be, and the less ductile the connection will be. This is the reason why the design standards have been reducing the maximum size of the bolt that is included in the design standard. While there is no restriction against designing with larger-diameter bolts, the LRFD and ASD manuals have 1-in. diameter bolts as the largest diameter included in any of the tables. If one uses the yield equations for bolts, lag screws, or dowels in the design standard, larger sizes can be used.

When determining the size of dowel to use in a given connection, a balancing act of choosing between a few large-diameter fasteners with high capacity and a larger number of fasteners with lower capacity must be performed by the designer. In both cases, the higher number of fasteners used in a given row of fasteners results in a group action effect. This means that the capacity of the connection is less than the sum of capacities of the individual fasteners. Large-diameter bolts do not share the load equally due to placement tolerances and variation in the material properties of the timber. The design specification provides an equation that determines the reduction factor associated with multiple-fastener connections, and should be used for all connections that utilize more than one fastener per row.

If one reviews Figure 7.15 again, it becomes clear that connections that yield in Modes I and II will fail in a brittle manner, since wood as a material fails in a brittle manner. By the same deduction process, connections that yield in Modes III and IV will behave in a more ductile manner and provide higher damping ability. Thus, connections that yield in Modes III and IV should be the preferred connection configurations for seismic design. This is not always possible, due to the fact that this implies that a larger number of smaller-diameter fasteners will be used, and space limitations may not allow the large number of fasteners to be used. If connections yielding in Modes I or II must be used in the design, and these connections are the controlling connections in the structure, then the designer should assume that the structure will remain essentially elastic so that the potential for a brittle failure is minimized.

Recent research by Heine (2001) and Anderson (2002) indicates that the current design specifications provide nonconservative results for multiple-bolt, single-shear connections. There are possibly two problems currently in the design specification. The first is that the minimum spacing requirement of four times the bolt diameter for full design strength may be insufficient to prevent the failure from occurring between the bolts. The second problem is that the current group action factors are derived based on an assumption of elastic response. Together these two issues result in connections that perform as much as 40% below the predicted strength. There is some evidence that the capacity of large connections (100 kip plus design loads) may not be able to achieve the anticipated design values. Figure 7.16 illustrates the experimental and theoretical results for connections made with $\frac{1}{2}$-in. diameter bolts. Note that the 4D spacing has a significant weakening effect on the connection. If one were to consider the LRFD group action factor for this same configuration, the smallest it

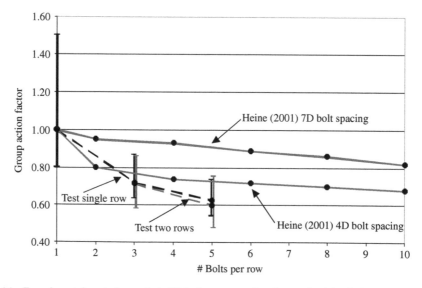

FIGURE 7.16 Experimental and theoretical (Heine) group action factor for $\frac{1}{2}$-in. bolts and variable spacing requirements (courtesy of Anderson, G.T. 2002. "Experimental Investigation of Group Action Factor for Bolted Wood Connections," M.S. thesis, Virginia Polytechnic Institute and State University, Blacksburg, VA).

would be is 0.88. Due to this problem, the 2001 edition of the ASD manual has a new appendix that provides guidance in designing multiple-bolt connections. The new appendix recommends additional checks on the capacity of the wood members to prevent block shear and other potential modes of failure in multiple-bolt connections.

This discussion of the yield modes and group action factor provides some guidance to the designer. First, the designer should try to use, when possible, connections that yield in Modes III and IV to provide the highest possible ductility and energy dissipation for the structure. Second, where possible, the spacing between bolts should be increased to at least seven times the diameter of the bolt (the minimum spacing for full design load that is required by the European design standard). This will minimize the group action effects in multiple-bolt connections. Third, the designer should follow the additional checks outlined in the new appendix for the 2001 ASD manual for preventing unintentional failure modes from occurring in multiple-bolt connections.

7.8.4 Heavy Timber Connectors

Shear plates and split rings are timber connectors that have very high capacities, but also tend to fail in more brittle modes. Historically, these connections have been used in heavy timber and glulam timber structures. Several applications have been made in large timber bow-string truss and glulam timber connections. Design of these connections is covered by the LRDF (American Forest and Paper Association 1996) and ASD (American Forest and Paper Association 2001) manuals, and the reader is referred to either of these documents for a clear description of how to design these connections. Additional guidance on design of these types of connections as well as general heavy timber structure design may be found in the *Timber Construction Manual* (American Institute of Timber Construction 2004). These connections should be considered in a similar class of connection to large-dowel connections. The connections are susceptible to group action and geometry effects in similar ways in which large-dowel connections experience the same phenomena.

While these connectors are good for situations where large numbers of smaller-dowel connections might be required, there are potential problems associated with their use. The first problem is that most contractors are not familiar with the installation of these specialized connectors. There are special tools for drilling and cutting the required surfaces to ensure proper bearing of the connectors in the connection, and the contractor would be advised to manufacture a couple of trial connections before beginning the real connections to learn how to use the cutting tools properly. Second, the connection cannot easily be inspected to ensure proper installation after the members are joined. The main components in these two types of connections need to bear uniformly around their perimeter. If the holes are overdrilled, not drilled circular, or both sides of the connection do not exactly line up for all of the connectors, the connection will not perform properly and will fail at load levels below the intended design level. Also, once the connection is assembled, there is no way to see whether the shear plate or split ring is even present, let alone installed properly. Therefore, these types of connections need to be constructed by a conscientious contractor who can be trusted to perform the installation properly, or continuous inspection of the installation by a responsible party is recommended.

Glossary

Anisotropic — Having different properties in different directions (i.e., not *isotropic*, which is to have the same properties in all directions).

Bow — The distortion of lumber in which there is a deviation, in a direction perpendicular to the flat face, from a straight line from end to end of the piece.

Box system — A type of lateral-force-resisting system (LFRS) in which forces are transmitted via horizontal diaphragms (i.e., roof and floors) to gravity load-bearing walls that act in shear, and thus form the main vertical elements of the LFRS.

Checking — A lengthwise separation of the wood that usually extends across the rings of annual growth and commonly results from stresses set up in wood during seasoning.

Creep — The continued increase in deflection that occurs in the viscoelastic material that sustains a constant load.

Cripple walls — Short stud walls between the foundation and the first-floor level.

Cup — A distortion of a board in which there is a deviation flat-wise from a straight line across the width of the board.

Diaphragm — A nearly horizontal structural unit that acts as a deep beam or girder when flexible relative to supports, and as a plate when its stiffness is higher than the associated stiffness of the walls.

Dowel — A wood connector, such as a nail, screw, or bolt.

Glulam — Glued–laminated lumber is an engineered product made by gluing together 50 mm (2 in.) or thinner pieces of lumber.

Hygroscopic — Readily absorbing moisture, as from the atmosphere.

I-joists — Wood joists that are structural members composed of an oriented strand board (OSB) or plywood web and two laminated vener lumber (LVL), oriented strand lumber (OSB), or solid lumber flanges, the newest and fastest growing engineered wood product, partially due to the declining availability of high-quality, large-dimension lumber for which it substitutes.

Laminated veneer lumber (LVL) — A structural composite lumber product made by adhesively bonding thin sheets of wood veneer oriented with the grain parallel in the long direction. Primary uses include headers, beams, rafters, and flanges for wood I-joists.

Lateral-force-resisting system (LFRS) — Any continuous load path potentially capable of resisting lateral (e.g., seismic) forces.

Lumber — Timber sawed into boards, planks, or other structural members of standard or specified dimensions.

Oriented strand board (OSB) — A structural panel made by adhesively bonding chips of small-diameter softwoods and previously underutilized hardwoods.

Orthotropic — Having different properties at right angles.

Plywood — A structural panel made from wood sheets (typically $\frac{1}{8}$-in. thick) peeled from tree trunks and adhesively laminated so as to orthogonally orient the wood grain in alternate plies.

Spaced column — Compression member made of two or more parallel single-member columns that are separated by spacers, located at specific locations along the column, and are rigidly tied together at the ends of the column.

Structural composite lumber (SCL) — A structural member, made by adhesively bonding thin strips of wood, oriented with the grain parallel to the long axis of the member. Primarily used for headers, heavily loaded beams, girders, and heavy timber construction.

Timber — Trees considered as a source of wood. Also, timbers used in the original round form, such as poles, piling, posts, and mine timbers.

Viscoelastic — A material that exhibits both viscous (time-dependent) and elastic responses to deformation.

Warp — Any variation from a true or plane surface. Warp includes bow, crook, cup, and twist, or any combination thereof.

References

American Forest and Paper Association. 1996. *Load and Resistance Factor Design (LRFD) Manual For Engineered Wood Construction*, American Forest and Paper Association, Washington, DC.

American Forest and Paper Association. 2001. *National Design Specification for Wood Construction*, American Forest and Paper Association, Washington, DC.

American Forest and Paper Association. 2001b. *ASD Allowable Stress Design Manual for Engineered Wood Construction and Supplements*, American Forest and Paper Association, Washington, DC.

American Forest and Paper Association/American Society of Civil Engineers. 2001. *Wood Frame Construction Manual (WFCM) for One- and Two-Family Dwellings*, American Forest and Paper Association/American Society of Civil Engineers, Washington, DC.

American Institute of Timber Construction. 2004. *Timber Construction Manual*, 5th ed., John Wiley & Sons, New York.

American Society of Civil Engineers. 1996. *Load and Resistance Factor Design (LRFD) for Engineered Wood Construction*, AF&PA/ASCE 16–95, American Society of Civil Engineers, New York.

American Society of Civil Engineers. 2002. *Minimum Design Loads for Buildings and Other Structures*, ASCE 7–02, American Society of Civil Engineers, New York.

Anderson, G.T. 2002. "Experimental Investigation of Group Action Factor for Bolted Wood Connections," M.S. thesis, Virginia Polytechnic Institute and State University, Blacksburg, VA.

APA, The Engineered Wood Association. 1997. *The Plywood Design Specification and Supplements*, APA, Tacoma, WA.

APA, The Engineered Wood Association. 2003. *Engineered Wood Construction Guide*, APA, Tacoma, WA.

Applied Technology Council. 1981. *Guidelines for the Design of Horizontal Wood Diaphragms*, ATC-7, Applied Technology Council, Redwood, CA.

Benuska, L., Ed. 1990. "Loma Prieta Earthquake Reconnaissance Report," *Earthquake Spectra*, 6 (Suppl.).

Breyer, D., Fridley, K.J., Pollock, D.G., and Cobeen, K.E. 2003. *Design of Wood Structures*, 5th ed., McGraw-Hill, New York.

Building Seismic Safety Council. 2003a. *NEHRP Recommended Provisions for Seismic Regulations for New Buildings and Other Structures. Part I: Provisions FEMA 368*, Building Seismic Safety Council, Washington, DC.

Building Seismic Safety Council. 2003b. *NEHRP Recommended Provisions for Seismic Regulations for New Buildings and Other Structures. Part 2: Commentary FEMA 369*, Building Seismic Safety Council, Washington, DC.

Council of American Building Officials. 1995. *CABO One- and Two-Family Dwelling Code*, Council of American Building Officials, Falls Church, VA.

Dolan, J.D. and Heine, C.P. 1997a. *Monotonic Tests of Wood-Frame Shear Walls with Various Openings and Base Restraint Configurations*, Timber Engineering Center Report no. TE-1997–001, Virginia Polytechnic Institute and State University, Blacksburg, VA.

Dolan, J.D. and Heine, C.P. 1997b. *Sequential Phased Displacement Cyclic Tests of Wood-Frame Shear Walls with Various Openings and Base Restraint Configurations*, Timber Engineering Center Report no. TE-1997–002, Virginia Polytechnic Institute and State University, Blacksburg, VA.

Dolan, J.D. and Johnson, A.C. 1997a. *Monotonic Performance of Perforated Shear Walls*, Timber Engineering Center Report no. TE-1996–001, Virginia Polytechnic Institute and State University, Blacksburg, VA.

Dolan, J.D. and Johnson, A.C. 1997b. *Sequential Phased Displacement (Cyclic) Performance of Perforated Shear Walls*, Timber Engineering Center Report no. TE-1996–002, Virginia Polytechnic Institute and State University, Blacksburg, VA.

Dolan, J.D., Bott, W., and Easterling, W.S. 2002. *Design Guidelines for Timber Diaphragms*, CUREE Publication W-XX, Consortium of Universities for Research in Earthquake Engineering, Richmond, CA.

Faherty, K. and Williamson, T. 1999. *Wood Engineering and Construction Handbook*, 3rd ed., McGraw-Hill, New York.

Federal Emergency Management Agency (FEMA). 1988. *Rapid Visual Screening of Buildings for Potential Seismic Hazards: A Handbook*, FEMA 154, Federal Emergency Management Agency, Washington, DC.

Fischer, D., Filliatrault, A., Folz, B., Uang, C.-M., and Seible, F. 2001. *Shake Table Tests of a Two-Story Woodframe House*, CUREE Publications W-06, Consortium of Universities for Research in Earthquake Engineering, Richmond, CA.

Heine, C.P. 1997. "Effect of Overturning Restraint on the Performance of Fully Sheathed and Perforated Timber Framed Shear Walls," M.S. thesis, Virginia Polytechnic Institute and State University, Blacksburg, VA.

Heine, C.P. 2001. "Simulated Response of Degrading Hysteretic Joints with Slack Behavior," Ph.D. dissertation, Virginia Polytechnic and State University, Blacksburg, VA.

Holmes, W.T. and Somers, P., Eds. 1996. "Northridge Earthquake Reconnaissance Report, Vol. 2," *Earthquake Spectra*, 11 (Suppl. C), 125–176.

International Code Council (ICC). 2003a. *International Building Code (IBC)*, Falls Church, VA.

International Code Council (ICC). 2003b. *International Residential Code (IRC)*, Falls Church, VA.

Lew, H.S., Ed. 1990. *Performance of Structures during the Loma Prieta Earthquake of October 17, 1989*, NIST Special Publication 778, National Institute of Standards and Technology, Gaithersburg, MD.

National Fire Prevention Association. 2002. *NFPA 5000 Building Code*, National Fire Prevention Association, Boston, MA.

Salenikovich, A.J. 2000. "The Racking Performance of Light-Frame Shear Walls," Ph.D. dissertation, Virginia Polytechnic Institute and State University, Blacksburg, VA.

Schierle, G.G. 2000. *Northridge Earthquake Field Investigations: Statistical Analysis of Woodframe Damage*, CUREE Publication W-02, Consortium of Universities for Research in Earthquake Engineering, Richmond, CA.

Steinbrugge, K.V., Schader, E.E., Bigglestone, H.C., and Weers, C.A. 1971. *San Fernando Earthquake February 9, 1971*, Pacific Fire Rating Bureau, San Francisco, CA.

Stewart, W.G. 1987. "The Seismic Design of Plywood Sheathed Shearwalls," Ph.D. dissertation, University of Canterbury, New Zealand.

Sugiyama, H. and Matsumoto, T. 1994. "Empirical Equations for the Estimation of Racking Strength of a Plywood Sheathed Shear Wall with Openings," *Mokuzai Gakkaishi*, 39, 924–929.

Tissell, J. 1990. "Performance of Wood-Framed Structures in the Loma Prieta Earthquake," in *Wind and Seismic Effects: Proceedings of the Twenty-Second Joint Meeting of the U.S.–Japan Cooperative Program in Natural Resources Panel on Wind and Seismic Effects*, NIST SP 796, National Institute of Standards and Technology, Gaithersburg, MD, September, pp. 324–330.

U.S. Department of Agriculture. 1999. *Wood Handbook: Wood as an Engineering Material*, Agriculture Handbook 72, Forest Products Laboratory, U.S. Department of Agriculture, Madison, WI (available online at http://www.fpl.fs.fed.us/documnts/FPLGTR/fplgtr113/fplgtr113.htm).

U.S. Department of Housing and Urban Development. 1995. *Preparing for the "Big One": Saving Lives through Earthquake Mitigation in Los Angeles, California, HUD-I511-PD&R*, January 17, U.S. Department of Housing and Urban Development, Washington, DC.

Yancey, C.W. et al. 1998. *A Summary of the Structural Performance of Single Family Wood Framed Housing, NISTIR 6224*, Building and Fire Research Laboratory, National Institute of Standards and Technology, Gaithersburg, MD.

Further Reading

The *Wood Handbook* (U.S. Department of Agriculture 1999) is a good introduction to the properties of wood and wood products. The several wood design handbooks (American Institute of Timber Construction 2004; American Forest and Paper Association 1996, 2001; APA 1997, 2003) are all required references for wood structure designers, and contain much useful background information. Breyer et al. (2003) is a good overall text for wood structure design. The Building Seismic Safety Council (2003a,b) provides the current consensus guidelines specific to seismic design of wood buildings. The CUREE project (www.curee.org) is a major research effort to better understand wood building performance and develop improved design data and practices. The CUREE Website provides publications and other information.

8

Aluminum Structures

Maurice L. Sharp
Consultant — Aluminum Structures,
Avonmore, PA

8.1 Introduction

8.1.1 The Material

8.1.1.1 Background

Of the structural materials used in construction, aluminum was the latest to be introduced into the market place even though it is the most abundant of all metals, making up about $\frac{1}{12}$ of the earth's crust. The commercial process was invented simultaneously in the United States and Europe in 1886. Commercial production of the metal started thereafter using an electrolytic process that economically separated aluminum from its oxides. Prior to this time aluminum was a precious metal. The initial uses of aluminum were for cooking utensils and electrical cables. The earliest significant structural use of aluminum was for the skins and members of a dirigible called the *Shenendoah* completed in 1923. The first structural design handbook was developed in 1930 and the first specification was issued by the industry in 1932 (Sharp 1994).

8.1.1.2 Product Forms

Aluminum is available in all the common product forms, flat-rolled, extruded, cast, and forged. Fasteners such as bolts, rivets, screws, and nails are also manufactured. The available thicknesses of flat-rolled products range from 0.006 in. or less for foil to 7.0 in. or more for plate. Widths to 17 ft are possible. Shapes in aluminum are extruded. Some presses can extrude sections up to 31 in. wide. The extrusion process allows the material to be placed in areas that maximize structural properties and joining ease. Because the cost of extrusion dies is relatively low, most extruded shapes are designed for specific applications.

   Castings of various types and forgings are possibilities for three-dimensional shapes and are used in some structural applications. The design of castings is not covered in detail in structural design books and specifications primarily because there can be a wide range of quality depending on the casting process. The quality of the casting affects structural performance.

8.1.1.3 Alloy and Temper Designation

The four-digit number used to designate alloys is based on the main alloying ingredients. For example, magnesium is the principal alloying element in alloys whose designation begins with a 5 (5083, 5456, 5052, etc.).

Cast designations are similar to wrought designations but a decimal is placed between the third and fourth digits (356.0). The second part of the designation is the temper, which defines the fabrication process. If the term starts with T, for example, -T651, the alloy has been subjected to a thermal heat treatment. These alloys are often referred to as heat-treatable alloys. The numbers after the T show the type of treatment and any subsequent mechanical treatment such as a controlled stretch. The temper of alloys that harden with mechanical deformation starts with H, for example, -H116. These alloys are referred to as non-heat-treatable alloys. The type of treatment is defined by the numbers in the temper designation. A 0 temper is the fully annealed temper. The full designation of an alloy has the two parts that define both chemistry and fabrication history, for example, 6061-T651.

8.1.2 Alloy Characteristics

8.1.2.1 Physical Properties

Physical properties usually vary only by a few percent depending on the alloy. Some nominal values are given in Table 8.1. The density of aluminum is low, about one third that of steel, which results in lightweight structures. The modulus of elasticity is also low, about one third of that of steel, which affects design when deflection or buckling controls.

8.1.2.2 Mechanical Properties

Mechanical properties for a few alloys used in general purpose structures are given in Table 8.2. The stress–strain curves for aluminum alloys do not have an abrupt break when yielding but rather have a gradual bend (see Figure 8.1). The yield strength is defined as the stress corresponding to a 0.002 in./in. permanent set. The alloys shown in Table 8.2 have moderate strength, excellent resistance to corrosion in the atmosphere, and are readily joined by mechanical fasteners and welds. These alloys often are employed in outdoor structures without paint or other protection. The higher-strength aerospace alloys are not shown. They usually are not used for general purpose structures because they are not as resistant to corrosion and normally are not welded.

8.1.2.3 Toughness

The accepted measure of toughness of aluminum alloys is fracture toughness. Most high-strength aerospace alloys can be evaluated in this manner; however, the moderate-strength alloys employed for

TABLE 8.1 Some Nominal Properties of Aluminum Alloys

Property	Value
Weight	$0.1 \, lb/in.^3$
Modulus of elasticity	
Tension and compression	10,000 ksi
Shear	3,750 ksi
Poisson's ratio	1/3
Coefficient of thermal expansion (68 to 212°F)	0.000013 per °F

Source: Gaylord, Gaylord, and Stallmeyer, *Structural Engineering Handbook*, McGraw-Hill, 1997.

TABLE 8.2   Minimum Mechanical Properties

Alloy and temper	Product	Thickness range, in.	Tension		Compression	Shear		Bearing	
			TS	YS	YS	US	YS	US	YS
3003-H14	Sheet and plate	0.009–1.000	20	17	14	12	10	40	25
5456-H116	Sheet and plate	0.188–1.250	46	33	27	27	19	87	56
6061-T6	Sheet and plate	0.010–4.000	42	35	35	27	20	88	58
6061-T6	Shapes	All	38	35	35	24	20	80	56
6063-T5	Shapes	to 0.500	22	16	16	13	9	46	26
6063-T6	Shapes	All	30	25	25	19	14	63	40

Note: All properties are in ksi. TS is the tensile strength, YS is the yield strength, and US is the ultimate strength.
Source: The Aluminum Association, *Structural Design Manual*, 2000.

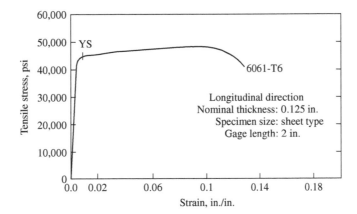

FIGURE 8.1   Stress–strain curve.

general purpose structures cannot be evaluated because they are too tough to get valid results in the test. Aluminum alloys also do not exhibit a transition temperature, their strength and ductility actually increase with decrease in temperature. Some alloys have a high ratio of yield strength to tensile strength (compared to mild steel) and most alloys have a lower elongation than mild steel, perhaps 8–10%, both considered to be negative factors for toughness. However, these alloys do have sufficient ductility to redistribute stresses in joints and in sections in bending to achieve full strength of the components. Their successful use in various types of structures, bridges, bridge decks, tractor trailers, railroad cars, building structures, and automotive frames, has demonstrated that they have adequate toughness. Thus far, there has not been a need to modify design based on toughness of aluminum alloys.

8.1.3   Codes and Specifications

Allowable stress design (ASD) for building, bridge, and other structures, which need the same factor of safety, and load and resistance factor design (LRFD) for building and similar type structures have been published by the Aluminum Association (2000). These specifications are included in a design manual that also has design guidelines, section properties of shapes, design examples, and numerous other aids for the designer.

The American Association of State Highway and Transportation Officials have published LRFD Specifications that cover bridges of aluminum and other materials (AASHTO 2002). The equations for strength and behavior of aluminum components are essentially the same in all of these specifications. The margin of safety for design differs depending on the type of specification and the type of structure.

Codes and standards are available for other types of aluminum structures. Lists and summaries are provided elsewhere (Sharp 1993; Aluminum Association 2000).

8.2 Structural Behavior

8.2.1 General

8.2.1.1 Behavior Compared to Steel

The basic principles of design for aluminum structures are the same as those for other ductile metals such as steel. Equations and analysis techniques for global structural behavior such as load–deflection behavior are the same. Component strength, particularly buckling, postbuckling, and fatigue, are defined specifically for aluminum alloys. The behavior of various types of components are provided in Sections 8.2.1.3–8.2.1.6 and Section 8.2.2. Strength equations are also given. The designer needs to incorporate appropriate factors of safety when these equations are used for practical designs.

8.2.1.2 Safety and Resistance Factors

Table 8.3 gives factors of safety as utilized for ASD. The calculated strength of the part is divided by these factors. This allowable stress must be less than the stress calculated using the total load applied to the part. In LRFD, the calculated strength of the part is multiplied by the resistance factors given in Table 8.4. This calculated stress must be less than that calculated using factored loads. Equations for determining the factored loads are given in the appropriate specifications discussed previously.

8.2.1.3 Buckling Curves for Alloys

The equations for the behavior of aluminum components apply to all thicknesses of material and to all aluminum alloys. Equations for buckling in the elastic and inelastic range are provided. Figure 8.2

TABLE 8.3   Factors of Safety for Allowable Stress Design

Component	Failure mode	Buildings and similar-type structures	Bridges and similar-type structures
Tension	Yielding	1.95	2.20
	Ultimate strength	1.65	1.85
Columns	Yielding (short column)	1.65	1.85
	Buckling	1.95	2.20
Beams	Tensile yielding	1.65	1.85
	Tensile ultimate	1.95	2.20
	Compressive yielding	1.65	1.85
	Lateral buckling	1.65	1.85
Thin plates in compression	Ultimate in columns	1.95	2.20
	Ultimate in beams	1.65	1.85
Stiffened flat webs in shear	Shear yield	1.65	1.85
	Shear buckling	1.20	1.35
Mechanically fastened joints	Bearing yield	1.65	1.85
	Bearing ultimate	2.34	2.64
	Shear str./rivets, bolts	2.34	2.64
Welded joints	Shear str./fillet welds	2.34	2.64
	Tensile str./butt welds	1.95	2.20
	Tensile yield/butt welds	1.65	1.85

Source: The Aluminum Association, *Structural Design Manual,* 2000.

TABLE 8.4   Resistance Factors for LRFD

Component	Limit state	Buildings	Bridges
Tension	Yielding	0.95	0.90
	Ultimate strength	0.85	0.75
Columns	Buckling	Varies with slenderness ratio	Varies with slenderness ratio
Beams	Tensile yielding	0.95	0.90
	Tensile ultimate	0.85	0.80
	Compressive yielding	0.95	0.90
	Lateral buckling	0.85	0.80
Thin plates in compression	Yielding	0.95	0.90
	Ultimate strength	0.85	0.80
Stiffened flat webs in shear	Yielding	0.95	0.90
	Buckling	0.90	0.80

Sources: Buildings — The Aluminum Association, *Structural Design Manual*, 2000; Bridges — American Association of State Highway and Transportation Officials, *AASHTO LRFD Bridge Design Specifications*, 2002.

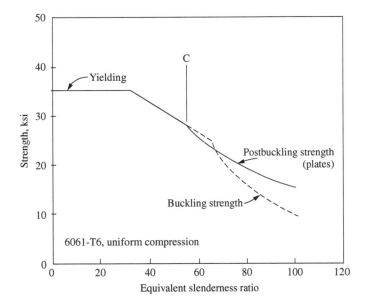

FIGURE 8.2   Buckling of components.

shows the format generally used for both component and element behavior. The strength of the component is normally considered to be limited by the yield strength of the material. For buckling behavior, coefficients are defined for two classes of alloys, those that are heat treated with temper designations -T5 or higher and those that are not heat treated or are heat treated with temper designations -T4 or lower. Different coefficients are needed because of the differences in the shapes of the stress–strain curves for the two classes of alloys.

8.2.1.4 Effects of Welding

In most applications, some efficiency is obtained by using alloys that have been thermally treated or strain hardened to achieve higher strength. The alloys are readily welded. However, welding partially anneals a narrow band of material (about 1.0 in. on either side of the weld) and thus this *heat-affected* material has a lower strength than the rest of the member. The lower strength is accounted for in the design equations presented in Equation 8.1.

If the strength of the heat-affected material is less than the yield strength of the parent material, the plastic deformation of the component at failure loads will be confined to that of the narrow band of lower-strength material. In this case, the component fails with only a small total deformation, thus exhibiting low structural toughness. For good structural toughness the strength of the heat-affected material should be well above the yield strength of the parent material. In the case of liquid natural gas containers an annealed temper of the plate, 5083-0, has been employed to achieve maximum toughness. The strength of the welded material is the same as that of the parent material and there is essentially no effect of welding on structural behavior.

8.2.1.5 Effects of Temperature

All of the properties important to structural behavior, static strength, elongation, fracture toughness, and fatigue strength, increase with decrease in temperature. Elongation increases but static and fatigue strengths decrease at elevated temperatures. Alloys behave differently but significant changes in mechanical properties can occur at temperatures over 300°F.

8.2.1.6 Effects of Strain Rate

Aluminum alloys are relatively insensitive to strain rate. There is some increase in mechanical properties at high strain rates. Thus, published strength data, based on conventional tests, are normally used for calculations for cases of rapidly applied loads.

8.2.2 Component Behavior

This section presents equations and discussion for determining the strength of various types of aluminum components. These equations are consistent with those employed in current specifications and publications for aluminum structures.

8.2.2.1 Members in Tension

As various alloys have different amounts of strain hardening, both yielding and fracture strength of members should be checked. The net section of the member is used in the calculation. The calculated net section stress is compared to the yield strength and tensile strength of the alloy as given in Table 8.2. Larger factors of safety are applied for ultimate than yield strength in both ASD and LRFD specifications as noted in Table 8.3 and Table 8.4.

The strengths across a groove weld are given for some alloys in Table 8.5. These properties are used for the design of tension members with transverse welds that affect the entire cross-section. For members

TABLE 8.5 Minimum Strengths of Groove Welds

Parent material	Filler metal	Tension TS[a]	Tension YS[b]	Compression YS[b]	Shear US
3003-H14	1100	14	7	7	10
5456-H116	5556	42	26	24	25
6061-T6	5356	24	20	20	15
6061-T6	4043	24	15	15	15
6063-T5, -T6	4043	17	11	11	11

[a] ASME weld-qualification values. The design strength is considered to be 90% of these values.
[b] Corresponds to a 0.2% set on a 10-in. gage length.
  Notes: All strengths are in ksi. TS is the tensile strength, YS is the yield strength, and US is the ultimate strength.
  Source: The Aluminum Association, *Structural Design Manual*, 2000.

with longitudinal welds in which only part of the cross-section is affected by welds the tensile or yield strength may be calculated using the following equation:

$$F_{pw} = F_n - \frac{A_w}{A}(F_n - F_w) \tag{8.1}$$

where F_{pw} is the strength of the member with a portion of cross-section affected by welding, F_n is the strength of the unaffected parent metal, F_w is the strength of the material affected by welding, A is the area of cross-section, and A_w is the area that lies within 1 in. of a weld.

8.2.2.2 Columns Under Flexural Buckling

The Euler column formula is employed for the elastic region and straight line equations in the inelastic region. The straight line equations are a close approximation to the tangent modulus column curve. The equations for column strength are as follows:

$$F_c = B_c - D_c \frac{KL}{r}, \quad \frac{KL}{r} \le C_c \tag{8.2}$$

$$F_c = \frac{\pi^2 E}{(KL/r)^2}, \quad \frac{KL}{r} > C_c \tag{8.3}$$

where F_c is the column strength (ksi), L is the unsupported length of column (in.), r is the radius of gyration (in.), K is the effective-length factor, E is the modulus of elasticity (ksi), and B_c, D_c, C_c are constants depending on mechanical properties (see Equations 8.4 to 8.9).

For wrought products with tempers starting with, -O, -H, -T1, -T2, -T3, and -T4, and cast products

$$B_c = F_{cy} \left[1 + \left(\frac{F_{cy}}{1000} \right)^{1/2} \right] \tag{8.4}$$

$$D_c = \frac{B_c}{20} \left(\frac{6B_c}{E} \right)^{1/2} \tag{8.5}$$

$$C_c = \frac{2B_c}{3D_c} \tag{8.6}$$

For wrought products with tempers starting with -T5, -T6, -T7, -T8, and -T9

$$B_c = F_{cy} \left[1 + \left(\frac{F_{cy}}{2250} \right)^{1/2} \right] \tag{8.7}$$

$$D_c = \frac{B_c}{10} \left(\frac{B_c}{E} \right)^{1/2} \tag{8.8}$$

$$C_c = 0.41 \frac{B_c}{D_c} \tag{8.9}$$

where F_{cy} is the compressive yield strength (ksi).

The column strength of a welded member is generally less than that of a member with the same cross-section but without welds. If the welds are longitudinal and affect part of the cross-section, the column strength is given by Equation 8.1. The strengths in this case are column buckling values assuming all parent metal and all heat-affected metal. If the member has transverse welds that affect the entire cross-section, and occur away from the ends, the strength of the column is calculated assuming that the entire column is a heat-affected material. Note that the constants for the heat-affected materials are given by Equations 8.4 to 8.6. If transverse welds occur only at the ends, the equations for parent metal are used but the strength is limited to the yield strength across the groove weld.

8.2.2.3 Columns Under Flexural–Torsional Buckling

Thin, open sections that are unsymmetrical about one or both principal axes may fail by combined torsion and flexure. This strength may be estimated using a previously developed equation that relates the combined effects to pure flexural and pure torsional buckling of the section. Equation 8.10 is in the form of effective and equivalent slenderness ratios and is in good agreement with test data (Sharp 1993). The equation must be solved by trial for the general case

$$\left[1 - \left(\frac{\lambda_c}{\lambda_y}\right)^2\right]\left[1 - \left(\frac{\lambda_c}{\lambda_x}\right)^2\right]\left[1 - \left(\frac{\lambda_c}{\lambda_\phi}\right)^2\right] - \left(\frac{y_0}{r_0}\right)^2\left[1 - \left(\frac{\lambda_c}{\lambda_x}\right)^2\right] - \left(\frac{x_0}{r_0}\right)^2\left[1 - \left(\frac{\lambda_c}{\lambda_y}\right)^2\right] = 0 \qquad (8.10)$$

where λ_c is the equivalent slenderness ratio for flexural–torsional buckling; λ_x, λ_y are the slenderness ratios for flexural buckling in the x and y directions, respectively; x_0, y_0 are the distances between centroid and shear center, parallel to principal axes, $r_0 = [(I_{xo}I_{yo})/A]^{1/2}$; I_{xo}, I_{yo} are the moments of inertia about axes through shear center; and λ_ϕ is the equivalent slenderness ratio for torsional buckling

$$\lambda_\phi = \sqrt{\frac{I_x + I_y}{(3J/8\pi^2) + (C_w/(K_\phi L)^2)}} \qquad (8.11)$$

where J is the torsion constant, C_w is the warping constant, K_ϕ is the effective length coefficient for torsional buckling, L is the length of the column, and I_x, I_y are the moments of inertia about the centroid (principal axes).

8.2.2.4 Beams

Beams that are supported against lateral–torsional buckling fail by excessive yielding or fracture of the tension flange at bending strengths above that corresponding to stresses reaching the tensile or yield strength at the extreme fiber. This additional strength may be accounted for by applying a shape factor to the tensile or yield strength of the alloy. Nominal shape factors for some aluminum shapes are given in Table 8.6. These factors vary slightly with alloy because they are affected by the shape of the stress–strain curve but the values shown are reasonable for all aluminum alloys.

This higher bending strength can be developed provided that the cross-section is compact enough so that local buckling does not occur at a lower stress. Limitations on various types of elements are given in Table 8.7. The bending moment for compact sections is as follows:

$$M = ZSF \qquad (8.12)$$

where M is the moment corresponding to yield or ultimate strength of the beam, S is the section modulus of the section, F is the yield or tensile strength of the alloy, and Z is the shape factor.

TABLE 8.6   Shape Factors for Aluminum Beams

Cross-section	Yielding, K_y	Ultimate, K_u
I and channel (major axis)	1.07	1.16
I (minor axis)	1.30	1.42
Rectangular tube	1.10	1.22
Round tube	1.17	1.24
Solid rectangle	1.30	1.42
Solid round	1.42	1.70

Sources: Gaylord, Gaylord, and Stallmeyer, *Structural Engineering Handbook*, McGraw-Hill, 1997 and The Aluminum Association, *Structural Design Manual*, 2000.

TABLE 8.7 Limiting Ratios of Elements for Plastic Bending

Element	Limiting ratio
Outstanding flange of I or channel	$b/t \leq 0.30 \ (E/F_{cy})^{1/2}$
Lateral buckling of I or channel	
Uniform moment	$L/r_y \leq 1.2 \ (E/F_{cy})^{1/2}$
Moment gradient	$L/r_y \leq 2.2 \ (E/F_{cy})^{1/2}$
Web of I or rectangular tube	$b/t \leq 0.45 \ (E/F_{cy})^{1/2}$
Flange of rectangular tube	$b/t \leq 1.13 \ (E/F_{cy})^{1/2}$
Round tube	$D/t \leq 2.0 \ (E/F_{cy})^{1/2}$

Source: Gaylord, Gaylord, and Stallmeyer, *Structural Engineering Handbook,* McGraw-Hill, 1997.

8.2.2.5 Effects of Joining

If there are holes in the tension flange the net section should be used for calculating the section modulus. Welding affects beam strength in the same way as it does tensile strength. The groove weld strength is used when the entire cross-section is affected by welds. Beams may not develop the bending strength as given by Equation 8.12 at the locations of the transverse welds. In these locations it is reasonable to use a shape factor equal to 1.0. If only part of the section is affected by welds, Equation 8.1 is used to calculate strength and compact sections can develop the moment as given by Equation 8.12. In the calculation the flange is considered to be the area that lies farther than two-thirds of the distance between the neutral axis and the extreme fiber.

8.2.2.6 Lateral Buckling

Beams that do not have continuous support for the compression flange may fail by lateral buckling. For aluminum beams an equivalent slenderness is defined and substituted in column formulas in place of KL/r. The slenderness ratios for buckling of I-sections, WF-shapes, and channels are as follows:

For beams with end moments only or transverse loads applied at the neutral axis:

$$\lambda_b = 1.4 \frac{L_b}{\sqrt{\dfrac{I_y d C_b}{S_c} \sqrt{1.0 + 0.152 \dfrac{J}{I_y} \left(\dfrac{L_b}{d}\right)^2}}} \tag{8.13}$$

For beams with loads applied to top and bottom flanges where the load is free to move laterally with the beam:

$$\lambda_b = 1.4 \frac{L_b}{\sqrt{\dfrac{I_y d C_b}{S_c}} \left[\pm 0.5 + \sqrt{1.25 + 0.152 \dfrac{J}{I_y} \left(\dfrac{L_b}{d}\right)^2}\right]} \tag{8.14}$$

where

λ_b = equivalent slenderness ratio for beam buckling (to be used in place of KL/r in the column formula)
C_b = coefficient depending on loading and beam supports, which is equal to
$\quad 12.5 M_{max}/(2.5 M_{max} + 3 M_A + 4 M_B + 3 M_C)$ for simple supports, wherein
$\quad M_{max}$ = absolute value of maximum moment in the unbraced beam segment
$\quad M_A$ = absolute value of moment at quarter-point of the unbraced beam segment
$\quad M_B$ = absolute value of moment at midpoint of the unbraced beam segment
$\quad M_C$ = absolute value of moment at three-quarter point of the unbraced beam segment
d = depth of beam

I_y = moment of inertia about the axis parallel to the web
S_c = section modulus for compression flange
J = torsion constant

A plus sign is to be used in Equation 8.14 if the load acts on the bottom (tension) flange and a minus sign if it acts on the top (compression) flange.

Equations 8.13 and 8.14 may also be used for cantilever beams of the specified cross-section by the use of the appropriate factor C_b. For a concentrated load at the end the factor is 1.28, and for a uniform lateral load the factor is 2.04.

These equations also may be applied to I-sections in which the tension and compression flanges are of somewhat different sizes. In this case the beam properties are calculated as though the tension flange is of the same size as that of the compression flange. The depth of the section is maintained.

Lateral buckling strengths of welded beams are affected similarly to that of flexural buckling of columns. For cases in which part of the compression flange has a heat-affected material, Equation 8.1 is used. The total flange area is that farther than two-thirds the distance from the neutral axis to the extreme fiber. If the beam has transverse welds away from the supports, the strength of the beam is calculated as though the entire beam is of a heat-affected material.

For other types of cross-sections and loadings not provided for above, and for cases in which the loads cause torsional stresses in the beam, other equations and analysis are needed. Some cases are covered elsewhere (Sharp 1993; Aluminum Association 2000).

8.2.2.7 Members Under Combined Bending and Axial Loads

The same interaction equations may be used for aluminum as for steel members. The following equations are for bending in one direction. Both formulas must be checked:

$$\frac{f_a}{F_{ao}} + \frac{f_b}{F_b} \leq 1.0 \tag{8.15}$$

$$\frac{f_a}{F_a} + \frac{C_b f_b}{F_b (1.0 - f_a/F_e)} \leq 1.0 \tag{8.16}$$

where f_a is the average compressive stress from the axial load, f_b is the maximum compressive bending stress, F_a is the strength of the member as a column, F_b is the strength of the member as a beam, F_{ao} is the strength of the member as a short column, and $F_e = \pi^2 E/(KL/r)^2$.

8.2.2.8 Buckling of Thin, Flat Elements of Columns and Beams Under Uniform Compression

The elastic buckling of plates is calculated using classical plate buckling theory. For inelastic stresses straight line formulas that approximate a secant–tangent modulus combination are used. These straight line formulas give higher stresses than those for columns that use tangent modulus, and they are in close agreement with test data. An equivalent slenderness ratio, $K_p b/t$, is utilized in the equations

$$F_p = B_p - \frac{D_p K_p b}{t}, \quad K_p \frac{b}{t} \leq C_p \tag{8.17}$$

$$F_p = \frac{\pi^2 E}{(K_p b/t)^2}, \quad K_p \frac{b}{t} > C_p \tag{8.18}$$

where F_p is the buckling stress of the plate (ksi), b is the clear width of the plate, t is the thickness of the plate, K_p is the coefficient depending on conditions of edge restraint of the plate (see Table 8.8), and B_p, D_p, C_p are the alloy constants defined in Equations 8.19 to 8.24.

For wrought products with tempers starting with -O, -H, -T1, -T2, -T3, and -T4, and cast products

$$B_p = F_{cy} \left[1 + \frac{(F_{cy})^{1/3}}{7.6} \right] \tag{8.19}$$

TABLE 8.8 Values of K_p for Plate Elements

Type of member	Stress distribution	Edge support	K_p
Column	Uniform compression	One edge free, one edge supported	5.1
		Both edges supported	1.6
Beam (flange)	Uniform compression	One edge free, one edge supported	5.1
		Both edges supported	1.6
Beam (web)	Varying from compression on one edge to tension on the other edge	Compression edge free, tension edge with partial restraint	3.5
		Both edges supported	0.67

Source: Gaylord, Gaylord, and Stallmeyer, *Structural Engineering Handbook*, McGraw-Hill, 1997.

$$D_p = \frac{B_p}{20}\left(\frac{6B_p}{E}\right)^{1/2} \tag{8.20}$$

$$C_p = \frac{2B_p}{3D_p} \tag{8.21}$$

For wrought products with tempers starting with -T5, -T6, -T7, -T8, and -T9

$$B_p = F_{cy}\left[1 + \frac{(F_{cy})^{1/3}}{11.4}\right] \tag{8.22}$$

$$D_p = \frac{B_p}{10}\left(\frac{B_p}{E}\right)^{1/2} \tag{8.23}$$

$$C_p = 0.41\frac{B_p}{D_p} \tag{8.24}$$

8.2.2.9 Buckling of Thin, Flat Elements of Beams Under Bending

For webs under bending loads, Equations 8.17 and 8.18 apply for buckling in the inelastic and elastic ranges. C_p is given by Equation 8.21. However, the values of B_p and D_p are higher than those of elements under uniform compression because they include a shape factor effect, the same as that defined for beams. The constants for the straight-line equation are as follows. They apply to all alloys and tempers

$$B_p = 1.3F_{cy}\left[1 + \frac{(F_{cy})^{1/3}}{7}\right] \tag{8.25}$$

$$D_p = \frac{B_p}{20}\left(\frac{6B_p}{E}\right)^{1/2} \tag{8.26}$$

8.2.2.10 Postbuckling Strength of Thin Elements of Columns and Beams

Most thin elements can develop strengths much higher than the elastic buckling strength as given by Equation 8.18. This higher strength is used in design. Elements of angle, cruciform, and channel (flexural buckling about the weak axis) columns may not develop postbuckling strength. Thus, the buckling strength should be used for these cases. For other cases the postbuckling strength (in the elastic buckling region) is given as follows.

$$F_{cr} = k_2\frac{\sqrt{B_pE}}{K_p b/t} \quad \text{for } \frac{b}{t} > \frac{k_1 B_p}{K_p D_p} \tag{8.27}$$

where F_{cr} is the ultimate strength of plate in compression (ksi), B_p, D_p are the coefficients defined in Equations 8.19 to 8.26, and k_1, k_2 are the coefficients ($k_1 = 0.5$ and $k_2 = 2.04$ for wrought products whose temper starts with -O, -H, -T1, -T2, -T3, and -T4, and castings; $k_1 = 0.35$ and $k_2 = 2.27$ for wrought products whose temper starts with -T5, -T6, -T7, -T8 and -T9).

8.2.2.11 Weighted Average Strength of Thin Sections

In many cases a component will have elements with different calculated buckling strengths. An estimate of the component strength is obtained by equating the ultimate strength of the section multiplied by the total area to the sum of the strength of each element times its area, and solving for the ultimate strength. This weighted average approach gives a close estimate of strength for columns and for beam flanges.

8.2.2.12 Effect of Local Buckling on Column and Beam Strength

If local buckling occurs at a stress below that for overall buckling of a column or beam, the strength of the component will be reduced. Thus, the elements should be proportioned such that they are stable at column or beam buckling strengths. There are methods for taking into account local buckling on column or beam strength provided elsewhere (Sharp 1993; Aluminum Association 2000).

8.2.2.13 Shear Buckling of Plates

The same equations apply to stiffened and unstiffened webs. Equivalent slenderness ratios are defined for each case. Straight line equations are employed in the inelastic range and the Euler formula in the elastic range. The equations are as follows:

$$F_s = B_s - D_s \lambda_s, \quad \lambda_s \le C_s \tag{8.28}$$

$$F_s = \frac{\pi^2 E}{\lambda_s^2}, \quad \lambda_s > C_s \tag{8.29}$$

For tempers -O, -H, -T1, -T2, -T3, -T4

$$B_s = F_{sy}\left(1 + \frac{F_{sy}^{1/3}}{6.2}\right) \tag{8.30}$$

$$D_s = \frac{B_s}{20}\left(\frac{6B_s}{E}\right)^{1/2} \tag{8.31}$$

$$C_s = \frac{2}{3}\frac{B_s}{D_s} \tag{8.32}$$

For tempers -T5, -T6, -T7, -T8, -T9

$$B_s = F_{sy}\left(1 + \frac{F_{sy}^{1/3}}{9.3}\right) \tag{8.33}$$

$$D_s = \frac{B_s}{10}\left(\frac{B_s}{E}\right)^{1/2} \tag{8.34}$$

$$C_s = 0.41\frac{B_s}{D_s} \tag{8.35}$$

where F_{sy} is the shear yield strength (ksi), λ_s is equal to $1.25h/t$ for unstiffened webs, $= 1.25a_1/t[1 + 0.7(a_1/a_2)^2]^{1/2}$ for stiffened webs, h is the clear depth of the web, t is the web thickness, a_1 is the smallest dimension of the shear panel, and a_2 is the largest dimension of the shear panel.

8.2.2.14 Web Crushing

One of the design limitations of formed sheet members in bending and thin-webbed beams is local failure of the web under concentrated loads. There also is interaction between the effects of the concentrated load and the bending strength of the web.

For interior loads

$$P = \frac{t^2(N + 5.4)(\sin \Theta)(0.46F_{cy} + 0.02\sqrt{EF_{cy}})}{0.4 + r(1 - \cos \Theta)} \tag{8.36}$$

where P is the maximum load on one web (kip), N is the length of the load (in.), F_{cy} is the compressive yield strength (ksi), E is the modulus of elasticity (ksi), r is the radius between web and top flanges (in.), t is the thickness (in.), and Θ is the angle between the plane of the web and the plane of the loading (flange).

For loads at the end of the beam

$$P = \frac{1.2t^2(N + 1.3)(\sin \Theta)(0.46F_{cy} + 0.02\sqrt{EF_{cy}}}{0.4 + r(1 - \cos \Theta)} \tag{8.37}$$

If there is significant bending stresses at the point of concentrated load, the interaction may be calculated using the following equation:

$$\left(\frac{M}{M_u}\right)^{1.5} + \left(\frac{P}{P_u}\right)^{1.5} \leq 1.0 \tag{8.38}$$

where M is the applied moment (in.-kip), P is the applied concentrated load (kip), M_u is the maximum moment in bending (in.-kip), and P_u is the web crippling load (kip).

8.2.2.15 Stiffeners for Flat Plates

The addition of stiffeners to thin elements greatly improves the efficiency of a material. This is especially important for aluminum components because there usually is no need for a minimum thickness based on corrosion, and thus parts can be thin compared to those of steel, for example.

8.2.2.16 Stiffening Lips for Flanges

The buckling strength of the combined lip and flange is calculated using the equations for column buckling given previously and the following equivalent slenderness ratio that replaces the effective slenderness ratio. Element buckling of the flange plate and lip must also be considered:

$$\lambda = \pi\sqrt{\frac{I_p}{\frac{3}{8}J + 2\sqrt{C_w K_\phi/E}}} \tag{8.39}$$

where

λ = equivalent slenderness ratio (to be used in the column buckling equations)
$I_p = I_{xo} + I_{yo}$ = polar moment of inertia of lip and flange about center of rotation, in.[4]
$\quad I_{xo}, I_{yo}$ = moments of inertia of lip and flange about center of rotation, in.[4]

K_ϕ = elastic restraint factor (the torsional restraint against rotation as calculated from the application of unit outward forces at the centroid of the combined lip and flange of a one unit long strip of the section), in. lb/in.

J = torsion constant, in.^{4}

$C_w = b^2(I_{yc} - bt^3/12)$ = warping term for lipped flange about center of rotation, in.^{6}

I_{yc} = moment of inertia of flange and lip about their combined centroidal axis (the flange is considered to be parallel to the y-axis), in.^{4}

b = flange width, in.

t = flange thickness, in.

E = modulus of elasticity, ksi

8.2.2.17 Intermediate Stiffeners for Plates in Compression

Longitudinal stiffeners (oriented parallel to the direction of the compressive stress) are often used to stabilize the compression flanges of formed sheet products and can be effective for any thin element of a column or the compression flange of a beam. The buckling strength of a plate supported on both edges with intermediate stiffeners is calculated using the column buckling equations and an equivalent slenderness ratio. The strength of the individual elements, plate between stiffeners, and stiffener elements also must be evaluated

$$\lambda = \frac{4Nb}{\sqrt{3}t}\sqrt{\frac{1 + A_s/bt}{1 + \sqrt{1 + 32I_e/3t^3b}}} \tag{8.40}$$

where λ is the equivalent slenderness ratio for the stiffener that replaces the effective slenderness ratio in the column formulas, N is the total number of panels into which the longitudinal stiffeners divide the plate (in.), b is the stiffener spacing (in.), t is the thickness of the plate (in.), I_e is the moment of inertia of the plate-stiffener combination about the neutral axis using an effective width of plate equal to b (in.^{4}), A_s is the area of the stiffener (not including any of the plates (in.^{2}).

8.2.2.18 Intermediate Stiffeners for Plates in Shear (Girder Webs)

Transverse stiffeners on girder webs must be stiff enough so that they remain straight during the buckling of the plate between stiffeners. The following equations are proposed for design:

$$I_s = \frac{0.46Vh^2}{E}(s/h), \quad \text{for } s/h \leq 0.4 \tag{8.41}$$

$$I_s = \frac{0.073Vh^2}{E}(h/s), \quad \text{for } s/h > 0.4, \tag{8.42}$$

where I_s is the moment of inertia of the stiffener (about face of web plate for stiffeners on one side of web only) (in.^{4}), s is the stiffener spacing (in.), h is the clear height of the web (in.), V is the shear force on web at stiffener location (kip), and E is the modulus of elasticity (ksi).

8.2.2.19 Corrugated Webs

Corrugated sheet is highly efficient in carrying shear loads. Webs of girders and roofs and side walls of buildings are practical applications. The behavior of these panels, particularly the shear stiffness, is dependent not only on the type and size of corrugation but also on the manner in which it is attached to edge members. Test information and design suggestions are published elsewhere (Sharp 1993). Some of the failure modes to consider are as follows:

1. *Overall shear buckling.* This primarily is a function of the size of corrugations, the length of the panel parallel to the corrugations, the attachment and the alloy.

2. *Local buckling.* Individual flat or curved elements of the corrugations must be checked for buckling strength.
3. *Failure of corrugations and/or fastening at the attachment to the edge framing.* If not completely attached at the ends, the corrugation may roll or collapse at the supports, or fastening may fail.
4. *Excessive deformation.* This characteristic is difficult to calculate and the best guidelines are based on test data. The shear deformation can be many times that of flat webs particularly for those cases in which the fastening at the supports is not continuous.

8.2.2.20 Local Buckling of Tubes and Curved Panels

The strength of these members for each type of loading is defined by an equation for elastic buckling that applies to all alloys and two equations for the inelastic region that are dependent on alloy and temper. The members also need to be checked for overall buckling.

8.2.2.21 Round Tubes Under Uniform Compression

The local buckling strength is given by the following equations:

$$F_t = B_t - D_t \sqrt{\frac{R}{t}}, \quad \frac{R}{t} \le C_t \tag{8.43}$$

$$F_t = \frac{\pi^2 E}{16(R/t)\left(1 + \dfrac{\sqrt{R/t}}{35}\right)^2}, \quad \frac{R}{t} > C_t \tag{8.44}$$

where F_t is the buckling stress for a round tube in end compression (ksi), R is the mean radius of the tube (in.), t is the thickness of the tube (in.), and C_t is the intersection of equations for elastic and inelastic buckling (determined by charting or trial and error).

The values of the constants, B_t and D_t are given by the following formulas:

For wrought products with tempers starting with -O, -H, -T1, -T2, -T3, and -T4, and cast products

$$B_t = F_{cy}\left(1 + \frac{F_{cy}^{1/5}}{5.8}\right) \tag{8.45}$$

$$D_t = \frac{B_t}{3.7}\left(\frac{B_t}{E}\right)^{1/3} \tag{8.46}$$

For wrought products with tempers starting with -T5, -T6, -T7, -T8 and -T9

$$B_t = F_{cy}\left(1 + \frac{F_{cy}^{1/5}}{8.7}\right) \tag{8.47}$$

$$D_t = \frac{B_t}{4.5}\left(\frac{B_t}{E}\right)^{1/3} \tag{8.48}$$

where F_{cy} is the compressive yield strength, ksi.

For welded tubes, Equations 8.45 and 8.46 are used along with the yield strength for welded material. The accuracy of these equations has been verified for tubes with circumferential welds and R/t ratios equal to or less than 20. For tubes with much thinner walls, limited tests show that much lower buckling strengths may occur (Sharp 1993).

8.2.2.22 Round Tubes and Curved Panels Under Bending

For curved elements of panels under bending, such as corrugated sheet, the local buckling strength of the compression flange may be determined using the same equations as given in the preceding section for tubes under uniform compression.

In the case of round tubes under bending a higher compressive buckling strength is available for low R/t ratios due to the shape factor effect. (Tests have indicated that this higher strength is not developed in curved panels.) The equations for tubes in bending for low R/t are given in Equation 8.49. Note that the buckling of tubes in bending is provided by two equations in the inelastic region, that defined in Equation 8.49 and that for intermediate R/t ratios, which is the same as that for uniform compression, and the equation for elastic behavior, which also is the same as that for tubes under uniform compression

$$F_{tb} = B_{tb} - D_{tb}\sqrt{R/t}, \quad R/t \le C_{tb} \tag{8.49}$$

where F_{tb} is the buckling stress for round tube in bending (ksi), R is the mean radius of the tube (in.), t is the thickness of the tube (in.), and C_{tb} is equal to $[(B_{tb} - B_t)/(D_{tb} - D_t)]^2$, the intersection of curves, Equations 8.49 and 8.43.

The values of the constants B_{tb} and D_{tb} are given by the following formulae:

For wrought products with tempers starting with -O, -H, -T1, -T2, -T3, and -T4, and cast products

$$B_{tb} = 1.5F_y\left(1 + \frac{F_y^{1/5}}{5.8}\right) \tag{8.50}$$

$$D_{tb} = \frac{B_{tb}}{2.7}\left(\frac{B_{tb}}{E}\right)^{1/3} \tag{8.51}$$

For wrought products with tempers starting with -T5, -T6, -T7, -T8, and -T9

$$B_{tb} = 1.5F_y\left(1 + \frac{F_y^{1/5}}{8.7}\right) \tag{8.52}$$

$$D_{tb} = \frac{B_{tb}}{2.7}\left(\frac{B_{tb}}{E}\right)^{1/3} \tag{8.53}$$

where F_y is the tensile or compressive yield strength, whichever is lower, ksi.

8.2.2.23 Round Tubes and Curved Panels Under Torsion and Shear

Thin walled curved members can buckle under torsion. Long tubes are covered in specifications of the Aluminum Association (2000) and provisions for stiffened and unstiffened cases are provided elsewhere (Sharp 1993).

8.2.3 Joints

8.2.3.1 Mechanical Connections

Aluminum components are joined by aluminum rivets, aluminum and steel (galvanized, aluminized, or stainless) bolts, and clinches. The joints are normally designed as bearing-type connections, because the as-received surfaces of aluminum products have a low coefficient of friction and slip often occurs at working loads. Some information has been developed for the amount of roughening of the surfaces and the limiting thicknesses of material for designing a friction-type joint, although current US specifications do not cover this type of design.

Table 8.9 presents strength data for a few of the rivet and bolt alloys available. Rivets are not recommended for applications that introduce large tensile forces on the fastener. The joints are proportioned based on the shear strength of the fastener and the bearing strength of the elements being joined. The bearing strengths apply to edge distances equal to at least twice the fastener diameter, otherwise reduced values apply. Steel bolts are often employed in aluminum structures. They are generally stronger than the aluminum bolts, and may be required for pulling together parts during assembly. They also have high fatigue strength, which is important in applications in which the fastener is subject to cyclic tension. The steel bolts must be properly coated or be of the 300 series stainless to avoid galvanic corrosion between the aluminum elements and the fastener.

Thin aluminum roofing and siding products are commonly used in the building industry. One failure mode is the pulling of the sheathing off the fastener due to uplift forces from wind. The pull-through strength is a function of the strength of the sheet, the geometry of the product, the location of the fastener, the hole diameter, and the size of the head of the fastener (Sharp 1993).

8.2.3.2 Welded Connections

The aluminum alloys employed in most nonaerospace applications are readily welded, and many structures are fabricated with this method of joining. Transverse groove weld strengths and appropriate filler alloys for a few of the alloys are given in Table 8.5. Because the weld strengths are usually less than those of the base material the design of aluminum welded structures is somewhat different from that of steel structures. Techniques for designing aluminum components with longitudinal and transverse welds are provided in the preceding section. If the welds are inclined to the direction of stress, either purely longitudinal or transverse, the strength of the connection more closely approximates that of the transversely welded case.

Fillet weld strengths are given in Table 8.10. Two categories are defined, longitudinal and transverse. These strengths are based on tests of specimens in which the welds were symmetrically placed and had no

TABLE 8.9 Strengths of Aluminum Bolts and Rivets

Alloy and temper	Minimum expected strength, ksi	
	Shear	Tension on net area
	Rivets	
6053-T61	20	—
6061-T6	25	—
	Bolts	
2024-T4	37	62
6061-T6	25	42
7075-T73	41	68

Source: The Aluminum Association, *Structural Design Manual*, 2000.

TABLE 8.10 Minimum Shear Strengths of Fillet Welds

Filler alloy	Shear strength, ksi	
	Longitudinal	Transverse
4043[a]	11.5	15
5356	17	26
5554	17	23
5556	20	30

[a] Naturally aged (2 to 3 months).
Source: Sharp, *Behavior and Design of Aluminum Structures*, McGraw-Hill, 1993.

large bending component. In the case of longitudinal fillets the welds were subjected to primarily shear stresses. The transverse fillet welds carried part of the load in tension. The difference in stress states accounts for the higher strengths for transverse welds. Aluminum specifications utilize the values for longitudinal welds for all orientations of welds, because many types of transverse fillet welds cannot develop the strengths shown because they have a more severe stress state than the test specimens, for example, more bending stress. Proportioning of complex fillet weld configurations is done using structural analysis techniques appropriate for steel and other metals.

8.2.3.3 Adhesive Bonded Connections

Adhesive bonding is not used as the only joining method for main structural components of non-aerospace applications. It is employed in combination with other joining methods and for secondary members. Although there are many potential advantages in the performance of adhesive joints compared to those for mechanical and welded joints, particularly in fatigue, there are too many uncertainties in design to use them in primary structures. Some of the problems in design are as follows:

1. There are no specific adhesives identified for general structures. The designer needs to work with adhesive experts to select the proper one for the application.
2. In order to achieve long-term durability proper pretreatment of the metal is required. There are little data available for long-term behavior, so the designer should supplement the design with durability tests.
3. There is no way to inspect the quality of the joint. Proper quality control of the joining process should result in good joints. However, a mistake can result in very low strengths, and the bad joint cannot be detected by inspection.
4. There are calculation procedures for proportioning simple joints in thin materials. Techniques for designing complex joints of thicker elements are under development, but are not adequate for design at this time.

8.2.4 Fatigue

Fatigue is a major design consideration for many aluminum applications, for example, aircraft, cars, trucks, railcars, bridges, and bridge decks. Most field failures of metal structures are by fatigue. The current design method used for all specifications, aluminum and steel structures, is to define categories of details that have essentially the same fatigue strength and fatigue curves for each of these categories. Smooth components, bolted and riveted joints, and welded joints are covered in the categories. For a new detail the designer must select the category that has a similar local stress.

Many of the unique characteristics related to the fatigue behavior of aluminum components have been summarized (Sharp et al. 1996). Some general comments from this reference follow:

1. Some cyclic loads, such as wind-induced vibration and dynamic effects in forced vibration, are nearly impossible to design for because stresses are high and the number of cycles build up quickly. These loads must be reduced or eliminated by design.
2. Good practice to eliminate known features of structures causing fatigue, such as sharp notches and high local stresses due to concentrated loads, should be employed in all cases. In some applications the load spectrum is not known, for example, light poles, and fatigue-resistant joints must be employed.
3. The fatigue strength of aluminum parts is higher at low temperature and lower at elevated temperature compared to that at room temperature.
4. Corrosion generally does not have a large effect on the fatigue strength of welded and mechanically fastened joints but considerably lowers that of smooth components. Protective measures such as paint improves fatigue strength in most cases.

5. Many of the joints for aluminum structures are unusual in that they are quite different from those of the fatigue categories provided in the specifications. Stress analysis to define the critical local stress is useful in these cases. Test verification is desirable if practical.

8.3 Design

Aluminum should be considered for applications in which life cycle costs are favorable compared to competing materials. The costs include

1. Acquisition, refining, and manufacture of the metal.
2. Fabrication of the metal into a useful configuration.
3. Assembly and erection of the components in the final structure.
4. Maintenance and operation of the structure over its useful life.
5. Disposal after the useful life.

The present markets for aluminum have developed because of life cycle considerations. Transportation vehicles, one of the largest markets, with aerospace applications, aircraft, trucks, cars, and railcars, are light weight, thus saving fuel costs and are corrosion resistant thus minimizing maintenance costs. Packaging, another large market, makes use of close loop recycling that returns used cans to rolling mills that produce sheet for new cans. Building and infrastructure uses developed because of the durability of aluminum in the atmosphere without the need for painting, thus saving maintenance costs.

8.3.1 General Considerations

8.3.1.1 Product Selection

Most aluminum structures are constructed of flat-rolled products, sheet and plate, and extrusions because they provide the least cost solution. The properties and quality of these products are guaranteed by producers. The flat-rolled products may be bent or formed into shapes and joined to make the final structure. Extrusions should be considered for all applications requiring constant section members. Most extruders can supply shapes whose cross-section fits within a 10-in. circle. Larger shapes are made by a more limited number of manufacturers and are more expensive. Extrusions are attractive for use because the designer can incorporate special features to facilitate joining, place material in the section to optimize efficiency, and consolidate number of parts (compared with fabricated sheet parts). Because die costs are low the designer should develop unique shapes for most applications.

Forgings are generally more expensive than extrusions and plate, and are employed in aerospace applications and wheels, where the three-dimensional shape and high performance and quality are essential. Castings are also used for three-dimensional shapes, but the designer must work with the supplier for design assistance.

8.3.1.2 Alloy Selection

For extrusions alloy 6061 is best for higher strength applications and 6063 is preferred if the strength requirements are less. 5XXX alloys have been extruded and have higher as-welded strength and ductility in structures but they are generally much more expensive to manufacture, compared to the 6XXX alloys.

6061 sheet and plate are also available and are used for many applications. For the highest as-welded strength the 5XXX alloys are employed.

Table 8.11 shows alloys that have been employed in some applications. Choice of specific alloy depends on cost, strength, formability, weldability, and finishing characteristics.

8.3.1.3 Corrosion Resistance

Alloys shown in Table 8.11, 3XXX, 5XXX, and 6XXX, have high resistance to general atmospheric corrosion and can be employed without painting. Tests of small, thin specimens of these alloys in

TABLE 8.11 Selection of Alloy

Application	Specific use	Alloys
Architecture		
Sheet	Curtain walls, roofing and siding, mobile homes	3003, 3004, 3105
Extrusions	Window frames, railings, building frames	6061, 6063
Highway		
Plate	Signs, bridge decks	5086, 5456, 6061
Extrusions	Sign supports, lighting standards, bridge railings	6061, 6063
Industrial		
Plate	Tanks, pressure vessels, pipe	3003, 3004, 5083, 5086, 5456, 6061
Transportation		
Sheet/plate	Automobiles, trailers, railcars, shipping containers, boats	5052, 5083, 5086, 6061, 6009
Extrusions	Stiffeners/framing	6061
Miscellaneous extrusions	Scaffolding, towers, ladders	6061, 6063

Source: Gaylord, Gaylord, and Stallmeyer, *Structural Engineering Handbook*, McGraw-Hill, 1997.

a seacoast or industrial environment for over 50 years of exposure have shown that the depth of attack is small and self-limiting. A hard oxide layer forms on the surface of the component which prevents significant additional corrosion.

If aluminum components are attached to steel components protective measures must be employed to prevent galvanic corrosion. These measures include painting the steel components and placing a sealant in the joint. Stainless steel or galvanized fasteners are also required.

Some of the 5XXX alloys with magnesium content over about 3% may be sensitized by sustained elevated temperatures and lose their resistance to corrosion. For these applications alloys 5052 and 5454 may be used.

8.3.1.4 Metal Working

All of the usual fabrication processes can be used with aluminum. Forming capabilities vary with alloy. Special alloys are available for automotive applications in which high formability is required. Aluminum parts may be machined, cut, or drilled and the operations are much easier to accomplish compared to steel parts.

8.3.1.5 Finishing

Aluminum structures may be painted or anodized to achieve a color of choice. These finishes have excellent long-term durability. Bright surfaces also may be accomplished by mechanical polishing and buffing.

8.3.2 Design Studies

Some specific design examples follow in which product form, alloy selection, and joining method are discussed. The Aluminum Association Specifications are used for calculations of component strength.

EXAMPLE 8.1

Lighting standard

Design requirements

1. Withstand wind loads for area.
2. Fatigue and vibration resistant.

3. Heat treatable after welding to achieve higher strength.
4. Base that breaks away under vehicle impact.

Alloy and product

Round, extruded tubes of 6063-T4 are selected for the shaft. This alloy is easily extruded and has low cost and excellent corrosion resistance. The -T4 temper is required so that the pole can be tapered by a spinning operation so that the structure can be heat treated and aged after welding. A permanent mold casting of 356-T6 is selected for the base. The shaft extends through the base. This base may be acceptable for break away characteristics. If not, a break away device must be employed.

Joining

MIG circumferential welds are made at the top and bottom of the base using filler alloy 4043. This filler alloy must be employed because of the heat treat operation after welding. The corrosion resistance of a 5XXX filler alloy may be lowered by the heat treatment and aging.

Design considerations

Wind-induced vibration can be a problem occasionally. The vibration involves both the standard and luminare. There is currently no accurate way to predict whether or not these structures will vibrate. Light pole manufacturers have dampers that they can use if necessary.

Calculation example — bending of welded tube

Determine the bending strength of a 8 in. diameter (outside) × 0.313 in. wall tube of 6063-T4, heat treated and aged after welding using ASD. Factors of safety corresponding to building-type structures apply.

For this special case of fabrication the specifications allow the use of allowable stresses for the welded construction equal to 0.85 times those for 6063-T6. Also, the allowable stresses can be increased one third for wind loading:

1. The allowable tensile stress (tensile properties are given in Table 8.2, shape factors in Table 8.6, and factors of safety in Table 8.3) is as follows:

$$\text{tensile strength:} \quad F_{tu} = 0.85(1.24)(1.33)(30)/(1.95) = 21.6 \text{ ksi}$$

$$\text{yield strength:} \quad F_{ty} = 0.85(1.17)(1.33)(25)/(1.65) = 20.0 \text{ ksi}$$

2. The allowable compressive strength is given by Equations 8.43 to 8.53: $R/t = (4.0 - 0.313)/(0.313) = 11.8$. Equation 8.49 applies because R/t is less than $69.6(C_{tb})$. Constants are determined from Equations 8.52 and 8.53:

$$F_{tb} = (0.85)(1.33)\left(45.7 - 2.8\sqrt{(R/t)}\right)\Big/1.65 = 24.7 \text{ ksi}$$

3. The lower of the three values, 20.0 ksi, is used for design. This bending stress must be less than that calculated from all loads.

EXAMPLE 8.2

Overhead sign truss

Design requirements

1. Withstand wind loads for locality (signs and truss are considered).
2. Prevent wind induced vibration of truss and members.
3. Provide structure that does not need painting.

Alloy and product

Extruded tubes of 6061-T651 are selected for the truss and end supports. This alloy is readily welded and has excellent corrosion resistance. It also is one of the lower-cost extrusion alloys.

Joining

The individual members will be machined at the ends to fit closely with other parts and welded together using the MIG process. 5356 filler wire is specified to provide higher fillet weld strength compared to that for 4043 filler.

Design considerations

Wind-induced vibration must be prevented in these structures. The trusses are particularly susceptible to the wind when they do not have signs installed. Vibration of the entire truss can be controlled by the addition of a suitable damper (at midspan) and individual members must be designed to prevent vibration by limiting their slenderness ratio. (Sharp 1993).

Calculation example — buckling of a tubular column with welds at ends

The diagonal member of the truss is a 4-in. diameter tube (outside diameter) of 6061-T651 with a wall thickness of 0.125 in. The radius of gyration is 1.37. Its length is 48 in. and it is welded at each end to chords using filler 5356. Use ASD factors of safety corresponding to bridge structures (Table 8.3). Assume that the effective length factor is 1.0. Allow one third increase in stress because of wind loading

$$KL/r = (1.0)(48)/1.37 = 35.0$$

For column buckling, Equation 8.2 applies ($KL/r \leq C_c$). The constants are calculated from Equations 8.7 to 8.9 and parent metal properties (Table 8.2)

$$F_c = (1.33)(39.4 - 0.246(35.0))/2.20 = 18.6 \text{ ksi}$$

For yielding at the welds (the entire cross-section is affected at the ends), the properties in Table 8.5 are employed

$$F_c = (1.33)(20)/(1.85) = 14.4 \text{ ksi}$$

The allowable stress is the lower of these values, 14.4 ksi.

Calculation example — tubular column with welds at ends and midlength

This is the same construction as described in the previous paragraph, except that the designer has specified that a bracket be circumferentially welded to the tube at midlength. This weld lowers the column buckling strength. The column is now designed as though all the material is heat affected. Equation 8.2 still applies but constants are now calculated using Equations 8.4 to 8.6 and properties from Table 8.5:

$$F_c = (1.33)[22.8 - (1.0)(35.0)(0.133)]/2.20 = 11.0 \text{ ksi}$$

This stress is less than that calculated previously for yielding (14.4 ksi) and now governs. This stress must be higher than that calculated using the total load on the structure.

EXAMPLE 8.3

Built-up highway girder

Design requirements

The loads to be used for static and fatigue strength calculations are provided in AASHTO specifications. Long time maintenance-free construction is also specified. The size of the girder is larger than the largest extrudable section.

Riveted construction

Alloy 6061-T6 is selected for the web plate, flanges, and web stiffeners. This alloy has excellent corrosion resistance, is readily available, and has the highest strength for mechanical joining. Rivets of 6061-T6 are used for the joining because they are a good match for the parts of the girder from strength and corrosion considerations. The extrusions for the flanges and stiffeners are special sections designed to facilitate fabrication and to achieve maximum efficiency of material. A sealant is placed in the faying surfaces to enhance fatigue strength and to prevent ingress of detrimental substances.

Welded construction

Alloy 5456-H116 is selected for the web plate and flange plate. This alloy has high as-welded strength compared to 6061-T6 and excellent corrosion resistance. Alloys 5083 and 5086 would also be satisfactory selections. The filler wire selected for MIG welding is 5556, to have high fillet weld strength.

Calculation example — strength of riveted joint

The 6061-T6 parts are assembled as received from the supplier, so that the joint must be designed as a bearing connection. Use ASD for design. The thickness of the web plate is $\frac{1}{2}$ in. and it is attached to the legs of angle flanges (two angles) that are $\frac{3}{4}$ in. thick. One inch diameter rivets (area is 0.785 in.^{2}) are used.

Allowable bearing load on the web (bearing area is $0.50 \times 1.0 = 0.50$ in.^{2}) for one fastener is (see Table 8.2 and Table 8.3):

$$\text{Based on yielding:} \quad P = (58)(0.50)/1.85 = 15.7 \text{ kip}$$
$$\text{Based on ultimate:} \quad P = (88)(0.50)/2.64 = 16.7 \text{ kip}$$

Allowable shear load on one rivet with double shear:

$$\text{Based on ultimate (see Table 8.3 and Table 8.9):} \quad P = (2)(0.785)(25)/2.64 = 14.9 \text{ kip}$$

The allowable load per rivet is the smaller of the three values or 14.9 kip.

Calculation example — fatigue life of welded girder with longitudinal fillet welds

The allowable tensile strength for a 5456-H116 girder is (see Table 8.2 and Table 8.3):

$$\text{Based on yield:} \quad F = 33/1.85 = 17.8 \text{ ksi (governs)}$$
$$\text{Based on ultimate:} \quad F = 46/2.2 = 20.9 \text{ ksi}$$

Calculate the number of cycles that the girder can sustain at a stress range corresponding to a stress of half the static design value (8.9 ksi).

Category B applies to a connection with the fillet weld parallel to the direction of stress. For this category the fatigue strength is

$$\text{The stress range:} \quad S = 130N^{-0.207} = 8.9 \text{ ksi}$$
$$\text{The number of cycles:} \quad N = 423,000 \text{ cycles}$$

(fatigue equations from Aluminum Association 2000).

Calculation example — intermediate stiffeners

Stiffeners on girder webs must be of sufficient size to remain straight when the web buckles. Stiffener sizes are given by Equations 8.41 and 8.42.

EXAMPLE 8.4

Roofing or siding for a building

Design requirements

1. Withstand wind loads (uplift as well as downward pressure).
2. Withstand concentrated loads from foot pressure or from reactions at supports.
3. Corrosion resistant so that painting is not needed.

Alloy and product

Sheet of alloy 3004-H14 is selected. This alloy and temper has sufficient formability to roll-form the trapezoidal shape desired. It also has excellent corrosion and reasonable strength. Other 3XXX alloys would also be good choices.

Design considerations

Attachment of the sheet panels to the supporting structure must be strong enough to resist uplift forces. The pull through strength of the sheet product as well as the fastener strength are considered. Sufficient overlap of panels and fasteners at laps are needed for watertightness.

Calculation example — web crushing load at an intermediate support

Consider the shape shown in Figure 8.3. The bearing length is 2 in. (width of flange of support). Use LRFD specifications for buildings. The material properties for 3004-H14 are given in Table 8.2 and the resistance factors are in Table 8.4. For an interior load use Equation 8.36. $\phi = 0.90$ for this case, the same as that for web buckling.

$$\phi P = \frac{(0.90)(0.032)^2(2.0 + 5.4)(0.866)(0.46 \times 14.0 + 0.02(10100 \times 14.0)^{1/2})}{0.4 + 0.032(1 - 0.5)}$$

$$= 0.198 \text{ kip per web}$$

This load must be higher than that calculated using the factored loads. (Equations for factored loads are given in the Aluminum Association Specifications.)

Calculation example — bending strength of section

To calculate the section strength, the strength of the flange under uniform compression and the strength of the web under bending are calculated separately, and then combined using a weighted average calculation. The area of the web used in the calculation is that area beyond two thirds of the distance from the neutral axis. The resistance factor for the strength calculations from Table 8.4 is 0.85. The radii

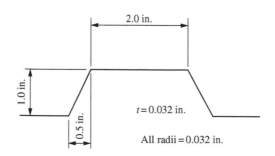

FIGURE 8.3 Example 8.4.

are neglected in subsequent calculations so that plate widths are to the intersection point of elements. The width to points of tangency of the corner radii is more accurate.

Strength of flange

Equation 8.27 governs because the b/t ratio (62.5) is larger than the b/t limit given for that equation. Values for B_p and D_p are given by Equations 8.19 and 8.20; the value of K_p is in Table 8.8

$$\phi F_{cr} = (0.85)(2.04)(18.4 \times 10100)^{1/2}/(1.6)(62.5) = 7.5 \text{ ksi}$$

Strength of web

The web is in bending and has a h/t ratio of 35 so Equation 8.17 governs. Values for B_p and D_p are given by Equations 8.25 and 8.26, and the value of K_p is in Table 8.8

$$\phi F_p = (0.85)(24.5 - (0.67)(0.147)(35)) = 17.9 \text{ ksi}$$

Strength of section

The bending strength of the section is between that calculated for the flange and the web. An accurate estimate of the strength is obtained from a weighted average calculation, which depends on the areas of the elements and the strength of each element. The area of the webs is that portion further than two thirds of the distance from the neutral axis

$$\phi F = [(2.0)t(7.5) + (2)(1.12)t(0.187)(17.9)]/(2.0 + 0.374)t = 9.5 \text{ ksi}$$

This stress must be higher than that calculated using factored loads.

Calculation example — intermediate stiffener

The bending strength of the section in Figure 8.3 can be increased significantly, with a small increase in material, by the addition of a formed stiffener at midwidth as illustrated in Figure 8.4. The strength of the stiffened panel is calculated using an equivalent slenderness ratio as given by Equation 8.40 and column buckling equations. The addition of a few percent more material as illustrated in Figure 8.4 can increase section strength by over 25%.

Calculation example — combined bending and concentration loads

The formed sheet product can experience high longitudinal compressive stresses and a high normal concentrated load at the same location such as an intermediate support. These stresses interact and must be limited as defined by Equation 8.38.

EXAMPLE 8.5

Orthotropic bridge deck

Design requirements

1. Withstand the static and impact loads as provided in an appropriate bridge design specification.
2. Withstand the cyclic loads provided in specifications.

FIGURE 8.4 Example 8.4.

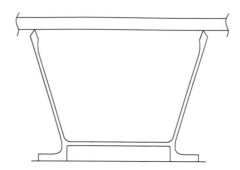

FIGURE 8.5   Example 8.5.

3. Fabricated by the use of welding.
4. Corrosion resistant so that painting is not needed.
5. Large prefabricated panels to shorten erection time.

Alloy and product

The selection depends on the type of construction desired. Figure 8.5 is a plate reinforced by an extruded closed stiffener. This construction has been used successfully. The plate is 5456-H116, chosen because of its high as-welded strength. The extrusion is 6061-T651, which has high strength and reasonable cost. The extrusion is designed to accommodate welding and attachment to supports to minimize fabrication costs. Both alloys have excellent corrosion resistance and will not need to be painted.

All extruded decks with segments either bolted or welded together, to achieve a shape similar to that in Figure 8.5, have also been used. 6061-T651 extrusions for all the segments are the choice in this case.

Joining

MIG welding with filler alloy 5556 is selected for attaching the extrusions to the plate. Fixturing is required to control the final shape of the panel.

Design considerations

Large panels, 11 × 28 ft or larger, complete with wearing surface have been fabricated. The panels must be attached tightly to the supporting structure to avoid fatigue failures of the fasteners.

Galvanized A356 bolts are suggested for the attachment, to obtain high static and fatigue strengths.

Calculation example — bending stresses in plate and section

Fatigue is the major design concern in a metal bridge deck. Wheel pressures cause bending stresses in the deck plate transverse to the direction of the stiffeners. These loads also cause longitudinal bending stresses in the stiffened panel. Fatigue evaluations are needed for both stresses. Deflection and static strength requirements of specifications also must be met.

EXAMPLE 8.6

Ship hull

Design requirements

1. Withstand pressures from operation in seas, including dynamic pressures from storms.
2. Withstand stresses from bending and twisting of entire hull from storm conditions.
3. The hull is of welded construction.

4. The hull must plastically deform without fracture when impacting with another object.
5. Employ joints that are proven to be fatigue resistant in other metal ship structures.
6. Corrosion resistant so that painting is not required even for salt water exposure.

Alloy and product

The hull is constructed of stiffened plate. Alloy 5456-H116 plate and 5456-H111 extruded stiffeners are selected. Main girders are fabricated using 5456-H116 plate. Other lower-strength 5XXX alloys are also suitable choices. This alloy is readily welded and has high as-welded strength. The 5456-H111 extrusions are more expensive than those of 6061-T6 but the welded 5XXX construction is much tougher using 5XXX stiffening, and thus would better accommodate damage without failure. This alloy has excellent resistance to corrosion in a salt water environment.

Joining

MIG welding using 5556 filler is specified. This is a high-strength filler and appropriate for joining parts of this high-strength alloy.

Design considerations

Loadings for hull or component design are difficult to obtain. The American Bureau of Shipping has requirements for the size of some of the components. Fireproofing is required in some areas.

Calculation example — buckling of stiffened panel

For a longitudinally framed vessel the hull plate and stiffeners will be under compression from bending of the ship. The stiffened panel must be checked for column buckling between major transverse members using Equations 8.2 and 8.3. For hull construction that is subjected to normal pressures, the stiffened panel will have lateral bending as well as longitudinal compression. Equations 8.15 and 8.16 are needed in this case.

Elements of the stiffened panel must be checked for strength under the compression loads. In addition, an angle or Tee stiffener can fail by a torsion about an enforced axis of rotation, the point of attachment of the stiffener to the plate. Equation 8.39 may be used for the calculation.

EXAMPLE 8.7

Latticed tower or space frame

Design requirements

1. Withstand wind, earthquake, and other imposed loads.
2. Corrosion resistant so that painting is not required.
3. Prevent wind-induced vibration.

Alloy and product

Extrusions of 6061-T651 are selected for the members because of their corrosion resistance, strength, and economy. 6063 extrusions can be more economical if the higher strength of 6061 is not needed. The designer should make full use of the extrusion process by designing features in the cross-section that will facilitate joining and erection, and that will result in optimal use of the material.

Joining

Mechanical fasteners are selected. Galvanized A325 or stainless steel fasteners are best for major structures because of their higher strength compared to those of aluminum.

Design considerations

Overall buckling of the system as well as the buckling of components must be considered. The manner in which the members are attached at their ends can affect both component and overall strength.

Special extrusions in the form of angles, Y-sections, and hat-sections have been used in these structures. Some of these sections can fail by flexural–torsional buckling under compressive loads. Equation 8.10 covers this case. These sections, because they are relatively flexible in torsion, can vibrate in the wind in torsion as well as flexure.

8.4　Economics of Design

There are two considerations that can affect the economy of aluminum structures: efficiency of design and life cycle costs. These considerations will be summarized briefly here.

Most structural designers are schooled in and are comfortable with design in steel. Although the design of aluminum structures is very similar to that in steel there are differences in their basic characteristics that should be recognized:

1. The density of aluminum alloys is about one third that of steel. Efficiently designed aluminum structures will weigh about one third to half those of efficiently designed steel structures, depending on failure mode. The lighter structures are governed by tensile or yield strength of the material and the heavier ones by fatigue, deflection, or buckling.
2. Modulus of elasticity of aluminum alloys is about one third that of steel. The size/shape of efficient aluminum components will need to be larger for aluminum structures as compared to those of steel for the same performance.
3. The fatigue strength of a joint of aluminum is one third to half that of steel, with identical geometry. The size/shape of the aluminum component will need to be larger than that of steel to have the same performance.
4. The resistance to corrosion from the atmosphere of aluminum is much higher than that of steel. The thickness of aluminum parts can be much thinner than those of steel and painting is not needed for most aluminum structures.
5. Extrusions are used for aluminum shapes of nonrolled sections as used for steel. The designer has much flexibility in the design to: (a) consolidate parts, (b) include features for welding to eliminate machining, (c) include features to snap together parts or to accommodate mechanical fasteners, and (d) to include stiffeners, nonuniform thickness, and other features to provide the most efficient placement of metal. Because die costs are low most extrusions are uniquely designed for the application.

Aluminum applications are economical generally because of life cycle considerations. In some cases, for example, castings, aluminum can be competitive on a first cost basis compared to steel. Light weight and corrosion resistance are important in transportation applications. In this case the higher initial cost of the aluminum structure is more than offset by lower fuel costs and higher pay loads. Closed loop recycling is possible for aluminum and scrap has high value. The used beverage can is converted into sheet to make additional cans with no deterioration of properties.

Glossary

Alloy — Aluminum in which a small percentage of one or more of other elements have been added primarily to improve strength.
Foil — Flat-rolled product that is less than 0.006 in. thick.
Heat-affected zone — Reduced strength material from welding measured 1 in. from centerline of groove weld or 1 in. from toe or heel of fillet weld.

Plate — Flat-rolled product that is greater than 0.25 in. thickness.
Sheet — Flat-rolled product between 0.006 and 0.25 in. thickness.
Temper — The measure of the characteristic of the alloy as established by the fabrication process.

References

Aluminum Association, *Aluminum Design Manual*, Washington DC, 2000.

AASHTO, American Association of State Highway and Transportation Officials, *AASHTO LRFD Bridge Design Specifications*, Washington DC, 2002.

Sharp, M.L., *Behavior and Design of Aluminum Structures*, McGraw-Hill, New York, 1993.

Sharp, M.L., Development of Aluminum Structural Technology in the United States, *Proceedings, 50th Anniversary Conference*, Structural Stability Research Council, Bethlehem, Pennsylvania, 21–22 June 1994.

Sharp, M.L., Nordmark, G.E., and Menzemer, C.C., *Fatigue Design of Aluminum Components and Structures*, McGraw-Hill, New York, 1996.

Further Reading

Aluminum Association, *The Aluminum Extrusion Manual*, Washington DC, 1998.

Aluminum Association, website for book store, www.aluminum.org

American Bureau of Shipping, *Rules for Building and Classing Aluminum Vessels*, New York, 1975.

American Welding Society, *ANSI/AWS D1.2/D1.2M:2003 Structural Welding Code Aluminum*, Miami, Florida, 2003.

Gaylord, E.H., Gaylord, C.N., and Stallmeyer, J.E., *Structural Engineering Handbook*, McGraw-Hill, New York, 1997.

Kissell, J.R. and Ferry, R.L., *Aluminum Structures, A Guide to Their Specifications and Design*, John Wiley & Sons, Inc., New York, 2002.

9
Reliability-Based Structural Design

Achintya Haldar

Department of Civil Engineering and Engineering Mechanics,
The University of Arizona,
Tucson, AZ

9.1 Introduction

Structural design consists of proportioning elements of a system to satisfy various criteria of performance, safety, serviceability, and durability under various demands. The presence of uncertainty cannot be avoided in every phase of structural engineering analysis and design, but it is not simple to satisfy design requirements in the presence of uncertainty. After three decades of extensive work in different engineering disciplines, several reliability evaluation procedures of various degrees of complexity are now available. First-generation structural design guidelines and codes are being developed and promoted worldwide using some of these procedures.

In civil engineering in particular, the structures group has been providing leadership in implementing the reliability-based design concept. The reliability-based structural design concept, in the form of the load and resistance factor design (LRFD) code, was formally introduced by the American Institute of Steel Construction (AISC) as early as 1986 [1]. The LRFD code was intended to be an alternative to the allowable stress design (ASD) [2] concept used exclusively before 1986. The third edition of the LRFD code was introduced in 2001 [1]. Similar design guidelines for concrete [3], masonry [4], and wood [5,6] are now available, reflecting the reliability-based design concept. Concrete, masonry, and wood codes do not explicitly state that they are based on the LRFD concept, but the intention is the same. Since this handbook will be used by all segments of the structural engineering profession, it would not be appropriate to use the LRFD concept to imply the reliability-based design concept. However, they are essentially the same and it would not be an exaggeration if the terms LRFD and the reliability-based design concept were used synonymously. Most of the worldwide codes applicable to structural engineering have already been modified or are in the process of being modified to implement the concept [7]. At present almost all civil engineering schools in the United States teach steel design using the LRFD concept. It is expected that reliability-based structural design will be the only design option worldwide in the near future. Under the sponsorship of the International Standard Organization (ISO), a standard (ISO 2394) [8] is now being considered for this purpose.

Since AISC's LRFD concept has almost two decades of history, it is used to make some observations here. AISC published the ninth edition of the ASD code in 1989 after the introduction of the LRFD code in 1986, indicating indecisiveness or uneasiness in using the LRFD concept in structural steel design. Even recent graduates who are taught steel design using only the LRFD concept are asked to use the ASD method in everyday practice after graduation. LRFD is calibrated with respect to ASD for a ratio of live load to dead load of 3, that is, the two methods will usually give identical sections. When the ratio is less than 3, as is expected in most design applications, the material savings could be as much as 18%. AISC [9] documented various amounts of savings for office buildings, parking structures, floor and framing systems, and industrial buildings, but the LRFD concept is still being ignored or over-looked by most of the profession. The author believes that the main cause of this nonuniform use of the LRFD code is not its difficulty, but the older generation's lack of familiarity with the concept, new terminologies, and the cost of changing from the old to new guidelines. Since LRFD will save on material cost in most designs, the extra cost of implementing LRFD can be recovered within a short period of time. In the near future, it is expected that the reliability-based design concept, similar to the LRFD concept, may not be an option but a requirement in structural analysis for steel, concrete, wood, and masonry structures.

It can be observed that steel design using the LRFD concept is not that different from concrete design using the ACI's ultimate strength design concept [10], and is not more difficult than using AISC's old ASD method. The ACI recently published strength design guidelines similar to the LRFD concept [3]. Design aids like figures, tables, and charts are very similar in the LRFD and ASD guidelines for steel. It is not difficult for a person familiar with one design concept to use the other with a minor effort. The use of ASD by recent graduates trained in LRFD attests to this statement. Unlike in Europe, a code is not a government document in the United States. It is developed by the profession, and its acceptance is voted by the users and developers. Since all major professional groups are advocating reliability-based design and following the worldwide trend, it would be very appropriate to use only the reliability-based approach in future designs.

9.2 Available Structural Design Concepts

Before introducing the reliability-based design concept in structural engineering, it may be informative to study different deterministic structural design concepts used in the recent past. The fundamental concept behind any structural design is that the resistance of a structural element, joint, or the structure as a whole should be greater than the load or combinations of loads that may act during its lifetime with some conservatism or safety factor built in. The level of conservatism is introduced in the design in several ways depending on the basic design concept being used. In the ASD approach, the basic concept is that the allowable stresses should be greater than the unfactored nominal loads or load combinations expected during its lifetime. The allowable stresses are calculated using a safety factor. In other words, the nominal resistance R_n is divided by a safety factor to compute the allowable resistance R_a, and safe design requires that the nominal load effect S_n is less than R_a. In the ultimate strength design method [10], the loads are multiplied by certain load factors to determine the ultimate load effects and the members are required to resist various design combinations of the ultimate load. In this case, the safety factors are used in the loads and load combinations.

It is well known that the loads that may act on a structure during its lifetime are very unpredictable and different levels of unpredictability or uncertainty exist for each load. The uncertainty associated with predicting dead load is expected to be lower than that of live, wind, or seismic load. Since different assumptions are made in developing the beam and column theories, the theoretical prediction of the resistance or strength of beams and columns is expected to have different levels of uncertainty. In the ASD approach, the safety factor is introduced in predicting resistance, and the loads are assumed at their nominal values ignoring the different levels of uncertainty in predicting them. In the ultimate strength design concept, conservatism is introduced by using different load factors. Conceptually, the use of load- and resistance-related safety factors may not assure a uniform underlying risk for different structural elements, for example, beams, columns, and slabs, or under different loading conditions (e.g., dead, live, wind, or seismic loads). In the LRFD or reliability-based design concept, conservatism is introduced by using both load and resistance factors and satisfying an underlying risk, combining the desirable features of both the ASD and ultimate design concepts. Essentially, the LRFD approach uses safety factors to estimate both the resistance and load under the constraint of an underlying risk. Since satisfying an underlying risk is the main objective of the reliability-based design concept, this assures uniform risk for the structure and may produce a more economical design than other deterministic design concepts.

9.3 Introduction of the Reliability-Based Structural Design Concept

Like the LRFD format, the reliability-based design concept can be represented in its basic form as

$$\sum \gamma_i Q_i \leq \phi R_n \tag{9.1}$$

where Q_i is the ith nominal load effect, γ_i is the load factor corresponding to Q_i, R_n is the nominal resistance, and ϕ is the resistance factor corresponding to R_n. The load factor γ_i and the resistance factor ϕ account for the uncertainties in the parameters related to the loads and resistance. The mathematical expressions for these factors will be derived in Section 9.4. These factors are derived based on reliability analysis of simple "standard" structures such as simple beams, centrally loaded columns, tension members, high-strength bolts, and fillet welds [11], and are calibrated to achieve levels of reliability similar to conventional ASD procedures such as the 1989 AISC specification [2].

One of the most important objectives of the reliability-based design approach is that it provides a reasonable platform to compare different design alternatives by considering the realistic behavior of structures and the uncertainty in the design variables satisfying some underlying design criteria. It tries to reduce the scatter in the underlying risk of different members designed according to the ASD concept. Thus, the risk-consistent load factors used in LRFD help to economize the design. Since dead load has less uncertainty than live load, it should have a smaller load factor. But in the ASD procedure, the dead and live loads have the same load factor of 1. In evaluating the strength or resistance, theoretically predicted values generally do not match experimental results. The strength of a beam is more predictable than column strength. The support conditions and the buckling behavior of a column make its strength prediction more uncertain, indicating that beams and columns should have different capacity reduction factors to satisfy the same underlying risk. This is common-sense logic, and reliability or risk-based design is expected to be superior to ASD.

ASD is essentially a deterministic design concept. A casual evaluation of the LRFD code indicates that it is also deterministic in nature; only the prescribed load and resistance factors are based on reliability analysis. This information may not be of any practical significance to a typical practicing engineer.

The design aids in AISC's LRFD and ASD guidelines are very similar. In both approaches, the isolated member approach was used to develop the design guidelines [11]. There are several advantages to the isolated-member approach: (1) in deterministic design methods that use safety factors, it is not practical to prepare detailed requirements for each structural configuration; (2) the characteristics of the individual members and connections are independent of the framework; and (3) most research has been devoted to the study of such elements, and theoretical and experimental verification of their performance is readily available. Nevertheless, the performance of a member is directly dependent on its location in a structural configuration and on its relationship or connection with the other members in the framework. Such dependence is not restricted to the computation of load effects through a deterministic analysis of the structure, but extends to the probabilistic variation of the load effect as well, which is influenced by the probabilistic characteristics of all the parameters of the structure [12]. Only a probabilistic structural analysis of the entire structure can account for this influence and accordingly determine the risk or reliability of any individual member, enabling an improved approach to reliability-based design. However, this approach could be very complicated [13] and may not be practical for everyday structural design.

The analytical procedures used to estimate the load effects are identical in ASD and LRFD approaches. Thus, the analytical procedures, computer programs, and other analysis aids used for ASD are also applicable for LRFD. Only the treatments of load effects are different in the two concepts. The nominal resistance is also calculated deterministically using a codified approach. The capacity reduction factor is used to address the level of uncertainty in estimating it. As pointed out earlier, for wider applications, LRFD or reliability-based design guidelines were calibrated with respect to time-tested ASD procedure. Thus, in many cases, the final selection of a structural member will be identical for ASD and the reliability-based design concept; however, the design procedures will be very different as outlined below. Using the reliability-based design concept, structural engineers will be more empowered to manage risk in a typical design.

9.4 Fundamental Concept of Reliability-Based Structural Design

It is assumed in the following sections that the readers are familiar with the basic concept of uncertainty analysis. If not, they are urged to refer to a recent book authored by Haldar and Mahadevan [14].

In general, R and S can be used to represent the resistance and load effect as random variables since they are functions of many other random variables. R is a function of material properties and the geometric properties of a structural element including cross-sectional properties. S is a function of the load effect that can be expected during the lifetime of the structural element. The uncertainty in R and S can be completely defined by their corresponding probability density functions (PDFs) denoted as $f_R(r)$ and $f_S(s)$, respectively. Then, the probability of failure of the structural element can be defined as the probability of the resistance being less than the load effect or simply $P(R > S)$. Mathematically, it can be expressed as [14]

$$P(\text{failure}) = P(R < S) = \int_0^\infty \left[\int_0^s f_R(r)\, dr \right] f_S(s)\, ds = \int_0^\infty F_R(s) f_S(s)\, ds \qquad (9.2)$$

where $F_R(s)$ is the cumulative distribution function (CDF) of R evaluated at s. Conceptually, Equation 9.2 states that for a particular value of the random variable $S = s$, $F_R(s)$ is the probability of failure. However, since S is also a random variable, the integration needs to be carried out for all possible values of S, with their respective likelihood represented by the corresponding PDF. Equation 9.2 can be considered as the fundamental equation of the reliability-based design concept.

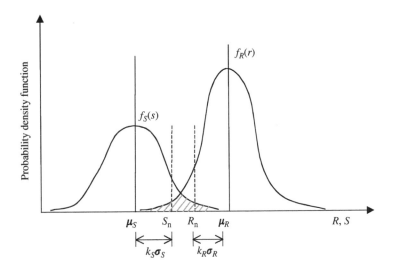

FIGURE 9.1 Reliability-based design concept. (Adopted from *Probability, Reliability and Statistical Methods in Engineering Design*, by Haldar and Mahadevan, 2000, with permission from John Wiley & Sons, Inc.)

The concept is shown in Figure 9.1. In Figure 9.1, the nominal values of resistance and load effect, denoted as R_n and S_n, respectively, and the corresponding PDFs of R and S are shown. The overlapped (dashed area) between the two PDFs provides a *qualitative measure* of the probability of failure. Controlling the size of the overlapped area is essentially the idea behind reliability-based design. Haldar and Mahadevan [14] pointed out that the area could be controlled by changing the relative locations of the two PDFs by separating the mean values of R and S (μ_R and μ_S), the uncertainty expressed in terms of their standard deviations (σ_R and σ_S), and the shape of the PDFs [$f_R(r)$ and $f_S(s)$].

Although conceptually simple, the evaluation of Equation 9.2 may not be easy except for some special cases. Consider $S = S_1 + S_2 + \cdots + S_n$, representing n statistically independent load effects (dead, live, wind loads, etc.), and R and S are independent normal random variables. Under these assumptions, the resistance and the ith load factors in Equation 9.1 can be shown to be [14]

$$\phi = \frac{1 - \varepsilon\beta\delta_R}{1 - k_R\delta_R} \tag{9.3}$$

and

$$\gamma_i = \frac{1 + \varepsilon\varepsilon_{nn}\beta\delta_{S_i}}{1 + k_{S_i}\delta_{S_i}} \tag{9.4}$$

where β is the reliability index, a measure of probability of failure, δ_R and δ_S are the coefficients of variation (COV) of R and S (a measure of uncertainty), k_R is the number of standard deviations below the mean resistance in selecting the nominal value of R (a measure of underestimation of the resistance), k_S is the number of standard deviations above the mean load in selecting the nominal value of the ith load (a measure of overestimation in the load), and

$$\varepsilon = \frac{\sqrt{\sigma_R^2 + \sigma_S^2}}{\sigma_R + \sigma_S} \tag{9.5}$$

$$\varepsilon_{nn} = \frac{\sqrt{\sigma_{S_1}^2 + \sigma_{S_2}^2 + \cdots + \sigma_{S_n}^2}}{\sigma_{S_1} + \sigma_{S_2} + \cdots + \sigma_{S_n}} \tag{9.6}$$

where σ_R and σ_{S_i} are the standard deviations of the resistance and the ith load, respectively. Equations 9.3 and 9.4 indicate that several parameters are needed to evaluate the reliability-based resistance and load factors and the nature of mathematical sophistication required to evaluate them, even when all the variables are assumed to be normal.

When R and S are independent lognormal random variables, expressions similar to Equations 9.3 and 9.4 can be derived [14]. In general, R and S are not independent normal or lognormal random variables, and the probability of failure using Equation 9.2 is expected to be very challenging. They are functions of many different random variables, and their exact probabilistic characteristics are very difficult to evaluate. Historically, several methods with various degrees of sophistication were proposed to evaluate the probability of failure or reliability [14]. Reliability is calculated as $(1.0 - \text{probability of failure})$. It is not possible to discuss the details of all the methods here. They are simply identified here for the completeness of discussion.

Initially, in the late 1960s, the first-order second-moment (FOSM) method, also known as the mean value first-order second-moment (MVFOSM), was proposed to calculate the probability of failure neglecting the distributional information on the random variables present in the problem. This deficiency was overcome by the advanced first-order second-moment method where all the variables are assumed to be normal and independent as proposed by Hasofer and Lind [15]. Rackwitz [16] proposed a more general formulation applicable to different types of distributions. Currently, it is the most widely used reliability evaluation technique. Using this concept, the probability of failure has been estimated using two types of approximations to the limit state at the design point (defined in the following section): first order (leading to the name first-order reliability method or FORM) and second order (leading to the name second-order reliability method or SORM). Since FORM is a commonly used reliability evaluation technique, it is discussed in more detail below. A person without a sophisticated background in probability and statistics can use simulation to evaluate the underlying risk or reliability, as discussed in Section 9.7.

9.4.1 First-Order Reliability Method

The basic idea behind reliability-based structural design is to design a structural member satisfying several performance criteria and considering the uncertainties in the relevant load- and resistance-related random variables, called the basic variables X_i. Since the R and S random variables in Equation 9.2 are functions of many other load- and resistance-related random variables, they are generally treated as basic random variables. The relationship between the basic random variables and the performance criterion, known as the performance or limit state function, can be mathematically represented as

$$Z = g(X_1, X_2, \ldots, X_n) \tag{9.7}$$

The failure surface or the limit state of interest can then be defined as $Z = 0$. The limit state equation plays an important role in evaluating reliability using FORM. It represents the boundary between the safe and unsafe regions and a state beyond which a structure can no longer fulfill the function for which it was designed. Assuming R and S are the two basic random variables, the limit state equation, and the safe and unsafe regions are shown in Figure 9.2. A limit state equation can be an explicit or implicit function of the basic random variables and can be linear or nonlinear. Reliability estimation using explicit limit state functions is discussed here. Haldar and Mahadevan [13] discussed reliability evaluation techniques for implicit limit state functions.

Two types of performance functions are generally used in structural engineering: strength and serviceability. Strength performance functions relate to the safety of the structures and serviceability performance functions are related to the serviceability (deflection, vibration, etc.) of the structure. The reliabilities underlying the strength and serviceability performance functions are expected to be different.

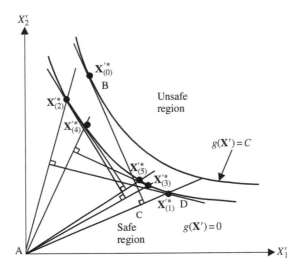

FIGURE 9.2 Limit state concept. (Adopted from *Probability, Reliability and Statistical Methods in Engineering Design*, by Haldar and Mahadevan, 2000, with permission from John Wiley & Sons, Inc.)
Note: A number in parenthesis indicates iteration number.

The limit state equation for the strength limit state in pure bending can be expressed as

$$g(\mathbf{X}) = 1.0 - \frac{M_{\mathrm{u}}}{M_{\mathrm{n}}} \tag{9.8}$$

where M_{u} is the unfactored applied moment and M_{n} is the nominal flexural strength of the member.

Using the AISC's LRFD design criteria, the limit state equation for the strength limit state for a beam–column element can be expressed as

$$g(\mathbf{X}) = 1.0 - \left(\frac{P_{\mathrm{u}}}{P_{\mathrm{n}}} + \frac{8}{9}\frac{M_{\mathrm{u}}}{M_{\mathrm{n}}}\right), \quad \text{if } \frac{P_{\mathrm{u}}}{\phi P_{\mathrm{n}}} \geq 0.2 \tag{9.9}$$

$$g(\mathbf{X}) = 1.0 - \left(\frac{P_{\mathrm{u}}}{2P_{\mathrm{n}}} + \frac{M_{\mathrm{u}}}{M_{\mathrm{n}}}\right), \quad \text{if } \frac{P_{\mathrm{u}}}{\phi P_{\mathrm{n}}} < 0.2 \tag{9.10}$$

where P_{u} is the unfactored tensile and compressive load effects, P_{n} is the nominal tensile and compressive strength, M_{u} is the unfactored flexural load effect, and M_{n} is the nominal flexural strength. Limit state equations can similarly be defined for other strength design criteria like tension, compression, shear, etc. All the strength-related parameters in Equations 9.8 to 9.10 are functions of the geometry, material, and cross-sectional properties of the members under consideration. The code also suggests analytical procedures to evaluate the load effects in some cases. The important point is that the loads effects in Equations 9.8 to 9.10 are to be unfactored in evaluating the reliability index.

For the serviceability limit state, the midspan deflection of beams under live load and the side sway (interstory drift and lateral deflection) of frames are commonly used. They can be represented as

$$g(\mathbf{X}) = 1.0 - \frac{\delta}{\delta_{\mathrm{limit}}} \tag{9.11}$$

$$g(\mathbf{X}) = 1.0 - \frac{u}{u_{\mathrm{limit}}} \tag{9.12}$$

where δ is the vertical deflection of a beam under unfactored live load, δ_{limit} is the allowable or prescribed vertical deflection, u is the lateral deflection under unfactored loads, and u_{limit} is the allowable or prescribed lateral deflection. The deflection limits δ_{limit} and u_{limit} are selected by the designer based on the performance requirements of the structure. They have to be preselected to evaluate the reliability index.

Incorporating the concept of limit state or performance function, the basic reliability evaluation formulation represented by Equation 9.2 can be rewritten as

$$P(\text{failure}) = \int \cdots \int_{g(\,)<0} f_X(x_1, x_2, \ldots, x_n)\, dx_1\, dx_2 \cdots dx_n \tag{9.13}$$

in which $f_X(x_1, x_2, \ldots, x_n)$ is the joint PDF of the basic random variables, and the integration is performed over the failure region. If the random variables are statistically independent, then the joint PDF can be replaced by the product of the individual PDFs in Equation 9.13. Reliability evaluation using Equation 9.13 is known as the full distributional approach.

Initially, all the basic random variables are considered to be statistically independent; they can have different types of distribution, and the limit state equation can be linear or nonlinear. FORM is an iterative technique. The final products of this technique are the reliability index β, the corresponding coordinates of the design or checking point or the most probable failure point $(x_1^*, x_2^*, \ldots, x_n^*)$, and the sensitivity indexes indicating the influence of the individual random variables on the reliability index. In the context of FROM, the reliability index β has a physical interpretation. It is the shortest distance from the origin to the limit state function at the checking point in the reduced standard normal variable space as shown in Figure 9.2 and Figure 9.3. As will be discussed further later, an optimization technique is used to estimate it iteratively. Once the information on β is available, the probability of failure can be obtained as

$$P(\text{failure}) = \Phi(-\beta) = 1.0 - \Phi(\beta) \tag{9.14}$$

where Φ is the CDF of the standard normal variable. If β is large, the probability of failure will be small. The coordinates on the limit state surface where the iteration converges represents the worst combination of the random variables that would cause failure and is appropriately named the design point or the most probable failure point $(x_1^*, x_2^*, \ldots, x_n^*)$. All these aspects of FORM are discussed in the following section.

The estimation of probability of failure using Equation 9.13 is expected to be complicated. As mentioned earlier, Rackwitz [16] suggested a solution strategy that can be used to evaluate the

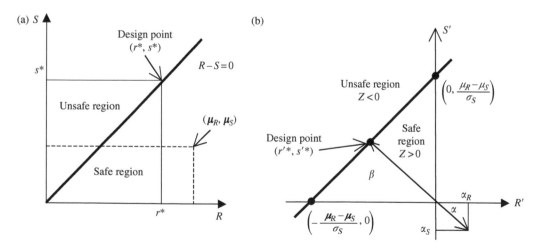

FIGURE 9.3 Reliability evaluation for linear performance function: (a) original coordinates and (b) reduced coordinates. (Adopted from *Probability, Reliability and Statistical Methods in Engineering Design,* by Haldar and Mahadevan, 2000, with permission from John Wiley & Sons, Inc.)

probability of failure for most problems of practical interest. The method requires that the limit state function be available in explicit form, and is discussed next. Haldar and Mahadevan [14] discussed a FORM method where the limit state function is implicit, but it will not be discussed here. Rackwitz's method [16] and some improvements suggested by Ayyub and Haldar [17] are discussed next. For ease of discussion, the method is presented in the form of eight tasks:

- *Task 1.* The limit state equation must be available or defined in terms of the basic random variables and performance criterion.
- *Task 2.* Assume an intelligent initial value of the reliability index β. An initial β value can be 3.0.
- *Task 3.* Assume the coordinates of the initial design or checking point in terms of all the basic random variables. The mean values of the random variables are generally assumed to be the coordinates of the checking point in the first iteration.
- *Task 4.* Not all the random variables in a limit state function are expected to be normal random variables. A nonnormal variable cannot be modified to a normal random variable for all possible values without compromising its underlying uncertainty. It is known that a normal random variable can be uniquely defined in terms of two parameters: its mean and standard deviation. For a nonnormal random variable, the two parameters of an equivalent normal distribution can be evaluated by using two constraints. The two constraints commonly used for this purpose are that the CDF and the PDF of the actual nonnormal variable be equal to the equivalent normal variable at the checking point $(x_1^*, x_2^*, \ldots, x_n^*)$. The coordinates of the checking point are either assumed or known at the beginning of each iteration, and thus the evaluation of two equivalent parameters is not expected to be complicated. Denoting the equivalent normal mean and standard deviation for the ith random variable as $\mu_{X_i}^N$ and $\sigma_{X_i}^N$, respectively, Haldar and Mahadevan [14] have shown that they can be estimated as

$$\mu_{X_i}^N = x_i^* - \Phi^{-1}[F_{X_i}(x_i^*)]\sigma_{X_i}^N \tag{9.15}$$

and

$$\sigma_{X_i}^N = \frac{\phi\{\Phi^{-1}[F_{X_i}(x_i^*)]\}}{f_{X_i}(x_i^*)} \tag{9.16}$$

where $F_{X_i}(x_i^*)$ and $f_{X_i}(x_i^*)$ are the CDF and the PDF of the nonnormal random variable evaluated at the checking point, and $\Phi^{-1}(\ )$ and $\phi(\ )$ are the inverse of the CDF and the PDF of the standard normal distribution.

- *Task 5.* The checking point coordinate for the ith random variable can be shown to be

$$x_i^* = \mu_{X_i}^N - \alpha_{X_i}\beta\sigma_{X_i}^N \tag{9.17}$$

where α_{x_i} is the direction cosine and can be evaluated as

$$\alpha_{X_i} = \frac{(\partial g/\partial X_i)^*\sigma_{X_i}^N}{\sqrt{\sum_{i=1}^n (\partial g/\partial X_i\sigma_{X_i}^N)^{*2}}} \tag{9.18}$$

Partial derivatives in Equation 9.18 are evaluated at the checking point x_i^*. If the random variables are normal, then their standard deviations can be used directly; otherwise, for nonnormal random variables, the equivalent standard deviations at the checking point need to be used.

- *Task 6.* Using Equation 9.17, the coordinates of the checking point can now be updated. These coordinates are expected to be different from the mean values assumed to start the iteration process. If necessary, Tasks 4 to 6 need to be repeated until the direction cosines α_{X_i}s converge with a predetermined tolerance. A tolerance level between 0.001 and 0.005 is common. Once the direction cosines converge, the coordinates of the checking point can be updated keeping the reliability index β as the unknown parameter.

- *Task 7.* Since the checking point must be on the limit state equation, an updated value for β can be obtained by substituting the updated coordinates of the checking point in terms of β.
- *Task 8.* Recalculate the coordinates of the checking point with the updated value for β as in Task 3. Tasks 3 to 7 need to be repeated until β converges to a predetermined tolerance level.

For hand calculation, a tolerance level of 0.01 is adequate. For computer applications, a tolerance level of 0.001 can be used. The algorithm converges in 5 to 10 cycles in most cases. A small computer program can be written to carry out the calculations.

The procedure for linear limit state equations is shown in Figure 9.3, and for nonlinear limit state equations consisting of two variables in Figure 9.2. In Figure 9.3a, the limit state equation, and the safe and unsafe regions are shown. As mentioned earlier, the limit state equation needs to be presented in the reduced coordinates system for the physical interpretation of the reliability index. The following transformation can be used for this purpose

$$X' = \frac{X_i - \mu_{X_i}}{\sigma_{X_i}}, \quad (i = 1, 2, \ldots, n) \tag{9.19}$$

The limit state equation, the safe and the unsafe regions, and the reliability index are shown in the reduced coordinates in Figure 9.3b. For linear limit state equations, no iteration is necessary and the reliability index can be obtained in one step. However, when the limit state is nonlinear as in Figure 9.2, an iterative procedure is necessary. The reliability index can be evaluated iteratively using the eight tasks discussed earlier. Since FORM is the most commonly used reliability evaluation technique at present, the eight steps are elaborated further with the help of an example.

EXAMPLE

A simply supported beam of span $L = 9.144$ m is loaded by a uniformly distributed load w in kN/m and a concentrated load P (in kN) applied at the midspan. The maximum deflection of the beam at the midspan can be calculated as

$$\delta_{\max} = \frac{5}{384} \frac{wL^4}{EI} + \frac{1}{48} \frac{PL^3}{EI} \tag{9.20}$$

where E is the Young's modulus and I is the moment of inertia of the cross-section of the beam. A beam with $EI = 182{,}262$ kN m^2 is selected to carry the load. Suppose w is a normal random variable with a mean of 35.03 kN/m and a standard deviation of 5.25 kN/m and P is a lognormal random variable with mean of 111.2 kN and a standard deviation of 11.12 kN. Further assume w and P are statistically independent. The allowable deflection, δ_a, is considered to be 38.1 mm. The task is to calculate the probability of failure of the beam in deflection satisfying the allowable deflection criterion.

Solution

This example is a problem of estimating the reliability index and the corresponding probability of failure when the limit state function contains independent nonnormal variables. This type of problem is very common in practice. The procedure is explained using the 8 tasks identified earlier.

Task 1 — Define the limit state function. Substituting deterministic values of all the parameters in Equation 9.20, the limit state function for the problem can be defined as

$$g() = 0.0381 - (4.99444 \times 10^{-4} w + 8.73919 \times 10^{-5} P) \tag{9.21}$$

where w is a normal random variable with specified mean and standard deviation values, that is, $w \sim N(35.03 \text{ kN/m}, 5.25 \text{ kN/m})$ and P is a lognormal random variable with parameters λ_P and ζ_P, that is, $P \sim \ln(\lambda_P, \zeta_P)$. These parameters can be calculated as [14]

$$\delta_P = \frac{11.12}{111.2} = 0.1 \cong \zeta_P \quad \text{and} \quad \lambda_P = \ln 111.2 - \frac{1}{2} \times 0.1^2 = 4.706$$

The probability of failure or the reliability index of the beam in deflection corresponding to Equation 9.21 is evaluated using the eight tasks identified earlier. The results are summarized in Table 9.1. For further clarification, the detail calculations for the third and the final iterations are given below.

Third iteration

Task 2. As shown in Table 9.1, an initial value of β of 3.0 is assumed to start the iteration.

Task 3. Using Equation 9.17, the coordinates of the new checking point for w and P can be shown to be

$$w^* = 35.03 + 0.926 \times 3 \times 5.25 = 49.615$$
$$p^* = 110.02 + 0.377 \times 3 \times 12.22 = 123.841$$

Task 4. Since P is a lognormal random variable, the equivalent normal mean and standard deviation at the checking point are calculated using Equations 9.15 and 9.16

$$f_P(123.841) = \frac{1}{\sqrt{2\Pi} \times 0.1 \times 123.841} \exp\left[-\frac{1}{2}\left(\frac{\ln 123.841 - 4.706}{0.1}\right)^2\right] = 0.0170$$

$$F_P(123.841) = P(P \le 123.841) = \Phi\left(\frac{\ln 123.841 - 4.706}{0.1}\right) = \Phi(1.13)$$

$$\Phi^{-1}[F_P(123.841)] = 1.13$$

$$\phi\{\Phi^{-1}[F_P(123.841)]\} = \frac{1}{\sqrt{2\Pi}} \exp\left[-\frac{1}{2} \times 1.13^2\right] = 0.21069$$

$$\sigma_P^N = \frac{0.21069}{0.0170} = 12.38$$

$$\mu_P^N = 123.841 - 1.13 \times 12.38 = 109.85$$

TABLE 9.1 Reliability Evaluation Using FORM with Uncorrelated Variables

Step 1	$g(\ ) = 0.0381 - (4.99444 \times 10^{-4} w + 8.73919 \times 10^{-5} P)$				
Step 2	β	3		3.88	
Step 3	w^*	35.03	49.803	49.615	53.852	53.77
	p^*	111.2	122.216	123.841	128.151	128.813
Step 4	μ_W^N	35.03	35.03	35.03	35.03	35.03
	σ_W^N	5.25	5.25	5.25	5.25	5.25
	μ_P^N	110.61	110.02	109.85	109.28	109.19
	σ_P^N	11.12	12.22	12.38	12.81	12.88
Step 5	$(\partial g/\partial W)^*$	-4.99444×10^{-4}	-4.99444×10^{-4}	-4.99444×10^{-4}	-4.99444×10^{-4}	-4.99444×10^{-4}
	$(\partial g/\partial P)^*$	-8.73919×10^{-5}	-8.73919×10^{-5}	-8.73919×10^{-5}	-8.73919×10^{-5}	-8.73919×10^{-5}
Step 6	α_W	-0.938	-0.926	-0.924	-0.920	-0.919
	α_P	-0.348	-0.377	-0.381	-0.393	-0.394
Step 7				3.88		3.88
Step 8						3.88

Note: In this transcription of Step 3 through Step 6 rows, the data columns align under five positions. Step 2 has value 3 in the first position and 3.88 in the fourth position. Step 7 has 3.88 in the third position and 3.88 in the fifth position. Step 8 has 3.88 in the fifth position.

Tasks 5 and 6

$$\alpha_w = \frac{-4.99444 \times 10^{-4} \times 5.25}{\sqrt{(-4.99444 \times 10^{-4} \times 5.25)^2 + (-0.873919 \times 10^{-4} \times 12.38)^2}} = \frac{-26.22081}{28.365193} = -0.924$$

$$\alpha_P = \frac{-10.819117}{28.365193} = -0.381$$

The direction cosines converged with a tolerance level of 0.005.

Task 7

$$w^* = 35.03 + 0.924 \times \beta \times 5.25 = 35.03 + 4.8510\beta$$

$$p^* = 109.85 + 0.381 \times \beta \times 12.38 = 109.85 + 4.71678\beta$$

$$0.0381 - 4.99444 \times 10^{-4} \times (35.03 + 4.8510\beta) - 0.873919 \times 10^{-4} \times (109.85 + 4.71678\beta) = 0$$

or

$$\beta = 3.88$$

Task 8. The reliability index did not converge with a tolerance level of 0.005. Go back to Task 3.

Final iteration

Task 3

$$w^* = 53.03 + 0.920 \times 3.88 \times 5.25 = 53.77$$

$$p^* = 109.28 + 0.393 \times 3.88 \times 12.31 = 128.813$$

Task 4. The equivalent normal mean and standard deviation for P at the checking point of 128.813 can be shown to be 109.19 and 12.88, respectively.

Tasks 5 and 6

$$\alpha_w = \frac{-4.9444 \times 10^{-4} \times 5.25}{\sqrt{(-4.9444 \times 10^{-4} \times 5.25)^2 + (-0.873919 \times 10^{-4} \times 12.88)^2}} = \frac{-26.22081}{28.534718} = -0.919$$

$$\alpha_P = \frac{-0.873919 \times 12.88}{28.534718} = -0.394$$

The direction cosines converged.

Task 7

$$w^* = 35.03 + 0.919 \times \beta \times 5.25 = 35.03 + 4.82475\beta$$

$$p^* = 109.19 + 0.394 \times \beta \times 12.88 = 109.19 + 5.07472\beta$$

$$0.0381 - 4.99444 \times 10^{-4} \times (35.03 + 4.82475\beta) - 0.873919 \times 10^{-4} \times (109.19 + 5.07472\beta) = 0$$

or

$$\beta = 3.88$$

Task 8. The reliability index converges with a tolerance level of 0.005.

This simple example outlines the calculation of the underlying risk of a specific design. The design of a structural member is discussed next if the acceptable risk is specified *a priori* for a specific design criterion or limit state.

9.5   Reliability-Based Structural Design Using FORM

In the previous section, procedures were presented to evaluate the reliablity of an already designed structural element using FORM. The next stage is to design a structural element using the reliability-based design procedure or FORM. The steps involved in reliability-based structural design are illustrated with the help of an example.

EXAMPLE

Suppose a simply supported steel beam of span 9.114 m needs to be designed. The beams are spaced 3.048 m apart and are subjected to nominal uniform dead and live loads of 4788.03 and 2394.02 Pa, respectively. Assume the beams are continuously laterally supported by the concrete slab, that is, the unbraced length, L_b, of the beam is zero. Suppose the same beam needs to be designed using the reliability-based design concept. The acceptable risk in terms of the reliability index β for strength, that is, for bending moment is suggested as 3.0.

In the United States, steel members are designed using British units, that is, pound, feet, and second units, and AISC's LRFD design aids were developed using the same units. For this example, all calculations were initially conducted using British units and then converted to SI units to satisfy the requirements of the book.

Solution

The beam is first designed using AISC's LRFD design criteria. The factored load can be calculated as

$$W = 1.2D + 1.6L = 1.2 \times 4788 + 1.6 \times 2394 = 9576 \text{ Pa}$$

The applied design bending moment for the beam can be shown to be

$$M_A = \frac{9576 \times 3.048 \times 9.114^2}{8 \times 1000} = 303 \text{ kN m}$$

Using AISC's LRFD design criteria [1] and Grade 50 steel, the most economical section of W18 × 35 is selected for the beam. The section has about 10% more plastic section modulus than required by the LRFD guidelines.

As mentioned earlier, in standard codified design the calculations are made using nominal values. However, for reliability-based design using FORM, the statistical descriptions of all the random variables must be known in terms of the underlying distributions and parameters to define them uniquely. For this example, suppose all the random variables are normal. A normal variable can be uniquely defined if the mean and the standard deviation values are known. Denoting μ_D, μ_L, and μ_R as the mean values of the dead and live loads and the resistance of the W section, and the corresponding nominal values as D_n, L_n, and R_n, respectively, Ellingwood et al. [18] showed that it might be reasonable to assume $D_n/\mu_D = 1.05$, $L_n/\mu_L = 1.4$, and $R_n/\mu_R = 0.9$. The uncertainty in the dead load, live load and resistance in terms of COV is generally considered to be 0.13, 0.37, and 0.13, respectively.

To start the design process, the limit state function in strength for the beam is

$$g() = R - D - L \tag{9.22}$$

where R, D, and L are the resistance, and dead and live loads effects, respectively. From the information given in the problem, the mean values of the dead and live loads are $4788.0/1.05 = 4560.0$ Pa and $2394.02/1.4 = 1710.0$ Pa and the corresponding COVs are 0.13 and 0.37, respectively. The mean value of the moment caused by the applied dead load can be calculated as

$$\mu_{M_D} = \frac{4560.0 \times 3.048 \times 9.114^2}{8 \times 1000} = 144.3 \text{ kN m}$$

The standard deviation of the bending moment due to the dead load M_D is

$$\sigma_{M_D} = 0.13 \times 144.3 = 18.76 \text{ kN m}$$

Similarly, the mean value of the moment caused by the applied live load is

$$\mu_{M_L} = \frac{1710.0 \times 3.048 \times 9.114^2}{8 \times 1000} = 54.1 \text{ kN m}$$

The corresponding standard deviation of the live load moment M_L is

$$\sigma_{M_L} = 0.37 \times 54.1 = 20.02 \text{ kN m}$$

Equation 9.17 can be used to find the checking points for R, D, and L for $\beta = 3$ as

$$r^* = \mu_R - \alpha_R \times \beta \times (0.13 \times \mu_R) = \mu_R - 0.13 \times \beta \times \alpha_R \times \mu_R$$
$$d^* = 144.3 - \alpha_D \times \beta \times 18.76$$

and

$$l^* = 54.1 - \alpha_L \times \beta \times 20.02$$

To calculate the direction cosines α_R, α_D, and α_L using Equation 9.18, the partial derivatives of the performance function with respect to R, D, and L evaluated at the checking point are

$$\frac{\partial g()}{\partial R} = 1, \quad \frac{\partial g()}{\partial D} = -1, \quad \frac{\partial g()}{\partial D} = -1$$

Using Equation 9.18, the corresponding direction cosines are

$$\alpha_R = \frac{\sigma_R}{\sqrt{\sigma_R^2 + \sigma_D^2 + \sigma_L^2}}$$
$$\alpha_D = \frac{\sigma_D}{\sqrt{\sigma_R^2 + \sigma_D^2 + \sigma_L^2}}$$

and

$$\alpha_L = \frac{\sigma_L}{\sqrt{\sigma_R^2 + \sigma_D^2 + \sigma_L^2}}$$

The checking point must satisfy the performance function represented by Equation 9.22; that is

$$g() = 0 = (\mu_R - 0.13 \times \beta \times \alpha_R \times \mu_R) - (144.3 - 18.76 \times \beta \times \alpha_D) - (54.1 - 20.02 \times \beta \times \alpha_L)$$

Substituting the direction cosine values in the preceding equation and simplifying will result in

$$\frac{\mu_R - 144.3 - 54.1}{\sqrt{(0.13 \times \mu_R)^2 + 18.76^2 + 20.02^2}} = \beta = 3$$

This is a quadratic equation in terms of the mean value of R. Solving the equation gives $\mu_R = 361.7$ kN m. To select a section, the nominal value of the resisting bending moment, R_n, is necessary and can be shown to be $0.9 \times 361.7 = 325.5$ kN m. A W18 × 35 of Grade 50 steel will satisfy this requirement. The same section was obtained using the codified LRFD approach.

On the other hand, suppose the beam needs to be designed for a reliability index of 4, implying that the underlying risk has to be much smaller than before or that the beam needs to be designed more conservatively. Following the same procedure discussed above, it can be shown that a larger size member of W21 × 44 of Grade 50 steel will be required. This result is expected.

This example clearly demonstrates the advantages of the reliability-based design procedure. It will not only suggest a section but also give the underlying risk in selecting the section. Thus, using the reliability-based design procedure, engineers are empowered to design a structure considering an appropriate acceptable risk different than that considered in the codified approach for a particular structure.

9.6 Reliability Evaluation with Nonnormal Correlated Random Variables

The previous section's discussion of reliability evaluation using FORM implicitly assumes that all the random variables in the performance function are uncorrelated. Considering the practical aspect of structural engineering problems, some of the random variables are expected to be correlated. Thus, the reliability evelution of a structure using FORM for correlated random variables is of considerable interest. Although this is considered to be an advanced topic, it is discussed very briefly below. More detailed information can be found elsewhere [14].

The correlation characteristics of random variables are generally presented in the form of the covariance matrix as

$$[\mathbf{C}] = \begin{bmatrix} \sigma_{X_1}^2 & \text{cov}(X_1, X_2) & \cdots & \text{cov}(X_1, X_n) \\ \text{cov}(X_1, X_2) & \sigma_{X_2}^2 & \cdots & \text{cov}(X_2, X_n) \\ \vdots & \vdots & & \vdots \\ \text{cov}(X_n, X_2) & \text{cov}(X_n, X_2) & \cdots & \sigma_{X_n}^2 \end{bmatrix} \quad (9.23)$$

The corresponding correlation matrix can be shown to be

$$[\mathbf{C}'] = \begin{bmatrix} 1 & \rho_{X_1,X_2} & \cdots & \rho_{X_1,X_n} \\ \rho_{X_2,X_1} & 1 & \cdots & \rho_{X_2,X_n} \\ \vdots & \vdots & & \vdots \\ \rho_{X_n,X_1} & \rho_{X_n,X_2} & \cdots & 1 \end{bmatrix} \quad (9.24)$$

where ρ_{X_i,X_j} is the correlation coefficient of the X_i and X_j variables.

Reliability evaluation for correlated nonnormal variables $\mathbf{X}$ requires the original limit state equation to be rewritten in terms of the uncorrelated equivalent normal variables $\mathbf{Y}$. Haldar and Mahadevan [14] showed that this can be done using the following equation:

$$[\mathbf{X}] = [\sigma_{\mathbf{X}}^N][\mathbf{T}][\mathbf{Y}] + \{\mu_{\mathbf{X}}^N\} \quad (9.25)$$

where $\mu_{X_i}^N$ and $\sigma_{X_i}^N$ are the equivalent normal mean and standard deviation of X, respectively, evaluated at the checking point using Equations 9.15 and 9.16, and $\mathbf{T}$ is a transformation matrix. Note that the matrix containing the equivalent normal standard deviation in Equation 9.25 is a diagonal matrix. The matrix $\mathbf{T}$ can be shown to be

$$[\mathbf{T}] = \begin{bmatrix} \theta_1^{(1)} & \theta_1^{(2)} & \cdots & \theta_1^{(n)} \\ \theta_2^{(1)} & \theta_2^{(2)} & \cdots & \theta_2^{(n)} \\ \vdots & \vdots & & \vdots \\ \theta_n^{(1)} & \theta_n^{(2)} & \cdots & \theta_n^{(n)} \end{bmatrix} \quad (9.26)$$

where $\{\theta^{(i)}\}$ is the normalized eigenvector of the ith mode of the correlation matric $[\mathbf{C}']$ and $\theta_1^{(i)}, \theta_2^{(i)}, \ldots, \theta_n^{(i)}$ are the components of the ith eigenvector. Eigenvalues are the variances of $\mathbf{Y}$. $\mathbf{Y}$ will have zero means [14]. Using Equation 9.25, the correlated $\mathbf{X}$ variables can be transformed into uncorrelated $\mathbf{Y}$ variables. Then it is straightforward to rewrite the performance function in terms of the $\mathbf{Y}$ variables. FORM can then be used to evaluate the corresponding risk and reliability. The additional steps required to evaluate the reliability corresponding to a performance function containing correlated random variables are elaborated further with the help of an example. The example considered in the previous section for uncorrelated variables is modifed for this purpose.

EXAMPLE

Consider the limit state function represented by Equation 9.21 in the previous example. w is considered to be a normal random variable and P is considered to be a lognormal variable with the same means and standard deviation. However, unlike the previous example, assume w and P are correlated. The correlation coefficient between them is 0.7. The task is to evaluate the risk or reliability index of the beam for this situation.

Solution

This is a problem on correlated nonnormal variables. The limit state equation for the problem is given by Equation 9.21. As before, the uniformly distributed load w is normal, that is, $w \sim N(35.03\,\text{kN/m}$, $5.25\,\text{kN/m})$, and P is lognormal with $\lambda_P = 4.706$ and $\zeta_P = 0.1$. However, they are now considered to be correlated. Assume that the the correlation coefficient $\rho_{w,P}$ is 0.7. The reliability index calculation is expected to be very complicated. As mentioned earlier, the limit state equation needs to be expressed in terms of the uncorrelated $\mathbf{Y}$ variables. However, since the equivalent normal mean and standard deviation need to be evaluated for P at the checking point and the coordinates of the checking point are expected to be different at each iteration, the limit state function will change for each iteration. Hand calculation is not recommended for this type of problem; a computer program is necessary. However, some important steps are discussed below for ease of comprehension.

The correlation matrix $[\mathbf{C'}]$ given by Equation 9.24 for the problem is

$$[\mathbf{C'}] = \begin{bmatrix} 1 & 0.7 \\ 0.7 & 1 \end{bmatrix}$$

The two eigenvalues for the correlation matrix can be shown to be 0.3 and 1.7 [14]. The corresponding normalized eigenvectors can be evaluated and the tranformation matrix $[\mathbf{T}]$ given by Equation 9.26 can be shown to be

$$[\mathbf{T}] = \begin{bmatrix} 0.707 & 0.707 \\ -0.707 & 0.707 \end{bmatrix}$$

The results using hand calculations are summarized in Table 9.2. Calculations for the first and the final iterations are shown in the following sections.

TABLE 9.2 Reliability Evaluation Using FORM for Correlated Nonnormal Variables

Step 1	$g(\ ) = 0.0381 - (4.99444 \times 10^{-4} w + 8.73919 \times 10^{-5} P)$ (original with uncorrelated variables)				
	$g(\ ) = 0.010938 - 1.166536 \times 10^{-3} Y_1 - 2.540336 \times 10^{-3} Y_2$ (first iteration)				
	$g(\ ) = 0.011356 - 9.594176 \times 10^{-4} Y_1 - 2.74745 \times 10^{-3} Y_2$ (final iteration)				
Step 2	β	3.0			3.136	
Step 3	w^*	35.03	50.438	50.306	50.132	50.957
	p^*	111.20	138.352	142.733	146.887	144.704
Step 4	μ_W^N	35.03	35.03	35.03	35.03	35.03
	σ_W^N	5.25	5.25	5.25	5.25	5.25
	μ_P^N	110.61	107.389	106.339	105.221	105.823
	σ_P^N	11.12	13.835	14.273	14.688	14.470
Step 5	$(\partial g/\partial Y_1)^*$	-1.166536×10^{-3}	-9.986566×10^{-4}	-9.715649×10^{-4}	-9.459591×10^{-4}	-9.594176×10^{-4}
	$(\partial g/\partial Y_2)^*$	-2.540336×10^{-3}	-2.708217×10^{-3}	-2.735308×10^{-3}	-2.760914×10^{-3}	-2.74745×10^{-3}
Step 6	α_{Y_1}	-0.1894	-0.1530	-0.1491	-0.1425	-0.1451
	α_{Y_2}	-0.9819	-0.9882	-0.9890	-0.9848	-0.9894
Step 7				3.136		3.137
Step 8	β					3.137

First iteration

Task 1. To define the appropriate limit state function, the following steps are followed. Since P is a lognormal random variable, the equivalent normal mean and standard deviation at the checking point are calculated using Equations 9.15 and 9.16 as

$$\mu_P^N = 110.61, \quad \sigma_P^N = 11.12$$

Using Equation 9.23, it can be shown that

$$\begin{Bmatrix} w \\ p \end{Bmatrix} = \begin{bmatrix} 5.25 & 0 \\ 0 & 11.12 \end{bmatrix} \begin{bmatrix} 0.707 & 0.707 \\ -0.707 & 0.707 \end{bmatrix} \begin{Bmatrix} Y_1 \\ Y_2 \end{Bmatrix} + \begin{Bmatrix} 35.03 \\ 110.61 \end{Bmatrix}$$

or

$$w = 3.711 Y_1 + 3.711 Y_2 + 35.03 \tag{9.27}$$
$$P = -7.86 Y_1 + 7.86 Y_2 + 110.61 \tag{9.28}$$

Thus, the modified limit state function for the first iteration can be written in terms of the uncorrelated normal **Y** variables as

$$g(\,) = 0.0381 - 4.99444 \times 10^{-4}[3.711(Y_1 + Y_2) + 35.03] - 8.73919 \times 10^{-5}[7.86(Y_2 - Y_1) + 110.61]$$

or

$$g(\,) = 0.010938 - 1.166536 \times 10^{-3} Y_1 - 2.540336 \times 10^{-3} Y_2 \tag{9.29}$$

Task 2. Assume $\beta = 3.0$.
Task 3. $w^* = 35.03$ and $p^* = 111.20$.
Task 4. The applicable mean and standard deviation values for w and P are given in Table 9.2.
Tasks 5 and 6

$$\frac{\partial g(\,)}{\partial Y_1} = -1.166536 \times 10^{-3} \quad \text{and} \quad \frac{\partial g(\,)}{\partial Y_2} = -2.540336 \times 10^{-3}$$

$$\alpha_{Y_1} = \frac{-1.166536 \times 10^{-3} \times \sqrt{0.3}}{\sqrt{(-1.166536 \times 10^{-3})^2 \times 0.3 + (-2.540336 \times 10^{-3})^2 \times 1.7}}$$

$$= \frac{-6.38938 \times 10^{-4}}{3.373257 \times 10^{-3}} = -0.1894$$

$$\alpha_{Y_2} = \frac{-2.540336 \times 10^{-3} \times \sqrt{1.7}}{3.373257 \times 10^{-3}} = -0.9819$$

During the first iteration, no comment can be made about the convergence of the direction cosines. As shown in Table 9.2, the direction cosines converged during the third iteration with a tolerance level of 0.005. Then, Task 7 can be carried out as shown below.

Task 7

$$y_1^* = 0.1491 \times \sqrt{0.3}\beta = 0.081665\beta$$

$$y_2^* = 0.9890 \times \sqrt{1.7}\beta = 1.289498\beta$$

The applicable limit state function, similar to Equation 9.29, can be shown to be

$$g(\,) = 0.01131131 - 9.715649 \times 10^{-4} Y_1 - 2.7353082 \times 10^{-3} Y_2 \tag{9.30}$$

Thus

$$0.01131131 - 9.715649 \times 10^{-4} \times 0.081665\beta - 2.7353082 \times 10^{-3} \times 1.289498\beta = 0$$

or

$$\beta = 3.136$$

Final iteration

As shown in Table 9.2, the operating value for β is 3.136 and the direction cosines values are -0.1425 and -0.9848.

Task 3

$$y_1^* = 0.1425 \times \sqrt{0.3} \times 3.136 = 0.24476$$

$$y_2^* = 0.9898 \times \sqrt{1.7} \times 3.136 = 4.0471$$

and

$$w^* = 3.711(0.24476 + 4.0471) + 35.03 = 50.957$$
$$p^* = 10.384(-0.24476 + 4.0471) + 105.221 = 144.704$$

Task 4. The applicable mean and standard deviation values for w and P are given in Table 9.2.

Tasks 5 and 6. The applicable limit state function, similar to Equation 9.30, can be shown to be

$$g(\ ) = 0.0113564 - 9.594176 \times 10^{-4}Y_1 - 2.74745 \times 10^{-3}Y_2 \tag{9.31}$$

$$\frac{\partial g(\ )}{\partial Y_1} = -9.594176 \times 10^{-4} \quad \text{and} \quad \frac{\partial g}{\partial Y_2} = -2.74745 \times 10^{-3}$$

$$\alpha_{Y_1} = \frac{-9.594176 \times 10^{-4} \times \sqrt{0.3}}{\sqrt{(-9.594176 \times 10^{-4})^2 \times 0.3 + (-2.74745 \times 10^{-3})^2 \times 1.7}}$$

$$= \frac{-5.2549466 \times 10^{-4}}{3.62057498 \times 10^{-3}} = -0.1451$$

$$\alpha_{Y_2} = \frac{-2.74745 \times 10^{-3} \times \sqrt{1.7}}{3.62057498 \times 10^{-3}} = -0.9894$$

The direction cosines converged with a tolerance level of 0.005.

Task 7

$$y_1^* = 0.1451 \times \sqrt{0.3}\beta = 0.07947\beta$$
$$y_2^* = 0.9894 \times \sqrt{1.7}\beta = 1.29\beta$$

Using Equation 9.31, it can be shown that

$$0.0113564 - 9.594176 \times 10^{-4} \times 0.07947\beta - 2.74745 \times 10^{-3} \times 1.29\beta = 0$$

or

$$\beta = 3.137$$

The reliability index converges with a tolerance level of 0.005. The reliability index for the correlated random variables case is considerably different than that observed for the uncorrelated case.

9.7 Reliability Evaluation Using Simulation

Reliability evaluation using sophisticated probability and statistical theories may not be practical for many practicing structural engineers. But a simple simulation technique makes it possible to calculate the risk or probability of failure without knowing the analytical techniques and with only a little background in probability and statistics. The advancement in computing power makes simulation an attractive option for risk evaluation at the present time.

International experts agree that simulation can be an alternative for implementing the reliability-based design concept in practical design [19]. Lewis and Orav [20] wrote, "Simulation is essentially a controlled statistical sampling technique that, with a model, is used to obtain approximate answer for questions about complex, multi-factor probabilistic problems." They added, "It is this interaction of experience, applied mathematics, statistics, and computing science that makes simulation such a stimulating subject, but at the same time a subject that is difficult to teach and write about."

Theoretical simulation is usually performed numerically with the help of computers, allowing a more elaborate representation of a complicated engineering system than can be achieved by physical experiments, and it is often less expensive than physical models. It allows a designer to know the uncertainty characteristics being considered in a particular design, to use judgment to quantify randomness beyond what is considered in a typical codified design, to evaluate the nature of implicit or explicit performance functions, and to have control of the deterministic algorithm used to study the realistic structural behavior at the system level.

The method commonly used for this purpose is called the Monte Carlo simulation technique. In the simplest form of the basic simulation, each random variable in a problem is sampled several times to represent the underlying probabilistic characteristics. Solving the problem deterministically for each realization is known as a simulation cycle, trial, or run. Using many simulation cycles will give the probabilistic characteristics of the problem, particularly when the number of cycles tends to infinity. Using computer simulation to study the presence of uncertainty in the problem is an inexpensive experiment compared to laboratory testing. It also helps evaluate different design alternatives in the presence of uncertainty, with the goal of identifying the optimal solution.

9.7.1 Steps in Simulation

The Monte Carlo simulation technique has six essential elements [14]: (1) defining the problem in terms of all the random variables; (2) quantifying the probabilistic characteristics of all the random variables in terms of their PDFs and the corresponding parameters; (3) generating values of these random variables; (4) evaluating the problem deterministically for each set of realizations of all the random variables; (5) extracting probabilistic information from N such realizations; and (6) determining the accuracy and efficiency of the simulation. The success of implementing the Monte Carlo simulation in design will depend on how accurately each element is addressed. All these steps are discussed briefly in the following sections.

Step 1: Defining the problem in terms of all the random variables. The function that needs to be simulated must be defined in terms of all the random variables present in the formulation. For example, if the uncertainty in the applied bending moment, M_a, at the midspan of a simply supported beam of span L loaded with a uniform load w per unit length and a concentrated load P at the midspan needs to be evaluated, the problem can be represented as

$$M_a = wL^2/8 + PL/4 \qquad (9.32)$$

In this equation, if the span is assumed to be a known constant but w and P are random variables with specified statistical characteristics, then the applied moment is also a random variable. Its probabilistic characteristics can be evaluated using simulation.

On the other hand, if the probability of failure of the same beam is of interest and M_R is denoted as its bending moment capacity, the corresponding function to be simulated is

$$g(\ ) = M_R - (wL^2/8 + PL/4) \qquad (9.33)$$

In this case, M_R is expected to be a random variable in addition to w and P. The probability of failure of the beam can be evaluated by studying cases where $g(\ )$ will be negative or where the applied moment is greater than the resisting moment.

Step 2: Quantifying the probabilistic characteristics of all the random variables. The uncertainties associated with most of the random variables used in structural engineering have already been quantified by their underlying distributions and the parameters needed to define them uniquely. The subject has been discussed in detail by Haldar and Mahadevan [14] and will not be discussed further here.

Step 3: Generating random numbers for all the variables. The generation of random numbers according to a specific distribution is the heart of Monte Carlo simulation. All modern computers have the capability to generate uniformly distributed random numbers between 0 and 1. The computer will produce the required number of uniform random numbers corresponding to an arbitrary seed value between 0 and 1. In most cases, these are known as pseudorandom numbers and provide a platform for all engineering simulations.

Since most random variables are not expected to be uniform between 0 and 1, it is necessary to transform a uniform random number u_i between 0 and 1 to another random number with the appropriate statistical characteristics. The inverse transformation technique [14] is commonly used for this purpose. In this approach, the CDF of a random variable X, $F_X(x_i)$ is equated to the generated random number u_i. Thus

$$F_X(x_i) = u_i \tag{9.34}$$

or

$$x_i = F_X^{-1}(u_i) \tag{9.35}$$

If x_i is a uniform random variable between a and b, and u_i is a uniform random number between 0 and 1, then it can be shown that

$$u_i = \frac{x_i - a}{b - a} \tag{9.36}$$

or

$$x_i = a + (b - a)u_i \tag{9.37}$$

when $a=0$ and $b=1$, $x_i = u_i$, which is obvious.

If X is a normal random variable with a mean of μ_X and a standard deviation of σ_X, then a normal random number x_i corresponding to a uniform number u_i between 0 and 1 can be shown to be

$$x_i = \mu_X + \sigma_X \Phi^{-1}(u_i) \tag{9.38}$$

where Φ^{-1} is the inverse of the CDF of a standard normal variable. Similarly, if X is a lognormal random variable with parameters λ_X and ς_X, then x_i can be generated according to the lognormal distribution as

$$x_i = \exp[\lambda_X + \varsigma_X \Phi^{-1}(u_i)] \tag{9.39}$$

Most computers will generate random numbers for commonly used distributions. If not, the above procedure can be used to generate random numbers for a specific distribution.

Step 4: Evaluating the problem deterministically for each set of realizations of all the random variables. N random numbers for each of the random variables present in the problem will give N sets of random numbers, each set representing a realization of the problem. Thus, deterministically solving the problem defined in Step 1 N times will give N sample points. The generated information will provide the uncertainty in the response variable. Using N sample points and standard procedures, all the necessary statistical information can be collected, as briefly discussed next.

Step 5: Extracting probabilistic information from N such realizations. Simulation can be used to evaluate the uncertainty in the response variable like M_a in Equation 9.32. However, if the objective is only to estimate the probability of failure, the following procedure can be used.

If the value of $g(\ )$ in Equation 9.33 is negative, it indicates failure. Let N_f be the number of simulation cycles when $g(\ )$ is negative and let N be the total number of simulation cycles. The probability of failure can be expressed as

$$p_f = \frac{N_f}{N} \tag{9.40}$$

Step 6: Determining the accuracy and efficiency of the simulation. The probability of failure using Equation 9.40 is a major concern. The estimated probability of failure will reach the true value when N approaches infinity. When p_f and/or N are small, a considerable amount of error is expected in the estimated value of p_f. Haldar and Mahadevan [14] discussed the related issues in great detail. The following recommendation can be followed. In many structural engineering problems, the probability of failure could be smaller than 10^{-5}, that is, on average only 1 out of 100,000 simulations would show a failure. At least 100,000 simulation cycles are required to predict this behavior. For a reasonable estimate, at least 10 times this minimum, that is, 1 million simulation cycles, is usually recommended to estimate the probability of failure of 10^{-5}. Thus, if n random variables are present in a formulation to be simulated, $n \times 10^6$ random numbers are required. Simulation could be cumbersome or tedious for structural reliability evaluation. However, simulation is routinely used to verify a new theoretical method.

9.7.2 Variance Reduction Techniques

The discussion of simulation will not be complete without discussing variance reduction techniques (VRTs). The concept behind simulation presented in the previous section is relatively simple. However, its application to structural engineering reliability analysis depends on the efficiency of the simulation. The attractiveness of the simulation method can be greatly improved if the probability of failure can be estimated with a reduced number of simulation cycles. This led to the development of many VRTs. The efficiency of simulation can be improved by using VRTs, which can be grouped in several ways [14]. One approach is to consider whether the variance reduction method alters the experiment by altering the input scheme, by altering the model, or by special analysis of the output. The VRTs can also be grouped according to description or purpose (i.e., sampling method, correlation methods, and special methods).

The sampling methods either constrain the sample to be representative or distort the sample to emphasize the important aspects of the function being estimated. Some of the sampling methods are systematic sampling, importance sampling, stratified sampling, Latin hypercube sampling, adaptive sampling, randomization sampling, and conditional expectation. The correlation methods employ strategies to achieve correlation between functions or different simulations to improve the efficiency. Some of the VRTs in correlation methods are common random numbers, antithetic variates, and control variates. Other special VRTs include partition of the region, random quadratic method, biased estimator, and indirect estimator. The VRTs can also be combined to further increase the efficiency of the simulation. The details of these VRTs cannot be presented here but can be found in Haldar and Mahadevan [14].

The type of VRT that can be used depends on the problem under consideration. It is usually impossible to know beforehand how much efficiency can be improved using a given technique. In most cases, VRTs increase the efficiency in the reliability estimation by using a smaller number of simulation cycles. Haldar and Mahadevan [14] noted that VRTs increase the computational difficulty for each simulation, and a considerable amount of expertise may be necessary to implement them. The most desirable feature of simulation, its basic simplicity, is thus lost.

9.7.3 Simulation in Structural Design

As mentioned earlier, simulation can be an attractive alternative to estimate the reliability of a structural system. Simulation will enable reliability estimation considering realistic nonlinear structural behavior, the location of a structural element in a complicated structural system, correlation characteristics of random variables, etc. Reliability evaluation using a classical method like FORM essentially evaluates the reliability at the element level. Thus, simulation has many attractive features. It also has some deficiencies. Like other reliability methods, if the reference or allowable values are not known, it will be unable to estimate the reliability. The outcome of the simulation could be different depending on the number of simulation cycles and the characteristics of the computer-generated random numbers. One fundamental drawback is the time or cost of simulation. Huh and Haldar [21] reported that simulating

100,000 cycles in a supercomputer (SGI Origin 2000) to estimate the reliability of a one-bay two-story steel frame subjected to only 5 s of an earthquake loading may take more than 23 h. Using an ordinary computer, it may take several years.

It is clear that the simulation approach provides a reasonable alternative to the commonly used codified approach. However, there are still some issues that need to be addressed before it can be adopted in structural design. Further evaluation is needed of issues related to the efficiency and accuracy of the deterministic algorithm to be used in simulations, appropriate quantification of randomness, defining the statistical characteristics and performance functions, the selection of reference or allowable values, evaluating the correlation characteristics of random variables in complex systems, simulation of random variables versus random field, simulation of multi-variate random variables, system reliability, the effect of load combinations, time-dependent reliability, available software to implement the simulation-based concept, etc. The documentation of case studies will help in this endeavor.

EXAMPLE

Since each computer is expected to give different sets of random numbers, it may not be practical to give a detailed example to demonstrate the application of the simulation method in reliability analysis. However, a simple example is given to illustrate its many desirable features.

Suppose the probability of failure of a steel beam needs to be evaluated. The limit state function can be defined as

$$g() = F_y Z - M_a \qquad (9.41)$$

where F_y is the yield stress, Z is the plastic section modulus, and M_a is the applied bending moment. The statistical characteristics of these variables are given in Table 9.3.

The probability of failure of the beam can be calculated in several ways [22] and the results are summarized in Table 9.4. If distributional information on the three random variables is ignored, then according to MVFOSM, the probability of failure is found to be 0.007183. However, if FORM is used, the corresponding probability of failure is found to be 0.023270. These probabilities of failure are quite different. Simulation can be used to establish which number is correct.

In the third column of Table 9.4, the probabilities of failure for several simulation cycles are given for direct Monte Carlo simulation. The results indicate that if the simulation cycles are relatively small, the

TABLE 9.3 Statistical Characteristics of Random Variables

Variables	Mean value	COV	Probability distribution
F_y	262.0 MPa	0.10	Normal
Z	$8.19 \times 10^{-4} \, m^3$	0.05	Lognormal
M_a	113.0 kN m	0.30	Type II

TABLE 9.4 Probability of Failure Evaluations Using Different Methods

		Monte Carlo simulation		
MVFOSM	FORM	Direct	Conditional expectation VRT	Conditional expectation plus antithetic variates VRT
$\beta = 2.448$	$\beta = 1.990$	$N = 10$ 0.000000[a]	0.021626	0.023266
$P(failure) = 0.007183$	$P(failure) = 0.023270$	$N = 50$ 0.000000[a]	0.022071	0.024660
		$N = 100$ 0.000000[a]	0.021425	0.024322
		$N = 500$ 0.018000	0.024250	0.024233
		$N = 1000$ 0.021000	0.025025	0.024489

[a] $N_f = 0$, for these cases.

direct Monte Carlo simulation method cannot accurately predict the probability of failure. However, as the number of simulation cycles increases, the probability of failure approaches the value obtained using FORM. The important conclusion is that MVFOSM should not be used for structural reliability evaluation.

Columns 4 and 5 give the probabilities of failure using the conditional expectation VRT and conditional expectation plus antithetic variates VRT. The results indicate the power of the VRTs.

In both VRT schemes, only ten simulations are necessary to predict the underlying probability of failure. This simple example indicates that a considerable amount of additional desirable information can be collected by intelligently solving the problem ten times deterministically instead of solving it only once.

9.8 Future Directions in Reliability-Based Structural Design

The discussion of reliability-based structural design would be incomplete without commenting on future trends. Some important directions are discussed below.

9.8.1 A New Hybrid Reliability Evaluation Method

As pointed out by Haldar and Marek [23], the available reliability evaluation techniques are not capable of estimating the risk of realistic structures. Simulation is an attractive alternative but it can be very inefficient. To address these concerns and combining the desirable features of theoretical and simulation-based techniques, Huh and Haldar [21,24] proposed a hybrid approach. The algorithm intelligently integrates the concepts of the response surface method, the finite element method, FORM, and an iterative linear interpolation scheme. In this algorithm, a real structure is represented as realistically as possible by finite elements. The behavior of the structure is traced considering all major sources of nonlinearity and uncertainty under static and dynamic loading conditions. The coordinates of the most probable failure point or design point are evaluated by FORM, and then a response surface is generated around this point by multiple deterministic analyses of the structure. Conceptually, this part of the algorithm is similar to the Monte Carlo simulation technique; however, the simulation is conducted around highly selective experimental design points required for the response surface method, removing the inefficiency in the algorithm.

Huh and Haldar have shown that instead of conducting 100,000 cycles on Monte Carlo simulation, the probability of failure of a realistic nonlinear steel frame can be obtained with only 50 runs without compromising the accuracy of the estimation. The unique feature of the algorithm is that actual dynamic loading including earthquake loading can be applied in the time domain. This enables the most realistic representation of the dynamic loading condition and provides an alternative to the classical random vibration approach. The method is expected to play a major role in the reliability analysis of real structures under static and dynamic loading conditions in the future.

9.8.2 Education

Lack of education could be a major reason for the profession's avoidance of the reliability-based structural design concept. However, there are some recent major developments in this area. In the Czech Republic, CSN 7314 01-1998 (Appendix A) [25] is one of the pilot codes that allows the Monte Carlo simulation as a design tool. It was pointed out [19] that in Canada, a 14 km long bridge with a span length of 250 m was recently built. The code did not cover this design, and simulation was used. Reliability-based design is very common for offshore structures. In Europe, highway and railway companies are using simulation for assessment purposes. In the United States, the general feeling is that we are safe if we design according to the design code but this is not entirely true. According to a judge, designers should use all available means to satisfy performance requirements. The automotive industry satisfied the code requirements in one case, but a judge ruled that they should have used simulation to address the problem more comprehensively.

Some of the developments in risk-based design using simulation are very encouraging. Simulation could be used in design in some countries, but it is also necessary to look at its legal ramifications. In some countries, code guidelines must be followed to the letter, and other countries permit alternative methods if they are better [19]. In Europe, two tendencies currently exist: Anglo-Saxon (more or less free to do anything) and middle-European (fixed or obligatory requirements). Current Euro-code is obligatory. We need to change the mentality and laws to implement simulation or the reliability-based design concept in addressing real problems.

In the context of education of future structural engineers, the presence of uncertainty must be identified in design courses. Reliability assessment methods can contribute to the transition from deterministic to a probabilistic way of thinking for students as well as designers. In the United States, the Accreditation Board of Engineering and Technology now requires that all civil engineering undergraduate students demonstrate knowledge of the application of probability and statistics to engineering problems, indicating its importance in civil engineering education.

Most of the risk-based design codes are the by-product of education and research at the graduate level. A pilot international project titled TERECO (TEaching REliability COncepts), sponsored by the Leonardo da Vinci Agency in Europe [26], was very successful.

In summary, the profession is moving gradually toward accepting the reliability-based design concept, and the structures group is providing leadership in this regard.

9.8.3 Computer Programs

As mentioned earlier, many theoretical procedures with various degrees of complexity have been developed over the last few decades; however, they are not popular with practicing engineers. One issue could be the lack of user-friendly software. Two types of issues need to be addressed. Reliability-based computer software should be developed for direct applications or the reliability-based design feature should be added to commercially available deterministic software. Some of the commercially available reliability based computer software are briefly discussed next.

NESSUS (Numerical Evaluation of Stochastic Structures Under Stress) was developed by the Southwest Research Institute [27,28] under the sponsorship of the NASA Lewis Research Center. It combines probabilistic analysis with a general-purpose finite element/boundary element code. The probabilistic analysis features an advanced mean value technique. The program also includes techniques such as fast convolution and curvature-based adaptive importance sampling.

PROBAN (PROBability ANalysis) was developed at Det Norske Veritas, Norway, through A.S. Veritas Research [29]. PROBAN was designed to be a general-purpose probabilistic analysis tool. It is capable of estimating the probability of failure using FORM and SORM for a single event, unions, intersections, and unions of intersections. It has a library of standard probability distributions. The approximate FORM/SORM results can be updated through an importance sampling simulation scheme. The probability of general events can be computed using Monte Carlo simulation and directional sampling.

CALREL (CAL-RELiability) is a general-purpose structural reliability analysis program designed to compute probability integrals in the form given by Equation 9.13. CALREL was developed at the University of California at Berkeley by Liu et al. [30]. It incorporates four general techniques for computing the probability of failure: FORM, SORM, directional simulation with exact or approximate surfaces, and Monte Carlo simulation. It has a library of probability distributions of independent and dependent random variables. Additional distributions can be included through a user-defined subroutine.

Under the sponsorship of the Pacific Earthquake Engineering Research (PEER) Center [31], a multiuniversity team developed a general-purpose finite element reliability code within the framework of OpenSees. A web address is given in the references for further information on the program [31].

Structural engineers without a formal education in reliability-based design may not be able to use these computer programs. They can be retrained with very little effort. They may be very knowledgeable using existing deterministic analysis software including commercially available finite element packages.

This expertise needs to be integrated with the reliability-based design concept. Thus, probabilistic features may need to be added to the deterministic finite element packages. Proppe et al. [32] discussed the subject in great detail. For proper interface with deterministic software, they advocated a graphical user interface, a communication interface that must be flexible enough to cope with different application programming interfaces and data formats, and the reduction of problem sizes before undertaking reliability analysis. COSSAN [33] software attempted to implement the concept.

The list of computer programs given here is not exhaustive. However, these programs are being developed and are expected to play a major role in implementing reliability-based structural analysis and design in the near future.

9.9 Concluding Remarks

The state of the art in reliability-based structural analysis and design has been presented in brief in this chapter. Some of the theoretical methods currently available were discussed. The use of the simulation technique is advocated as an alternative to theoretical models. A hybrid method proposed by the author combining the desirable features of theoretical methods and simulation technique was discussed.

All major structural design codes for concrete, masonry, steel, and wood are now using the LRFD concept, which is essentially a reliability-based design concept. In spite of significant developments in the reliability-based structural design concept, it is not popular with practicing structural engineers. Issues related to this were discussed. Since reliability-based structural design may be the only design option worldwide in the near future, it is hoped that this discussion will help readers stay ahead of the curve.

References

[1] American Institute of Steel Construction (AISC), AISC, *Manual of Steel Construction Load and Resistance Factor Design*, 1st, 2nd, and 3rd Editions, Chicago, IL, 1986, 1994, 2001.

[2] American Institute of Steel Construction (AISC), *Manual of Steel Construction Allowable Stress Design*, 9th Edition, Chicago, IL, 1989.

[3] American Concrete Institute (ACI), *Building Code Requirements for Structural Concrete (318-02)*, Farmington Hills, MI, 2002.

[4] American Concrete Institute (ACI), *Building Code Requirements for Masonry Structures and Specification for Masonry Structures — 2002*, ACI 530-02/ASCE 5-02/TMS 402-02, reported by the Masonry Standards Joint Committee, 2002.

[5] American Society of Civil Engineers (ASCE), *Standard for Load and Resistance Factor Design (LRFD) for Engineered Wood Construction*, ASCE 16–95, 1995.

[6] American Wood Council, *The Load and Resistance Factor Design (LRFD) Manual for Engineered Wood Construction*, Washington, DC, 1996.

[7] American Society of Civil Engineers (ASCE), A panel session on "Past, present and future of reliability-based structural engineering worldwide," *Civil Engineering Conference and Exposition*, Washington, DC, 2002.

[8] International Standard Organization (ISO), *ISO 2394 — General Principles on Reliability for Structures*, 2nd Edition, 1998-06-01.

[9] American Institute of Steel Construction (AISC), *Economy in Steel ASD vs. LRFD*, AISC Lecture Series, 1988/1989.

[10] American Concrete Institute (ACI), *Building Code Requirements for Structural Concrete (318-99)*, Farmington Hills, MI, 1999.

[11] Bjorhovde, R., Galambos, T.V., and Ravindra, M.K., LRFD criteria for steel beam-columns, *J. Struct. Eng., ASCE*, 104(9), 1943, 1982.

[12] Mahadevan, S. and Haldar, A., Stochastic FEM-based validation of LRFD, *J. Struct. Eng., ASCE*, 117(5), 1393, 1991.

[13] Haldar, A. and Mahadevan, S., *Reliability Assessment Using Stochastic Finite Element Analysis*, John Wiley & Sons, New York, NY, 2000.

[14] Haldar, A. and Mahadevan, S., *Probability, Reliability, and Statistical Methods in Engineering Design*, John Wiley & Sons, New York, NY, 2000.

[15] Hasofar, A.M. and Lind, N.C., Exact and invariant second moment code format, *J. Eng. Mech., ASCE*, 100(EM1), 111, 1974.

[16] Rackwitz, R., Practical probabilistic approach to design, *Bulletin No. 112*, Comite European du Beton, Paris, France, 1976.

[17] Ayyub, B.M. and Haldar, A., Practical structural reliability techniques, *J. Struct. Eng., ASCE*, 110(8), 1707, 1984.

[18] Ellingwood, B., Galambos, T.V., MacGregor, J.G., and Cornell, C.A., Development of a probability based load criterion for American standard A58: Building Code Requirements for Minimum Design Loads in Buildings and Other Structure, *Special Publication 577*, National Bureau of Standards, Washington, DC, 1980.

[19] Marek, P., Haldar, A., Guštar, M., and Tikalsky, P., Editors, *Euro-SiBRAM 2002 Colloquium Proceedings*, ITAM Academy of Sciences of Czech Republic, Prosecka 76, 19000 Prague 9, Czech Republic, 2002.

[20] Lewis, P.A.W. and Orav, E.J., *Simulation Methodology for Statisticians, Operations Analysts, and Engineers*, Vol. 1, Wadsworth & Brooks/Cole Advanced Books & Software, Pacific Grove, CA, 1989.

[21] Huh, J. and Haldar, A., Stochastic finite element-based seismic risk evaluation for nonlinear structures, *J. Struct. Eng., ASCE*, 127(3), 323, 2001.

[22] Ayyub, B.M. and Haldar, A., Improved simulation techniques as structural reliability models, in *Proc. 4th Int. Conf. on Structural Safety and Reliability*, Konishi, I., Ang, A.H.-S., and Shinozuka, M., Eds., IASSR, Japan, 1985, I-17.

[23] Haldar, A., and Marek, P., Role of simulation in engineering design, in *Proc. 9th Int. Conf. on Applications of Statistics and Probability (ICASP9-2003)*, 2, 945, 2003.

[24] Huh, J. and Haldar, A., Seismic reliability of nonlinear frames with PR connections using systematic RSM, *Probab. Eng. Mech.*, 17(2), 177, 2002.

[25] Czech Institute of Standards, *CSN 73 1401-1998 Design of Steel Structures*, Prague, Czech Republic, 1998.

[26] Marek, P., Brozzetti, J., and Gustar, M., *Probabilistic Assessment of Structures using Monte Carlo Simulation*, Academy of Sciences of the Czech Republic, Praha, Czech Republic, 2001.

[27] Cruse, T.A., Burnside, O.H., Wu, Y.-T., Polch, E.Z., and Dias, J.B., Probabilistic structural analysis methods for select space propulsion system structural components (PSAM), *Comput. Struct.*, 29(5), 891, 1988.

[28] Southwest Research Institute, *NEUSS*, San Antonio, Texas, 1991.

[29] Veritas Sesam Systems, *PROBAN*, Houston, Texas, 1991.

[30] Liu, P.-L., Lin, H.-Z., and Der Kiureghian, A., *CALREL*, University of California, Berkeley, CA, 1989.

[31] McKenna, F., Fenves, G.L., and Scott, M.H., *Open System for Earthquake Engineering Simulation*, http://opensees.berkeley.edu/, Pacific Earthquake Engineering Research Center, Berkeley, CA, 2002.

[32] Proppe, C., Pradlwarter, H.J., and Schueller, G.I., Software for stochastic structural analysis — needs and requirements, in *Proc. 4th Int. Conf. on Structural Safety and Reliability*, Corotis, R.B., Schueller, G.I., and Shinizuka, M., Eds., 2001.

[33] COSSAN (Computational Stochastic Structural Analysis) — Stand-Alone Toolbox, *User's Manual*, IfM-Nr: A, Institute of Engineering Mechanics, Leopold-Franzens University, Innsbruck, Austria, 1996.

10

Structure Configuration Based on Wind Engineering

Yoshinobu Kubo
*Department of Civil Engineering,
Kyushu Institute of Technology,
Tobata, Kitakyushu,
Japan*

10.1 Introduction

When a structure is immersed in an air stream, aerodynamic forces are induced in the structures by the air stream and the aerodynamic forces apply on the structure as wind load generating deflection and vibration. Flexible structures like high-rise buildings and long-span bridges are susceptible to deflection and vibration under wind action. In order to reduce the deflection or to control the vibration of the structure, the effect of aerodynamic/aeroelastic forces should be reduced. The best method for the reduction of the aerodynamic force effect is to adapt a structural configuration that can reduce the aerodynamic/aeroelastic force effect. For this purpose, when designing a flexible structure susceptible to vibration under wind action, it is necessary and useful for structural engineers to understand the mechanism of the process generating aerodynamic forces and the relationship between structural configuration and aerodynamic forces.

The following steps are an outline of the design for the structure susceptible to vibration under wind action:

1. Decide the design wind speed for the structure by using data measured previously at meteorological observatories close to the construction site.
2. Estimate the wind load applying on the structure under the design wind speed.
3. When the wind load exceeds the design wind load, improve the proposed structural configuration so as to possess a smaller wind load based on the data previously measured or the results of wind tunnel tests conducted to find a better configuration.
4. Check the occurrence of aeroelastic vibrations, after confirmation that the wind load is smaller than the design wind load. If the occurrence of aeroelastic vibrations is predicted, the structural configuration should be further improved to be the vibration amplitude less than the allowable value so as not to induce the aeroelastic vibrations.

This chapter deals with the mechanism of the process generating aerodynamic forces and the relationship between the structural configuration and wind load or aeroelastic vibration induced in the structure.

10.2   Effects of Wind Load

Aerodynamic forces are drag force, lift force, and aerodynamic moment. The drag force is a force parallel to the wind direction, the lift force is a force perpendicular to the wind direction, and the aerodynamic moment is a rotating force around a specified point. When the terminology of wind load is used, the wind load usually indicates the drag force.

10.2.1   Mechanism of Wind Load

A cross-sectional shape of structural member is usually a nonstreamline shape, which is called a "bluff body." The representative shapes of a bluff body used in a structure are circular and rectangular cross-sections.

Figure 10.1 shows flow visualization around a circular cross-section structure [1]. The approaching flow separates at an angle θ of about $80°$, which is measured at the center of the circular section from the stagnation point to downstream direction along the surface. The separated flows on upper and lower sides roll up from both separation points and vortex streets are generated in a wake of the bluff body. The vortex vibrates with frequency linearly proportional to the wind velocity as discovered by Strouhal [2]. Figure 10.2 shows the mean pressure distribution around the circular structure [3]. Positive pressure is induced on the upstream surface and negative pressure on the downstream surface and on both upper and lower side surfaces. The drag force is calculated by subtracting pressure on downstream surfaces from pressure on upstream surface of the structure. That is, the drag force is generated by the pressure difference between upstream and downstream surfaces. The mechanism generating drag force is simple for a single bluff body as mentioned above, but complicated for multiple bluff bodies. In this section, the drag force coefficients for a single body and multiple bodies are introduced for structural engineers.

Coefficients for drag and lift forces and aerodynamic moment are C_D, C_L, and C_M, respectively, and defined as follows:

$$C_D = \frac{F_D}{\frac{1}{2}\rho U^2 A}, \quad C_L = \frac{F_L}{\frac{1}{2}\rho U^2 A}, \quad C_M = \frac{F_M}{\frac{1}{2}\rho U^2 BA} \tag{10.1}$$

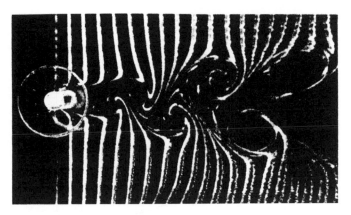

FIGURE 10.1 Flow visualization of the wake of a circular structure [1].

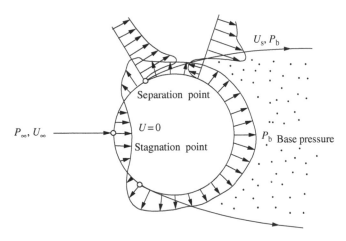

FIGURE 10.2 Rough sketch of the pressure distribution of a circular structure [3].

where ρ, U, A, and B are air density, wind velocity, representative area (usually projected area perpendicular to wind direction), and representative width, respectively. F_D, F_L, and F_M are the drag and the lift forces and the aerodynamic moment, respectively.

Strouhal number St is useful to predict the onset wind velocity of vortex-excited vibration. The definition of Strouhal number is

$$St = \frac{f_v D}{U} \tag{10.2}$$

where f_v, D, and U are frequency of vortex in the wake, representative length (usually height projected perpendicularly to wind direction), and wind velocity, respectively.

In some following figures, Strouhal number is indicated with drag force coefficients.

10.2.2 Configuration Effect for Single Bluff Body

10.2.2.1 Side Ratio Effect of Rectangular Cross-Section Structure

The drag force of a rectangular cross-section structure (rectangular structure) changes with a variety of side ratios as shown in Figure 10.3 [3]. When side ratio $B/D = 0$, the rectangular structure is

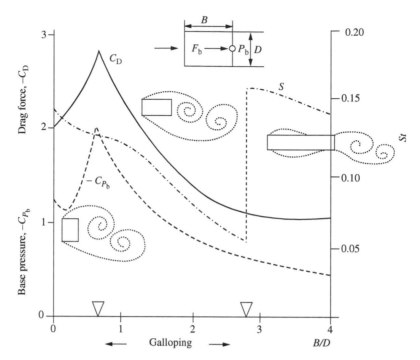

FIGURE 10.3   Base pressure, drag force, and Strouhal number to side ratio B/D of a rectangular cross-section structure [3].

a flat plate installed perpendicular to the wind direction; when the side ratio is large to infinity, the rectangular structure is a flat plate installed parallel to the wind direction.

The drag force of a rectangular structure is strongly related to flow pattern around the rectangular structure:

1. In the case of $B/D < 0.6$, separation flow is completely separated. The separated shear layers from both upstream edges flow down without reattaching to the side surfaces and roll up alternately in the wake to construct Karman vortex streets. In this region, the distance between the rolling up position of the separated flow and the downstream surface of the structure becomes shorter with increment of B/D and the base pressure on the downstream surface decreases remarkably. As a result, the drag force increases with increment of B/D. At about $B/D = 0.6$, the rolling up position becomes nearest to the downstream surface of the structure to be a peak value of the drag force in variety of B/D.

2. In the case of $0.6 < B/D < 2.8$, the approaching flow is also completely separated. In this region, the separated flow interferes with the downstream corners of the structure to increase the distance between the rolling up position and the downstream surface of the body and increases the pressure on the downstream surface (which is called pressure recovery). As a result, the drag force decreases with increment of B/D. At the same time, the interval of vortex occurrence increases with increment of B/D and Strouhal number decreases as shown in Figure 10.3.

3. In the case of $B/D > 2.8$, the separation flow steadily reattaches to the side surface and separates again at a further downstream position than the reattachment position on the side surface to construct the Karman vortex streets in the wake of the body. The separation flow periodically reattaches to the surfaces and constructs vortex as on the surfaces. This region is called the complete reattachment region of separation flow.

10.2.2.2 Drag Reduction Method

10.2.2.2.1 Corner Shape Effect

The drag force of the rectangular cross-section structure can be reduced by making corners round or by cutting corners. Figure 10.4 shows one example for the reduction of the drag force of a square structure [4]. The horizontal axis indicates the ratio of radius of rounded corner to side length and the vertical axis indicates the drag force coefficient. Type 2 is the square structure with a rounded corner at upstream corners, Type 3 has four rounded corners, and Type 4 has a rounded corner at downstream corners. The drag force of Type 2 decreases with increase in the radius of the rounded corner. The drag force of Type 4 takes a little larger value than the value of a square structure without rounded corners. Type 3 takes the value between Type 2 and Type 4. Referring to the results, it is clear that it is very important to reduce the drag force to make upstream corners round or of cutting shape. On the other hand, making downstream corners round has no effect in the reduction of drag force.

10.2.2.2.2 Roughness Effect

Figure 10.5 shows the drag force of a circular cross-section structure to Reynolds number [5]. Reynolds number (Re) is defined as the ratio of inertial force to viscous force of fluid:

$$Re = \frac{UD}{v} \qquad (10.3)$$

where U, D, v are wind velocity, representative length, and kinetic viscosity, respectively.

The circular cross-section structure is susceptible to the influence of Re as shown in Figure 10.5. The drag force coefficient of a circular cross-section structure with a smooth surface takes a constant value of 1.2 in region of $Re < 1.5 \times 10^5$ and the drag force coefficient decreases with increment of Re up to $Re = 3.5 \times 10^5$ (which is the critical Reynolds number) and increases gradually with increment of Re. In the region larger than the critical Reynolds number (which is the supercritical region), the drag force is about less than half of the value in the subcritical region. If the same condition of fluid around the circular cross-section structure is realized in the subcritical region as in the supercritical region by using some method, the drag force can be reduced to half the value of the drag force of a cylindrical member in a subcritical region. In the supercritical region, the fluid around the structure is in the state of turbulent flow. By attaching artificial roughness on the surface of the cable, the fluid around the structure can be made turbulent and the drag force reduced. The result in Figure 10.6 was obtained from the work to

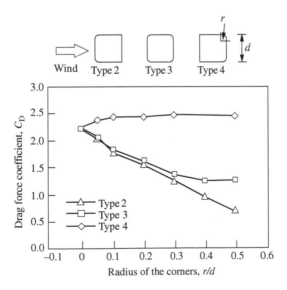

FIGURE 10.4 Drag force coefficient of a square cylinder with a rounded corner [4].

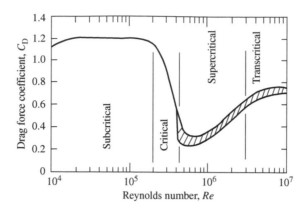

FIGURE 10.5 Drag force coefficient of a circular structure to Reynolds number [5].

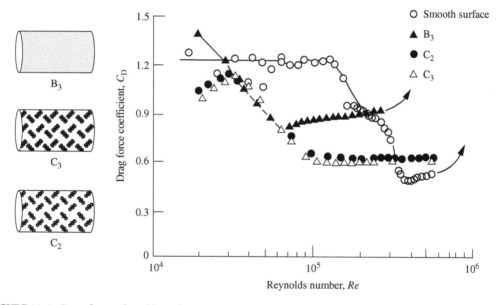

FIGURE 10.6 Drag force of a cable with various surface roughnesses to Reynolds number [6].

reduce the drag force of a cable for a long-span cable-stayed bridge [6]. Referring to the results, the cable with indented surface (C_2 and C_3) takes a smaller drag force than the cable with uniform roughness (B_3). The indented cable was adopted for the cable of Tatara Bridge in Japan, which is the world's longest cable-stayed bridge, and opened to traffic in 1999.

The Strouhal number and the drag force for various structural members are shown in Table 10.1 and Table 10.2 in Section 10.4.

10.2.3 Vicinity Arrangement Effect of Multiple Bluff Bodies [48]

When structures are built or placed in the vicinity area, their wind loads vary through the arrangement of the structures, the distance between structures, and the number of structures. The flow condition varies based on the relative distance between structures and wind load takes various values corresponding to the vicinity arrangement condition.

10.2.3.1 Side by Side Arrangement

The definition of wind load is indicated in Figure 10.7 for a side by side arrangement. In the case of side by side arrangement, since the fluid around a body flows in an asymmetric flow pattern, the lift force is also induced along with the drag force. The wakes of both structures interfere and the lift and drag forces take various values corresponding to the distance between the two structures because of bistable flow running in the gap between both structures. In the following explanations, interval parameter T/D is used. T and D are spacing distance between two structures and representative length, respectively.

Figure 10.8 shows the lift and drag force coefficients of circular structures of side by side arrangement [7]. It is seen that both coefficients take two values against the interval parameter T/D and this fact

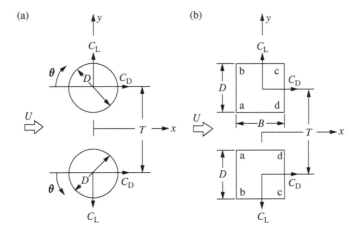

FIGURE 10.7 Side by side arrangement of two structures with: (a) circular and (b) rectangular cross-sections.

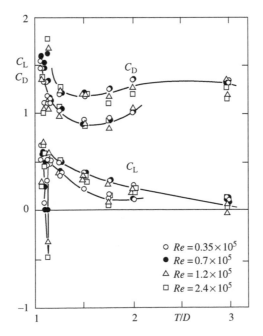

FIGURE 10.8 Lift and drag force coefficients for the side by side arrangement to T/D [7].

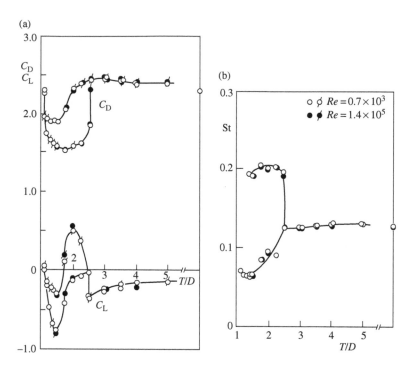

FIGURE 10.9 (a) Drag and lift force coefficient and (b) Strouhal number [8] of two square structures ($B/D = 1$) in the side by side arrangement to T/D.

supports the idea that the flow around the structures is in a bistable state. For a very small value of the spacing ratio $T/D = 1.125$, one structure shows negative lift, that is, the attractive force acts between the two structures. On the other hand, for a spacing ratio larger than $T/D = 1.125$, the repulsive lift force between the two structures acts. The lift force takes the extreme value at T/D just larger than 1.125 and decreases with increment of T/D in the region of T/D larger than 1.125. The drag force takes the extreme value at a very small spacing ratio of T/D and gradually decreases to the value of a single body with increment of T/D. The bistable flow disappears in the region of T/D larger than 3.

Figure 10.9 shows the drag and lift forces of square structures arranged side by side [8]. In the region of T/D less than 3, a bi-stable flow is generated also in the case of circular structures arranged side by side. The drag and lift forces take two values against each T/D. Since the lift force takes a negative value over the whole region of T/D, the attractive force acts on both structures. The life force takes a positive value only in the region $1.8 < T/D < 2.4$. Different from the case of circular structures, the square structures generally take negative values of lift force.

Figure 10.10 shows drag and lift force coefficients of rectangular structures arranged side by side for various side ratios [8]. In the region of side ratio larger than 4, the lift force is positive. Therefore, the lift force changes its sign at side ratio around $B/D = 4$ for the cases of $T/D = 2$ and 3. The drag force has a similar tendency as the drag force of single rectangular structure, that is, the drag force decreases with increment of side ratio.

10.2.3.2 Tandem Arrangement

A tandem arrangement is as practical as a side by side arrangement. In tandem arrangement, the downstream structure is immersed in the wake of the upstream structure. Interference of flow between structures is closely connected with the properties and behavior of the wake of an upstream structure.

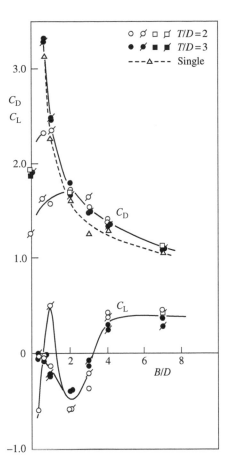

FIGURE 10.10 Variation of drag and lift force coefficients to the side ratio *B/D* of rectangular structures for *T/D* = 2 and 3 [8].

Figure 10.11 shows the drag force coefficients to *L/D* (spacing ratio, *L* is the spacing distance between the two structures) for tandem arrangement of circular structures [9–18]. In the figure, C_{D1} and C_{D2} indicate drag force coefficients of upstream and downstream structures, respectively. C_{D1} is positive over the whole spacing ratio and C_{D2} is negative in the spacing ratio less than 3.6. In the gap between two structures, negative pressure is induced by the wake of the upstream structure. Therefore, focussing on the upstream structure, positive pressure acts on the upstream surface and negative pressure is induced on the downstream surface by the structure's own wake. As a result, the drag force of the upstream structure takes a positive value. On the downstream structure, in case of space distance less than 3.6, negative pressure acts on the upstream surface, induced by the wake of the upstream structure, and negative pressure on the downstream surface. Since the absolute value of pressure for the upstream surface is larger than for the downstream surface, the drag force becomes negative for the downstream structure. The direction of the drag force of the downstream structure is the upstream direction. As shown in Figure 10.11, the drag force coefficients of both structures jump at around *L/D* = 3.6. This is called "critical spacing distance." In the spacing distance larger than the critical spacing distance, the effect of wake interference is weakened and both drag force coefficients take positive values. The flow pattern for a tandem arrangement of circular structures is shown in Figure 10.12 [19]. This figure provides useful information to understand the relationship between spacing distance and flow patterns or wake interference.

Figure 10.13 shows drag force coefficients of a tandem arrangement of rectangular structures with various side ratios [20–22]. The jumps of drag force coefficients are seen in rectangular structures

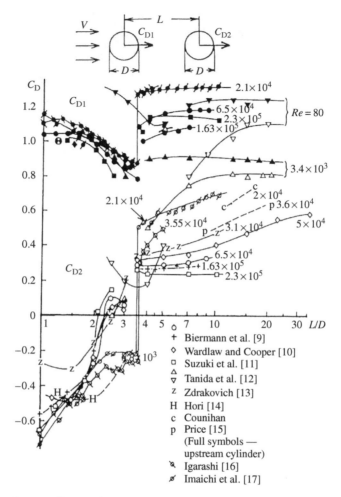

FIGURE 10.11   Drag force coefficient of two tandem circular cylinders [12,16–18,48].

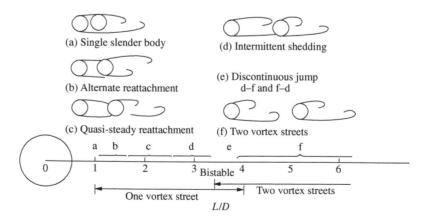

FIGURE 10.12   Classification of flow regimes in a tandem arrangement. (Zdrakovich 1985. Reprinted with permission from Elsevier.)

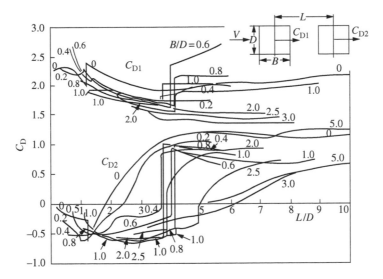

FIGURE 10.13 Drag force coefficients of two tandem rectangular cylinders with various side ratios, $B/D = 0$, 0.2, 0.4, 0.6, 0.8, 1.2, 2.5, 3, and 5 [20–22,48].

with side ratio between 0.4 and 2.0 at a spacing distance of around $L/D = 4$. As seen in the tandem arrangement of circular structures, C_{D2} changes its own sign from negative to positive against increment of L/D. Therefore, the wake interference is supposed to be weakened in the region of spacing distance over $L/D = 4$. Both drag force coefficient curves of the rectangular structures show only small changes to a variety of spacing distances for the cases except a side ratio between 0.4 and 2.0. In the single rectangular structure, it should be pointed out that the side ratio of 0.6 is a critical value. But in the tandem arrangement of rectangular structures the critical value is not found.

10.2.3.3 Staggered Arrangement

When practical structures are considered, the wind direction cannot be fixed. Therefore, even if two structures stand side by side or in tandem arrangement, the arrangement of structures becomes a staggered arrangement depending on the wind direction. Although it is very difficult to find the universality about the behavior of aerodynamic forces for staggered arrangement of the structures, which is induced by the complicated flow conditions, it will be useful for structural engineers to understand the behavior of aerodynamic forces of the staggered arrangement. But the amount of experimental data is restricted and the data presented here are only for circular structures and viaducts.

10.2.3.3.1 Circular Structures

Figure 10.14 shows the definition of location and the direction of aerodynamic forces between two structures. Figure 10.15 and Figure 10.16 show the plot of constant lift and drag force coefficient curves of the downstream structure, respectively [13]. Figure 10.15 shows a positive repulsive force in the vicinity side by side arrangement and negative lift force directed toward the wake axis of the upstream structure in the remainder of the staggered arrangement. The most remarkable feature is that there are two different lines of maximum lift force, shown with a chain-dotted line. One is inside the upstream structure wake for L/D up to 3. The other is nearer to the wake boundary for L/D larger than 2.7.

When the side surface of the downstream structure approaches closely to the axis of the wake of the upstream structure, the gap flow between the two structures generates the negative lift force, and its absolute value increases with decrement of T/D. The gap flow with high velocity induces a very low

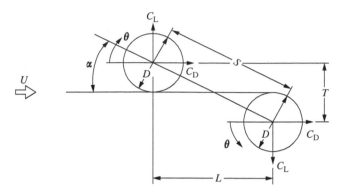

FIGURE 10.14 Definition of location and the direction of aerodynamic forces between two structures.

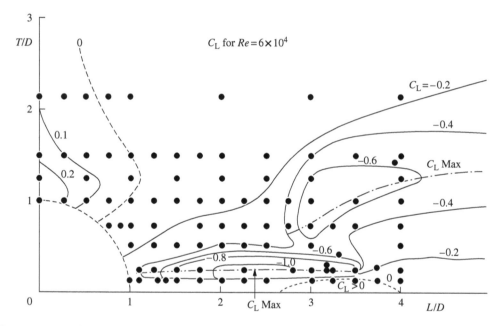

FIGURE 10.15 Lift force coefficient of a downstream structure to *L/D*. (Zdrakovich and Pridden 1977. Reprinted with permission from Elsevier.)

pressure region around both the inner and the lower sides of the downstream structure as shown in Figure 10.17 [13]. Such a pressure distribution of the downstream structure causes a very higher lift force.

On the other hand, Figure 10.15 and Figure 10.16 show that the slight asymmetry of the flow around the downstream structure has an immediate effect on the lift force, but very little effect on the drag force. A very high lift force is thus generated, but the negative drag force in a tandem arrangement is generated by a no flow condition in the gap.

10.2.3.3.2 *Viaducts in Vicinity Arrangement*

Figure 10.18 shows the model arrangement to measure the drag and lift forces of staggered arrangement of the viaducts [23]. Wind tunnel tests were conducted to obtain data for wind load to design viaducts with a noise barrier in the staggered arrangement. Figure 10.19 and Figure 10.20 show the drag forces of the model with a relative arrangement. On comparing drag force coefficients with

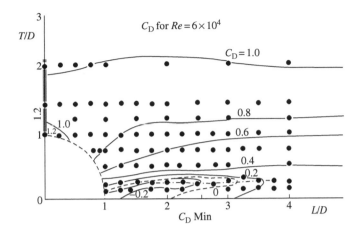

FIGURE 10.16 Drag force coefficient of a downstream structure to L/D. (Zdrakovich and Pridden 1977. Reprinted with permission from Elsevier.)

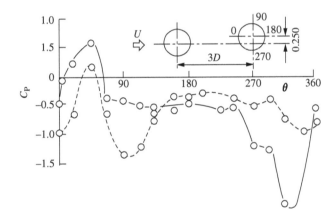

FIGURE 10.17 Pressure distribution around the downstream structure in the staggered arrangement with $L/D = 3$ and $T/D = 0.25$. (Zdrakovich and Pridden 1977. Reprinted with permission from Elsevier.)

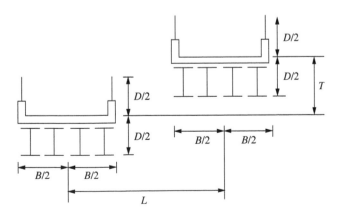

FIGURE 10.18 Definition of location of the model for measurement to the reference model in the staggered arrangement of viaducts.

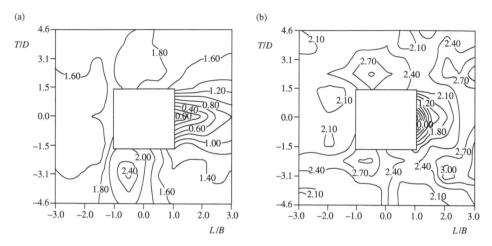

FIGURE 10.19 Drag force coefficient distribution of the model for measurement in the staggered arrangement of viaducts without noise barrier: (a) mean value and (b) mean value plus fluctuation (root mean square).

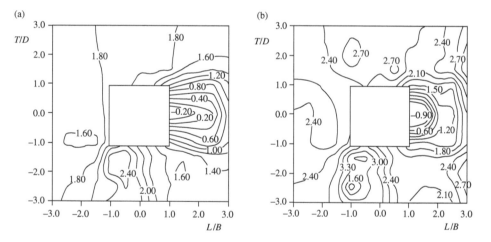

FIGURE 10.20 Drag force coefficient distribution of the model for measurement in the staggered arrangement of viaducts without noise barrier: (a) mean value and (b) mean value plus fluctuation (root mean square).

and without noise barrier, the mean value for both cases is almost equal. But the drag force in the case of including fluctuation takes the value of 1.5 to 2 times the mean value. This means that in the staggered arrangement of the viaducts, the fluctuating force applying on the downstream viaduct is very large. Therefore, in the design of staggered arrangement viaducts, the wind load should be decided by considering the addition of the fluctuating component to the mean value.

10.3 Control of Aeroelastic Responses

Flexible structures are easy to vibrate under wind action. By the degree of intensity of the vibration, the structure is destroyed, or it becomes fatigued. Therefore, methods should be developed to reduce the degree of intensity of the vibration and to keep it safe. One of the methods is to adopt a structural configuration able to reduce the aeroelastic forces. It is, however, necessary to understand the mechanism of aeroelastic vibrations before investigating the method.

10.3.1 Mechanism of Aeroelastic Vibration of Structures

Aeroelastic vibrations of flexible structures are classified into four types of vibrations (Figure 10.21): (1) vortex-excited vibration, (2) self-excited vibrations including galloping and flutter, (3) buffeting in natural wind, and (4) aerodynamic interference vibration. Vertical and horizontal axes indicate amplitude of vibration and wind velocity, respectively. In the following sections, each vibration will be discussed.

10.3.1.1 Vortex-Excited Vibration

The vortex alternately sheds from one side to another in the wake of the bluff body as shown in Figure 10.1. The shedding frequency is linearly proportional to the wind velocity. The pressure fluctuation occurs on the surface of the bluff body corresponding to the shedding frequency. When the shedding frequency is equal to the natural frequency of the structure, the force generated by the fluctuation of the surface pressure becomes resonant with the vibration of the structure. Another definition is that the vortex-excited vibration is the vibration whose frequency coincides with the vortex shedding frequency. The vortex-excited vibration occurs usually at a relatively lower wind velocity as shown in Figure 10.21. As already mentioned, each bluff body possesses an intrinsic value of Strouhal number. Incidentally, $St = 0.2$ for a circular structure, $St = 0.12$ for a square structure.

Figure 10.22 shows a typical example of vortex-excited vibration of a circular structure elastically supported with springs as indicated in the figure. The frequency of the wake of the circular structure in stationary state increases with increment of wind velocity as an inclined line in the figure. But the frequency of the wake of the circular structure supported with elastic springs takes a constant value almost equal to the natural frequency of the vibration system during the vibration of the structure. The region is called "locking-in region" or "locked-in region," which is a kind of synchronization phenomenon that the wake synchronizes with the natural frequency of vibration system. This is called "vortex-excited vibration." In the condition of vortex-excited vibration, the frequencies of both wake and bluff body coincide with each other. Since the amplitude and the region of wind velocity for vortex-excited vibration are limited, this vibration has another name, "limited vibration."

Onset wind velocity of vortex-excited vibration can be estimated from the Strouhal number. In the equation defining Strouhal number (Equation 10.2), all that is needed is to replace f_v as the frequency of the wake with the natural frequency f_0 of the vibration system. The onset wind velocity U_{cr} (critical wind velocity) can be estimated by the following equation:

$$U_{cr} = \frac{f_0 d}{St} \tag{10.4}$$

The expression by reduced wind velocity is as follows:

$$U_{r,cr} = \frac{U_{cr}}{f_0 d} = \frac{1}{St} \tag{10.5}$$

Therefore, Strouhal number directly gives the onset wind velocity by the reduced wind velocity expression for the vortex-excited vibration. $U_{r,cr} = 5$ is for a circular structure and $U_{r,cr} = 8.33$ is for a square structure. Other expressions for the onset wind velocity are $U_{cr} = 5f_0 d$ for the circular structure and $U_{cr} = 8.33 f_0 d$ for the square structure.

10.3.1.2 Self-Excited Vibration

Although almost aeroelastic vibration has characteristics of self-excited vibration, galloping and flutter are regarded to be representative self-excited vibrations. The term "galloping" is usually used for

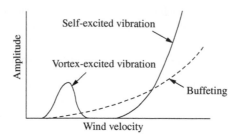

FIGURE 10.21 Types of aeroelastic response.

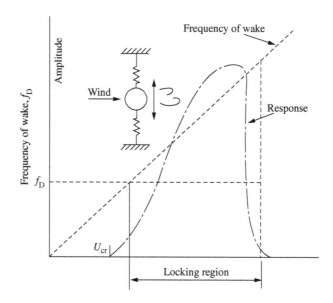

FIGURE 10.22 Vortex-excited vibration of a circular cross-section structure elastically supported with springs.

self-excited vibration including heaving vibration mode, and the term "flutter" is used for self-excited vibration including torsional vibration mode. Self-excited vibration is a vibration that increases the vibration amplitude by adding force induced by the motion of the body in the air as shown in Figure 10.23. Therefore, once a vibration occurs, the vibration grows to a large amplitude to collapse the structure.

The explanation for self-excited vibration by the equation of motion is as follows: m is the mass of the vibration system, c is the structural damping constant, k is the spring constant, y is the displacement, F_0 is the fluctuation amplitude of the aeroelastic force, ω is the frequency of the aeroelastic force, and β is the phase lag of the aeroelastic force to displacement. The equation of motion is

$$m\ddot{y} + c\dot{y} + ky = F_0 \sin(\omega t + \beta)$$
$$= F_0 \cos \beta \sin \omega t + F_0 \sin \beta \cos \omega t$$

by putting $y = y_0 \sin \omega t$, $\dot{y} = \omega y_0 \cos \omega t$

$$= F_0 \, \cos \, \beta \frac{y}{y_0} + F_0 \, \sin \, \beta \frac{\dot{y}}{\omega y_0} \tag{10.6}$$

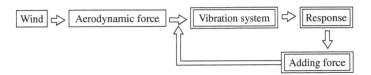

FIGURE 10.23 Explanation of self-excited vibration.

rewriting the equation by moving the right-hand side to the left-hand side

$$m\ddot{y} + \left(c - \frac{F_0 \sin \beta}{\omega y_0}\right)\dot{y} + \left(k - \frac{F_0 \cos \beta}{y_0}\right)y = 0 \tag{10.7}$$

The required condition for vibration growth is that a term proportional to velocity of displacement is negative. This is the condition for the occurrence of self-excited vibration

$$\left(c - \frac{F_0 \sin \beta}{\omega y_0}\right) \leq 0 \tag{10.8}$$

when $c = 0$, $\sin \beta \geq 0$. That is, when $\beta \geq 0$, self-excited vibration occurs. The meaning of $\beta \geq 0$ is that external force F_0 works before displacement increases the amplitude.

An interesting expression is used to explain vibration. Figuratively speaking, a horse beaten to run is forced vibration. A horse is chasing a carrot to eat which is hung down from a bar fixed on her back. The horse will increase its running speed with the intention of bringing the carrot into its mouth. This is literally self-excited phenomenon.

The following phenomena are classified into self-excited vibration.

10.3.1.2.1 Galloping

A square section and D-type section induce galloping. Figure 10.24 shows a spring-mass vibration system under wind action. When the mass goes downward, the mass receives the upward wind. Considering the moment when the mass is going down, the aerodynamic forces act on the mass as shown in Figure 10.25. The upward force F_y from drag and lift forces can be regarded as a driving force to heave the mass up and down. F_y can be expressed as follows:

$$F_y = L\cos \alpha + D\sin \alpha \tag{10.9}$$

By using Taylor's expansion at around $\alpha = 0$ and expressed in first order

$$F_y = \left.\left(\frac{dL}{d\alpha} + D\right)\right|_{\alpha=0} \alpha = \frac{1}{2}\rho U^2 A \left.\left(\frac{dC_L}{d\alpha} + C_D\right)\right|_{\alpha=0} \alpha \tag{10.10}$$

Assuming that α is very small and satisfies $\tan \alpha \approx \alpha$, $\alpha \approx -\dot{y}/U$ can be assumed. The equation of motion is

$$m\ddot{y} + c\dot{y} + ky = -\frac{1}{2}\rho U^2 A \left.\left(\frac{dC_L}{d\alpha} + C_D\right)\right|_{\alpha=0} \frac{\dot{y}}{U} \tag{10.11}$$

by putting $\omega^2 = k/m$, $2\zeta\omega = c/m$, and rearranging the equation

$$\ddot{y} + \left\{2\zeta\omega + \frac{\rho U A}{2m}\left.\left(\frac{dC_L}{d\alpha} + C_D\right)\right|_{\alpha=0}\right\}\dot{y} + \omega^2 y = 0 \tag{10.12}$$

The condition toward oscillatory instability is that the second term is negative. If $\zeta = 0$

$$\left.\left(\frac{dC_L}{d\alpha} + C_D\right)\right|_{\alpha=0} < 0 \tag{10.13}$$

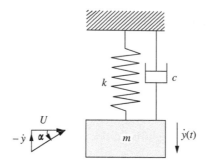

FIGURE 10.24   Spring-mass system subjected to wind action.

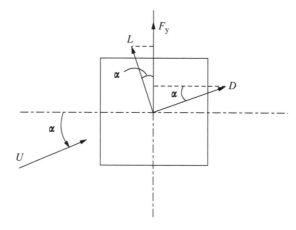

FIGURE 10.25   Heaving force by drag and lift forces and relative wind velocity.

becomes the condition for oscillatory instability. This is the well known Den Hartog criterion. The criterion teaches us that it is possible to estimate the occurrence of galloping from the lift forces to the angle of attack.

Figure 10.26 shows the lift force coefficient to the angle of attack of a square structure [4]. The meaning of the negative slope at zero angle of attack is as follows. When the square structure is going down, the square structure receives upward wind with the positive angle of attack and the downward force F_y is generated through the negative lift force. Therefore, F_y acts to help increase the downward displacement, that is, the force acting on the structure has the same direction as the movement of the structure, and the amplitude of vibration increases infinitely in linear vibration system. This is the mechanism of galloping. The method to estimate the dynamic behavior by using static parameters is called as the quasisteady theory. The galloping of a square structure is a representative example to which the quasisteady theory is applicable.

10.3.1.2.2 Stall Flutter
Shallow rectangular sections and shallow H-sections induce stall flutter, which is the torsional type self-excited vibration with single degree of freedom. Since this type of flutter is induced by the separation flow from the leading edge of the shallow body, it is also called "separation flow flutter."

10.3.1.2.3 Classical Flutter
A flat plate induces classical flutter, which is the self-excited vibration with two degrees of freedom. Heaving and torsional vibrations are coupled and this is the most intense self-excited vibration.

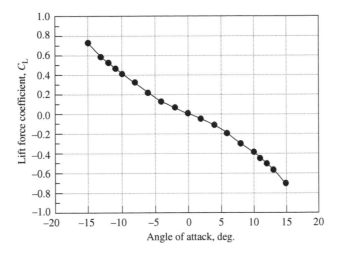

FIGURE 10.26 Lift force coefficient to angle of attack of a square structure [4].

Other names are "coupled flutter" and "potential flutter." In this flutter, the flow around the flat plate is the potential flow without separation from the leading edge.

10.3.1.3 Buffeting and Aerodynamic Interference Vibration

10.3.1.3.1 Buffeting

When the approaching flow to a structure is turbulence, the structure is forced to vibrate randomly due to the random external force induced by the turbulent flow. Buffeting is inherently a random vibration due to the aerodynamic random force induced by upstream turbulence which is generated by the existence of the upstream structure.

10.3.1.3.2 Aerodynamic Interference Vibration

When the structures or structural members are arranged in the vicinity area, an upstream structure or member affects the approaching flow to a downstream structure and the behavior of the downstream structure. For example, in bundle cable of cable-stayed bridges, the downstream cable intensely vibrates, affected by the wake of the upstream cable. As the structure for city life becomes complicated, the number of buildings and viaducts built in the vicinity area increases and the problems associated with aerodynamic interference tend to increase. Since the phenomena are complicated and various, as introduced later, it is very difficult to solve the problem by the deterministic method.

10.3.1.4 Mechanism of aeroelastic vibration of shallow section

Figure 10.27 shows the vortex arrangement on the surface in the motion of torsional and heaving vibrations [24]. Left and right rows show the flow situation around the shallow rectangular structure in torsional and heaving vibration, respectively, the middle row shows sketches of vortex arrangement on the surface, and the arrow indicates force direction acting on the surface. According to the figure, the number of vortices on the surface of the shallow rectangular structure is the same for the identical reduced wind velocity regardless of kinds of motion.

Do inherent differences exist between the vortex-excited vibration and self-excited vibration? According to the research associated with a shallow rectangular structure, there is no inherent difference between them. Figure 10.28 is a summarized expression concerning the mechanism of aeroelastic vibration. The fundamental concept is that aeroelastic vibration is controlled by the vortex

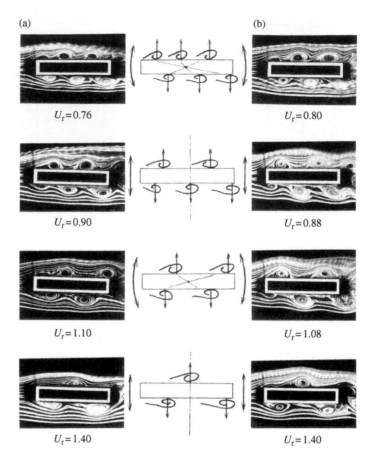

FIGURE 10.27 Vortex arrangement at various wind velocities [24]: (a) torsional vibration and (b) heaving vibration.

arrangement on both upper and lower surfaces. The assumption of the fundamental concept is that the separation flow takes a certain time length to form a vortex on the surface and the formed vortex flows down with the speed equal to wind velocity. Figure 10.28 also shows aeroelastic vibrations to reduced wind velocity for torsional and heaving vibrations. The top panel indicates that torsional vibration and heaving vibration appear alternately to increment of wind velocity, if both vibration modes have the same natural frequency. The torsional vibration mode appears under the condition that the number of vortices on the upper surface is equal to lower surface and the heaving vibration occurs under the condition that there is difference of one vortex between the number of vortices on upper and lower surfaces. From the point of view mentioned above, flutter is regarded as torsional vibration III. In this sense, inherently there is no difference between vortex-excited vibration and stall flutter.

10.3.2   Control of Aeroelastic Vibration of a Single Bluff Body

Since the vortex-excited vibration of a single bluff body is seen at a relatively lower wind velocity, the occurrence frequency is highest among the various aeroelastic vibrations. Various methods have been developed to suppress the vortex-excited vibration of tower-like structures and bridge girders. In the following, methods to control aeroelastic vibrations are introduced as countermeasures for tower-type or bridge girder-type structures.

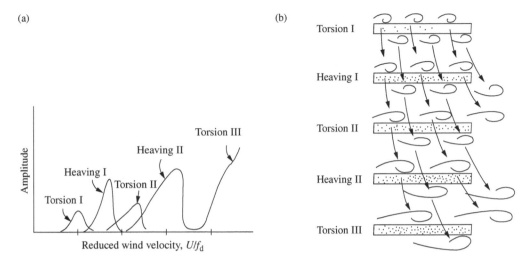

FIGURE 10.28 Summarized expression for aeroelastic vibration and the corresponding vortex arrangement [24]: (a) sketch of aeroelastic response of a shallow rectangular structure and (b) expected vortex arrangement.

10.3.2.1 Tower-Type Structure

The tower-type structures are usually composed of typical bluff bodies that are circular and square structures or shapes nearer to these. A method to suppress the vortex-excited vibrations of circular structures has been developed. The concept of the method is to reduce the space correlation of surface pressure in the axial direction of the bluff body. Another method is to remove the simultaneous fluctuation of surface pressure along the longitudinal axis, or to remove the simultaneous separation of the flow to generate vortices in axial direction of the bluff bodies.

The methods for structures with circular cross-section are helical winding of wire, helical strake, and perforated shroud as shown in Figure 10.29 [25,26]. The wind tunnel test result is shown in Figure 10.30 for a chimney with helical strake [27]. Although the vortex-excited vibration occurs in the case of chimney without helical strakes, it is completely suppressed by attaching the helical strakes.

Since octagonal and square shape cross-sections have the negative slope of lift force around zero degrees of angle of attack as shown in Figure 10.26 and Figure 10.31 [28], they have the possibility of galloping as mentioned in the previous section. It is necessary to develop a method for suppressing the galloping of both shape cross-sections. The methods invented up to now are shown in Figure 10.32. They are corner cut-outs, corner vanes, gap, and protuberant plates. The concept for corner vanes is to suppress the separation from the corner by inducing the approaching flow along the surface. For the corner cuts, corner cut-outs, and protuberant plates, having two corners in close proximity is the concept to control the separation flow from the upstream separation point by contacting it to the downstream separation point. And for gap the concept is to reduce the pressure difference between front and back surfaces. Figure 10.33 and Figure 10.34 show the examples of wind tunnel results for corner vanes and protuberant plates [29,30]. In each figure, the results are compared for both the cases with and without the countermeasures. It is understood that both methods are very useful for suppressing the galloping. Especially interesting is that the angle between two separation points affects the performance of suppression of galloping and that there is an optimum angle around $30°$ (corresponding to $p/H = 0.3$). The result is obtained from the wind tunnel tests for protuberant plates. These countermeasures were used for steel towers of cable-stayed and suspension bridges. In these bridges, before the cable is installed, the steel tower is in a free standing state with low structural damping and is easy to induce vortex-excited vibration and galloping.

There is another aeroelastic problem on a tower of a cable-stayed bridge. When the tower is in the A-shape, the in-plane vibration is sometimes induced by the wind with bridge axis direction. Since the

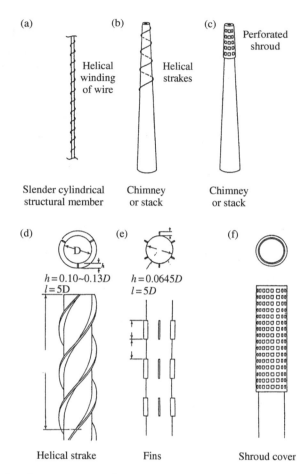

FIGURE 10.29 Examples of countermeasure for vortex-excited vibration of a circular cylinder [25,26].

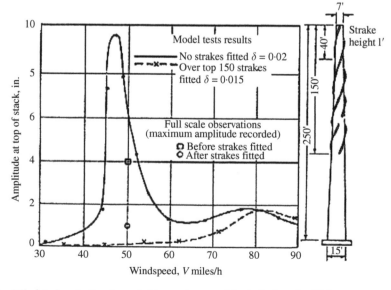

FIGURE 10.30 Wind-induced vibration of chimney improved by helical strake [27].

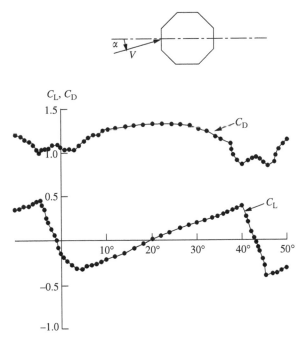

FIGURE 10.31 Drag and lift force coefficients for octagonal cylinder at $Re = 1.2 \times 10^6$ [28].

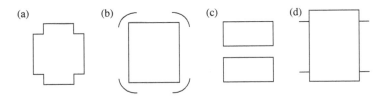

FIGURE 10.32 Types of countermeasure for bridge tower: (a) corner cut-outs; (b) corner vanes; (c) gap; and (d) protuberant plates.

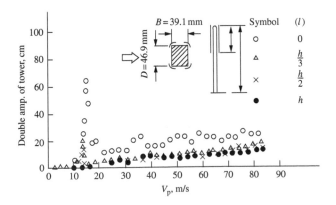

FIGURE 10.33 Effect of corner vanes for bridge tower on reduction of aeroelastic vibration amplitude [29].

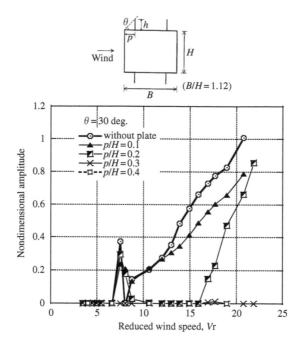

FIGURE 10.34 Aeroelastic response of protuberant plates with various plate locations ($H = 50$ mm, $h = 9$ mm) [30].

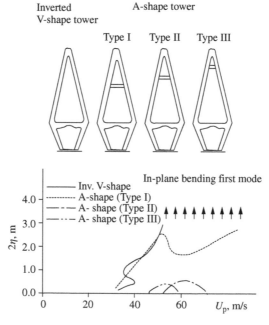

FIGURE 10.35 Aeroelastic response of A-shape tower and inverted V-shape tower with various locations of horizontal beam [31].

in-plane stiffness of the tower is not improved even after cable installation, being different from the out-of-plane rigidity, it is improved after the installation of the cable. Therefore, the in-plane vibration by the wind in bridge axis direction is still a severe problem even after completion for a cable-stayed bridge with an A-shape tower. Figure 10.35 shows the results of wind tunnel tests for the A- and the inverted V-shape

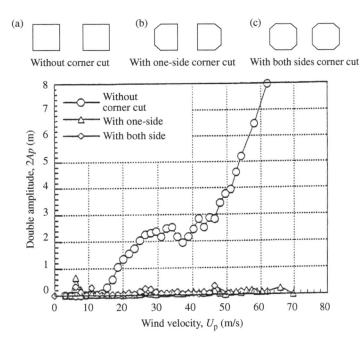

FIGURE 10.36 Aeroelastic response of A-shape tower with/without corner cut column [32].

towers in a free standing state [31,32]. The location of a horizontal beam remarkably affects the aeroelastic vibration in the out-of-plane direction. The key point to suppress the aeroelastic vibration is the selection of mounting position for horizontal beam of the tower.

Figure 10.36 shows the effect of corner-cut for the tower to the wind with perpendicular direction to the bridge axis [32]. When the corner-cuts are conducted at up- and downstream corners, the galloping is not suppressed; however, when the corner-cuts are conducted at only upstream corners, the aeroelastic performance is remarkably improved.

10.3.2.2 Bridge Girder Structures

The relatively shallower cross-section is usually used for bridge girders. Especially since a cable-stayed bridge and a suspension bridge as the representative of longer span bridge and girder bridge with longer span than 80 m have generally low frequency and small structural damping, they are susceptible to vibration under wind action. Various countermeasures have been developed to suppress aeroelastic vibrations induced in bridge girders.

10.3.2.2.1 Vortex-Excited Vibration

Figure 10.37 shows typical examples of countermeasure for vortex-excited vibration of bridge girders. The concept to suppress the vibration is to control the separation flow from the leading edge of the girders. In the figure, (a) horizontal plates were mounted to project outboard from the lower chord of each girder. It is expected for the plate to suppress the separation of flow from the under side of the plate girder. (b) The open girder induced vortex-excited vibration with large amplitude. Figure 10.38 shows the aeroelastic responses during process up to finding the final solution [33]. The figure indicates that the amplitude of vortex-excited vibration is reduced in cases when a soffit plate is added and when a soffit plate with wider fairing width is added. For an open deck, it is useful to suppress the vortex-excited vibration by closing the open side by the soffit plate. The more effective means besides adding the soffit plate is to mount the fairing with wider width. (c) Baffle plates are mounted in the inner part of the open deck. The role of the baffle plates is to prevent the separation flow of the under side of the girder from rolling into the open space of the girder. (d) Flaps are also useful for suppression of vortex-excited vibration. The role of the flap is to suppress the separation at the tip of the upper surface of the deck and

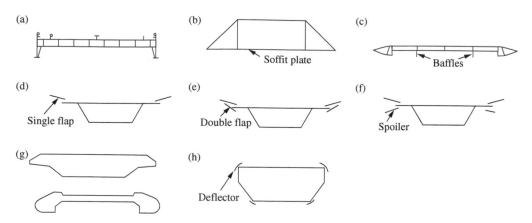

FIGURE 10.37 Countermeasures for vortex-excited vibration of shallow cross-section girder: (a) horizontal plate at lower flange; (b) soffit plate; (c) baffle plate; (d) single flap; (e) double flap; (f) flap + spoiler; (g) blunt fairing; and (h) deflector.

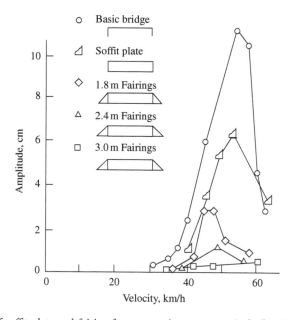

FIGURE 10.38 Effect of soffit plate and fairing for suppressing vortex-excited vibration of open girder [33].

to prevent the flow from rolling on the upper surface of the deck. (e) Deflector has a similar role as the corner vanes for the square-shaped tower introduced in the previous section. The deflector has the role of making the fluid at the leading corner run along the girder shape. (f) Blunt fairing was developed as the fairing for a cable-stayed bridge made of prestressed concrete. At the present time, the blunt fairing is also used in a steel bridge. The concept for the blunt fairing is as follows [34]. The required conditions for a concrete bridge is to eliminate a sharp edge corner, because it is difficult to make the sharp edge corner and the tip of the sharp edge corner is easy to break. After studying aeroelastic performance of various blunt fairings, the upper and lower slopes of the blunt fairing is decided in the shape as shown in Figure 10.39. Figure 10.40 shows the results during the process to decide angle of the lower slope. The lower angle as shown in the figure was changed from 20° to 50° in every 5°. Referring to the figure, the aeroelastic response is remarkably influenced by the lower angles of the blunt fairing. After comparison

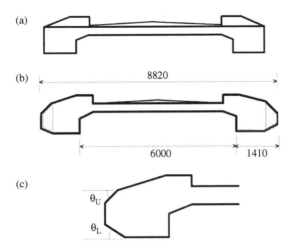

FIGURE 10.39 Open deck girder with blunt fairing made of prestressed concrete: (a) open deck girder without fairing; (b) fundamental cross-section for open deck girder; and (c) definition of θ_U and θ_L. (Kubo et al. 1993. Reprinted with permission from Elsevier.)

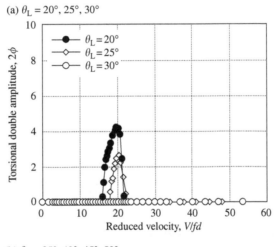

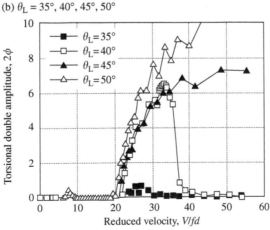

FIGURE 10.40 Aeroelastic response of bridge girder with blunt fairing with various lower slope angles θ_L. (Kubo et al. 1993. Reprinted with permission from Elsevier.)

of the wind tunnel test results, the optimum lower angle of 30° was observed. Figure 10.41 shows the flow visualization around the girders with the blunt fairing with various angles of lower slope. In the case of 20°, the separation flow from the lower side of the leading edge reattaches on the lower surface of the downstream edge. In the case of 30°, the separation flow flows down without reattachment. And in the cases of larger than 40°, the separation flow rolls in the inner space of the open deck. As a result, in the cases of less than 30°, vortex-excited vibration is induced, and in the cases of larger than 40°, stall flutter is induced.

10.3.2.2.2 Galloping

Since a box-girder bridge is usually constructed by using a beam with a rectangular section with a small side ratio, the galloping is easily induced. For suppression of the galloping, it is very useful to mount the horizontal plate on the surface of the lower part of the girder as shown in Figure 10.42 [3]. Figure 10.43 shows one of the examples of the effectiveness of the horizontal plate for suppressing galloping [35]. The top figure is the graph of the aeroelastic response of the original section. Galloping occurs at high wind velocity. The bottom figure is the graph of the aeroelastic response of the improved section by attaching a horizontal plate on the surface of the lower part of the the girder. The figure shows that the galloping which appears in original section is suppressed in the improved section.

10.3.2.2.3 Flutter

The shallow cross-section is susceptible to flutter that is self-excited vibration with torsional vibration. Figure 10.44 shows countermeasures to suppress flutter [3]. The girders with open grating (a) and central opening (b) are used to reduce the pressure difference between upper and lower surfaces of the girder, which is a main cause of occurrence of flutter. The vertical plates (a) called as "stabilizer" and "center barrier" have a role to prevent the separation flow from the upstream upper chord of the truss-stiffened girder from smoothly flowing down and to raise the flutter onset wind velocity.

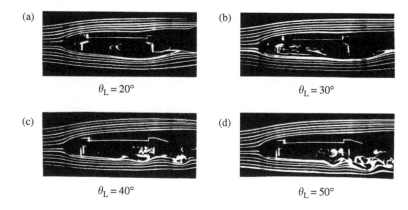

FIGURE 10.41   Flow visualization of bridge girder with blunt fairing with various lower slope angles. (Kubo et al. 1993. Reprinted with permission from Elsevier.)

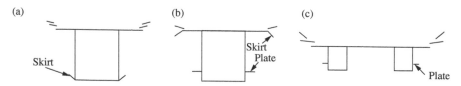

FIGURE 10.42   Countermeasures for galloping of box-girder [3]: (a) double flap + skirt; (b) flap + skirt + plate; and (c) double flap + plate.

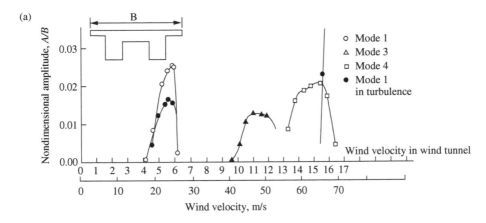

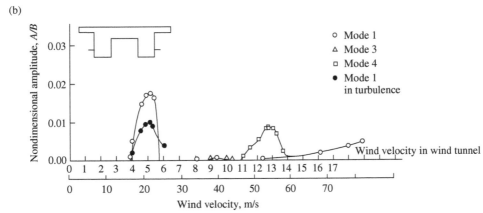

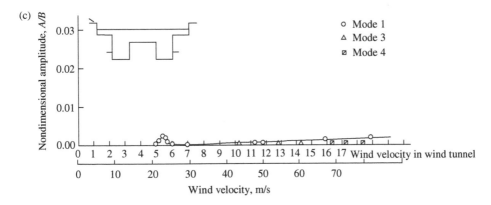

FIGURE 10.43 Effect of horizontal plates and double flap to suppress galloping and vortex-excited vibration [35]: (a) aeroelastic response of the original steel box-girder bridge; (b) aeroelastic response improved by horizontal plates at lower portion of the girder (galloping is suppressed); and (c) aeroelastic response improved by horizontal plate and double flap (galloping and vortex-excited vibration are suppressed).

Figure 10.45 shows the stabilizer for the truss-stiffened girder and the results of sectional model tests [36]. The flutter onset wind velocity of the truss-stiffened girder is remarkably raised by using a stabilizer. Wind-nose (c), deflector (d), and spoiler (e) help make the approaching flow run along the girder shape without separating the flow at the tip of the girder.

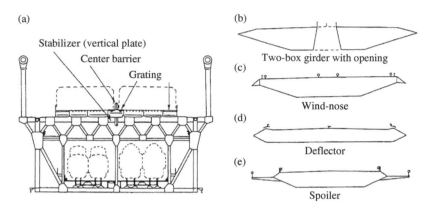

FIGURE 10.44 Countermeasures to suppress flutter [3]: (a) truss-stiffened girder (gratings, center barrier, stabilizer); (b) two-box girder with opening; (c) wind-nose; (d) deflector; and (e) spoiler.

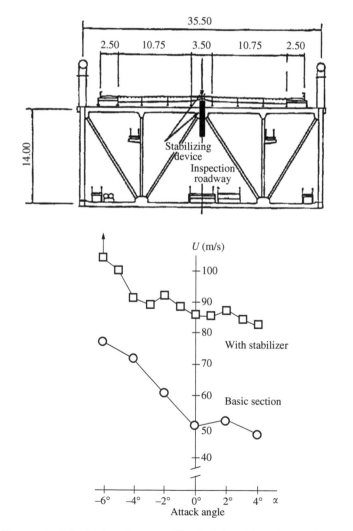

FIGURE 10.45 Flutter onset wind velocity of a truss-stiffened girder with/without stabilizer [36].

The ultimate countermeasure for suppressing aeroelastic vibration should be enabled by using not additive devices as deflector and fairing but only structural members. Figure 10.46 shows a bridge deck with two plate girders as one of the examples for trying to improve aeroelastic performance by only structural members [37]. In order to suppress the aeroelastic vibrations of the bridge girder with two plate girders, investigations were done to find how the locations of the plate girders under the deck influence the aeroelastic performance in both torsional and heaving vibration modes. Figure 10.47 shows the results of wind tunnel tests [37]. Referring to the experimental results, the inner location of the plate girder gives a better performance with regards to aeroelastic behavior. Before deciding on the structural details of a bridge, if the aeroelastic behavior is investigated, more economical and reasonable bridges can be realized without using additive devices.

10.3.2.3 Cables

Transmission lines, cables for cable-stayed bridges, and hanger for suspension bridge induce vortex-excited vibration, rain-wind-induced vibration, and galloping by adhering ice and snow. In order to suppress the vortex-excited vibration, helical winding of wire has been used in towers with circular cross-sections.

For rain-wind-induced vibration, axial grooves and indent shapes have been applied on the surface of the cable for a long-span cable-stayed bridge as shown in Figure 10.6 and Figure 10.48 [2,6]. The axial grooves and indent shapes prevent water rivulets from moving freely on the cable surface to simply induce the aeroelastic vibration. A diagonal hanger of a suspension bridge is susceptible to inducing aeroelastic vibration, explained as the axial flow along the diagonal hanger inducing the vibration. Cable fin is proposed as the countermeasure to prevent the occurrence of axial flow.

It is very difficult to suppress the galloping caused by adhering ice and snow by means of adopting an adequate configuration, because the form by ice and snow is not determined.

10.3.3 Control of Aerodynamic Interference Caused by Multiple Structure

The aerodynamic interference is induced by multiple structure or by multiple structural members and the degree of the interference is different by the difference among the arrangements. An intense vibration is induced by the wake of the upstream structure, which is called as "wake galloping" and "wake flutter."

10.3.3.1 Bundle Cables

In the cable-stayed bridge of prestressed concrete, multicables are sometimes used at one anchorage point from an economical viewpoint. When two cables in a bundle cable are placed closely in tandem arrangement against the wind direction, wake galloping is induced in the downstream cable. In the wake galloping, the downstream cable is vibrated by the wake generated by the upstream cable [38].

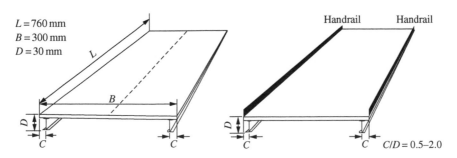

FIGURE 10.46 Bridge girder composed of two plate girders [37].

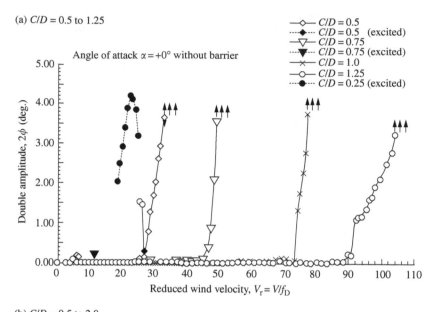

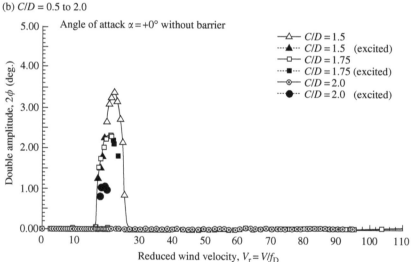

FIGURE 10.47 Torsional response of the bridge girder with various locations of two plate girders [37].

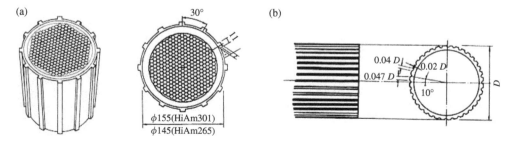

FIGURE 10.48 Cables developed for suppression of rain-wind-induced vibration [3]: (a) protuberant axial groove and (b) U-groove.

The response behavior varies according to the spacing distance between both cables as shown in Figure 10.49. The response behavior drastically changes as the boundary with respect to the critical spacing distance $L/D = 3.6$, which was introduced in Section 10.2.

Aeroelastic performance of three and four bundle cables were also investigated as shown in Figure 10.50 [39]. In the figure, the circle with oblique lines indicates the measured cable. In any of the cases, the downstream cable is vibrated with large amplitude by the wake of the upstream cables. Wake galloping is difficult to be generated in a lateral triangle arrangement, because any cable does not exist in the wake of other cables.

Wake flutter was observed in hanger cables of the Akashi Strait Bridge with large spacing distance of $10D$ (ten times of hanger diameter D) [40]. The wake galloping occurs intensively in tandem arrangement and the direction of vibration of downstream cable is almost perpendicular to the wind direction; the wake flutter, however, occurs intensively in a slightly deviated position from tandem arrangement and the locus of vibration of the downstream cable is an ellipse. Although the mechanism is still not clear, the helical winding of wire was proposed as a countermeasure to suppress the wake flutter. The experimental data are shown in Figure 10.51.

10.3.3.2 Parallel Bridges

A new bridge is sometimes constructed parallel to an old bridge to reduce traffic congestion in a large city. This fact causes a new problem concerning the aerodynamic interference caused by constructing the new bridge close to and along the old bridge.

Figure 10.52 shows the example of the aerodynamic interference caused by the parallel arrangement of a box three-span box-girder bridge [41]. The figure shows clearly that the relative distance among the bridges gives a remarkable effect for the aeroelastic performance of the bridge. The aeroelastic forces applied on the downstream bridge are induced by the wake of upstream bridge and generate the intense vibration with larger amplitude than 3% of deck height. It is very difficult to suppress the vibration induced by aerodynamic interference by the aerodynamic countermeasure by selection of a bridge cross-section providing better aeroelastic performance. Finding the solution is the future subject in the field for suppression of aeroelastic vibrations. One of the solutions is to construct a new bridge in a place away from old bridge where both bridges do not interfere with each other with respect to aerodynamic performance. It is very difficult to find the appropriate place in almost all cases from the viewpoint of effective land use. If possible, it should be investigated at the urban planning stage.

10.4 Wind Design Data

In this section, data for design of structures subjected to wind load and the method to read the experimental data are introduced.

10.4.1 Estimation of Wind Load and Onset Wind Velocity

Table 10.1 and Table 10.2 show Strouhal numbers and aerodynamic coefficients for various cross-sections [5]. The wind load is calculated by Equation 10.1. The stress intensity caused by wind load is compared with the allowable stress intensity of the material used for the structure and confirmed to be less than the allowable stress intensity.

The onset wind velocity of vortex-excited vibration is estimated by Equation 10.5. The following is a summary about the estimation of the onset wind velocity of a bridge girder in *The Wind-Resistant Design Handbook in Japan* [42]. The unit of onset wind velocity is m/s. Onset wind velocity should be larger than the design wind velocity as the design criterion.

(a) $S_H = 2d, 2.5d, 3d$

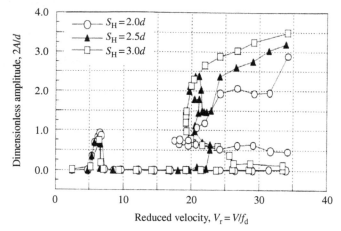

(b) $S_H = 3.5d, 4.0d, 5.0d$

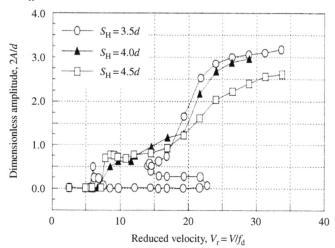

(c) $S_H = 6.0d, 7.0d, 8.0d$

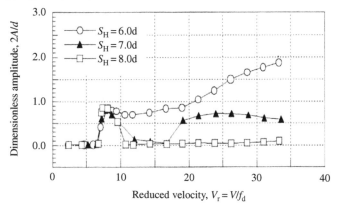

FIGURE 10.49 Aeroelastic response of a downstream cable of a bundle cable in wake-galloping [38].

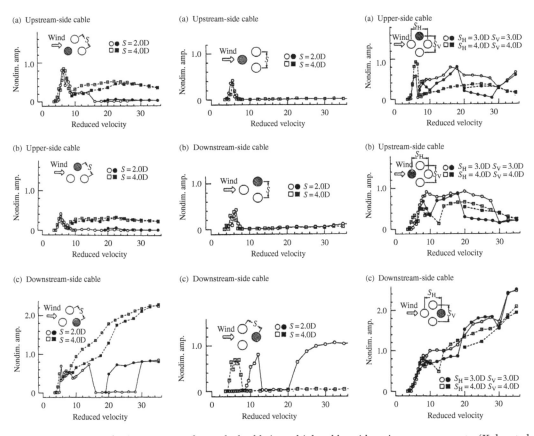

FIGURE 10.50 Aeroelastic responses of a marked cable in multiple cables with various arrangements. (Kubo et al. 1995. Reprinted with permission from Elsevier.)

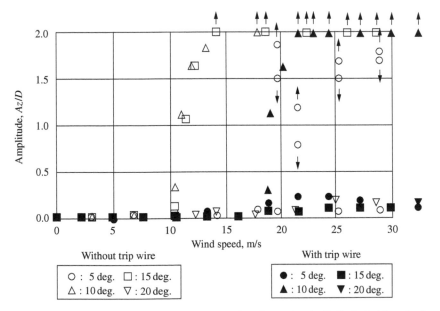

FIGURE 10.51 Aeroelastic response of downstream cable with/without trip wire (helical wire) in wake flutter against various wind directions. In the figure, deg. means wind direction from tandem arrangement of the cable [40].

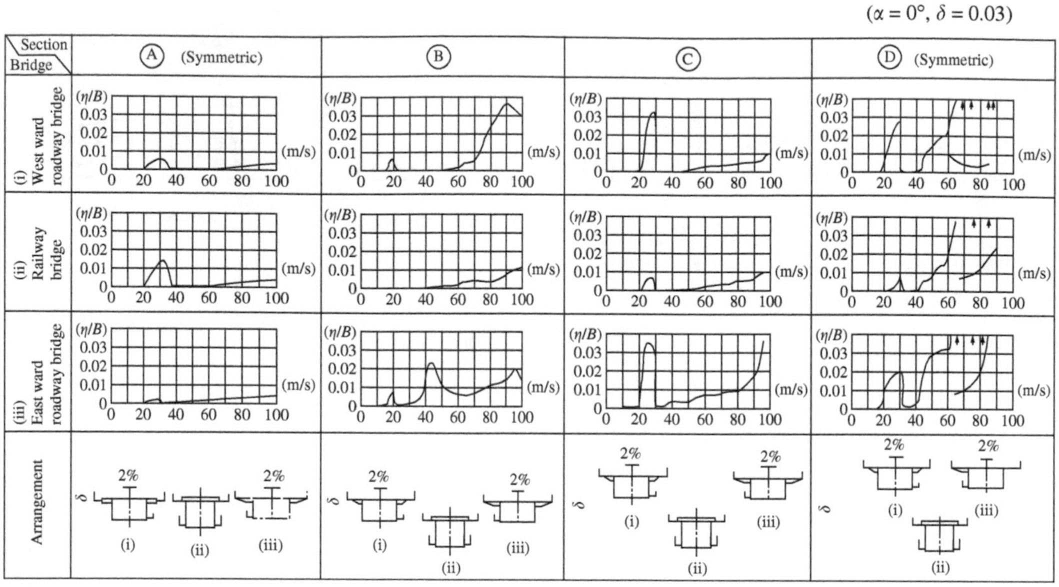

$(\alpha = 0°, \delta = 0.03)$

FIGURE 10.52 Responses of girders in parallel arrangement [41].

1. *For vortex-excited vibration*

$$U_{vh} = 2.0 f_h B \quad \text{heaving mode} \tag{10.14}$$

$$U_{v\theta} = 1.33 f_\theta B \quad \text{torsional mode} \tag{10.15}$$

where f_h and f_θ are natural frequencies (Hz) of heaving and torsional motion, respectively, and B is the projected width of girder in unit of meter.

2. *For galloping*

$$U_g = 8 f_h B \tag{10.16}$$

3. *For flutter*

$$U_f = 2.5 f_\theta B \tag{10.17}$$

10.4.2 Estimation of Amplitude

The equations for the estimation of maximum amplitude in vortex-excited vibration are given in *The Wind-Resistant Design Handbook in Japan*.

1. *Heaving vibration*

$$h = (E_h \cdot E_{th} \cdot B)/(m_r \cdot \delta_h) \tag{10.18}$$

where

h = heaving mode amplitude (m)
$E_h = 0.065 \beta_{ds} (B/D)^{-1}$
$E_{th} = 1 - 15 \beta_t (B/D)^{1/2} I_u^2 \geq 0, m_r = m/(\rho B^2)$
    D = effective height of girder
    β_{ds} = 2 or 1 (2 for vertical web, 1 for others), correction factor for girder form

TABLE 10.1 Strouhal Number of Various Cross-Sections of Structural Members [5]

Wind	Profile dimensions, in mm	Value of $\mathscr{S}$	Wind	Profile dimensions, in mm	Value of $\mathscr{S}$
→ / ↓	$t=2.0$; 50, 50, t (I-section)	0.120 / 0.137	↓	$t=1.0$; 12.5, 12.5, 25, 50 (Z-section)	0.147
→	$t=0.5$; 25, 25 (H-section)	0.120	↓	$t=1.0$; 12.5, 12.5, 12.5, 50	0.150
↓	$t=1.0$; 25, 50 (channel)	0.144	← / ↑ / ↗	$t=1.0$; 50, 50 (L-section)	0.145 / 0.142 / 0.147
↓	$t=1.5$; 12.5, 50 (T-section)	0.145	← / ↑ / ↗	$t=1.0$; 25, 25 (L-section)	0.131 / 0.134 / 0.137
↓ / ↑	$t=1.0$; 25, 50 (channel)	0.140 / 0.153	→ / ↓	$t=1.0$; 25, 25, 25, 25	0.121 / 0.143
↓ / ↑	$t=1.0$; 12.5, 50 (T-section)	0.145 / 0.168	→	$t=1.0$; 25, 25, 25, 12.5	0.135
→ / ↓	$t=1.5$; 50 (angle)	0.156 / 0.145	→	$t=1.0$; 50, 100 (angle)	0.160
Cylinder 11,800 < $\mathscr{R}e$ < 91,100; 25		0.200	→ / ↑	$t=1.0$; 25, 50 (angle)	0.114 / 0.115

TABLE 10.2 Drag and Lift Force Coefficients of Structural Members with Various Cross-Sections

Profile and wind direction	C_D	C_L
	2.01	0
	2.04	0
	1.81	0
	2.0	0.3
	1.83	2.07
	1.99	−0.09
	1.62	−0.48
	2.01	0
	1.99	−1.19
	2.19	0

β_t = 1 or 0 (0 for hexagonal cross-section, 1 for others), correction factor for influence of turbulence

I_u = intensity of turbulence

m = mass of girder per unit length ($\text{kg fs}^2/\text{m}^2$)

ρ = air density

δ_h = logarithmic damping decrement of the structure in heaving vibration mode

2. *Torsional vibration*

$$\theta = (E_\theta \cdot E_{t\theta} \cdot B)/(I_{pr} \cdot \delta_\theta) \tag{10.19}$$

where

θ = torsional mode amplitude (degree)

$E_\theta = 17.16\beta_{ds}(B/D)^{-3}$

$E_{t\theta} = 1 - 20\beta_i(B/D)^{1/2}I_u^2 \geq 0$

$I_{pr} = I_p/(\rho B^4)$

   D = effective height of girder

   I_p = polar inertia moment of mass of girder per unit length (kg fs^2)

δ_θ = logarithmic damping decrement of the structure in torsional vibration mode

3. *Amplitude estimation from experimental data:* As indicated in Equation 10.8, the occurrence of self-excited vibration is assessed as the sum of structural damping and aerodynamic damping. In wind tunnel tests, at specified wind velocity, aerodynamic during experiment from measured total damping. The aerodynamic damping usually is as shown in Figure 10.53. That is called the "A–δ curve."

When A–δ curve is obtained, the amplitude of the vibration at specified wind velocity can be estimated for the structure with any structural damping. In the case where the structural damping is δ_s, setting the value of $-\delta_s$ on the vertical axis and obtaining the intersection point on the curve, the corresponding amplitude on the horizontal axis is the estimated amplitude A_s of the steady-state vibration at the specified wind velocity.

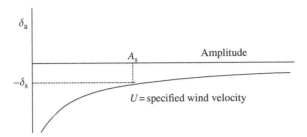

FIGURE 10.53　Aerodynamic damping to amplitude at specified wind velocity (A–δ curve).

10.5　Examples of Real Bridges

In this section, some bridges that had aeroelastic vibration problems are introduced. It is considered that it is very important for bridge engineers to know the damage caused to the bridge by wind action, which would be helpful when they design new bridges.

10.5.1　Collapse of Tacoma Narrows Bridge [43]

The collapse of Tacoma Narrows Bridge in Figure 10.54 is very famous in the bridge engineering field as the accident was caused by a wind action. The accident occurred at a wind speed of 19 m/s in 1940. The Tacoma Narrows Bridge was constructed by utilizing the most advanced analytical method of suspension bridge in those days. Although the effect of wind load as the drag force of the girder was taken into account in the design of the bridge, the occurrence of aeroelastic vibration of the girder was not taken into consideration. The girder configuration was similar to the shallow H-shape section shown in Figure 10.55. This type of girder section induces vortex-excited vibration in heaving and torsional modes and the flutter as the torsional self-excited vibration as explained in Section 10.4. As is well known, the bridge collapsed in the torsional vibration with large amplitude caused by the flutter. It was reported that the Tacoma Narrows Bridge was vibrating from the beginning of the erection and investigation was done to control the vibration by attaching fairings on the girder ends so as not to generate the separation flow from the leading edge of the girder as shown in Figure 10.56. Unfortunately, the bridge collapsed before the fairing could be attached. After the collapse, the cause was investigated from various points of view. The girder configuration chosen from various investigations was a truss-stiffened girder reducing the wind load and the possible occurrence of aeroelastic vibrations. The truss-stiffened girder developed in the investigation of the collapse of the Tacoma Narrows Bridge has been used mainly in United States and Japan as the girder of long-span suspension bridges.

10.5.2　Vortex-Excited Vibration of the Great-Belt Bridge [44]

The Great-Belt Bridge (Figure 10.57) was constructed as the second long-span suspension bridge in 1998. The aeroelastic performance was investigated in the wind tunnel by using a full span $\frac{1}{100}$-scaled model. The occurrence of vortex-excited vibration in the smooth flow was predicted through the wind tunnel tests. The vibration, however, was not predicted in the turbulent flow. Therefore, at the beginning of erection, the girder was constructed without attaching a device to suppress the vortex-excited vibrations. Just before being opened to traffic, however, the bridge girder vibrated in the vortex-excited vibration. It was considered that the device to suppress the vortex-excited vibration should have been attached on the bridge girder, when the vortex-excited vibration was observed. After the observation of the vibration, a vane-type device was prepared and attached on the lower corners of the bridge girder as shown in Figure 10.58. As a result, the vortex-excited vibration was suppressed.

FIGURE 10.54 Collapse of the Tacoma Narrows Bridge [43].

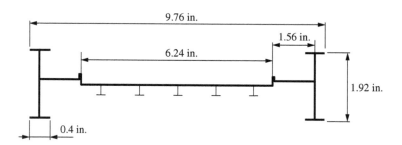

FIGURE 10.55 Cross-section of the girder of the Tacoma Narrows Bridge [43].

FIGURE 10.56 Fairing planned to be attached on the Tacoma Narrows Bridge before the collapse [43].

10.5.3 Vortex-Excited Vibration of the Box Girder Bridge in Trans-Tokyo Bay Highway [45,46]

This is an example of the vortex-excited vibration occurring in a continuous steel box girder bridge. The outline of the bridge is shown in Figure 10.59. The longest span of 240 m vibrated in the wind with direction perpendicular to the bridge axis under a wind velocity of 20 m/s. Up to this example, it was considered that the vortex-excited vibration is difficult to be induced in a box girder bridge. During the design of the bridge, the wind tunnel test was conducted to check the aeroelastic performance of the bridge.

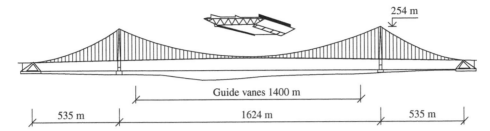

FIGURE 10.57 The Great-Belt Bridge [44].

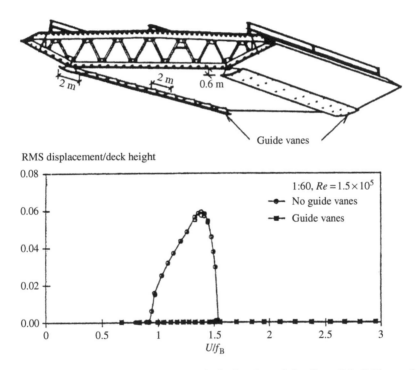

FIGURE 10.58 Corner vane to suppress vortex-excited vibration of the Great-Belt Bridge and comparison between the aeroelastic response with/without corner vane [44].

The result was that the vortex-excited vibration would be induced in both smooth and turbulent flows. Aerodynamic countermeasure of fairing, double flap, and skirt (as shown in Figure 10.60) was examined to suppress the vortex-excited vibration. The vortex-excited vibration could not be suppressed (as shown in Figure 10.61). This means that it is very difficult to suppress the aeroelastic vibration of a steel box girder bridge by aerodynamic countermeasure through changing the structure configuration. Therefore, in this example, a tuned mass damper (TMD) was used for suppression of vortex-excited vibration.

Figure 10.62 shows the comparison of aerodynamic responses of first mode in both the wind tunnel tests and the field measurement before being opened to the public. In the figure, solid lines show the results of wind tunnel tests for two structural damping decrements of $\delta = 0.028$ and 0.044. The other points are the field measurement results measured in the wind for bridge axis right angle within $\pm 20°$ and the turbulent intensity of 4 to 10%. These results prove that the accuracy of the wind tunnel test is high. The maximum amplitude was measured around a wind velocity of 16 m/s and the result was

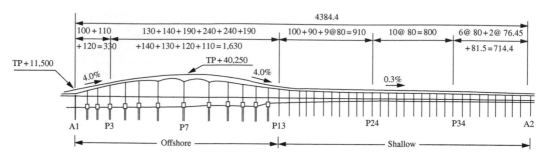

FIGURE 10.59   Outline of the Trans-Tokyo Bay Highway [45].

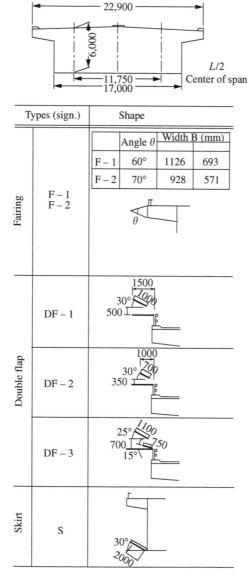

FIGURE 10.60   Bridge girder section and aerodynamic countermeasures tested for the bridge of the Trans-Tokyo Bay Highway to suppress the vortex-excited vibration [45].

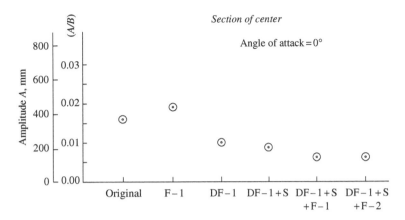

FIGURE 10.61 Comparison of estimated amplitudes of vortex-excited vibration of the box-girder with various aerodynamic countermeasures [45].

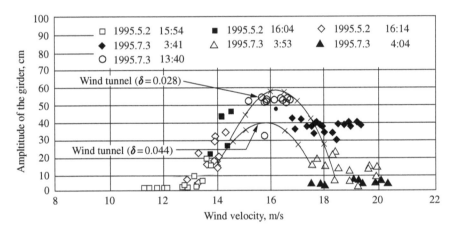

FIGURE 10.62 Comparison between the wind tunnel test results and the field observation results for the continuous steel box girder bridge in the Trans-Tokyo Bay Highway [45].

60 cm; that is, 10% of girder height at the center of the longest span. The vortex-excited vibration was suppressed by TMD and the bridge is now open to traffic.

10.5.4 Vortex-Excited Vibration of the Tower of the Akashi-Strait Bridge During Self-Standing State [47]

During the erection of a suspension bridge and a cable-stayed bridge, there is a period in which the tower is self-supporting. In this period, the tower is in the aeroelastically unstable state, because the structural damping is very small ($\delta = 0.025$) and is susceptible to the vibration under the wind action. The height of the tower is 280 m. When the tower vibrates, a large bending moment occurs at the foundation of the tower. The corner cut-outs introduced in 3–2–1 was chosen as the tower configuration to suppress the galloping of the tower. Even this tower configuration induced the vortex-excited vibration of bending mode in out of plane. Figure 10.63 shows the comparison of results of the field measurement and the wind tunnel tests. The solid and chain lines indicate the results of the wind tunnel tests in three wind directions of 0, 5, and 10° from the right-angled direction at the bridge axis. The dots by the symbol are the measuring results corresponding to the three wind directions. These results show the

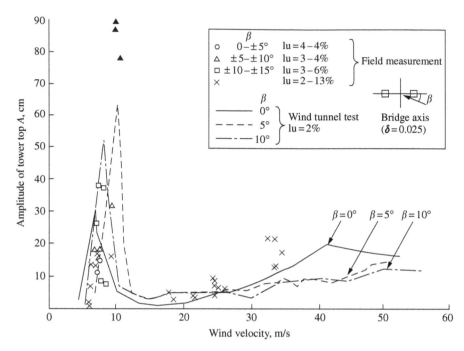

FIGURE 10.63 Comparison between the aeroelastic responses of wind tunnel test results and field measurement results of the tower of the Akashi-Kaikyo Bridge during self-standing [47].

good coincidence of the field measurement value and the wind tunnel test value. The vortex-excited vibration of the tower is now suppressed by applying the TMD.

10.6 Summary

It has been explained that structure configuration is very important in wind engineering. The structure configuration controls the separation flow around the structure. That is, the phenomenon occurring in the structure under wind action is induced by the behavior of the separation flow from the leading edge of the structure and the reattachment of the separation flow on the structure surface. Both the deflection of the structure and the aeroelastic vibration of the structure are induced by the behavior of the separation flow around the structure.

The deflection under wind action is cause by wind load. The wind load is proportional to the square of wind velocity. The reduction of wind load is one of the themes in wind engineering. It was explained that the reduction in the wind load is achieved to some extent by changing the corner configuration of the structure. Another theme is how the aeroelastic vibration of the structure is suppressed. The aeroelastic vibration problem is a peculiar problem in the flexible structure exposed to wind. The suppression of vibration has been achieved by the control of the separation flow and the structural method of adding a damping device including TMD. From the viewpoints of maintenance of the structure with long life, aerodynamic control by structural configuration is rather preferable than the structural method. Because the structural method has movable parts in the damper system, the movable parts should be replaced periodically with new parts.

Until now, it was usual to carry out the examination of the wind-resistant design after the structural design ends. When there was a problem on aeroelastic vibration, the aeroelastic performance was improved by attaching some aerodynamic devices, for example, fairing, deflector, and guide vane so on. The ideal form of the bridge design is the form to fuse the structural design with the wind-resistant design. If it is so done, the ideal bridge cross-section is realized without attaching aerodynamic devices on the bridge girder. In the bridge design, when the lowest natural frequency of the bridge is under 1 Hz, there is the possibility of occurrence of aeroelastic vibration under the design wind velocity. The following equation is very useful to roughly estimate the lowest natural frequency of the girder bridge:

$$f = 100/\text{span length (m)}, \quad \text{in Hz}$$

Therefore, in the case the lowest frequency is under 1 Hz, the design should be proceeded with respect to structural and wind-resistant designs. Then the ideal bridge section will be found also from the view-point of both aesthetic and economical reasons.

For the bridge engineers who are interested in wind engineering of structures, some reference books are introduced in the Further Reading section.

References

[1] Y. Kubo and K. Kato, "The role of end plates in two dimensional tests," *Proc. JSCE*, vol. 368/I-5, pp. 179–186.

[2] V. Strouhal, "Uber eine besondere Art der Tonerregung," *Ann. Phys.*, vol. 5, pp. 216–250, 1878.

[3] Japan Society of Steel Construction, "Wind Engineering on Structures" (in Japanese), Tokyo Denki University Press, Tokyo, 1997.

[4] Y. Kubo, E. Yamaguchi, S. Kawamura, K. Tou, and K. Hayashida, "Aerodynamic Characteristics of Square Prism with Rounded Corners," Proceedings of the 14th National Symposium on Wind Engineering in Japan, pp. 281–286, 1996 (in Japanese).

[5] Y. Nakamura, "Aerodynamic characteristics of fundamental structures (part 1)," *J. Wind Eng.*, *JAWE*, no. 36, pp. 49–84; E. Simiu and R.H. Scanlan, "*Wind Effects on Structures (Third Edition)*," John Wiley & Sons, Inc., New York, 1996.

[6] T. Miyata, H. Yamada, and T. Hojo, "Aerodynamic Response of PE Stay Cables with Pattern-Indented Surface," Proceedings of Cable-Stayed and Suspension Bridges (Deauville), 1994.

[7] A. Okajima, K. Sugitani, and T. Mizota, "Flow around a pair of circular cylinders arranged side by side at high reynolds numbers," *Trans. JSME*, vol. 52, no. 480, pp. 2844–2850, 1986 (in Japanese).

[8] A. Okajima, K. Sugitani, and T. Mizota, "Flow around two rectangular cylinders in the side-by-side arrangement," *Trans. JSME*, vol. 51, no. 472, pp. 3877–3886, 1985 (in Japanese).

[9] D. Biermann and W.H. Herrnstein, "The Interference Between Struts in Various Combinations," National Advisory Committee for Aeronautics, Tech. Rep. 468, pp. 515–524, 1933.

[10] R.L. Wardlaw and K.R. Kooper, "A Wind Tunnel Investigation of the Steady Aerodynamic Forces on Smooth and Standard Twin Bundled Power Conductors for the Aluminum Company of America," National Aeronautical Establishment, Canada, LTR-LA-117, 1973.

[11] N. Suzuki, H. Sato, M. Iuchi, and S. Yamamoto, "Aerodynamic Forces Acting on Circular Cylinders Arranged in a Longitudinal Row," Proceedings of the 3rd International Symposium on Wind Effects on Buildings and Structures, Tokyo, pp. 377–387, 1971.

[12] Y. Tanida, A. Okajima, and Y. Watanabe, "Stability of circular cylinder oscillating in uniform or in a wake," *J. Fluid Mech.*, vol. 61, pp. 769–784, 1973.

[13] M.M. Zdrakovich and D.L. Pridden, "Interference between two circular cylinders; series of unexpected discontinuities," *J. Ind. Aero.*, vol. 2, pp. 255–270, 1977.

[14] E. Hori, "Experiments on Flow around a Pair of Parallel Circular Cylinders," Proceedings of the 9th Japan National Congress for Applied Mechanics, Tokyo, pp. 231–234, 1959.

[15] S.J. Price and M.P. Paidoussis, "The aerodynamic forces acting on groups of two and three circular cylinders when subject to a cross-flow," *J. Wind Eng. Ind. Aero.*, vol. 17, pp. 329–347, 1984.

[16] T. Igarashi, "Characteristics of the flow around two circular cylinders arranged in tandem (1st Report), *Bull. JSME*, vol. 24, no. 188, pp. 323–331, 1981.

[17] K. Imaichi, Preprint for JSME, no. 734–5, pp. 104–106, 1973 (in Japanese).

[18] M.M. Zdrakovich, "Review of flow interference between two circular cylinders in various arrangement," *Trans. ASME, J. Fluids Eng., Ser. I*, vol. 99, no. 4, pp. 618–633, 1977.

[19] M.M. Zdrakovich, "Flow-induced oscillations of two interfering circular cylinders," *J. Sound Vibration*, vol. 101, no. 4, pp. 511–521, 1985.

[20] A. Takano, I. Arai, and M. Matsuzaka, "An experiment on the flow around a group of square prisms," *Trans. JSME*, vol. 47, no. 417, pp. 982–991, 1981 (in Japanese).

[21] K. Hirano, H. Ohsako, and A. Kawashima, "Interference between two normal flat plates in the various staggered arrangements (1st, Drag and Vortex Shedding Frequency)," *Trans. JSME*, vol. 49, no. 2363–2370, 1983.

[22] A. Okajima, Preprint for JSME, no. 7867–1, pp. 15–16, 1986 (in Japanese).

[23] Y. Kubo, E. Yamaguchi, S. Kawamura, K. Tou, and K. Hayashida, "Aerodynamic Characteristics of Square Prism with Rounded Corners," Proceedings of the 16th National Symposium on Wind Engineering in Japan, 2000 (in Japanese).

[24] Y. Kubo, K. Hirata, and K. Mikawa, "Mechanism of aerodynamic vibration of shallow bridge girder sections," *J. Wind Eng. Ind. Aerodyn.*, vol. 41–44, pp. 1297–1308, 1992.

[25] D.E. Walshe and L.R. Wootton, "Preventing Wind-Induced Oscillation of Structures of Circular Section," Proc. ICE, Paper 7289, 1970.

[26] H. Liu, *Wind Engineering, A HandBook for Structural Engineers*, Prentice Hall, New York p. 125, 1991.

[27] C. Scruton and A.R. Flint, "Wind-Excited Oscillations of Structures," Proc. I.C.E., paper no. 6758, 1964.

[28] R.H. Scanlan and R.L. Wardlaw, "Reduction of Flow-Induced Structural Vibrations," Isolation of Mechanical Vibration, Impact, and Noise, AMD, vol. 1, sec. 2, ASME, 1973.

[29] S. Ohno, S. Sano, and C. Morimoto, "Study of aerodynamic stability of cable-stayed curved bridge and its countermeasures against wind-induced vibration," *J. Wind Eng.*, no. 37, pp. 567–579, 1988.

[30] Y. Kubo, A. Koishi, K. Tasaki, and H. Nakagiri, "Mechanism of Separation Interference on Bridge Sections," Proceedings of the 13th National Symposium on Wind Engineering in Japan, pp. 353–358, 1994 (in Japanese).

[31] K. Ogawa, "On 3-D aerodynamic characteristics of bridge tower," *J. Wind Eng.*, JAWE, no. 59, pp. 53–54, 1994.

[32] Y. Kubo, K. Kato, E. Yamaguchi, H. Nakagiri, and K. Okumura, "Suppression Mechanism of Aerodynamic Vibration of A-Shape Tower due to Horizontal Beam," Proceedings of the 14th National Symposium on Wind Engineering in Japan, pp. 593–598, 1996 (in Japanese).

[33] R.L. Wardlaw, "Improvement of Aerodynamic Performance," Aerodynamics of Large Bridges edited by A. Larsen, A.A. Balkema, pp. 59–70, 1992.

[34] Y. Kubo, K. Honda, K. Tasaki, and K. Kato, "Improvement of aerodynamic instability of cable-stayed bridge deck by separated flows mutual interference method," *J. Wind Eng. Ind. Aerodyn.*, vol. 49, pp. 553–564, 1993.

[35] K. Nagai, M. Oyadomari, and R. Inamuro, "Aeroelastic Response of 3 Span Double Steel Box Girder Bridge," Proceedings of the 40th Annual Conference on the Japan Society of Civil Engineers, 1, pp. 481–482, 1985.

[36] T. Miyata, K. Yokoyam, M. Yasuda, and Y. Hikamai, "Akashi Kaikyo Bridge: Wind Effects and Full Model Wind Tunnel Tests," Aerodynamics of Large Bridges edited by A. Larsen, A.A. Balkema, pp. 217–236, 1992.

[37] K. Sadashima, Y. Kubo, T. Koga, Y. Okamoto, E. Yamaguchi, and K. Kato, "Aeroelastic responses of 2-edge girder for cable-stayed bridges," *J. Struct. Eng.*, JSCE, vol. 46A, pp. 1073–1078, 2000 (in Japanese).

[38] Y. Kubo, K. Kato, H. Maeda, K. Oikawa, and T. Takeda, "New Concept on Mechanism and Suppression of Wake-Galloping of Cable-Stayed Bridges," Proceedings of the Cable-stayed and Suspension Bridges, pp. 491–498, 1994.

[39] Y. Kubo, T. Nakahara, and K. Kato, "Aerodynamic behavior of multiple elastic circular cylinders with vicinity arrangement," *J. Wind Eng. Ind. Aerodyn.*, vol. 54/55, pp. 227–237, 1995.

[40] R. Toriumi, N. Furuya, M. Takeguchi, M. Miyazaki, and Y. Saito, "A Study on Wind-Induced Vibration of Parallel Suspenders Observed at the Akashi-Kaikyo Bridge," Proceedings of the 3rd International Symposium on Cable Dynamics, pp. 177–182, 1999.

[41] A. Honda, N. Shiraishi, and S. Motoyama, "Aerodynamic instability of Kansai international airport access bridge," *J. Wind Eng., JAWE*, no. 37, pp. 521–528, 1988.

[42] *The Wind-Resistant Design Handbook for Road Bridges*, Japan Society for Roads, 1991.

[43] F.B. Farquarson (ed.), "Aerodynamic Stability of Suspension Bridges," Parts I–V, Bulletin No. 116, University of Washington Engineering Experiment Station, Seattle, 1949–1954.

[44] A. Larsen, S. Esdahl, J.E. Andersen, and T. Vejrum, "Vortex Shedding Excitation of the Great Belt Suspension Bridge," Proceedings of the 10th ICWE, pp. 947–954, 1999.

[45] Y. Yoshida, Y. Fujino, H. Tokita, and A. Honda, "Wind tunnel study and field measurement of vortex-induced vibration of continuous steel box girder in Trans-Tokyo Bay highway," *J. Struc. Mech. Earthquake Eng., JSCE*, no. 633, I-49, pp. 103–117, 1999.

[46] Y. Yoshida, Y. Fujino, H. Sato, H. Tokita, and S. Miura, "Control of vortex-induced vibration of continuous steel box girder in Trans-Tokyo Bay highway," *J. Struc. Mech. Earthquake Eng., JSCE*, no. 633, I-49, pp. 119–134, 1999.

[47] H. Shimodoi, T. Kanesaki, K. Hata, and N. Sasaki, "Aeroelastic Response of Tower of the Akashi Kaikyo Bridge during Self Standing State," Proceedings of the 40th Annual Conference on the Japan Society of Civil Engineers, 1, pp. 976–977, 1994.

[48] N.P. Cheremisnoff, "Encyclopedia of Fluid Mechanics: Aerodynamics and Compressible Flows," Chapter 10: Wake Interference and Vortex Shedding, Gulf Publishing, 1989.

Further Reading

R.D. Blevins, "Flow-Induced Vibration (Second Edition)," Krieger Publishing Company, New York, 1994.

C. Dyrbye and S.O. Hansen, "Wind Loads on Structures," John Wiley & Sons, Inc., New York, 1996.

Japan Society of Steel Construction, "Wind Engineering on Structures" (in Japanese), Tokyo Denki University Press, Tokyo, 1997.

H. Liu, "Wind Engineering, A Hand Book for Structural Engineers," Prentice Hall, New York, 1991.

P. Sachs, "Wind Forces in Engineering," Pergamon Press, New York, 1972.

E. Simiu and R.H. Scanlan, "Wind Effects on Structures (Third Edition)," John Wiley & Sons, Inc., New York, 1996.

Index

Note: Italicized page numbers refer to figures and tables.

Milton Keynes UK
Ingram Content Group UK Ltd.
UKHW052025071024
449327UK00027B/2427